Biotechnological Production Of Natural Ingredients For Food Industry

First Edition

Edited By

Juliano Lemos Bicas
Mário Roberto Maróstica Jr.
Glaucia Maria Pastore
School of Food Engineering
University of Campinas-Brazil

Biotechnological production of natural ingredients for food industry

Editors:

Juliano Lemos Bicas

Mário Roberto Maróstica Jr.

Glaucia Maria Pastore

ISBN (eBook): 978-1-68108-265-3

ISBN (Print): 978-1-68108-266-0

First published in 2016.

Bentham Science Publishers Ltd.
Executive Suite Y - 2
PO Box 7917, Saif Zone
Sharjah, U.A.E.
Email: subscriptions@benthamscience.org

BENTHAM SCIENCE

CONTENTS

FOREWORD

Hence, depending to whom the question is addressed, the assessment about the necessity of additives in food will turn out extremely different. Industry, on one hand, aims to produce qualitatively high ranking food with minimal costs. Beside palatability (aroma, taste, texture and color) the food should offer extended shelf life and high nutritional impact. The consumers, on the other hand, want to have all thesebenefits without the addition of any food additive. In terms of food the "all without mania" (flavors, taste enhancers, colorants, preservatives, antioxidants, thickeners, gelling and stabilizing agents *etc.*) is prevalent in developed countries. However, consumer's empathy of preparing their daily meal on the basis of fresh ingredients is more and more decreasing nowadays.

Therefore processed food, especially convenient food, purchased in the supermarket and not on the farmers market represents the dominating basis of modern nutrition. Hence, quality, nutritional value and shelf life must be guaranteed for longer and longer storage times. Moreover, nowadays food should offer additional benefits regarding lifestyle and health. To comply with all of these requests, food additives are indispensable in industrial food processing. For a long time no economical alternative to chemical synthesis of food additives existed. Not at the least because of the depletion of the fossil fuels reserves together with several concerns about climate changeand the all over chemophobia, the production of natural food additives seems to cut this *Gordianknot* of different interests. Natural additives serve consumers wish on natural nutrition and approve food industry the application of indispensable additives. Whether recovered from natural resources or produced biotechnologicallyroutes natural additives are more accepted in public mind. However, natural resources are often limited which calls for biotechnological alternatives.

The aim of the book is to provide coverage of natural food additives and their production. Within this textbook, comprising 13 chapters, international well accepted experts in the field give a prevailing and comprehensive overview on food additives such as flavors colors, sweeteners, amino and organic acids, vitamins, prebiotics, edible oils, antimicrobial compounds, biosurfactantsand enzymes. This book secures the experts in the field as well as interested consumers to inform themselves about the current state of the art of biotechnological processes for the production of natural food additives. Furthermore legal, economic and ecological aspects are also addressed. It is demonstrated that biotechnology alongside the attribute "natural" can compete against chemical production processes because of improved production strains, the use of stable and often immobilized used enzymes, disposition of cheap waste streams of food producing processes as precursors and/or as the nutrient medium for producing microorganisms. Last but not least, it is more and more evident that a liable application of genetic engineering is an indispensable part of white and green biotechnology as it is already the case for red biotechnology. The conclusion of all chapters is that biotechnology, particularly genetic engineering, is a powerful tool which will help to cope at least with some of humankinds future challenges in the light of limited resources and a fast growing world population.

Ulrich Krings
Institute of Food Chemistry, Leibniz University of Hanover
Hanover
Germany

PREFACE

After the advent of Organic Chemistry (1800's), the supply for organic chemicals achieved industrial scale. This was particularly important for Food Industry, which soon incorporated synthetic substances for the preparation of food products: saccharin, benzoate and indigotine, for instance, were already in use at the beginning of 20th century and are still applied today as food additives. However, public awareness involving the possible toxic effects of artificial ingredients has increased the demand for new alternative sources of their natural counterparts. Moreover, the impact of human activity on environment has been considered a major issue to be overcome, leading to intense search for sustainable or alternative production processes.Therefore, biotechnology has emerged as an important tool to supply natural ingredients for food industry, since they occur under controlled conditions, may use renewable sources and are recognized as an environmentally friendly technique.

Although such approach has been used empirically for the production of fermented food (bread, wine, beer, cheeses *etc.*) aiming at either preservation or modification of their sensory attributes, it was only recently that science begun to able to understand and explain the phenomenology behind these biotechnological processes, which reflected in an increased number of R&D projects for the production of food ingredients by microorganisms, enzymes of cell cultures.The fact is that, nowadays, the so-called White Biotechnology is considered an inextinguishable resource of natural food ingredients. Additionally, food biotechnology remains a vigorousresearch field and many fundamental studieson this subject are still needed. This may be evidenced by the intense growth, during the last decades, of articles and patents covering all aspects of biotechnological production of food ingredients. Most of these processes are already found in commercial scale, but others are still waiting for further developments.

This e-Book aims to be a fundamental reference for people who want to deepen into this field, particularly those students, scientists, researchers and professionals working with Food Science and Technology, Food Chemistry, Food Biotechnology, Food Engineering, Bioprocess Engineering, Biotechnology, Applied Microbiology, Nutrition and others. It is divided in 12 chapters. The first one presents a brief overview of food biotechnology, particularly those aspects involving and historical perspective and some examples of commercially relevant products and processes currently available. All other chapters are devoted each one to a particular class of products with potential interest for food or feed industries: sweeteners (Ch. 2), aminoacids and nucleotides (including flavor enhancers) (Ch. 3), organic acids (Ch. 4), vitamins and nutraceuticals (Ch. 5), aroma compounds (Ch. 6), colorants (Ch. 7), edible oils (Ch. 8), hydrocolloids (Ch. 9), antimicrobial compounds (Ch.

10), biosurfactants (Ch. 11) and enzymes (Ch. 12).

We hope you enjoy it!

Juliano Lemos Bicas
Mário Maróstica Jr.
Glaucia Maria Pastore
School of Food Engineering
University of Campinas
Brazil

List of Contributors

Ana M. O. Finco	Bioprocess Engineering and Biotechnology Department, Federal University of Paraná, Curitiba, Brazil
André Lebert	Clermont Université, Université Blaise Pascal, Institute Pascal, BP20206, F-63174 AUBIERE cedex, France CNRS, UMR 6602, Institute Pascal, F-63171 Aubière, France Clermont Université, Université Blaise Pascal, Université Blaise Pascal, F-63171 AUBIERE cedex, France
Antônio I. Magalhães	Bioprocess Engineering and Biotechnology Department, Federal University of Paraná, Curitiba, Brazil
Adriane B. P. Medeiros	Bioprocess Engineering and Biotechnology Department, Federal University of Paraná, Curitiba, Brazil
André Luís Lopes da Silva	Bioprocess Engineering and Biotechnology Department, Federal University of Paraná, Curitiba, Brazil
Adenise L. Woiciechowski	Bioprocess Engineering and Biotechnology Department, Federal University of Paraná, Curitiba, Brazil
Christian Larroche	Clermont Université, Université Blaise Pascal, Institute Pascal, BP20206, F-63174 AUBIERE cedex, France CNRS, UMR 6602, Institute Pascal, F-63171 Aubière, France Clermont Université, Université Blaise Pascal, Université Blaise Pascal, F-63171 AUBIERE cedex, France
Cristine Rodrigues	Bioprocess Engineering and Biotechnology Department, Federal University of Paraná, Curitiba, Brazil
Carlos R. Soccol	Bioprocess Engineering and Biotechnology Department, Federal University of Paraná, Curitiba, Brazil
Camilo Barroso Teixeira	Bioprocess Laboratory, School of Food Engineering, University of Campinas, Campinas, Brazil
Dorit Eberhardt	Faculty of Biology & CeBiTec, Bielefeld University, D-33615 Bielefeld, Germany
Daniel E. R. Fernández	Bioprocess Engineering and Biotechnology Department, Federal University of Paraná, Curitiba, Brazil
Elia N. M. Ramírez	Bioprocess Engineering and Biotechnology Department, Federal University of Paraná, Curitiba, Brazil
Gwendoline Christophe	Clermont Université, Université Blaise Pascal, Institute Pascal, BP20206, F-63174 AUBIERE cedex, France CNRS, UMR 6602, Institute Pascal, F-63171 Aubière, France Clermont Université, Université Blaise Pascal, Université Blaise Pascal, F-63171 AUBIERE cedex, France

Gustavo Bernardes Fanaro	Department of Food and Nutrition, School of Food Engineering, University of Campinas, Campinas, Brazil
Gabriela Alves Macedo	Bioprocess Laboratory, School of Food Engineering, University of Campinas, Campinas, Brazil
Gustavo Molina	Bioprocess Laboratory, School of Food Engineering, University of Campinas, Campinas, Brazil Institute of Science and Technology, Food Engineering – UFVJM, Diamantina, Brazil Clermont Université, Université Blaise Pascal, Institut Pascal, BP20206, F-63174 AUBIERE cedex, France
Haroldo Yukio Kawaguti	Department of Food Science, School of Food Engineering, University of Campinas, Campinas, Brazil
Hélia Harumi Sato	Department of Food Science, School of Food Engineering, University of Campinas, Campinas, Brazil
Ibrahim M. Banat	School of Biomedical Sciences, Faculty of Life and Health Sciences, University of Ulster, Northern Ireland, UK
Juliano Lemos Bicas	Department of Food Science, School of Food Engineering, University of Campinas, Campinas, Brazil
Jenyffer M. Campos	Departamento de Nutrição, Universidade Federal de Pernambuco, Recife, Brazil
Júlio C. de Carvalho	Bioprocess Engineering and Biotechnology Department, Federal University of Paraná, Curitiba, Brazil
Jane Mary L. N. Gelinski	Bioprocess Engineering and Biotechnology Department, Federal University of Paraná, Curitiba, Brazil
Jaide V. K. Jensen	Faculty of Biology & CeBiTec, Bielefeld University, D-33615 Bielefeld, Germany
Juliana M. Luna	Centro de Ciências e Tecnologia , Universidade Católica de Pernambuco, Recife, Brazil
José Valdo Madeira	Bioprocess Laboratory, School of Food Engineering, University of Campinas, Campinas, Brazil
Juliana de Oliveira	Bioprocess Engineering and Biotechnology Department, Federal University of Paraná, Curitiba, Brazil
K. Kamireddy	Plant Cell Biotechnology Department, CSIR-Central Food Technological Research Institute, Mysore - 570 020, India
Luiz A. J. Letti	Bioprocess Engineering and Biotechnology Department, Federal University of Paraná, Curitiba, Brazil

Luiz D. G. Mamani	Bioprocess Engineering and Biotechnology Department, Federal University of Paraná, Curitiba, Brazil
Leonie A. Sarubbo	Centro de Ciências e Tecnologia , Universidade Católica de Pernambuco, Recife, Brazil
Luciana P. S. Vandenberghe	Bioprocess Engineering and Biotechnology Department, Federal University of Paraná, Curitiba, Brazil
Marius Herbst	Faculty of Biology & CeBiTec, Bielefeld University, D-33615 Bielefeld, Germany
Mário Roberto Maróstica	Department of Food and Nutrition, School of Food Engineering, University of Campinas, Campinas, Brazil
Pierre Fontanille	Clermont Université, Université Blaise Pascal, Institut Pascal, BP20206, F-63174 AUBIERE cedex, France CNRS, UMR 6602, Institut Pascal, F-63174 AUBIERE cedex, France Clermont Université, Université Blaise Pascal, LABEX IMobS3, BP80026, F-63171 AUBIERE cedex, France
P Giridhar	Plant Cell Biotechnology Department, CSIR-Central Food Technological Research Institute, Mysore - 570 020, India
P.S. Priyanka	Plant Cell Biotechnology Department, CSIR-Central Food Technological Research Institute, Mysore - 570 020, India
Paula Speranza	Bioprocess Laboratory, School of Food Engineering, University of Campinas, Campinas, Brazil
Rodrigo Malanski	Bioprocess Engineering and Biotechnology Department, Federal University of Paraná, Curitiba, Brazil
Raquel Diniz Rufino	Centro de Ciências e Tecnologia , Universidade Católica de Pernambuco, Recife, Brazil
R. K. Saini	Plant Cell Biotechnology Department, CSIR-Central Food Technological Research Institute, Mysore - 570 020, India
Sidnei Emilio Bordignon	Bioprocess Engineering and Biotechnology Department, Federal University of Paraná, Curitiba, Brazil
S. Sravan Kumar	Plant Cell Biotechnology Department, CSIR-Central Food Technological Research Institute, Mysore - 570 020, India
Vanessa Béligon	Clermont Université, Université Blaise Pascal, Institut Pascal, BP20206, F-63174 AUBIERE cedex, France CNRS, UMR 6602, Institut Pascal, F-63171 Aubière, France
Vanete Thomaz Soccol	Bioprocess Engineering and Biotechnology Department, Federal University of Paraná, Curitiba, Brazil

Volker F. Wendisch Chair of Genetics of Prokaryotes, Faculty of Biology & CeBiTec, Bielefeld University, D-33615 Bielefeld, Germany

Biotechnological Production Of Natural Ingredients For Food Industry
First Edition

CHAPTER 1

Introductory Overview of Biotechnological Additives

Gustavo Molina[a,b,*] and **Gustavo Bernardes Fanaro**[c]

[a] *Laboratory of Bioflavors, Department of Food Science, School of Food Engineering, University of Campinas, Campinas, São Paulo, Brazil*

[b] *Institute of Science and Technology, Food Engineering, UFVJM, Diamantina, Minas Gerais, Brazil*

[c] *Department of Food and Nutrition, School of Food Engineering, University of Campinas, Campinas, São Paulo, Brazil*

Abstract: The use of biotechnology in the manufacture of food and beverages has been practiced for many years. Because of this important developments over the years, biotechnology can be considered as a significant part of human life and industrial development, enabling the creation of breakthrough products and technologies to combat diseases, protect the environment, increase crop yields and to produce feed, fuels, renewable energy, industrial additives and several other useful products.

Keywords: Enzymes, Food ingredients, Fuels, Genetic engineering, Microbial fermentation.

INTRODUCTION

Biotechnology can be briefly defined as "any technological application that uses biological systems, living organisms, or derivatives, to make or modify products or processes for specific use" [1]. In this sense, biotechnology involves the application of tools based on biotechnology in traditional industrial processes ("bioprocess") and the manufacturing of biobased products (such as fuels, chemi-

* **Corresponding author Gustavo Molina:** Institute of Science and Technology, UFVJM, Diamantina, Minas Gerais, Brazil; E-mail:gustavomolinagm@gmail.com.

Juliano Lemos Bicas, Mário Roberto Maróstica, Glaucia Maria Pastore (Eds.)

cals and plastics) from renewable feedstock [2].

The biocatalysts used in these processes, such as bacteria, yeast and fungi or microalgae, are considered as an inexhaustible source of a diverse range of important compounds and industrial additives. Currently, microbial biotechnology plays an important role in the production of food additives, fine and bulk chemicals, solvents, enzymes, agrochemicals and biopharmaceuticals, and many others [3, 4].

Biotechnology presents several advantages compared to conventional chemical production models and also to the direct extraction of a desired additive from nature [5]. This happens mainly because microbial biocatalysts display desirable chirality and are biodegradable and the reactions are conducted in mild conditions, with lower energy consumption and lower environmental impact [4].

Furthermore, the industrial roles of biotechnology have been considerably expanded in the current scenario to produce renewable chemicals for industrial and economical purposes, also aim at reducing the use of petrochemical derivatives and the depletion of fossil fuels, in this way producing biofuels and bioenergy as a primary product through a 'biorefinery' concept [2].

Thus, biotechnology presents unique opportunities to produce natural food ingredients with industrial and economical interest. In this sense, the aim of this chapter is to present the historical development of biotechnology and also to illustrate some of the major products from this important industrial sector.

HISTORICAL ASPECTS OF BIOTECHNOLOGY FOR MODERN DEVELOPMENTS

First, it is important to understand the difference between the "traditional" and the so-called "modern biotechnology" [6].

The traditional biotechnology can be considered as the fermentation process used to produce beer, wine, cheese, soy sauce and others [7], and the biotechnology process in agriculture started with the history of agriculture itself. With the emerging development of agriculture, humankind began to select the plants with the best yields and resistance according to its needs [8]. Therefore, the biotechnological techniques are not new, considering that the manufacture of food and beverages, for example, has been practiced for more than 14,000 years with vinegar, alcoholic beverages, sourdough and cheese [9]. In fact, Food Biotechnology has been developed empirically since Ancient History and, ever since, the fermentation technology has been applied as the main tool to preserve food products or improve aroma, flavor and texture. Despite its long history, food science and technology has only recently understood the phenomenon involved in such biotechnological processes, and, today, the use of microbial and enzymatic processes for the production of food ingredients is highly developed [10].

Modern biotechnology, in turn, is based on recombinant DNA techniques, which started with the creation of the first recombinant gene, a couple of decades ago, and is currently helping to improve food, beverages, medicine and fuels [11]. The major examples of application of this technique are genetic modified organisms (GMO), metabolic engineered microorganisms and several breakthrough for the creation of crucial products for human use, such as new drugs, healthier foods and so on [11, 12]. Two major case studies will be presented in sequence to illustrate the development of biotechnology over the years.

Enzymes

For several years, enzymes have played an important role in many industries (food/feed, detergent, biofuels, textile and others). Currently, most food products have at least one ingredient produced with enzyme technology. Some examples produced with the enzymatic process include: sweeteners, syrups, bakery products, alcoholic beverages, precooked cereals, baby food, fish meal, cheese and dairy products, egg products, fruit juice, soft drinks, vegetable oil and puree, candy, spice and flavor extracts, liquid coffee, flavors and tenderized meat [13].

The industrial production of enzymes for use in food processing started in 1874, when Christian Hansen extracted chymosin (or rennin) from calf stomachs for the clotting of milk for cheese production [14, 15]. However, the mechanism of enzymes was unknown until 1877, when Moritz Traube proposed a "protein-like material that catalyzes fermentation and other chemical reactions".

The first patent for the industrial use of enzymes was named Taka-Diastase, an amylolytic produced by the fungi *Aspergillus oryzae* grown on rice. The patent was lodged in USA by Dr. Jokichi Takamine, a Japanese immigrant in 1884 [16].

In 1897, Buchner showed that an alcoholic fermentation could be performed using a cell-free yeast extract. The word 'zymase' was used to describe this cell-free extract, and this term evolved to the current 'enzyme' [17].

In 1930s, pectinases were used for juice clarification, and in World War II invertase was used for the production of inverted sugar syrup in a pioneered process with immobilized enzymes. However, the "modern" enzyme production by microbes can be observed only in the 1960s, when the acid hydrolysis of starch for glucose production was replaced by enzymatic hydrolysis using the fungal-derived amyloglucosidase [18].

During the 1990s, a growing use of enzymes was observed, mainly in the baking and animal feed industries. The estimated value of the worldwide use of industrial enzymes grew in this decade from US$ 1 billion in 1995 to US$ 1.5 billion in 2000 [19]. In 2007 the estimated was US$ 4 billion and the increase was projected to US$ 7 billion in 2013 [20, 21].

The several applications proposed by researchers in the 1990s are now being reinvestigated to make them more efficient, increase yield and decrease costs [22].

Thus, the modern enzyme industry is the result of this rapid development which happened over the last four decades [19]. The major worldwide enzyme industry association, the Association of Manufacturers and Formulators of Enzyme Products (AMFEP), lists approximately 160 enzymes manufactured commercially, and at least 36 of them are produced with genetically modified microorganisms [14]. These enzymes are mostly used in baking, beverages and brewing, dairy, dietary supplements, as well as fats and oils [20].

Microbial enzymes used in food processing are normally sold as enzyme preparations containing not only a desired enzyme (or a blend of enzymes) but also other metabolites of the production strain, as well as preservatives and stabilizers, and these must be food-grade and meet the standards of regulatory policy where the enzyme is used [23].

The most important industrial enzymes today include protease, carbohydrases, lipases, pectinases and amylases. Compared with conventional chemical catalysts, enzyme catalysis is highly specific [24] and functions under mild temperatures, pressures and pH [25]. Unlike the many chemical synthesis processes, enzymes require nontoxic and noncorrosive conditions. Approximately 60% of the enzymes used commercially come from modern biotechnology.

One of the main applications is in the detergent industry [26]. In this regard, Kumar [27] reports that the development of a genetically modified alkaline protease from *Bacillus lentus* (BLAP) is estimated to reduce environmental pollution associated with detergents by more than 65%.

Recent advances in DNA, proteins and bioinformatics techniques provide access to a great information base that facilitates the choice of microorganisms and enzymes for bioconversions [28]. New biocatalysts are being discovered by studying microorganisms isolated from extreme environments [29] and also by using metagenomics [30]. The evolution of these enzymes to be applied in a particular process can be possible with *in vitro* studies at substantially higher rates than achieved in nature [31, 32]. In this sense, the development of enzyme technologies has the potential to provide these products at low costs [33].

Genetic Engineering

Genetic engineering is described as the science that studies the intentional changes of the characteristics of a specific organism by the manipulation of its DNA, creating new variations of an organism. Thus, it is now possible to introduce specifics features from almost any organism to a plant, bacteria, virus or animal by manipulating the DNA and its transfer from one organism to another (the so-called recombinant DNA technique). It is also possible to manipulate an organism's genetic characteristic by introducing, modifying or eliminating specific genes [8, 34].

The first genetically engineered food was produced in 1967, when a new variety of potato called Lenape was bred for its high solids content, which made it useful for the production of potato chips. However, this potato presented high concentrations of a toxin called solanine and after two years this potato was withdrawn from the market by the USDA (United States Department of Agriculture) [34].

The first successful genetically engineered plant was produced only in 1983, when an antibiotic resistance gene was inserted into a tobacco plant [8], and the first approved GM crop, a tobacco with herbicide tolerance, was cultivated in China in 1990. The commercialization of a GM food took place in 1994 in the USA (Flavr Savr tomato), and, after ten years, more than 25,000 field studies were performed in more than 45 countries [35].

One of the most famous uses of genetic engineering in agriculture was the introduction of a gene in maize from the common soil bacterium *Bacillus thuringiensis* (Bt). This bacterium was considered a natural insecticide, since it was able to express a class of insecticidal proteins called Cry δ-endotoxins. This group of toxins is very effective against a certain type of insect, but it is harmless to birds and mammals, including humans. In 1996, the so-called Bt corn was described as "the most important technological advancement in insect pest management since the development of synthetic insecticides" because of its resistance to infestation by the microorganism *Ostrinia nubilalis*, which is known as the European corn borer, one of the most serious corn pests [36, 37].

The main characteristic introduced to commercialized food products by the genetic engineering in crops is the herbicide resistance/tolerance and insect resistance. Some of the genetic modified plants used in agriculture are described in Table **1**.

Table 1. Some genetic modified crops and their manufacturer.

Crop	Characteristic	Manufacturer	Year
Soybean			
Roundup®	herbicide resistance	Monsanto	1996
A2704-12	herbicide tolerance	Bayer	2008
Maize			
T25	herbicide tolerance	Bayer	1998
MON 810	insect resistance	Monsanto	1998
Bt11	herbicide tolerance insect resistance	Syngenta	1998

(Table 1) contd.....

Crop	Characteristic	Manufacturer	Year
Rapeseed			
GT73	herbicide resistance	Monsanto	1997
T45	herbicide tolerance	Bayer	1998
Cotton			
MON531	herbicide resistance	Monsanto	1997
MON1445	insect resistance	Monsanto	1997
Sugar beet			
H7-1	herbicide resistance	KWS SAAT AG/Monsanto	2007

[Adapted from Batista & Oliveira [8]].

In the last years, the GMO industry was focused on the production of plants resistant and tolerant to insects (*e.g. Bacillus thuriengensis*, or Bt plants, mainly cotton and maize) and herbicide (as Roundup ReadyTM soybeans), amounting to 99% of all GMO plants engineered. Moreover, plants resistant to viruses (New Leaf potato) and sterile (InvigorTM hybrid system) have been produced and developed [35].

In 2003, global acreage planted with biotechnological crops already amounted to 167 million acres in 18 countries. In US, for example, the main transgenic crops cultivated were 40% for corn, 81% for soybeans, 73% for cotton and 70% for canola [38].

During 2007, globally GM crops were cultivated on 114.3 million hectares (282.4 million acres), 12% more than in 2006 [8].

Since GMOs were first introduced, an intense debate took place in several countries. While in some countries, *e.g.* the USA, there was some acceptability by the consumers, in others, *e.g.* EU countries, the consumers rejected the use of this technology mainly because of the allegation of lack of food security [39 - 41].

Independently of creeds and opinion, the use of genetic engineer in agriculture can produce cheaper products, higher yield of production and wider variety of new products without allergenic proteins. However, it is also possible to create products with antibiotic resistance, food with some degree of allergenicity and limited access to seeds through the patenting of GM crops [42].

The major GM crops are soybean, maize, cotton, canola, squash and papaya, and many other transgenic crops will be commercialized over the next few years. Some examples among the companies that invest in research in plant biotechnology, and which are the main traders of GM crops, are Syngenta, Monsanto, Bayer CropScience, DuPont/Pioneer Hi-Bred, Dow AgroSciences and BASF [43].

MAIN BIOTECHNOLOGICAL INGREDIENTS: FROM RESEARCH TO INDUSTRY

White biotechnology or industrial biotechnology comprises several processes used for the manufacture of commodities and chemicals, from the use of whole cells or enzymes as catalysts [43]. Biocatalysis has more commonly been directed towards the production of high-value products in large-scale for the fine chemical and pharmaceutical industries [33].

Biotechnology has been studied extensively over the past two decades and efforts have been mainly invested in the search for innovation strategies [44 - 49]. The result of this investment drives the biotechnology industry, which is a dynamic and diverse industry searching for new technologies, applications and products in several industrial areas including the pharmaceutical, agriculture, chemical, computer, medical device and environmental industries [45].

In the chemical industry, for example, biotechnology enables the environmentally acceptable production of goods and services with safer, rapid and effective processes [43]. This industrial segment has used traditional biotechnological processes such as the microbial production of enzymes, antibiotics, amino acids, ethanol, vitamins, enzyme catalysis and others for many years [50, 51]. In addition, traditional biotechnology is widely used in the production of fermented foods and treatment of waste [43, 52].

The biotechnology impact on the chemical industry is expected to growth over 20% by 2025, corresponding to an important fraction of the total value of this industry from the use of biocatalysts (enzyme and whole-cell catalysts) and biotechnological products [53].

In Europe, for example, the high demand for chemicals reached approximately 28% of the worldwide demand, and it identified the industrial biotechnology as a key emerging technology area [54]. By 2010, the share of biotechnological processes in the production of various chemical products was expected to rise from 5 to 20%. Since then, the greatest impact has been in the fine chemical sector, where up to 60% of the products might be based on biotechnology [33, 55]. Also in 2010, the German chemical industry obtained 87% of feedstock from fossil sources, with oil (72%), gas (14%) and coal (1%) being the most important components of this supply [56].

In fact, in some cases, microbial fermentation may be the only viable process for obtaining some products and ingredients in an industrial scale and with the possibility of commercial exploitation [57 - 59]. Table **2** shows some of the established fermentation products.

Table 2. Industrial products produced by biotechnology.

Product	**Annual production (tons)**
Bioethanol	26,000,000
Glutamic acid	1,000,000
Citric acid	1,000,000
Lysine	350,000
Lactic acid	250,000
Food enzymes	100,000
Vitamin C	80,000
Gluconic acid	50,000
Antibiotics	25,000
Xanthan	20,000

[Adapted from Gavrilescu & Chisti [43]].

The production of fuels through biotechnology continues to attract a lot of attention, as they offer many advantages over petroleum-based fuels [60], particularly in the case of bioethanol [61], biogas, biodiesel [62] and biohydrogen [63].

Although the production of several biofuels is already established in industrial scale or as an emerging technology, some processes still require many advances, as in the case of biohydrogen production.

Bioethanol can be highlighted as one of the best optimized processes in the industry and as the most widely used biofuel for transportation worldwide. This alcohol can be produced from different types of raw materials and the main examples are corn, used in United States and Europe, and sugarcane, used in Brazil, for example [64]. The production of bioethanol in these countries reached 13,300, 1,371 and 6,267 millions of gallons in 2013, respectively [65].

However, recent trends indicate the great potential of production of this biofuel from agro-industrial residues and lignocellulosic biomass as raw materials, such as sugarcane and sweet sorghum bagasse, coffee ground, wheat straw and others derived from fruits, legumes or cereals [66]. Brazil is recognized as the largest single producer of sugarcane with approximately 31% of global production [67], and a large volume of bagasse is generated during sugarcane processing [66]. In this sense, much effort has been directed to the use of biomass from this raw material, such as bagasse and straw, aiming at the production of second generation ethanol. In this perspective, in 2014, Brazil launched the first industrial plant in the southern hemisphere for commercial scale of 2G ethanol, named Bioflex®, with an investment of approximately US$ 260 million, and it is expected that the production will reach 82 million liters of anhydrous bioethanol by 2015 [68].

The above example corroborates the fact that lignocellulosic waste will become the main feedstock for ethanol production in the near future. However, for now, biotechnology still faces some challenges, such as the requirement of tailored recombinant microorganisms for the implementation of consolidated bioprocessing of different feedstock into ethanol, and others, for a future qualitative improvement in the industrial production of ethanol [66].

Biotechnological processes also offer various additives on an industrial scale, being the organic acids among the highest annual production scales. These compounds are extremely useful as starting materials for the industry as they constitute a key group among the building block chemicals [43].

Annual microbial production of the main organic acids are approximately

1,600,000 and 150,000 t for citric acid and lactic acid, respectively, while for acetic acid the production is approximately 190,000 t (from 7,000,000 t produced) [69].

Over the last decades, there has been an expanding interest in the production of microbial polysaccharides for food, pharmaceutical and medical use, and they may act as viscosity or gelling agents, stabilizers or emulsifiers [70]. Because of their broad spectrum of application, prices could vary from US$ 14 per kg, for xanthan used in food application, to approximately US$ 50 per kg, for cyclic dextrans used in pharmaceutical products [71]. The market value for these polysaccharides was evaluated in approximately US$ 4 million in 2008, with xanthan as the most significant bacterial exopolysaccharide (EPS) in this market [72].

Meanwhile, in the healthcare and pharmaceutical segments, modern biotechnology industries have generated more than 100 new drugs and vaccines since the mid-1970s [73]. In the year 2000, for example, worldwide investment in biotechnologies amounted to US$ 37 billion, and it was expected to increase by 30%/year over the years. In addition, the antibiotics market alone exceeds US$ 30 billion and includes approximately 160 antibiotics and derivatives [43].

Biopharmaceutical is an industrial segment mainly focused on recombinant proteins, vaccines and monoclonal antibodies. Examples of this class of products include tissue plasminogen activator (tPA), insulin and recombinant hepatitis B vaccine. The global market for biopharmaceutical products already exceeds US$ 40 billion, and it has grown by more than 300% compared to the beginning of the 20th century. In the past decade, for example, it is worth noting that the market of erythropoietin (indicated for anemia) reached US$ 6.8 billion while insulin (indicated for diabetes) reached US$ 4.0 billion [74].

Some other important pharmaceutical products obtained through biotechnological processes include antitumor compounds, immune suppressants, cholesterol lowering agents and enzyme inhibitors [50]. In fact, biotechnological processes considerably boost the growth of the pharmaceutical market and continues to grow faster than the average economy supported, exceeding US$ 400 billion in recent years [43].

Another sector of interest is the production of biopesticides, mainly because these compounds produce a lowr overall impact on the environment than conventional chemical pesticides, do not leave toxic residues, reduce the risk of resistance development in the target species and tend to be highly target specific [72]. In 2000, approximately US$ 160 million worth of biopesticides were commercialized, and over 90% represented sales relating to Bt products [75]. At present, biopesticides amount to less than 2% of the global pesticides market but this is expected to increase significantly in the future.

On the other hand, biofertilizers and inoculants are attracting attention as inexpensive and safe alternatives to chemical fertilizers that are used to deliver nitrogen, phosphorus, potassium, sulfur and certain other inorganic nutrients required for crop growth [76]. Some significant examples used currently in agriculture include *Rhizobium* spp., Cyanobacteria and *Azospirllum* for N_2 fixation [77].

As discussed, biotechnological additives are transforming the pharmaceutical, cosmetic and food industries, leading to significant technological changes in products and introducing more benefits related to this area. The biotechnology world market has grown, more than 10% per year, as well as the number of products available in the US and UE [78]. The development of these products has a positive impact on the additive industry, in which there is a strong trend to increase the active ingredients and nutrient levels of products derived from animals and plants on special diets [79].

CONCLUDING REMARKS

Many challenges for the biotechnology process still exists, from laboratory to an industrial scale. Some of the previous challenges have been overcome with the use of specific engineered biocatalysts and, with recent progresses, industrial biotechnology is changing the way energy, chemicals and other products are produced, and all this is being achieved with reduced environmental impact and enhanced sustainability. Finally, biotechnology significantly impacts several industrial sectors, and in some cases it is the only viable technology for obtaining products and ingredients in an industrial scale. The application of biotechnology has consistently led to economic advantages and enhanced product quality, and sustainability considerations are playing an increasing role in the increasing acceptance of bioprocesses when compared to conventional processes.

CONFLICT OF INTEREST

The authors confirms that they have no conflict of interest to declare for this publication.

ACKNOWLEDGEMENTS

The editors acknowledge Espaço da Escrita (Coordenadoria Geral - State University of Campinas) for the English corrections in in this text.

REFERENCES

[1] *FAO Statement on Biotechnology*; Food and Agriculture Organization: Rome, Italy, **2000**.

[2] Erickson, B.; Nelson, ; Winters, P. Perspective on opportunities in industrial biotechnology in renewable chemicals. *Biotechnol. J.*, **2012**, *7*(2), 176-185.
[http://dx.doi.org/10.1002/biot.201100069] [PMID: 21932250]

[3] Demain, A.L. REVIEWS: The business of biotechnology. *Ind. Biotechnol. (New Rochelle N.Y.)*, **2007**, *3*(3), 269-283.
[http://dx.doi.org/10.1089/ind.2007.3.269]

[4] Vandamme, E.J. Agro-Industrial residue utilization for industrial biotechnology products. In: *Biotechnology for agro-industrial residues utilisation*; Nigam, P.S.; Ashok, P., Eds.; Springer, **2009**; pp. 3-13.
[http://dx.doi.org/10.1007/978-1-4020-9942-7_1]

[5] Molina, G.; Bicas, J.L.; Moraes, E.A.; Marostica, M.R., Jr; Pastore, G.M. Recent developments and industrial perspectives in the microbial production of bioflavors. In: *Applications of Microbial Engineering*; Gupta, V.K.; Schmoll, M.; Maki, M.; Tuohy, M.; Mazutti, M.A., Eds.; CRC Press, **2013**; pp. 122-157.
[http://dx.doi.org/10.1201/b15250-6]

[6] Prieto, P.A. Modern biotechnology for the production of dairy products. In: *Bioprocesses and biotechnology for functional foods and nutraceuticals*; German, J.B.; Neeser, J.R., Eds.; CRC Press, **2004**; pp. 22-44.
[http://dx.doi.org/10.1201/9780203026380.ch3]

[7] Krings, U.; Berger, R.G. Biotechnological production of flavours and fragrances. *Appl. Microbiol. Biotechnol.*, **1998**, *49*(1), 1-8.
[http://dx.doi.org/10.1007/s002530051129] [PMID: 9487704]

[8] Batista, R.; Oliveira, M.M. Facts and fiction of genetically engineered food. *Trends Biotechnol.,* **2009**, *27*(5), 277-286.
[http://dx.doi.org/10.1016/j.tibtech.2009.01.005] [PMID: 19324440]

[9] Knorr, D.; Sinskey, A.J. Biotechnology in food production and processing. *Science,* **1985**, *229*(4719), 1224-1229.
[http://dx.doi.org/10.1126/science.229.4719.1224] [PMID: 17770800]

[10] Bicas, J.L.; Marostica, M.R., Jr; Barros, F.F.; Molina, G.; Pastore, G.M. Bioadditives produced by fermentation. In: *Fermentation processes engineering in the food industry*; Soccol, C.R.; Pandey, A.; Larroche, C., Eds.; CRC Press, **2013**; pp. 371-403.
[http://dx.doi.org/10.1201/b14070-16]

[11] Jube, S.; Borthakur, D. Recent advances in food biotechnology research. In: *Food biochemistry and food processing*; Hui, Y.H.; Nip, W-K.; Nollet, L.M.; Paliyath, G.; Sahlstrøm, S.; Simpson, B.K., Eds.; Blackwell Publishing, **2006**; pp. 35-70.
[http://dx.doi.org/10.1002/9780470277577.ch3]

[12] Ruane, J.; Sonnino, A. Agricultural biotechnologies in developing countries and their possible contribution to food security. *J. Biotechnol.,* **2010**, *156*(4), 356-363.
[http://dx.doi.org/10.1016/j.jbiotec.2011.06.013] [PMID: 21723334]

[13] *Enzymes. A primer on use and benefits today and tomorrow*; Enzyme Technical Association: Washington, DC, **2001**.

[14] Ramos, O.S.; Malcata, F.X. Food-Grade enzymes. In: *Comprehensive Biotechnology*; Moo-Young, M., Ed.; Pergamon, **2011**; pp. 555-569.
[http://dx.doi.org/10.1016/B978-0-08-088504-9.00213-0]

[15] Hjort, C. Industrial enzyme production for food applications. In: *Novel enzyme technology for food applications*; Rastall, R., Ed.; Woodhead Publishing: Cambridge, **2007**; pp. 43-59.

[16] Headon, D.R.; Walsh, G. The industrial production of enzymes. *Biotechnol. Adv.,* **1994**, *12*(4), 635-646.
[http://dx.doi.org/10.1016/0734-9750(94)90004-3] [PMID: 14545919]

[17] Jegannathan, K.R.; Nielsen, P.H. Environmental assessment of enzyme use in industrial production – a literature review. *J. Clean. Prod.,* **2013**, *42*, 228-240.
[http://dx.doi.org/10.1016/j.jclepro.2012.11.005]

[18] MacCabe, A.P.; Orejas, M.; Tamayo, E.N.; Villanueva, A.; Ramón, D. Improving extracellular production of food-use enzymes from *Aspergillus nidulans*. *J. Biotechnol.,* **2002**, *96*(1), 43-54.
[http://dx.doi.org/10.1016/S0168-1656(02)00036-6] [PMID: 12142142]

[19] Kirk, O.; Borchert, T.V.; Fuglsang, C.C. Industrial enzyme applications. *Curr. Opin. Biotechnol.,* **2002**, *13*(4), 345-351.
[http://dx.doi.org/10.1016/S0958-1669(02)00328-2] [PMID: 12323357]

[20] Fernandes, P. Enzymes in food processing: a condensed overview on strategies for better biocatalysts. *Enzyme Res.,* **2010**, *2010*, 862537.
[http://dx.doi.org/10.4061/2010/862537]

[21] Lievonen, J., Ed. *Technological opportunities in biotechnology*; VTT, Group for Technological Studies: Finland, **1999**.

[22] Romero-Guido, C.; Belo, I.; Ta, T.M.; Cao-Hoang, L.; Alchihab, M.; Gomes, N.; Thonart, P.; Teixeira, J.A.; Destain, J.; Waché, Y. Biochemistry of lactone formation in yeast and fungi and its utilisation for the production of flavour and fragrance compounds. *Appl. Microbiol. Biotechnol.,* **2011**, *89*(3), 535-547.
[http://dx.doi.org/10.1007/s00253-010-2945-0] [PMID: 20981417]

[23] Olempska-Beer, Z.S.; Merker, R.I.; Ditto, M.D.; DiNovi, M.J. Food-processing enzymes from recombinant microorganisms--a review. *Regul. Toxicol. Pharmacol.,* **2006**, *45*(2), 144-158.
[http://dx.doi.org/10.1016/j.yrtph.2006.05.001] [PMID: 16769167]

[24] Bommarius, A.S., Ed. *Biocatalysis: fundamentals and applications*; Wiley: New York, **2004**.
[http://dx.doi.org/10.1002/3527602364]

[25] Phyhian, S.; Todd, C.; Wiggins, K., Eds. *Biocatalysts for fine chemical synthesis; Roberts, S.M., Casy, G., Nielsen, M.-S*; WILEY-VCH: Weinheim, **1999**.

[26] Maurer, K.H.; Kottwitz, B., Eds. *Enzyme und Waschmittel Henkel informiert*; Henkel Waschmittel GmbH: Düsseldorf, **1999**.

[27] Kumar, C.G.; Takagi, H. Microbial alkaline proteases: from a bioindustrial viewpoint. *Biotechnol. Adv.,* **1999**, *17*(7), 561-594.
[http://dx.doi.org/10.1016/S0734-9750(99)00027-0] [PMID: 14538129]

[28] Chotani, G.; Dodge, T.; Hsu, a.; Kumar, M.; LaDuca, R.; Trimbur, D.; Weyler, W.; Sanford, K. The commercial production of chemicals using pathway engineering. *Biochim. Biophys. Acta - Protein Struct. Mol. Enzymol.,* **2000**, *1543*, 434-455.

[29] Madigan, M.T.; Marrs, B.L. Extremophiles. *Sci. Am.,* **1997**, *276*(4), 82-87.
[http://dx.doi.org/10.1038/scientificamerican0497-82] [PMID: 11536798]

[30] Lorenz, P.; Liebeton, K.; Niehaus, F.; Eck, J. Screening for novel enzymes for biocatalytic processes: accessing the metagenome as a resource of novel functional sequence space. *Curr. Opin. Biotechnol.,* **2002**, *13*(6), 572-577.
[http://dx.doi.org/10.1016/S0958-1669(02)00345-2] [PMID: 12482516]

[31] Turner, N.J. Directed evolution of enzymes for applied biocatalysis. *Trends Biotechnol.,* **2003**, *21*(11), 474-478.
[http://dx.doi.org/10.1016/j.tibtech.2003.09.001] [PMID: 14573359]

[32] Zhao, H.; Chockalingam, K.; Chen, Z. Directed evolution of enzymes and pathways for industrial biocatalysis. *Curr. Opin. Biotechnol.,* **2002**, *13*(2), 104-110.
[http://dx.doi.org/10.1016/S0958-1669(02)00291-4] [PMID: 11950559]

[33] Hatti-Kaul, R.; Törnvall, U.; Gustafsson, L.; Börjesson, P. Industrial biotechnology for the production of bio-based chemicals--a cradle-to-grave perspective. *Trends Biotechnol.,* **2007**, *25*(3), 119-124.
[http://dx.doi.org/10.1016/j.tibtech.2007.01.001] [PMID: 17234288]

[34] Uzogara, S.G. The impact of genetic modification of human foods in the 21st century: a review. *Biotechnol. Adv.,* **2000**, *18*(3), 179-206.
[http://dx.doi.org/10.1016/S0734-9750(00)00033-1] [PMID: 14538107]

[35] Phillips, P.W. Biotechnology in the global agri-food system. *Trends Biotechnol.,* **2002**, *20*(9), 376-381.
[http://dx.doi.org/10.1016/S0167-7799(02)02039-5] [PMID: 12175768]

[36] Falk, M.C.; Chassy, B.M.; Harlander, S.K.; Hoban, T.J., IV; McGloughlin, M.N.; Akhlaghi, A.R. Food biotechnology: benefits and concerns. *J. Nutr.,* **2002**, *132*(6), 1384-1390.
[PMID: 12042463]

[37] Eck, P. Recombinant DNA Technologies in Food. In: *Biochemistry of Foods*; Eskin, N.A.; Shahidi, F., Eds.; Academic Press, **2012**; pp. 504-553.

[38] Runge, C.F.; Ryan, B. HThe economic status and performance of plant biotechnology in 2003: adoption, research and development: United States, **2003**.

[39] Cichocka, D.; Claxton, J.; Economidis, I.; Högel, J.; Venturi, P.; Aguilar, A. European Union research and innovation perspectives on biotechnology. *J. Biotechnol.,* **2010**, *156*(4), 382-391.
[http://dx.doi.org/10.1016/j.jbiotec.2011.06.032] [PMID: 21745504]

[40] Costa-Font, M.; Gil, J.M.; Traill, W.B. Consumer acceptance, valuation of and attitudes towards genetically modified food: Review and implications for food policy. *Food Policy,* **2008**, *33*, 99-111.
[http://dx.doi.org/10.1016/j.foodpol.2007.07.002]

[41] Pardo, R.; Midden, C.; Miller, J.D. Attitudes toward biotechnology in the European Union. *J. Biotechnol.,* **2002**, *98*(1), 9-24.
[http://dx.doi.org/10.1016/S0168-1656(02)00082-2] [PMID: 12126802]

[42] Azadi, H.; Ho, P. Genetically modified and organic crops in developing countries: a review of options for food security. *Biotechnol. Adv.,* **2010**, *28*(1), 160-168.
[http://dx.doi.org/10.1016/j.biotechadv.2009.11.003] [PMID: 19913085]

[43] Gavrilescu, M.; Chisti, Y. Biotechnology-a sustainable alternative for chemical industry. *Biotechnol. Adv.,* **2005**, *23*(7-8), 471-499.
[http://dx.doi.org/10.1016/j.biotechadv.2005.03.004] [PMID: 15919172]

[44] Organisation for Economic Co-operation and Development. The application of biotechnology to industrial sustainability. *Available from: http://www.keepeek.com/Digital-Asset-Management/oecd/environment/the-application-of-biotechnology-to-industrial-sustainability_9789264195639-en#page1,* **2001**.

[45] Hall, L.a.; Bagchi-Sen, S. An analysis of firm-level innovation strategies in the US biotechnology industry. *Technovation,* **2007**, *27*, 4-14.
[http://dx.doi.org/10.1016/j.technovation.2006.07.001]

[46] Amir-Aslani, A.; Negassi, S. Is technology integration the solution to biotechnology's low research and development productivity? *Technovation,* **2006**, *26*, 573-582.
[http://dx.doi.org/10.1016/j.technovation.2005.06.001]

[47] Terziovski, M.; Morgan, J.P. Management practices and strategies to accelerate the innovation cycle in the biotechnology industry. *Technovation,* **2006**, *26*, 545-552.
[http://dx.doi.org/10.1016/j.technovation.2004.10.016]

[48] Oinas, P.; Malecki, E.J. The Evolution of Technologies in Time and Space: From National and Regional to Spatial Innovation Systems. *Int. Reg. Sci. Rev.,* **2005**, *25*(1), 102-131.

[http://dx.doi.org/10.1177/016001702762039402]

[49] Hall, L.; Bagchi-Sen, S. An analysis of R&D, innovation and business performance in the US biotechnology industry. *Int. J. Biotechnol.,* **2001**, *3*, 267-286. [http://dx.doi.org/10.1504/IJBT.2001.000165]

[50] Demain, A.L. Small bugs, big business: the economic power of the microbe. *Biotechnol. Adv.,* **2000**, *18*(6), 499-514. [http://dx.doi.org/10.1016/S0734-9750(00)00049-5] [PMID: 14538099]

[51] Schmid, R.D., Ed. *Pocket guide to biotechnology and genetic engineering*; WILEY-VCH: Weinheim, **2003**.

[52] Winter, J., Ed. *Environmental biotechnology: concepts and applications; Jördening, H.-J*; WILEY-VCH: Weinheim, **2004**.

[53] Conway, R. *US Biobased Products Market Potential and Projections Through,* **2008**, *2025*, 12-18.

[54] Lorenz, P.; Zinke, H. White biotechnology: differences in US and EU approaches? *Trends Biotechnol.,* **2005**, *23*(12), 570-574. [http://dx.doi.org/10.1016/j.tibtech.2005.10.003] [PMID: 16253362]

[55] Bachmann, R.; Riese, J. Industry overview: from promise to profit. *Ind. Biotechnol. (New Rochelle N.Y.),* **2005**, *1*(1), 9-15. [http://dx.doi.org/10.1089/ind.2005.1.9]

[56] Kircher, M. How to turn industrial biotechnology into reality. *N. Biotechnol.,* **2012**, *29*(2), 243-247. [http://dx.doi.org/10.1016/j.nbt.2011.11.002] [PMID: 22100431]

[57] Strohl, W.R., Ed. *Biotechnology of antibiotics*; Dekker: New York, **1997**. [http://dx.doi.org/10.1201/b14856]

[58] Liese, A.; Seelbach, K.; Wandrey, C., Eds. *Industrial biotransformations. A collection of processes*; WILEY-VCH: Weinheim, **2000**. [http://dx.doi.org/10.1002/9783527614165]

[59] Schreiber, S.L., Ed. *Biosynthesis: aromatic polyketides, isoprenoids, alkaloids*; Springer: Berlin, **2000**.

[60] Demirbas, A. Biofuels sources, biofuel policy, biofuel economy and global biofuel projections. *Energy Convers. Manage.,* **2008**, *49*, 2106-2116. [http://dx.doi.org/10.1016/j.enconman.2008.02.020]

[61] Roehr, M., Ed. *The biotechnology of ethanol: classical and future applications*; WILEY-VCH: Weinheim, **2001**.

[62] Graboski, M.S.; McCormick, R.L. Combustion of fat and vegetable oil derived fuels in diesel engines. *Pror. Energy Combust. Sci.,* **1998**, *24*(97), 125-164. [http://dx.doi.org/10.1016/S0360-1285(97)00034-8]

[63] Nandi, R.; Sengupta, S. Microbial production of hydrogen: an overview. *Crit. Rev. Microbiol.,* **1998**, *24*(1), 61-84. [http://dx.doi.org/10.1080/10408419891294181] [PMID: 9561824]

[64] Balat, M. Production of bioethanol from lignocellulosic materials *via* the biochemical pathway: A review. *Energy Convers. Manage.,* **2011**, *52*(2), 858-875.

[http://dx.doi.org/10.1016/j.enconman.2010.08.013]

[65] United States Department of Agriculture – Foreign Agriculture Service. *Available from: http://www.fas.usda.gov/ [accessed Dec 1, **2014**].*,

[66] Bhatia, L.; Johri, S.; Ahmad, R. An economic and ecological perspective of ethanol production from renewable agro waste: a review. *AMB Express,* **2012**, *2*(1), 65. [http://dx.doi.org/10.1186/2191-0855-2-65] [PMID: 23217124]

[67] Hartemink, A.E. Chapter 3 Sugarcane for Bioethanol: Soil and Environmental Issues. In: *AMB Express*; Academic Press: Agronomy, D.L.S.B.T.-A, **2008**; 99, p. 125, 182.

[68] GranBio. *Available from: http://www.granbio.com.br/.,*

[69] Sauer, M.; Porro, D.; Mattanovich, D.; Branduardi, P. Microbial production of organic acids: expanding the markets. *Trends Biotechnol.,* **2008**, *26*(2), 100-108. [http://dx.doi.org/10.1016/j.tibtech.2007.11.006] [PMID: 18191255]

[70] Prajapat, JB; Food, A.P. **2013**, Food and Health Applications of Exopolysaccharides produced by Lactic acid Bacteria. *Adv. Dairy Res.,* **2013**, *1*(2), 1-7.

[71] Morris, G.; Harding, S. *Polysaccharides, Microbial*; Encycl. Microbiol, **2009**, pp. 482-494.

[72] Imeson, A., Ed. *Food stabilisers thickening and gelling agents*; Wiley Blackwell Publishing: Weinheim, **2010**.

[73] Hulse, J.H. Biotechnologies: past history, present state and future prospects. *Trends Food Sci. Technol.,* **2004**, *15*(1), 3-18. [http://dx.doi.org/10.1016/S0924-2244(03)00157-2]

[74] Melmer, G. Biopharmaceuticals and the industrial environment. In: *Production of recombinant proteins: novel microbial and eukaryotic expression systems*; Gellissen, G., Ed.; WILEY-VCH: Weinheim, **2005**; pp. 361-383. [http://dx.doi.org/10.1002/3527603670.ch16]

[75] Vega, O.F A review of Bacillus thuringiensis (Bt) production and use in Cuba. *Biocontrol News Inf.,* **1999**, *20*, 47-48.

[76] Subba Rao, N.S. *Biofertilizers in agriculture*; Balkema: Rotterdam, **1982**.

[77] Pimentel, D., Ed. *Encyclopedia of pest management*; Dekker: New York, **2002**. [http://dx.doi.org/10.1201/NOE0824706326]

[78] Vialta, A. Ingredients - New functionalities. In: *Brasil Food Trends 2020*; Fiesp, I.T., Ed.; São Paulo, **2010**; pp. 98-109.

[79] Bhamra, T. The Role of Innovation in the Move Towards Sustainable Design of Packaging. *Available from: http://www.idspackaging.com/common/paper/paper_6/ [accessed Jun 1, **2014**].*,

CHAPTER 2

Alternative Sweeteners: Current Scenario and Future Innovations for Value Addition

R. K. Saini, S. Sravan Kumar, P. S. Priyanka, K. Kamireddy and **P. Giridhar***

Plant Cell Biotechnology Department, CSIR-Central Food Technological Research Institute, Mysore - 570 020, India

Abstract: Sustainable growth and consistent demand for zero or low-calorie alternative sweeteners by the global market are mainly attributable to public consciousness about health impact of artificial sugar substitutes. Despite limited market for natural sweeteners, a spurt in preference to plant derived-alternative sweeteners is known. Sugar substitutes, such as non-nutritional artificial sweeteners, low calorie or zero calorie natural sweeteners that include sugar alcohols and plant derived non-saccharide sweeteners find use in making various types of foods and beverages. From an industry point of view, approval for usage of sugar substitutes in food products by the regulatory agencies can initiate major trends. These trends can contribute to the safety and health consciousness of consumers and also to food and beverage industries to get better market and price. There is a need to further refine the available technologies for the production of alternative sweeteners, especially natural sweeteners through a plant-derived or microbial cell based production platform with the intervention of metabolic engineering to produce novel sweeteners.

Keywords: Acesulfame, Alitame, Artificial sweeteners, Aspartame, Erythritol, Isomalt, Lactitol, Low-Calorie sweeteners, Maltitol, Mannitol, Natural sweeteners, Neotame, Non-saccharide sweeteners, Polyols, Rebaudioside, Reduced-Calorie Sweeteners, Sorbitol, Stevioside, Sucralose, Xylitol.

* **Corresponding author P. Giridhar:** Plant Cell Biotechnology Department, CSIR-Central Food Technological Research Institute, Mysore - 570020, India; Telephone: 91-821-2516501; Fax: 91-821-2517233; E-mail:parvatamg@yahoo.com.

Juliano Lemos Bicas, Mário Roberto Maróstica, Glaucia Maria Pastore (Eds.)

INTRODUCTION

A wide range of food products containing natural sweeteners with an emphasis to bring down number of calories has gained momentum, especially to address obesity and diabetes *etc.*, which are prevalent in consumers and attributable to changing dietary habits and sedentary lifestyle. Moreover these alternative sweeteners are helpful to manufacturers of food products and also to consumers, in a high sugar price environment that prevails today. Sweetener can be defined as any substance added to food or beverage to make it taste sweeter. Sugar substitutes as a food additive mimic the effect of sugars in taste usually with fewer calories. Sugar substitutes can be classified as natural, synthetic or artificial on the basis of by their production [1]. The first category (synthetic) includes some of the important artificial sweeteners, such as aspartame, neotame, acesulfame, saccharin, sucralose *etc.* (Fig. **1**). Many of them are known as high-intensity sweeteners which are much sweeter and have a minimal energy contribution in food compared to sucrose (Fig. **2**). Food containing high-intensity sweeteners prevent excessive calorie intake and are claimed to be helpful in weight loss and in diabetes management [2]. The second category of sugar substitutes is natural sweeteners that occur naturally in certain fruits and vegetables, but can also be manufactured artificially. These natural sweeteners can be grouped in two major categories, comprising saccharides and non-saccharides. Saccharides based on natural sweeteners, also known as nutritive sweeteners or carbohydrates, contain polyhydroxy aldehydes or ketones, such as sucrose, glucose, trehalose *etc.* Non-saccharides based on natural sweeteners can be grouped in five major classes, such as terpenoids, proteins, flavonoids, steroidal saponins and polyols. Polyols or Sugar alcohols are compounds with multiple hydroxyl functional groups and commonly added to foods because polyols have lower calorie than sugars. Maltitol, lactitol, sorbitol, xylitol, erythritol, and isomalt are some of the more common examples of polyols (Fig. **3**). Another major class of non-saccharide sweeteners includes flavonoids and their derivatives (Fig. **4**), such as neohesperidin, phyllodulcin, naringin *etc.* Steroidal saponins are another class of non-saccharide sweeteners composed of rhamnopyranosyl units such as osladin (Fig. **5**). Though the above mentioned categorization is acceptable for demarcation of some sugar substitutes, it is also important to have knowledge about their alternative source whether natural or a derivative, processed or refined, or chemically derived from herbs or sugar. All these alternative sweeteners include non-nutritive, low calorie, low glycemic or saccharide-derived and non-saccharide sweeteners.

Sweeteners

- Synthetic- Acesulfame K, Alitame Aspartame, Neotame, Cyclamate
- Natural
  - Non-saccharides
    - **Terpenoids**- Abrusosides & glycyrrhizin, Gaudichaudioside-A, Carnosiflosides-V, VI Periandrin V, Mogroside V, Cucurbitane, Stragole, Hernandulcin, Perillartine, Mogroside, Steviosides
    - **Steroidal saponins**- Cyclocaryoside, Polypodoside, Osladin
    - **Proteins**- Cynarin, Monellin, Miraculin, Thaumatin, Mabinilin, Neoculin
    - **Flavonoids**- Hesperidin dihydrochacones, Naringin dihydrochalcones, Neohesperidin, Phyllodulcin, Naringin, Selligueain A
    - **Polyols**- Xylitol, Erythritol, Maltitol, Sorbitol, Isomalt
  - Saccharide- Sucrose, Glucose, Honey

Fig. (1). Classification of sweeteners.

In the USA, the Food and Drug Administration regulates artificial sweeteners as food additives that must be approved as GRAS (Generally Recognized as Safe). Most of the sugar substitutes approved for food use belong to artificial sweeteners category such as acesulfame, aspartame, saccharin, neotame, and sucralose. Food Safety and Standards Authority of India (FSSAI) also approved four artificial sweeteners, including aspartame, acesulfame K2, saccharin and sucralose to be used in food industry. The food and beverage industries are replacing sugar with low calorie sweeteners in a variety of food products. Low calorie sweeteners increase a fraction of the cost in food production. Many reviews on sweeteners focused mainly on chemistry, biosynthesis, production, characterization and application in foods [3 - 9]. In view of the ever-increasing demand for alternative

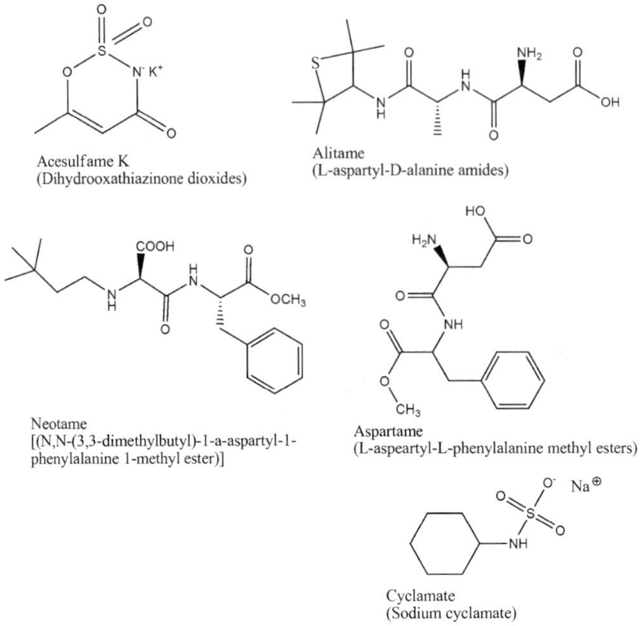

Fig. (2). Chemical structures of synthetic sweeteners.

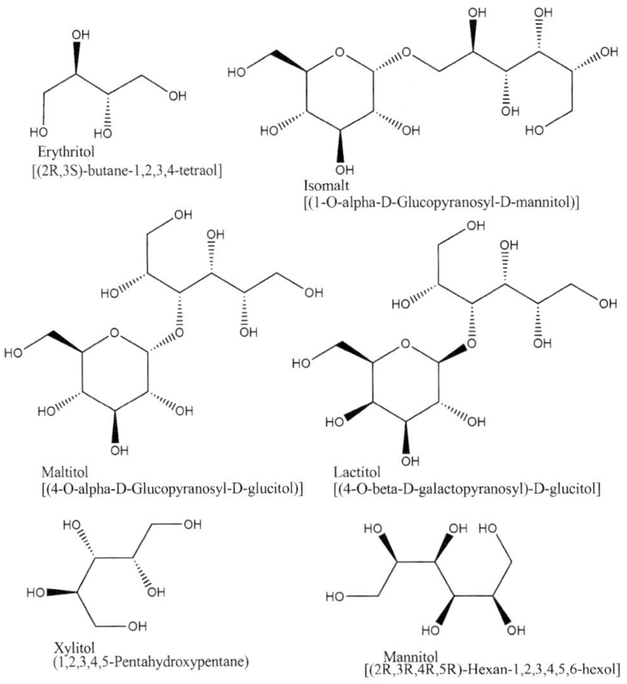

Fig. (3). Chemical structures of polyol class of natural sweeteners.

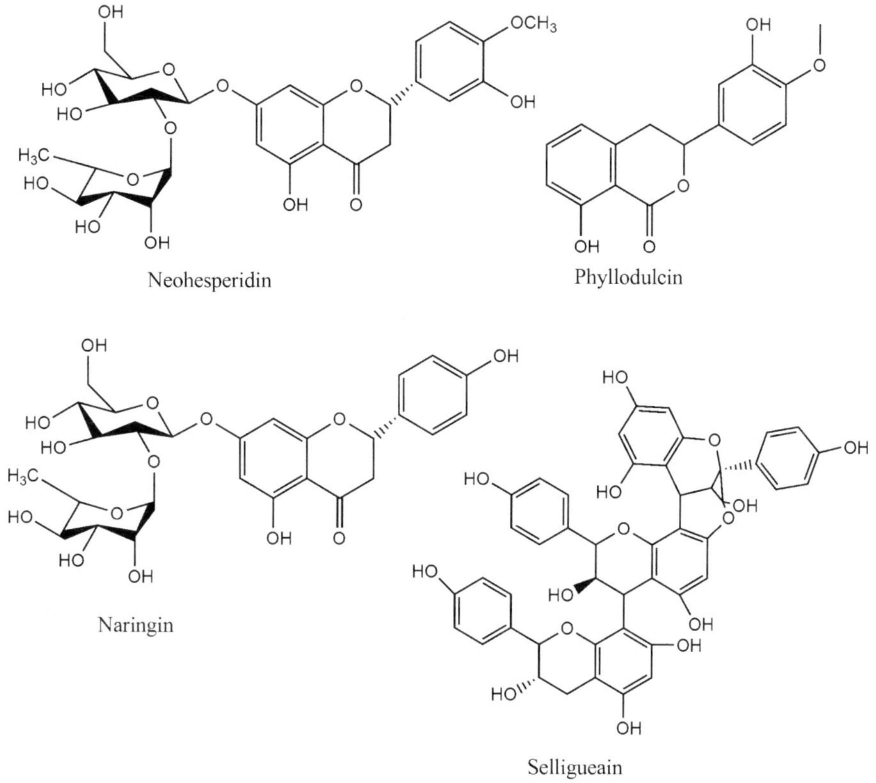

Fig. (4). Chemical structures of flavonoid group sweeteners.

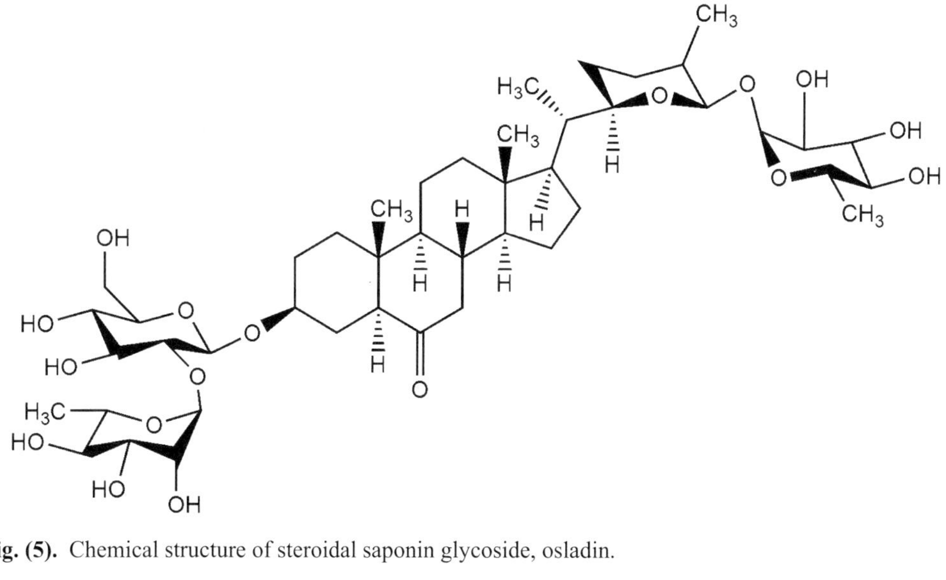

Fig. (5). Chemical structure of steroidal saponin glycoside, osladin.

sweeteners in food industry, a glance at recent developments in biotechnological production of sweeteners, especially natural sweeteners, is warranted. In this chapter, the global sweeteners market, a brief up about different types of alternative sweeteners, the emergence of new trends in the sweeteners world and their role in the food industry and potential areas for future research are discussed.

NATURAL SWEETENERS

At present, globally there is a growing consensus about drinks and beverages loaded with artificial sweeteners and their adverse effects on consumer's health. In this context, health and wealth have become the heat wave towards natural sweeteners. Natural sugar substitutes as alternative sweeteners gained importance as consumers have been preferring food products with their addition, instead of the conventional artificial sweeteners. The greater demand for such surge in products fortified as alternative natural sweeteners is due to preference for sugar-free foods, diet with reduced calories, diet with low calories, or healthy diet. Natural sweeteners are currently been marketed as consumer-friendly products that are likely to replace 20-30% of sugars in food products without compromising the taste. The European Commission (EU) approved glycosides from *Stevia rebaudiana* as a sweetener in food ingredient. Recently, Food and Drug Administration (FDA) of the USA also approved stevia for use in food products. In India, FSSAI also approved stevia usage in soft drink concentrates, chewing gums, carbonated water *etc.* Stevia has become gradually popular in the last few years, as an alternative to artificial sweeteners. Foods contain several different types of natural sugar in various forms, including dextrose, maltose, sorbitol, mannitol, xylitol, levulose, fructose and maltitol *etc.*, which are also common.

GLOBAL SWEETENERS MARKET

In terms of consumption, sugar has dominant share (83% to 85%) in the global sweeteners market [10]. Global market for sugar and sweeteners in 2012 was nearly US$77.5 billion and is forecasted to grow to US$97.2 billion by 2017. Thus, sugar alone accounts for 175 x 10^6 metric tonnes per year (~US$65 billion). The non-nutritive category has been growing at a faster pace at a compound annual growth rate (CAGR) of 5.2%. The market for high-intensity sweeteners is expected to reach nearly US$1.9 billion in 2017 [10]. Global market for alternative sweetener in food and beverage industry was 2.8% since 2012 and expected to grow 9.7% in the next three years to reach a figure of US$ 1.4 billion in 2017 (<http://nextgenfoodtech.com/sweetener-market-scenario.html> accessed on May, 20, 2014). The North American market is expected to reach US$ 6 million in 2016 followed by European market which represents the 2nd largest market category (US$ 2 billion in 2016 for CGAR of 1.5%) (<http://www.bccresearch.com/market-research/food-and-beverage/non-sugar-sweeteners-market-fod044a.html> accessed on May 20, 2014). The sweetener market is mainly dominated by high intensity sweetener (HIS) and high fructose syrup (HFS).

According to a report (ISO, 2012), saccharine is still dominant among high intense sweeteners, and the USA remains the strong hold for aspartame. Similarly, cyclamate is mostly used in Asia, though there is a stable market for acesulfame-K. In addition to these high-intense sweeteners, neotame gains ground in food process sector as an alternative sweetener.

Presently, acesulfame-K holds lion's share of 40% followed by aspartate (30%). However, due to growing consensus about aspartame on health, European Food Safety Authority (EFSA) predicted a decline in the aspartame market to US$ 360 million in 2017 from US$ 480 million in 2013. In India the total artificial sweetener market is valued at US$ 12.5 of which 95% of market is dominated by aspartame (<http://nextgenfoodtech.com/sweetener-market-scenario.html> accessed on May, 20, 2014).

Among the alternative sweeteners, being a natural sweetener, stevia was the 5th best selling non-caloric sweetener behind sucralose, aspartate, saccharine and cyclamate. At present there is a sustainable growth of natural sweeteners in sweetener industry because of recent developments in this area. Stevia has become particularly popular, reaching a global market of US$ 110 million in 2013-2014 and is expected to reach US$ 275 million in 2017 (forecasted by Mintell and Leatherhead Food Research, <http://nextgenfoodtech.com/sweetener-market-scenario.html> accessed on May 20, 2014). Steviol glycosides are popular in view of their high sweetening potential relative to sucrose. They are stevioside (150-300), Rebaudioside-A (200-400), Rebaudioside B (300-350), Rebaudioside-C (50-120), Rebaudioside-D (200-300), Rebaudioside-E (250-300), Rebaudioside (110), Steviolbioside-H (100-120), and Steviolbioside (50-120), respectively. Stevia is projected to drive the European market to grow at 5% CAGR. In Brazil and Asia-Pacific countries there is also a growing demand for low intensity sweeteners (LIS), high intensity sweeteners (HIS) and high fructose syrups (HFS) especially for beverage application (<http://www.preparedfoods.com/articles/112661-size-of-the-worlds-sweetener-market> accessed on May, 20, 2014) in food industry.

USES OF SWEETENERS IN FOOD INDUSTRY

Sweeteners of all categories have pivotal role in nowadays global food industry including beverages. Quality and consistency of sweetener compounds are vital, though there are varied sources of sweeteners, and wide range of process technologies. Irrespective of their nutrition status, the bulk sweeteners used in food industry are mainly based on cane, corn and beetroot followed by polyols and hydrogenated sugars. However, for beverages and confectionary, both high potency artificial sweeteners and low calorie nutritive sweeteners are in use. The International Numbering System (INS) for sweeteners is given in Table **1**. Very recently, the FDA approved the sixth sugar substitute ‘advantame’, an ultrahigh potency sweetener from Ajinomoto Co. Inc. (AJINY) which is about 20.000 times sweeter than sucrose and gives a sugar-like taste (<http://www.bidnessetc.com/21899-general-mills-inc-nyse-gis-the-coca-cola-company-nyse-ko-and-mondelez-international-nasdaq-mdlz-news-analysis-advantame-is-good-news-for-food-and-beverage-makers> accessed on July 17, 2014).

Table 1. The International Numbering System (INS) for major sweeteners.

S/No.	Name of Sweetener	INS Number
1	Sorbitol	420
2	Mannitol	421
3	Glycerol	422
4	Isomalt	953

(Table 1) contd.....

S/No.	Name of Sweetener	INS Number
5	Maltitol	965
6	Lactitol	966
7	Xylitol	967
8	Erythritol	968
9	Acesulfame K	950
10	Aspartame	951
11	Cyclamates	952
12	Saccharins	954
13	Sucralose	955
14	Alitame	956
15	Thaumatin	957
16	Glycyrrhizin	958
17	Neohesperidin DC	959
18	Stevioside	960
19	Neotame	961
20	Aspartame-acesulfame Salt	E962

Acesulfame K, aspartame and erythritol are allowed in the USA in table-top sweeteners, carbonated beverages, unstandardized beverage concentrates and mixes, unstandardized dairy beverages, filling mixes, fillings, topping mixes, toppings, unstandardized salad dressings, unstandardized dessert mixes, unstandardized desserts, yogurt, breath freshener products (except chewing gum), unstandardized fruit spreads, unstandardized confectionery, baking mixes; unstandardized bakery products, chewing gum and syrups (especially erythritol). Neotame is also used in the above-mentioned foods along with breakfast cereals. Other permitted sweeteners such as hydrogenated starch hydrolysates, isomalt, lactitol, maltitol and maltitol syrup, mannitol are permitted in unstandardized foods. Monk fruit extract is permitted only in table top sweeteners. Sorbitol is permitted to sweeten a blend of prepared fish and prepared meat. Similarly, saccharin is permitted as sweetener in breath freshener products, chewing gum, unstandardized frozen foods, syrups, beverages *etc.*; steviol glycosides and sucralose finds extensive use in bakery, beverages, cereal products, desserts, confectionary, sauces, purees, fruit spreads, toppings, dental creams, and chewing gums.

Xylitol finds use in unstandardized foods that come under good manufacturing practice. Regarding approvals for plant derived non-saccharide sweeteners, Japan is the only country so far that has approved monellin and curculin for food applications. Compared to these, other category of sweeteners such as 'sweet proteins' obtained moderate attention in commercial sweeteners market. Among the seven identified sweet proteins such as thaumatin, molenin, mobinilin, pentadine, brazzein, carculin and miraculin, only two are in the market. They are Talin (thaumatin from Naturex, France) and Cweet (Brazzein from Natur Research Ingradients, Losangeles, USA). Almost all these sweet proteins are obtained from Tropical rainforest plants. Product development and regulatory status aspects of thaumatin is best advanced among sweet proteins. Miraculin has approval in Japan and some other countries, and it has novel food status in the EU. Glycyrrhizin is approved in the USA as GRAS and approved in the EU as E958. Luo Han Guo (monk fruit) has GRAS approval by FDA in the USA. Pentadin comes from the climbing plant Oubli of West Africa and do not have any approval so far.

NATURAL SWEETENERS OF PLANT ORIGIN

A search for non-nutritive or low calorie intense sweeteners remains a potential area of research. Significant work has been done to explore plant secondary metabolites as sweeteners. A sweet taste of the identified molecule is contributed by diverse chemical structures as evident from ~ 80 sweet compounds reported (other than monosaccharides, disaccharides, polyols) that derived from natural sources including vascular plants [11, 12]. Natural sweeteners production is mainly based on saccharides or non-saccharides. In the following items of this chapter, we will focus on non-saccharide natural sweeteners.

Non-Saccharide Natural Sweeteners

These compounds are obtained from Apiaceae, Asteraceae, Capparidaceae, Fabaceae, Illiaceae, Lamiaceae, Marantaceae, Menispermaceae, Pinaceae, Piperaceae, Rutaceae *etc.* These sweetener compounds include small molecular secondary metabolites (flavonoids, terpenoids, caumarins, steroidal saponins *etc.*,

(Table **2**) and also macromolecular proteins that have intense sweetening property [12 - 14]. However, chemical property of many of these sweetener molecules is not well understood. A list of such sweetener compounds, their source and sweetening potential compared with sugar were summarized (Table **2**).

All these sweetener molecules exhibit diverse functional properties with varied stability properties, low to high water solubility nature and find use in liquor, medicines, beverages, confectionary and solid foods as flavouring and sweetening agents.

Table 2. List of some non-saccharide natural sweeteners.

Non-saccharide natural sweeteners	Source	Sweetness folds compared to sucrose	Reference
Proteins			
Mabinlin	*Fruits of Capparis masaikai*	100	[11]
Monelin	Fruit pulp of *Dioscoriophyllum cumminsii*	2000	[15 - 17]
Thaumatin I and Thaumatin II	fruit aril portion of *Thaumatococcus daniellii*	10.000	[18]
Pentadine and Brazzein	*Pentadiplandra brazzeana*	500	[19]
Curculin and neoculin	Fruits of *Curculigo latifolia*	Sweet tasting and taste modifying ability	[20]
Terpenoids			
Abrousoside A-E (triterpene glycosides) and glycyrrhizin	Leaves and root of Indian liquorice - *Abrus pricatorius*	30 to 100 times sweeter than 2% sucrose	[21]
Labdane diterpene glycosides and Gaudichaudioside-A	Aerial parts of *Baccharis gaudichaudiana*	55 times sweeter than 2% sucrose solution	[22]
Pentacyclic triterpene (glycyrrhizin) saponin glycoside	Root stalks and stolons of Licorice - *Glycyrrhizin glabra*	100 and 50 times sweeter than sucrose	[23]
Trans anithole and estragole	Dry fruits of *Illicium verum*	13-15	[24]
Sesquiterpene, hernandulcin	Aerial parts of *Lippia dulcus*	1000	[25]

(Table 2) contd.....

Non-saccharide natural sweeteners	Source	Sweetness folds compared to sucrose	Reference
Triterpene glycoside, periandrum I, II and V	Rhizomes of *Periandra dulcis*	90-200 times sweeter than 2% sucrose	[26]
Monoterpenoids perillartine	Whole plant of *Perilla frutescens*	200	
Diterpene acids	Resin of *Pinus* species	2000	
Diterpene tricyclic glycosides, steviosides, rebaudiosides A to E, dulcosides A to B	Leaves of *Stevia rebaudiana*	100 to 400	
Cucurbitane	Aerial parts of climber *Thladiantha grozavesti*	150	[27]
Triterpenoid glycoside	Fresh whole plant of *Myrrhis adorata* Dried leaves of *Piper marginatum*		[23, 26, 28]
Mogroside-V	Fresh roots of *Osmorhiza longistylis*		
Flavonoids			
Neohesperidin Dihydrochalcone	Fruit peel of *Citrus aurantium*	1000	[23, 26]
Naringin	Fruit peel of *Citrus paradise*	1000	
Trilobatin	Leaves of *Symplocos paniculata*	400-1000	
Hesperidine dihydrochalcone glucoside	Fruits of *Citrus sinensis*, *Citrus limon*	300	
Dihydrochalcone glycoside-glycyphyllin	All parts of *Smilax glycyphylla*	100-200	[29]
Trimeric proanthocyanidin selligueain-A	Rhizomes of *Polypodium feii*	35 times sweeter than 2% sucrose	[30]
Dihydroflavonol (+ dihydroquercetin---acetate)	Young shoots of *Tessaria dodoneiofla*	80 times sweeter than 2% sucrose	[31]
Dihydroisocoumarin, phylodelcin	Fermented crushed leaves of *Hydrangea macrophylla*	300 to 400	[32]
Steroid group			
Steroidal saponin glycosides such as polypodoside-A	*Polypodium glycyrrhiza*	100 times sweeter than 1% sucrose	[33]
Osladin	Rhizomes of *Polypodium vulgarae*	300-3000	[34]
Pterocarioside-A and B	Leaves and stems of *Pterocarya paliurus*	50 to 100 times sweeter than 2% sucrose	[35]

There are some other natural sweeteners which stand as healthier option for a touch of sweetness such as Lo Han Kuo extracts of Chinese fruit (*Siraitia grosvenori* or *Momordica grosvenori*) which is 250 times sweeter than sugar [36 - 40] with antioxidant property and erythritolerythritol that is a naturally processed sugar alcohol exhibiting 70% as sweet as sugar. There are low glycemic options such as maple syrup - a boiled sap of sugar from maple tree, barley malt syrup that contains mostly maltose which does not directly activate an insulin response, agave, a nectar of cactus species due to its high ratio of fructose to glucose [41], brown rice syrup, which is a mild sweetener and acts as healthier option, as this is prepared from cultured cooked brown rice upon enzymatic conversion of sugars to maltose *etc.*

ECONOMICAL IMPORTANCE AND COMMERCIAL RELEVANCE OF BIOTECHNOLOGICALLY PRODUCED SWEETENERS

During the last decade, globally attempts tried to create commercialization prospects for biotechnology-based methods for production of sweeteners [9]. However, limitations are imminent, such as low yielding, which is not good enough for laboratory level analysis, lack of sustainable sources for large scale extraction, and lack of cost efficiency. Moreover, artificial sweeteners are ruling the roost in global food processing industry, hence venturing into commercialization of biotech processes pertaining to natural sweeteners is lagging behind. Nevertheless, changes in public perception have provoked researchers to continue their efforts to further refine the available technology to value addition with noteworthy contributions recently made [8, 30, 42 - 44].

NATURAL SACCHARIDE-DERIVED SWEETENERS

Sweeteners based on saccharide or sugars such as sugar alcohols are not considered non-nutritive or non-caloric sweeteners, as they provide some calories though it is less than the regular table sugar, hence considered an attractive alternative. Regardless of their name sugar alcohols are not alcoholic. Along with sweet taste they also add bulk and texture to food. They are mainly used in many processed foods and other products, including chewing gum, candy, frozen desserts, chocolate, toothpaste, mouthwash, baked goods *etc.*

Polyols

Global market for polyols or polyhydric alcohols is of about US$1009 million, where sorbitol accounts for 83%, xylitol for 14% and mannitol for about 3% (<www.specchemonline.com, Sugaring the pill>, October 21, 2004). Polyols vary in sweetness from about 50% less sweet to that of sucrose. They are manufactured from corn starch, birch wood, sucrose and whey. In a decreasing order the calories per gram of sweeteners is: sucrose (4) > polyglycitol (3) > sorbitol (2.6) > xylitol (2.4) > maltitol (2.1) > isomalt (2) > lactitol (2) > mannitol (1.6) > erythritol (0.2). Similarly, compared to sucrose, sweetness of polyols is 90% for maltitol powder, 70% for erythritol and mannitol, 45-65% for isomalt, 40-90% for maltitol syrups and polyglycitol.

The demand for polyols in various end-use sectors such as industrial application, food and confectionary will improve the overall market revenues. Developing technology to minimize cost and maximize environmental benefit in the lowest cost, in the most environmentally sustainable way, should be pre-requisite for industrial scale production of polyols.

Mannitol

Mannitol is produced by hydrogenation of fructose-glucose syrup in presence of hydrogen gas and Raney nickel catalyst at high temperature and pressure [45] that leads to 30% mannitol production which needs to be purified by chromatography and crystallized at low temperature. But this has some limitations such as low production of mannitol and higher cost of production. In view of this, to meet the market demand and also to have cost effective methods, biological mannitol production processes are actively being investigated [45 - 47]. Mannitol has existed as commercial product for more than six decades and is widely used as a component of sugar-free gum and candy, and also in pharmacy as approved by FDA. It is produced by range of microorganisms, algae, fungi, yeast, lichens and plants. Microbial fermentation is an alternative process for the production of mannitol. This could be achieved either by heterofermentative lactic acid bacteria, which have the ability to convert fructose into mannitol or by homofermentative organisms, which are unable to produce mannitol directly from fructose, but they are mainly present in fermented food products. Engineering homofermentative organisms will help add extra value to the product and also to increase the viability of starter culture. As lactic acid bacteria (LAB) are fastidious organisms which add up to the cost of production. Hence, engineering other organisms, such as yeasts and *Escherichia coli* will be a cost effective alternative (Table **3**).

Aarnikunnas *et al.*, [53] used metabolic engineering of *L. fermentum* for production of mannitol through inactivating the lactate dehydrogenase D (ldhD) and L-(+)-lactate dehydrogenase (ldhL) coding genes. Kaup *et al.,* [54] constructed *E. coli* strain for production of mannitol from fructose in a whole cell biotransformation, wherein it produced up to 66 g mannitol from 90 g fructose l^{-1} within 8 h. The strain was constructed to express NAD^+-dependent MDH, NAD^+-dependent formate dehydrogenase and glucose facilitator for the uptake of fructose without concomitant phosphorylation. Recently, kinetic models of *Lactococcus lactis* have been developed for enhanced production of mannitol and 2.3-butanediol, through genetic engineering of lactate dehydrogenase (LDH) and mannitol-1-phosphate dehydrogenase (MPD) [54].

Table 3. Some important strategies for production of mannitol.

	Organism used	**Strategy**	**References**
lactic acid bacteria (LAB)	*L. lactis* Homofermentative	Inactivation of lactate dehydrogenase	[48, 49]
		Deleting *mtlF*(encoding EIIA mtl in the PTS mtl) or mtlA genes	[50]
		Double mutant (Δ*ldh* Δ*mtlA* and Δ*ldh* Δ*mtl F*)	
		Lactobacillus plantarum mtlD gene (encoding M1P-DH) and the *Eimeria tenella* mannitol 1-phosphatase (M1Pase) gene were co-expressed.	[9]
		Single (Δ*ldh B*) and double mutant strains (Δ*ldh B* Δ*ldh X*) by blocking *ldh* genes (*ldhX* and *ldh B*)	[51]
		lactococcal *mtl D* gene and MTLP from *Eimeria tenella* cloned under a nisin-inducible promoter	
	L. pseudomesenteroides	Fructo kinase inactivation	[52]
	L. fermentum Heterofermentative	disrupting both genes encoding D-and L-specific LDH (*ldh*)	[53]

(Table 3) contd.....

	Organism used	Strategy	References
	Leuconostoc pseudomesenteroides (random mutant)	*fru K* gene mutant	[52]
	Lactococcus lactis	Overproduction of mannitol-1-phosphate dehydrogenase (MPD)	[54]
Other organisms	*Escherichia coli*	Co-expressed the *L. Pseudomesenteroides mdh* gene (MDH); the *Mycobacterium vaccae fdh*gene (FDH); and the *Zymomonas mobilis glf gene* (GLF).	[55]
	Bacillus megaterium	co-expressed *mdh* gene (MDH) and *fdh gene* (FDH)	[56]
	Corynebacterium glutamicum	Expression of glf Zm	[57]

Sorbitol

Sorbitol, also known as D-glucitol is a noncyclic hexitol derived from D-glucose. It is used as sweetener in many food products like chewing gum, candies, desserts, ice-cream, food for diabetic patients, pharmaceutical compounds and for the production of sorbose and ascorbic acid [58]. Traditionally it was produced by hydrogenation of glucose syrup at high temperature and pressure in presence of Raney nickel catalyst, after which it is purified by ion exchange chromatography. Several yeasts and bacteria naturally synthesize sorbitol, while LAB do not, but are able to metabolize it [59, 60] (Table **4**).

Table 4. Some important strategies for production of Sorbitol.

Organism used	Strategy	Reference
Zymomonas mobilis	Permeabilized by freezing at –20 °C and thawing at room temperature	[61]
	CTAB-permeabilized	[62]
	Z. mobilis treated with toluene and Co-immobilized invertase in both chitin and calcium alginate.	[63]
L. casei BL232	*gut F* gene encoding S6PDH (enzyme responsible for the interconversion of sorbitol 6-P to fructose 6-P) was integrated in the chromosomal lactose operon.	[64]
	ldhL (LDH enzyme) inactivated	[65]
	inactivation of the *mtl D* gene (mannitol 1-P dehydrogenase)	

(Table 4) contd.....

Organism used	Strategy	Reference
L. plantarum D- and L-LDH-deficient	WCFS1 enzymes S6PDH (srlD1 andsrlD2) translationally fused with the expression signals of the *L. plantarum ldh L* gene	[58]

Tani and Vongsuvanlert [59] screened one hundred and eleven strains of methanol- utilizing yeast for sorbitol production from glucose using an intact cell system. Among these strains, *Candida boidinii* (*Kloeckera* sp.) 2201 was found best with respect to sorbitol production. Later Erzinger and Vitolo [60] studied the comparative ability to free (CTAB-treated or not) and immobilized *Zymomonas mobilis* in batch reactor for the production of gluconic acid and sorbitol from glucose and fructose, and maximum yield of sorbitol was obtained with free CTAB-treated cells. Ro and Kim [63] attempted simultaneous production of gluconic acid and sorbitol in a continuous stirred-tank reactor (CSTR) operation from sucrose using invertase and a glucose-fructose oxidoreductase of *Zymomonas mobilis*. Invertase was immobilized on chitin, and chitin-immobilized invertase was co-immobilized with permeabilized cells of *Z. mobilis* for maximum production. Recombinant strain of *L. casei* constructed with D-sorbito--6-phosphate dehydrogenase (gutF) was found to be capable of producing sorbitol from glucose. Inactivation of the L-lactate dehydrogenase gene further enhanced sorbitol production, suggesting that the engineered route provides an alternative pathway for NAD^+ regeneration [64]. Subsequently, De Boeck *et al.,* [65] described a sorbitol-producing *Lactobacillus casei* (strain BL232) with expression of sorbitol-6-phosphate dehydrogenase (gut F) gene. A further improvement in sorbitol production was achieved by inactivation of mannitol-1-phosphate dehydrogenase (mtlD).

Xylitol

Xylitol has been used in human food since 1960s. This five-carbon sugar is found in little amount in a variety of vegetables and fruits. On industrial scale it is produced by chemical conversion of xylan followed by purification using chromatography [66]. *Saccharomyces cerevisiae* or *Candida* has been studied as an alternative to chemical method of production. Candida can be efficiently used to produce xylitol but, the fact that some species of it are pathogenic restricts its use in food industry. Xylitol production by biotechnological methods for commercial applications is well known [67]. As it has applications and potentials in food, odontological and pharmaceutical industries [68 - 71], effective technologies for production were made in past decades. In view of this, xylitol market has already crossed US$ 340 million per year in 2009. Due to cost intensive chemical processes involved in xylitol production, alternative ecofriendly strategies by bioconversion of renewable biomass were a good success [72]. Xylitol producing microbial strains such as bacteria [71], fungi [73] and yeast [72, 74 - 76] helped researchers to develop effective microbial fermentation technologies including construction of recombinant yeast and bioreactor process strategies (Table **5**).

Table 5. Microbes based methods for production of xylitol.

Organism used	Strategy	References
Saccharomyces cerevisiae	Xylose reductase gene from *Pichia stipites*	[77]
S. cerevisiae 2805	Xylose reductase genes (XYL1) of *Pichia stipitis* (40 copies of gene on chromosome)	[75]
S. cerevisiae	Recombinant xylose reductase, a recombinant cellodextrin transporter, a recombinant intracellular β-glucosidase,	[78]
S.cerevisiae strain (transketolase-deficient)	XYL2 gene (xylitol dehydrogenase) of *Pichia stipitis* and deletion of the endogenous XKS1 gene (xylulokinase)	[79]
*Candida guilliermondii*FTI20037	Immobilized cell system	[80]
Kluyveromyces marxianus	Xylitol dehydrogenase and xylulose kinase activity, are reduced or eliminated	[81]
*Lactococcus lactis*NZ9800	D-xylose reductase and xylose transporter	[82]
Bacillus subtilis	Xylitol-phosphate dehydrogenase (XPDH) genes (from gram positive bacteria)	[83]

Hallborn *et al.,* [77] obtained highly efficient conversion of xylose to xylitol by transforming *S. cerevisiae* with xylose reductase (XR) coding gene of *Pichia stipitis*. In another study, Toivari *et al.,* [79] described the recombinant *S. cerevisiae* strains that can produce xylitol, ribitol, D-ribose from D-glucose in a single fermentation step. A transketolase-deficient *S. cerevisiae* strain accumulates D-xylulose 5-phosphate and releases ribitol, D-ribose, D-ribulose, and D-xylulose. Expression of xylitol dehydrogenase (XYL2) of *Pichia stipitis* in the transketolase-deficient strain result in an 8.5-fold enhanced production of ribitol and xylitol. As an addition to this, co-expression of 2-deoxy-glucose 6-

phosphate phosphatase (DOG 1) and XYL2 into the transketolase-deficient strain result in a further 1.6-fold increase in ribitol production. Deletion of the endogenous xylulokinase gene (XKS1) is also beneficial to increase the amount of xylitol. Recently, Sarrouh and da Silva [80] studied the biotechnological production of xylitol from immobilized cells of *Candida guilliermondii* FTI20037 using sugarcane bagasse hydrolysate in a repeated batch fermentation system. In this study, seven repeated batches were performed in a fluidized bed bioreactor using immobilized cell system, and these cells can be reused for six successive batches maintaining an average xylitol yield of 0.7 g/L and a volumetric productivity of 0.42 g/L h at the end of 432 h of fermentation.

Erythritol

Erythritol is unique member of polyol family and belongs to acyclic alcohols. It contains four carbon atoms with a small linear sugar. It is found in grapes, pears, melons, and mushrooms and in fermented food such as sake, soy sauce, and wine. It was first isolated from algae, lichens and grass. It is excreted through urine as it is not metabolized. It can be produced by a chemical process where dialdehyde starch is converted into erythritol at high-temperature in presence of a nickel catalyst [84]. However this chemical process did not reach industrialization due to low yields. It can also be produced through fermentation using *Aspergillus niger* and *Penicillium herquei*. Commercially it is produced using biotechnological approach [85] (Table **6**).

Highest yield of erythritol is obtained by *Torula corallina*, and it is used for industrial production. The increased production of erythritol by *T. corallina* with trace elements such as Cu^{+2} has been well studied. Supplementation of Cu^{+2} reduced the production of fumarate, a strong inhibitor of erythrose reductase, which led to less inhibition of erythrose reductase and high yield of erythritol [86]. Recently, Mirończuk *et al.,* [87] studied repeated batch cultures (RBC) system to improve the productivity of erythritol from pure and crude glycerol. An acetate negative mutant of *Yarrowia lipolytica* Wratislavia K1 is capable of producing high amount erythritol in RBC. In 20 % of fresh replaced medium, the strain Wratislavia K1 is capable to produce 220 gl^{-1} erythritol.

Microbial production of polyols especially erythritol was extensively investigated, but these microbial methods were not widely used for industrial scale due to formation of various by-products such as glycerol and ribitol. The yield and rate of erythritol production by *Aureobasidium* sp.-SN G42 was 47% in a 1.00,000 L fermenter, under optimized process conditions to minimize by-product formation [88]. Similarly Park *et al.,* [89] demonstrated the advantage of feed batch culture over batch fermentation, for which the production of erythritol was 23% erythritol greater for a 5 L-fermenter for cultivating *Trichoderma* sp.

Table 6. Various methods for production of erythritol.

Method/source	Organism	Strategy	Reference
Wild strain	*Aureobasidium* sp. SN-G42	Improved oxygen transfer rate	[86]
	Ustilaginomycetes sp. 618A-01	No by-product formation	[90]
	Trichosporon sp.	Repeated fed-batch culture	[89]
Mutation	*Candida magnolia* strain was *Aureobasidium* sp. SN124A	Mutated by UV irradiation and NTG treatment. Mutant by UV irradiation and nitrosoguanidine (NTG) treatment	[86, 91]
	Penicillium sp. KJUV29	Reduced byproduct and foam formation	[92]
	Torula sp. Mutant	No byproduct formation	[93]
	Candida magnolia M2	Osmophilic mutant	[94]
Optimizing glucose concentration	Osmophilic microorganisms	High initial concentration of glucose	[88, 89]
	Aureobasidium sp.	Mutant able to tolerate high glucose concentration (40% wv^{-1} glucose)	[90]
	Trichosporon sp.	220 g l^{-1} of glucose concentration	[91]
	Pseudozyma tsukubaensis KN75	Grown aerobically in a fed-Batch culture with glucose as a carbon source	[95]
Other carbon sources	*Yarrowia lipolytica* Wratislavia K1 *Yarrowia lipolytica*	Repeated batch cultures (RBC) Raw glycerol	[87, 96, 97]
	Y. lipolytica (acetate-negative mutant) *Y. lipolytica* Wratislavia K1 (acetate-negative mutant)	Raw glycerol 300 g l^{-1} at pH 3 Media containing glycerol, at pH 5.5	[97, 98]
	Yarrowia lipolytica CICC 1675	Two-stage osmotic pressure control fed-batch strategy	[99]

(Table 6) contd.....

Method/source	Organism	Strategy	Reference
Supplementation with co- factors			
Inositol	*Torula* sp.	Increases erythrose reductase in cells	[100]
phytic acid		Not known	
phosphate		Increases erythrose reductase in cells	
Mn^{2+}		Increased the permeation of the intracellular erythritol	[93]
Cu^{2+}		Increased the activity of erythrose reductase	
Removal of inhibitors and byproducts			
Tricyclazole biosynthesis)	*T. coralline*	Inhibit DHN-melanin: inhibitor of erythrose reductase	[101]

Isomalt

Isomalt or Isomaltulose (palatinose) is sugar-free bulk sweetener with low calories, presently, being used as a precursor for the production of isomalt. Some organisms, *e.g., Leuconostoc mesenteroides*, are able to produce Isomaltulose, whereas *Protaminobacter rubrum* is able to transglycosylate sucrose to isomaltulose [102]. Using immobilized non-viable cells is the cost effective method for production of isomaltulose [103, 104]. Isomalt is produced by hydrogenation of isomaltulose at moderate temperature and pressure at neutral pH in presence of Raney-Nickel in a fixed bed reactor [103, 104] (Table **7**).

Table 7. Some of the strategies attempted for production of isomalt.

Organisms/ System	Strategy	Reference
E. coli	Expressing sucrose isomerase (SIase) MDH Gene from *Pseudomonas fluorescens* DSM 50106	[105, 106]
Potato tubers	palI gene(sucrose isomerase) from *Erwinia rhapontici*	[107]
Potato tubers	Sucrose isomerases from the microorganisms *Protaminobacter rubrum* and *Erwinia rhapontici*	[108]

Recombinant *E. coli* with sucrose isomerase (SIase) gene can convert sucrose into isomaltulose from economical industrial medium composed of untreated cane molasses, yeast extract and corn steep liquor. Immobilizations of recombinant *E. coli* in calcium alginate gel can further improve the efficiency of recycling. The immobilization is most effective with 2% sodium alginate (w/v) and 3% calcium chloride (w/v). The immobilized *E. coli* cells exhibited good stability up to 30 batches of cultures with the productivity of 0.45 g isomaltulose g pellet^{-1} h^{-1}. A continuous isomaltulose formation process using a column reactor remained stable for 40 days with 83% isomaltulose yield, which is potentially beneficial for commercial production of isomaltulose [105].

Maltitol

Maltitol is produced by hydrogenation of maltose in presence of nickel and then purified followed by crystallization. Its sweetness is of about 85-95% that of sucrose. It is used as fat replacer, crystalline sugar, bulk sweetener, confectionary, pie fillings, ice-cream, salad dressing, cookies and cakes [109].

Similarly, natural sweeteners bulk production has gained popularity especially sorbitol and xylitol from berries, fruits, vegetables and mushrooms. But due to lack of commercial viability, catalytic hydrogenation of the appropriate reducing sugars has been employed to obtain these sweeteners.

Industrial Relevance

If we look at the key drivers for worldwide sweetener industries, second generation polyol sweeteners (maltitols, xylitols and lactitols) are currently replacing sorbitol, which is a first generation of polyol. Among these, xylitol (XIVIA of Danisco, Denmark) is approved in over 50 countries in food use. China is the largest single consumer for sorbitol accounting for over 1/3 of the world consumption in 2010. Now it is obvious that erythritol production by microbial methods using osmophilic yeasts is a success and also has been produced commercially using mutant strains of *Aureobasidium* sp. and *Pseudozyma tsukubaensis*. Various approaches for the efficient erythritol production, strategies used to enhance erythritol productivity in microbes, and the potentially biotechnological applications of erythritol [110] and mannitol [111] were reviewed. Although the currently available processes could be helpful to augment erythritol and mannitol production by fed-batch process, the newly isolated-strains or improved strains of respective microbes would still be very competitive, economically viable in the industrial production [95]. To achieve this, developing robust microbial systems by recombination of the cloned erythrose reductase gene or mannitol reductase gene respectively is a best alternative. Production of polyols such as erythritol, mannitol and xylitol are mainly carried out by chemical processes. However, due to low yield, chemical processes were not preferred for industrial production of polyols. Many companies such as Mitsubishi Chemical Corporation (Tokyo, Japan), Bolak Corporation (Hwasung, Kyunggi Do, Korea), Cargill Food and Pharma Specialty (Blair, Nebraska, USA) are involved in commercial production of erythritol. Similarly IPI industry, Singapore has developed a green technology for obtaining 0.870 g/g of xylose corresponding to 94.9 % theoretical yield by using a flocculation yeast strain *Candida athensensis* SB18 (<http://www.ipi-singapore.org/cos/o.x?ptid=1071682&c=/ipi/ipcat&func=preview&rid=719> accessed on July, 18, 2014). Xylitol production by fermentation process using yeast strain was well optimized [112] to get 65 to 85% xylitol (theoretical value from xylose). Compared to chemical process this fermentative process involves less cost due to non-necessity of xylose purification. However, it has its own limitations. Recently, a technology demonstration with feasibility of process scale-up for production of xylose from cellulose has been demonstrated by Xylitol Canada (<http://www.biofuelsdigest.com/bdigest/2013/06/18/xylitol-canada-completes-technology-demonstration-of-xylose-production-process/>). Similarly, a process for conversion of hemi cellulose to xylitol under mild condition was reported [113]. Recently, Rafiqul *et al.,* (2012) [114] reviewed the potential and prospects for bio production of xylitol by enzyme technology as a promising alternative to chemical process and fermentative process.

RESEARCH ADVANCES IN PLANT DERIVED NON-SACCHARIDE SWEETENERS

Biotechnological intervention for natural sweeteners was initiated in the 1980s to produce thaumatin by using recombinant DNA technology in laboratory by anticipating more yields compared to natural thaumatin, a protein extracted from the fruit of a West African plant *Thaumatococcus daniellii.* In 1985-1986 the intensely-sweet thaumatin protein was successfully cloned by Unilever (The

Netherlands) and Ingene (The USA) [115] and sequencing of two commercial proteins (thaumatin I and thaumatin II) was completed in 1990. By using this recombinant DNA technology very low to high (7 mg/L) recombinant thaumatin was produced by using various bacterial and fungal strains [116 - 119]. Subsequently efforts were made to express in transgenic plants of *Solanum tuberosum*. However, yield was very low [120]. As in the case of thaumatin, researchers have attempted to express monellin (from *Dioscoriophyllum cuminsii*) in microorganisms [33, 121] and also in transgenic plants [122] wherein, the yields were quite high (23.9 µg/g wet weight) in tomato compared to very low yields of *E.coli*.

Another protein sweetener, mabinilin, from leaves of Chinese plant *Capparis masakai*, was investigated and four variants of mabinilin (I, II, III and IV) were identified. Later, the cDNA representing the four mabinilin isoforms were cloned and sequenced [123]. Also, attempts were made to express mabinilin II in transgenic potato tubers, but no significant reports have been reported [124]. Expression of Brazzein protein (from fruits of *Pentadiplandra brazzeana*) in recombinant host [124, 125] and its expressions in *E. coli* were reported [126]. Sporadic reports on cDNA clones for Curculin (from *Curculigo latifolia*) and miraculin (from *Richadella dulcinea*) were isolated and sequenced [127, 128]. Similarly production of recombinant miraculin was reported by inserting a synthetic gene encoding miraculin into an *E. coli* expression vector [129].

Besides these sweetener proteins, several other sweetener compounds exist in nature and such compounds have to be isolated, purified and characterized followed by gene cloning and development of recombinant expression of natural proteins by using available and emerging technology. Noteworthy, achievements on *Stevia rebaudiana* were reported to identify the steviosides biosynthetic pathway genes apart from commercially viable processes and patents for steviosides production [130 - 133].

IMPENDING AREAS OF RESEARCH

Current knowledge pertains to some plant secondary metabolites as sweeteners is very limited and it refers to sporadic reports about the nature, chemistry and some basic characteristic features of sweetener molecules. Unlike natural sweetener alcohols, many of these non-saccharide sweeteners are still to be explored further for their commercial feasibility for production, yield and economics. Apart from this, plant source is also essential for its accessibility, sustainable production systems for large scale extraction and production of respective sweetener molecules. Plant tissue culture methods are highly helpful for effective propagation of natural sweeteners producing plants such as *Curculigo latifolia, Pentadiplanadra brazzeana, Abrus precatorius, Baccharis gaudichaudiana, Glycyrrhiza glabra, Illicium verum, Piper marginatum, Periandra dulcis, Perilla frutescens etc.* Such propagation methods not only help generate biomass and also facilitate to identification of elite clones. Similarly, molecular biological methods are vital for marker assisted selection to identify the high metabolite yielding germplasm. As most of the sweetener principles are extracted from fruits, leaves and rhizomes of different plants as explained earlier, unraveling their bioactive principles and nutraceutical potential will give value addition.

Miraculin is a taste-modifying protein from the red berries of *Richadella dulcifica*. This compound by itself is not sweet, but, it is able to turn a sour taste into a sweet taste. This unique property has led to increasing interest in this protein by developing genetic engineering approaches, wherein the high-yield production of miraculin in transgenic tomato plants (up to 102.5 μg/g fresh weights) was reported [123]. In spite of these developments in plant derived natural sweetener production, only a very few tasting natural products were commercialized as sweeteners but the number may rapidly increase [134]. Especially great potential for commercial production has been envisaged for steviol glycosides from *Stevia rebaudiana* and mogrosides from *Siraitia grosvenorii* [9] in view of their high potency taste and high stability [135, 136]. Biotransformation studies using *in vitro* plant cell cultures have enormous potential for producing high value low yielding compounds in a cost effective manner. For example, rebaudioside-D from stevia presents more sweetening potential than steviosides and rebaudioside-A, but its concentration is very low in stevia leaves. This can be addressed by biotransformation methods. As there are certain limitations of plant cell cultures for production of desired secondary metabolites (sweeteners), metabolic engineering approach to some extent aid for enhanced production of these value added compounds [137, 138]. In this regard, the advantages of metabolic engineering have been recently reviewed [9]. Irrespective of the methods of production of plant-derived natural sweeteners, *i.e.,* conventional, biotechnological, microbial fermentation methods, ultimately efficient and cost-effective downstream processing for recovery of biosynthetic products is warranted as it will decide economic viability of the process. Hence, a holistic approach is required for sustainable processes, which include identification of gene, metabolic and protein engineering, fermentation and downstream processes.

As the stability during processing, and food formulation, sensory acceptability along with nontoxic nature at permitted levels are required, investigations on these aspects still need for further refining the existing knowledge on these natural sweeteners. It is a known fact that different sweeteners have different taste profiles, so to meet consumer taste and expectations for a particular product several sweeteners are always blended to pass a sensory evaluation. A person's desire towards sweetness is a basic biological taste and the sweet receptors found on the tongue have been extensively studied and needs further validation [139, 140]. Recently, Brahmachari *et al.*, [141] have reviewed the pharmacokinetics along with an insight into the structure–sweetness relationship, safety evaluation and clinical trials of these ent-kaurene glycosides. Similarly, there was incited interest among researchers about the versatile bioactive properties of stevioside to synthesize several stevioside analogues such as sulphopropyl and sodiosulphopropyl esters (chemically modified structures) which help improve its bioactive properties such as organoleptic activity [142]. However, such studies on other natural sweeteners are sparse and needs to be explored.

THE FUTURE OF INNOVATION IN ALTERNATIVE SWEETENERS

The future of innovation in natural sweetener depends on various factors and developments, followed by researching that drive the market for natural sweetener by looking at end-users preferences and also technology innovations. To attend consumers concerns about health and safety, sufficient R&D backup should be developed to make natural sweeteners as valuable alternatives for synthetic sweeteners. The focus should be to further retrospect the sweetness potential of different, types of sweeteners and multidirectional biotechnological and pharmacological potentials of natural sweeteners. Once the feasible technologies are developed to facilitate biotechnological production of natural sweeteners, that will have far-reaching industrial and scientific implications for the production of other small molecules.

Similarly, research areas such as maintaining the taste and labeling of food products containing these sweeteners are vital from quality and safety aspects. Through more investment inflows into the chemical and biotech research industries, these processes can pace up and result in more substitutes being produced. With advanced technical compilations it is essential to explore novel natural sweeteners as there is a great scope for R&D labs to unravel wide ranges of unexplored plants and fruits to produce high intensity sweeteners.

CONFLICT OF INTEREST

The authors confirms that they have no conflict of interest to declare for this publication.

ACKNOWLEDGEMENTS

The editors acknowledge Espaço da Escrita (Coordenadoria Geral - State University of Campinas) for the English corrections in in this text.

REFERENCES

[1] Sardesai, V.M.; Waldshan, T.H. Natural and synthetic intense sweeteners. *J. Nutr. Biochem.,* **1991**, *2*, 236-244.
[http://dx.doi.org/10.1016/0955-2863(91)90081-F]

[2] Lemus-Mondaca, R.; Vega-Galvez, A.; Zura-Bravo, l.; Ah-Hen, K. *Stevia rebaudiana* Bertoni, source of a high-potency natural sweetener: a comprehensive review on the biochemical, nutritional and functional aspects. *Food Chem.,* **2012**, *132*, 1121-1132.
[http://dx.doi.org/10.1016/j.foodchem.2011.11.140]

[3] Butchko, H.H.; Stargel, W.W.; Comer, C.P.; Mayhew, D.A.; Benninger, C.; Blackburn, G.L.; de Sonneville, L.M.; Geha, R.S.; Hertelendy, Z.; Koestner, A.; Leon, A.S.; Liepa, G.U.; McMartin, K.E.; Mendenhall, C.L.; Munro, I.C.; Novotny, E.J.; Renwick, A.G.; Schiffman, S.S.; Schomer, D.L.; Shaywitz, B.A.; Spiers, P.A.; Tephly, T.R.; Thomas, J.A.; Trefz, F.K. Aspartame: review of safety. *Regul. Toxicol. Pharmacol.,* **2002**, *35*(2 Pt 2), S1-S93.
[http://dx.doi.org/10.1006/rtph.2002.1542] [PMID: 12180494]

[4] Grotz, V.L.; Munro, I.C. An overview of the safety of sucralose. *Regul. Toxicol. Pharmacol.,* **2009**, *55*(1), 1-5.

[http://dx.doi.org/10.1016/j.yrtph.2009.05.011] [PMID: 19464334]

[5] Mattes, R.D.; Popkin, B.M. Nonnutritive sweetener consumption in humans: effects on appetite and food intake and their putative mechanisms. *Am. J. Clin. Nutr.,* **2009**, *89*(1), 1-14. [http://dx.doi.org/10.3945/ajcn.2008.26792] [PMID: 19056571]

[6] Soffritti, M.; Padovani, M.; Tibaldi, E.; Falcioni, L.; Manservisi, F.; Belpoggi, F. The carcinogenic effects of aspartame: The urgent need for regulatory re-evaluation. *Am. J. Ind. Med.,* **2014**, *57*(4), 383-397. [http://dx.doi.org/10.1002/ajim.22296] [PMID: 24436139]

[7] Kant, R. Sweet proteins--potential replacement for artificial low calorie sweeteners. *Nutr. J.,* **2005**, *4*, 5. [http://dx.doi.org/10.1186/1475-2891-4-5] [PMID: 15703077]

[8] Masuda, T.; Kitabatake, N. Developments in biotechnological production of sweet proteins. *J. Biosci. Bioeng.,* **2006**, *102*(5), 375-389. [http://dx.doi.org/10.1263/jbb.102.375] [PMID: 17189164]

[9] Philippe, R.N.; De Mey, M.; Anderson, J.; Ajikumar, P.K. Biotechnological production of natural zero-calorie sweeteners. *Curr. Opin. Biotechnol.,* **2014**, *26*, 155-161. [http://dx.doi.org/10.1016/j.copbio.2014.01.004] [PMID: 24503452]

[10] Global markets for sugars and sweeteners in processed foods and beverages (FOD018D). *Available from: http://www.bccresearch.com/pressroom/fod/market-high-intensity-sweeteners-expected-reach-nearly-$1.9-billion-2017,* **2013**,

[11] Kim, N.C.; Kinghorn, A.D. Highly sweet compounds of plant origin. *Arch. Pharm. Res.,* **2002**, *25*(6), 725-746. [http://dx.doi.org/10.1007/BF02976987] [PMID: 12510821]

[12] Rao, R.S.; Jyothi, ChP.; Prakasham, R.S.; Sarma, P.N.; Rao, L.V.; Rao, L.V. Xylitol production from corn fiber and sugarcane bagasse hydrolysates by *Candida tropicalis. Bioresour. Technol.,* **2006**, *97*(15), 1974-1978. [http://dx.doi.org/10.1016/j.biortech.2005.08.015] [PMID: 16242318]

[13] Gibbs, B.F.; Alli, I.; Mulligan, C. Sweet and taste-modifying proteins: a Review. *Nutr. Res.,* **1996**, *16*, 1619-1630. [http://dx.doi.org/10.1016/0271-5317(96)00175-3]

[14] Douglas, K.; Djendoel, S. Discovery of terpenoid and phenolic sweeteners from plants. *Pure Appl. Chem.,* **2002**, *74*, 1169-1179.

[15] Nirasawa, S.; Nishino, T.; Katahira, M.; Uesugi, S.; Hu, Z.; Kurihara, Y. Structures of heat-stable and unstable homologues of the sweet protein mabinlin. The difference in the heat stability is due to replacement of a single amino acid residue. *Eur. J. Biochem.,* **1994**, *223*(3), 989-995. [http://dx.doi.org/10.1111/j.1432-1033.1994.tb19077.x] [PMID: 8055976]

[16] Kohmura, M.; Mizukoshi, T.; Nio, N.; Suzuki, E.I.; Ariyoshi, Y. Structure–taste relationships of the sweet protein monellin. *Pure Appl. Chem.,* **2002**, *74*, 1235-1242. [http://dx.doi.org/10.1351/pac200274071235]

[17] Tlili, N.; Elfalleh, W.; Saadaoui, E.; Khaldi, A.; Triki, S.; Nasri, N. The caper (*Capparis* L.):

ethnopharmacology, phytochemical and pharmacological properties. *Fitoterapia,* **2011**, *82*(2), 93-101. [http://dx.doi.org/10.1016/j.fitote.2010.09.006] [PMID: 20851750]

[18] van Der Wel, H.; Wiersma, A.; Brouwer, J.N. Tritium labelling of Thaumatin I, a sweet-tasting protein from *Thaumatococcus daniellii* Benth, by reductive methylation. *J. Labelled Comp. Radiopharm.,* **1978**, *14*, 735-740. [http://dx.doi.org/10.1002/jlcr.2580140511]

[19] Ming, D.; Hellekant, G. Brazzein, a new high-potency thermostable sweet protein from *Pentadiplandra brazzeana* B. *FEBS Lett.,* **1994**, *355*(1), 106-108. [http://dx.doi.org/10.1016/0014-5793(94)01184-2] [PMID: 7957951]

[20] Shirasuka, Y.; Nakajima, K.; Asakura, T.; Yamashita, H.; Yamamoto, A.; Hata, S.; Nagata, S.; Abo, M.; Sorimachi, H.; Abe, K. Neoculin as a new taste-modifying protein occurring in the fruit of *Curculigo latifolia. Biosci. Biotechnol. Biochem.,* **2004**, *68*(6), 1403-1407. [http://dx.doi.org/10.1271/bbb.68.1403] [PMID: 15215616]

[21] Kaneda, N.; Lee, I.S.; Gupta, M.P.; Soejarto, D.D.; Kinghorn, A.D. (+)-4 beta-hydroxyhernandulcin, a new sweet sesquiterpene from the leaves and flowers of *Lippia dulcis. J. Nat. Prod.,* **1992**, *55*(8), 1136-1141. [http://dx.doi.org/10.1021/np50086a016] [PMID: 1431936]

[22] Compadre, C.M.; Pezzuto, J.M.; Kinghorn, A.D.; Kamath, S.K. Hernandulcin: an intensely sweet compound discovered by review of ancient literature. *Science,* **1985**, *227*(4685), 417-419. [http://dx.doi.org/10.1126/science.3880922] [PMID: 3880922]

[23] Hussain, R.A.; Poveda, L.J.; Pezzuo, J.H.; Seejarto, D.D.; Kinghorn, A.D. Sweetening agents of plant origin: Phenylpropanoid constituents of seven sweet-tasting plants. *Econ. Bot.,* **1990**, *44*, 174-182. [http://dx.doi.org/10.1007/BF02860485]

[24] Takasaki, M.; Konoshima, T.; Murata, Y.; Sugiura, M.; Nishino, H.; Tokuda, H.; Matsumoto, K.; Kasai, R.; Yamasaki, K. Anticarcinogenic activity of natural sweeteners, cucurbitane glycosides, from *Momordica grosvenori. Cancer Lett.,* **2003**, *198*(1), 37-42. [http://dx.doi.org/10.1016/S0304-3835(03)00285-4] [PMID: 12893428]

[25] Suzuki, M.; Kurimoto, E.; Nirasawa, S.; Masuda, Y.; Hori, K.; Kurihara, Y.; Shimba, N.; Kawai, M.; Suzuki, E.; Kato, K. Recombinant curculin heterodimer exhibits taste-modifying and sweet-tasting activities. *FEBS Lett.,* **2004**, *573*(1-3), 135-138. [http://dx.doi.org/10.1016/j.febslet.2004.07.073] [PMID: 15327988]

[26] Surana, S.J.; Gokhale, S.B.; Rajmane, R.A.; Jadhav, R.B. Non-saccharides natural intense sweetners An overview of current status. *Natural product radiance,* **2006**, *5*(4), 270-278.

[27] Kinghorn, A.D.; Soejarto, D.D. Discovery of terpenoid and phenolic sweeteners from plants. *Pure Appl. Chem.,* **2002**, *74*, 1169-1179. [http://dx.doi.org/10.1351/pac200274071169]

[28] Tillequin, F.; Paris, M.; Jacquemin, H.; Paris, R.R. *Flavonoids from Piper marginatum-Isolation of a new flavonoid, marginatoside*; Planta Medica: German, FR, **1978**. Pepper

[29] Nanayakkara, N.P.; Hussain, R.A.; Pezzuto, J.M.; Soejarto, D.D.; Kinghorn, A.D. An intensely sweet dihydroflavonol derivative based on a natural product lead compound. *J. Med. Chem.,* **1988**, *31*(6),

1250-1253.
[http://dx.doi.org/10.1021/jm00401a030] [PMID: 3373494]

[30] Hussain, R.A.; Lin, Y.M.; Poveda, L.J.; Bordas, E.; Chung, B.S.; Pezzuto, J.M.; Soejarto, D.D.; Kinghorn, A.D. Plant-derived sweetening agents: saccharide and polyol constituents of some sweet-tasting plants. *J. Ethnopharmacol.,* **1990**, *28*(1), 103-115.
[http://dx.doi.org/10.1016/0378-8741(90)90067-4] [PMID: 2314108]

[31] Kim, J.; Pezzuto, J.M.; Soejarto, D.D.; Lang, F.A.; Kinghorn, A.D. Polypodoside A, an intensely sweet constituent of the rhizomes of *Polypodium glycyrrhiza. J. Nat. Prod.,* **1988**, *51*(6), 1166-1172.
[http://dx.doi.org/10.1021/np50060a019] [PMID: 3236009]

[32] Lee, H.J.; Kim, M.C.; Um, B.H.; Kim, C.Y. Determination of phyllodulcin from sweet hydrangea leaves by subcritical water extraction and high performance liquid chromatographic analysis. *J. Liquid Chromatogr. Relat. Technol.,* **2008**, *31*, 1306-1313.
[http://dx.doi.org/10.1080/10826070802019764]

[33] Kim, S.H.; Kang, C.H.; Kim, R.; Cho, J.M.; Lee, Y.B.; Lee, T.K. Redesigning a sweet protein: increased stability and renaturability. *Protein Eng.,* **1989**, *2*(8), 571-575.
[http://dx.doi.org/10.1093/protein/2.8.571] [PMID: 2813335]

[34] Kennelly, E.J.; Cai, L.; Long, L.; Shamon, L.; Zaw, K.; Zhou, B-N.; Pezzuto, J.M.; Kinghorn, A.D. Novel Highly Sweet Secodammarane Glycosides from *Pterocarya paliurus. J. Agric. Food Chem.,* **1995**, *43*, 2602-2607.
[http://dx.doi.org/10.1021/jf00058a009]

[35] Kennelly, E.J.; Suttisri, R.; Kinghorn, A.D. Novel sweet-tasting saponins of the cycloartane, oleanane, secodammarane, and steroidal types. *Adv. Exp. Med. Biol.,* **1996**, *405*, 13-24.
[http://dx.doi.org/10.1007/978-1-4613-0413-5_2] [PMID: 8910692]

[36] Makapugay, H.C.; Dhammika Nanayakkara, N.P.; Soejarto, D.D.; Kinghorn, A.D. High-performance liquid chromatographic analysis of the major sweet principle to Lo Han Kuo fruits. *J. Agric. Food Chem.,* **1985**, *33*, 348-350.
[http://dx.doi.org/10.1021/jf00063a007]

[37] Matsumoto, K.; Kasai, R.; Ohtani, K.; Tanaka, O. Minor cucurbitane glycosides from fruits of *Siraitia grosvenori* (Cucurbitaceae). *Chem. Pharm. Bull. (Tokyo),* **1990**, *38*, 2030-2032.
[http://dx.doi.org/10.1248/cpb.38.2030]

[38] Li, D.; Ikeda, T.; Matsuoka, N.; Nohara, T.; Zhang, H.; Sakamoto, T.; Nonaka, G. Cucurbitane glycosides from unripe fruits of Lo Han Kuo (*Siraitia grosvenori*). *Chem. Pharm. Bull. (Tokyo),* **2006**, *54*(10), 1425-1428.
[http://dx.doi.org/10.1248/cpb.54.1425] [PMID: 17015982]

[39] Li, D.; Ikeda, T.; Nohara, T.; Liu, J.; Wen, Y.; Sakamoto, T.; Nonaka, G. Cucurbitane glycosides from unripe fruits of *Siraitia grosvenori. Chem. Pharm. Bull. (Tokyo),* **2007**, *55*(7), 1082-1086.
[http://dx.doi.org/10.1248/cpb.55.1082] [PMID: 17603208]

[40] Xia, Y.; Riverohuguet, M.; Hughes, B.; Marshall, W. Isolation of the sweet compounds from *Siraitia grosvenori. Food Chem.,* **2008**, *107*, 1022-1030.
[http://dx.doi.org/10.1016/j.foodchem.2007.09.014]

[41] Nagel, R. Agave: Nector of the Gods. , **2008**; pp. 46-51.

[42] Solange, I.M.; Es, I.; Roberto, C. Acid hydrolysis and fermentation of brewer's spent grain to produce xylitol. *J. Sci. Food Agric.,* **2005**, *85*, 2453-2460. [http://dx.doi.org/10.1002/jsfa.2276]

[43] Monedero, V.; Pérez-Martínez, G.; Yebra, M.J. Perspectives of engineering lactic acid bacteria for biotechnological polyol production. *Appl. Microbiol. Biotechnol.,* **2010**, *86*(4), 1003-1015. [http://dx.doi.org/10.1007/s00253-010-2494-6] [PMID: 20180114]

[44] Ortiz, M.E.; Bleckwedel, J.; Raya, R.R.; Mozzi, F. Biotechnological and *in situ* food production of polyols by lactic acid bacteria. *Appl. Microbiol. Biotechnol.,* **2013**, *97*(11), 4713-4726. [http://dx.doi.org/10.1007/s00253-013-4884-z] [PMID: 23604535]

[45] Soetaert, W.; Vanhooren, P.; Vandamme, E.J. The production of mannitol by fermentation. In: *Carbohydrate Biotechnology Protocols*; Bucke, C., Ed.; Humana: Totowa, **1999**; pp. 261-275. [http://dx.doi.org/10.1007/978-1-59259-261-6_21]

[46] Racine, F.M.; Saha, B.C. Production of mannitol by *Lactobacillus intermedius* NRRL B-3693 in fed-batch and continuous cell recycle fermentations. *Process Biochem.,* **2007**, *42*, 1609-1613. [http://dx.doi.org/10.1016/j.procbio.2007.09.001]

[47] Silveira, M.M.; Jonas, R. The biotechnological production of sorbitol. *Appl. Microbiol. Biotechnol.,* **2002**, *59*(4-5), 400-408. [http://dx.doi.org/10.1007/s00253-002-1046-0] [PMID: 12172602]

[48] Neves, A.R.; Ramos, A.; Shearman, C.; Gasson, M.J.; Santos, H. Catabolism of mannitol in *Lactococcus lactis* MG1363 and a mutant defective in lactate dehydrogenase. *Microbiology,* **2002**, *148*(Pt 11), 3467-3476. [http://dx.doi.org/10.1099/00221287-148-11-3467] [PMID: 12427938]

[49] Wisselink, H.W.; Moers, A.P.; Mars, A.E.; Hoefnagel, M.H.; de Vos, W.M.; Hugenholtz, J. Overproduction of heterologous mannitol 1-phosphatase: a key factor for engineering mannitol production by *Lactococcus lactis. Appl. Environ. Microbiol.,* **2005**, *71*(3), 1507-1514. [http://dx.doi.org/10.1128/AEM.71.3.1507-1514.2005] [PMID: 15746354]

[50] Gaspar, P.; Neves, A.R.; Ramos, A.; Gasson, M.J.; Shearman, C.A.; Santos, H. Engineering *Lactococcus lactis* for production of mannitol: high yields from food-grade strains deficient in lactate dehydrogenase and the mannitol transport system. *Appl. Environ. Microbiol.,* **2004**, *70*(3), 1466-1474. [http://dx.doi.org/10.1128/AEM.70.3.1466-1474.2004] [PMID: 15006767]

[51] Gaspar, P.; Neves, A.R.; Gasson, M.J.; Shearman, C.A.; Santos, H. High yields of 2,3-butanediol and mannitol in Lactococcus lactis through engineering of NAD□ cofactor recycling. *Appl. Environ. Microbiol.,* **2011**, *77*(19), 6826-6835. [http://dx.doi.org/10.1128/AEM.05544-11] [PMID: 21841021]

[52] Helanto, M.; Aarnikunnas, J.; von Weymarn, N.; Airaksinen, U.; Palva, A.; Leisola, M. Improved mannitol production by a random mutant of *Leuconostoc pseudomesenteroides. J. Biotechnol.,* **2005**, *116*(3), 283-294. [http://dx.doi.org/10.1016/j.jbiotec.2004.11.001] [PMID: 15707689]

[53] Aarnikunnas, J.; Von Weymarn, N.; Rönnholm, K.; Leisola, M.; Palva, A. Metabolic engineering of

Lactobacillus fermentum for production of mannitol and pure L-lactic acid or pyruvate. *Biotechnol. Bioeng.,* **2003**, *82*(6), 653-663.
[http://dx.doi.org/10.1002/bit.10615] [PMID: 12673764]

[54] Costa, R.S.; Hartmann, A.; Gaspar, P.; Neves, A.R.; Vinga, S. An extended dynamic model of *Lactococcus lactis* metabolism for mannitol and 2,3-butanediol production. *Mol. Biosyst.,* **2014**, *10*(3), 628-639.
[http://dx.doi.org/10.1039/c3mb70265k] [PMID: 24413179]

[55] Kaup, B.; Bringer-Meyer, S.; Sahm, H. Metabolic engineering of *Escherichia coli*: construction of an efficient biocatalyst for D-mannitol formation in a whole-cell biotransformation. *Appl. Microbiol. Biotechnol.,* **2004**, *64*(3), 333-339.
[http://dx.doi.org/10.1007/s00253-003-1470-9] [PMID: 14586579]

[56] Bäumchen, C.; Roth, A.H.; Biedendieck, R.; Malten, M.; Follmann, M.; Sahm, H.; Bringer-Meyer, S.; Jahn, D. D-mannitol production by resting state whole cell biotrans-formation of D-fructose by heterologous mannitol and formate dehydrogenase gene expression in *Bacillus megaterium. Biotechnol. J.,* **2007**, *2*(11), 1408-1416.
[http://dx.doi.org/10.1002/biot.200700055] [PMID: 17619232]

[57] Bäumchen, C.; Bringer-Meyer, S. Expression of glf Z.m. increases D-mannitol formation in whole cell biotransformation with resting cells of *Corynebacterium glutamicum. Appl. Microbiol. Biotechnol.,* **2007**, *76*(3), 545-552.
[http://dx.doi.org/10.1007/s00253-007-0987-8] [PMID: 17503033]

[58] Ladero, V.; Ramos, A.; Wiersma, A.; Goffin, P.; Schanck, A.; Kleerebezem, M.; Hugenholtz, J.; Smid, E.J.; Hols, P. High-level production of the low-calorie sugar sorbitol by *Lactobacillus plantarum* through metabolic engineering. *Appl. Environ. Microbiol.,* **2007**, *73*(6), 1864-1872.
[http://dx.doi.org/10.1128/AEM.02304-06] [PMID: 17261519]

[59] Tani, Y.; Vongsuvanlert, V. Sorbitol production by a methanol yeast *Candida boidinii* (*Kloeckera sp.*) No. 2201. *J. Ferment. Technol.,* **1987**, *65*, 405-411.
[http://dx.doi.org/10.1016/0385-6380(87)90136-1]

[60] Erzinger, G.S.; Vitolo, M. Zymomonas mobilis as catalyst for the biotechnological production of sorbitol and gluconic acid. *Appl. Biochem. Biotechnol.,* **2006**, *131*(1-3), 787-794.
[http://dx.doi.org/10.1385/ABAB:131:1:787] [PMID: 18563654]

[61] Bringer-Meyer, S.; Sahm, H. Process for obtaining sorbitol and gluconic acid by fermentation, and cell material suitable for this purpose US5017485, **1991**.

[62] Jang, K.H.; Park, C.J.; Chun, U.H. Improvement of oxidoreductase stability of cethyltrimethylammoniumbromide permeabilized cells of *Zymomonas mobilis* through glutaraldehyde crosslinking. *Biotechnol. Lett.,* **1992**, *14*, 311-316.
[http://dx.doi.org/10.1007/BF01022330]

[63] Ro, H.S.; Kim, H.S. Continuous production of gluconic acid and sorbitol from sucrose using invertase and an oxidoreductase of *Zymomonas mobilis. Enzyme Microb. Technol.,* **1991**, *13*, 920-924.
[http://dx.doi.org/10.1016/0141-0229(91)90109-N]

[64] Nissen, L.; Pérez-Martínez, G.; Yebra, M.J. Sorbitol synthesis by an engineered *Lactobacillus casei* strain expressing a sorbitol-6-phosphate dehydrogenase gene within the lactose operon. *FEMS*

Microbiol. Lett., **2005**, *249*(1), 177-183.
[http://dx.doi.org/10.1016/j.femsle.2005.06.010] [PMID: 16002237]

[65] De Boeck, R.; Sarmiento-Rubiano, L.A.; Nadal, I.; Monedero, V.; Pérez-Martínez, G.; Yebra, M.J. Sorbitol production from lactose by engineered *Lactobacillus casei* deficient in sorbitol transport system and mannitol-1-phosphate dehydrogenase. *Appl. Microbiol. Biotechnol.,* **2010**, *85*(6), 1915-1922.
[http://dx.doi.org/10.1007/s00253-009-2260-9] [PMID: 19784641]

[66] Härkönen, M.; Nuojua, P. Eri tekijöiden vaikutus ksyloosin katalyyyttiseen hydraukseen ksylitoliksi. *Kemia-Kemi,* **1979**, *6*, 445-447.

[67] Prakasham, R.S.; Sreenivas Rao, R.; Hobbs, P.J. Current trends in biotechnological production of xylitol and future prospects. *Curr. Trends Biotechnol. Pharm.,* **2009**, *3*, 8-36.

[68] Hyvonen, L.; Koivistoinen, P.; Voirol, F. Food technological evaluation of xylitol. In: *Advances in Food Research*; Chichester, C.O.; Mrak, E.M.; Stewart, G., Eds.; Academic Press: New York, **1982**; Vol. 28, pp. 373-403.

[69] Mäkinen, K.K. Can the pentitol-hexitol theory explain the clinical observations made with xylitol? *Med. Hypotheses,* **2000**, *54*(4), 603-613.
[http://dx.doi.org/10.1054/mehy.1999.0904] [PMID: 10859647]

[70] Mäkinen, K.K. The rocky road of xylitol to its clinical application. *J. Dent. Res.,* **2000**, *79*(6), 1352-1355.
[http://dx.doi.org/10.1177/00220345000790060101] [PMID: 10890712]

[71] Yoshitake, J.; Obiwa, H.; Shimamurs, M. Production of polyalcohol by *Cornebacterium sp.* I. Production of pentitol from aldopentose. *Agric. Biol. Chem.,* **1971**, *35*, 905-911.
[http://dx.doi.org/10.1271/bbb1961.35.905]

[72] Sreenivas Rao, R.S.; Prakasham, R.S.; Krishna Prasad, K.; Rajesham, S.; Sharma, P.N.; Venkateswar Rao, L. Xylitol production by *Candida sp.* Parameter optimization using Taguchi approach. *Process Biochem.,* **2004**, *39*, 951-956.
[http://dx.doi.org/10.1016/S0032-9592(03)00207-3]

[73] Dahiya, J.S. Xylitol production by *Petromyces albertensis* grown on medium containing D-xylose. *Can. J. Microbiol.,* **1991**, *37*, 14-18.
[http://dx.doi.org/10.1139/m91-003]

[74] Kim, S.Y.; Kim, J.H.; Oh, D.K. Improvement of xylitol production by controlling oxygen supply in *Candida parapsilosis. J. Ferment. Bioeng.,* **1997**, *83*, 267-270.
[http://dx.doi.org/10.1016/S0922-338X(97)80990-7]

[75] Kim, Y.S.; Kim, S.Y.; Kim, J.H.; Kim, S.C. Xylitol production using recombinant *Saccharomyces cerevisiae* containing multiple xylose reductase genes at chromosomal δ-sequences. *J. Biotechnol.,* **1999**, *67*(2-3), 159-171.
[http://dx.doi.org/10.1016/S0168-1656(98)00172-2] [PMID: 9990733]

[76] Leathers, T.D.; Gupta, S.C. Xylitol and riboflavin accumulation in xylose growth cultures of *Pichia guilliermondii. Appl. Microbiol. Biotechnol.,* **1997**, *47*, 58-61.
[http://dx.doi.org/10.1007/s002530050888]

[77] Hallborn, J.; Walfridsson, M.; Airaksinen, U.; Ojamo, H.; Hahn-Hägerdal, B.; Penttilä, M.; Keräsnen, S. Xylitol production by recombinant *Saccharomyces cerevisiae. Biotechnology (N. Y.),* **1991**, *9*(11), 1090-1095.
[http://dx.doi.org/10.1038/nbt1191-1090] [PMID: 1367625]

[78] Ha, S-J.; Galazka, J.M.; Joong Oh, E.; Kordić, V.; Kim, H.; Jin, Y-S.; Cate, J.H. Energetic benefits and rapid cellobiose fermentation by *Saccharomyces cerevisiae* expressing cellobiose phosphorylase and mutant cellodextrin transporters. *Metab. Eng.,* **2013**, *15*, 134-143.
[http://dx.doi.org/10.1016/j.ymben.2012.11.005] [PMID: 23178501]

[79] Toivari, M.H.; Ruohonen, L.; Miasnikov, A.N.; Richard, P.; Penttilä, M. Metabolic engineering of *Saccharomyces cerevisiae* for conversion of D-glucose to xylitol and other five-carbon sugars and sugar alcohols. *Appl. Environ. Microbiol.,* **2007**, *73*(17), 5471-5476.
[http://dx.doi.org/10.1128/AEM.02707-06] [PMID: 17630301]

[80] Sarrouh, B.; da Silva, S.S. Repeated batch cell-immobilized system for the biotechnological production of xylitol as a renewable green sweetener. *Appl. Biochem. Biotechnol.,* **2013**, *169*(7), 2101-2110.
[http://dx.doi.org/10.1007/s12010-013-0127-0] [PMID: 23397324]

[81] Apajalahti, J.; Leisola, M. *Yeast strains for the production of xylitol,* **2001**. US6271007

[82] Nyyssölä, A.; Pihlajaniemi, A.; Palva, A.; von Weymarn, N.; Leisola, M. Production of xylitol from D-xylose by recombinant *Lactococcus lactis. J. Biotechnol.,* **2005**, *118*(1), 55-66.
[http://dx.doi.org/10.1016/j.jbiotec.2005.03.014] [PMID: 15916828]

[83] Povelainen, M.; Miasnikov, A.N. Production of D-arabitol by a metabolic engineered strain of *Bacillus subtilis. Biotechnol. J.,* **2006**, *1*(2), 214-219.
[http://dx.doi.org/10.1002/biot.200500035] [PMID: 16892251]

[84] Pfeifer, V.; Sohens, V.; Conway, H.; Lancaster, E.; Dabic, S.; Griffin, E. Two stage process for dialdehyde starch using electrolytic regeneration of periodic acid. *Ind. Eng. Chem. Res.,* **1960**, *52*, 201-206.
[http://dx.doi.org/10.1021/ie50603a020]

[85] Embuscado, M.E.; Patil, S.K. Erythritol. In: *Food science and technology, Alternative sweeteners*; Dekker, M., Ed.; Marcel Dekker: New York, **2001**; pp. 235-254.

[86] Lee, J.K.; Koo, B.S.; Kim, S.Y. Fumarate-mediated inhibition of erythrose reductase, a key enzyme for erythritol production by *Torula corallina. Appl. Environ. Microbiol.,* **2002**, *68*(9), 4534-4538.
[http://dx.doi.org/10.1128/AEM.68.9.4534-4538.2002] [PMID: 12200310]

[87] Mirończuk, A.M.; Furgała, J.; Rakicka, M.; Rymowicz, W. Enhanced production of erythritol by *Yarrowia lipolytica* on glycerol in repeated batch cultures. *J. Ind. Microbiol. Biotechnol.,* **2014**, *41*(1), 57-64.
[http://dx.doi.org/10.1007/s10295-013-1380-5] [PMID: 24281394]

[88] Sawada, K.; Taki, A.; Yamakawa, T.; Seki, M. Key role for transketolase activity in erythritol production by *Trichosporonoides megachiliensis* SN-G42. *J. Biosci. Bioeng.,* **2009**, *108*(5), 385-390.
[http://dx.doi.org/10.1016/j.jbiosc.2009.05.008] [PMID: 19804861]

[89] Park, J.; Seo, B.; Kim, J.; Park, Y. Production of erythritol in fed-batch cultures of *Trichosporon sp. J.*

Ferment. Bioeng., **1998**, *86*, 577-580.
[http://dx.doi.org/10.1016/S0922-338X(99)80010-5]

[90] Hirata, Y.; Igarashi, K.; Ezaki, S.; Atomi, H.; Imanaka, T. High-level production of erythritol by strain 618A-01 isolated from pollen. *J. Biosci. Bioeng.,* **1999**, *87*(5), 630-635.
[http://dx.doi.org/10.1016/S1389-1723(99)80126-1] [PMID: 16232530]

[91] Ishizuka, H.; Wako, K.; Kasumi, T.; Sasaki, T. Breeding of a mutant of *Aureobasidium sp.* with high erythritol production. *J. Ferment. Bioeng.,* **1989**, *68*, 310-314.
[http://dx.doi.org/10.1016/0922-338X(89)90003-2]

[92] Lee, K.J.; Lim, J.Y. Optimized conditions for high erythritol production by *Penicillium sp.* KJ-UV29, mutant of *Penicillium sp.* KJ81. *Biotechnology and Bioprocess Engineering,* **2003**, *8*, 173-178.
[http://dx.doi.org/10.1007/BF02935892]

[93] Lee, J.K.; Ha, S.J.; Kim, S.Y.; Oh, D.K. Increased erythritol production in *Torula sp.* by Mn2+ and Cu2+. *Biotechnol. Lett.,* **2000**, *22*, 983-986.
[http://dx.doi.org/10.1023/A:1005672801826]

[94] Yang, S.W.; Park, J.B.; Han, N.S.; Ryu, Y.W.; Seo, J.H. Production of erythritol from glucose by an osmophilic mutant of *Candida magnoliae. Biotechnol. Lett.,* **1999**, *21*, 887-890.
[http://dx.doi.org/10.1023/A:1005566420982]

[95] Jeya, M.; Lee, K.M.; Tiwari, M.K.; Kim, J.S.; Gunasekaran, P.; Kim, S.Y.; Kim, I.W.; Lee, J.K. Isolation of a novel high erythritol-producing *Pseudozyma tsukubaensis* and scale-up of erythritol fermentation to industrial level. *Appl. Microbiol. Biotechnol.,* **2009**, *83*(2), 225-231.
[http://dx.doi.org/10.1007/s00253-009-1871-5] [PMID: 19169680]

[96] Koutinas, A.A.; Wang, R-H.; Webb, C. The biochemurgist–bioconversion of agricultural raw materials for chemical production. *Biofuels Bioprod. Biorefin.,* **2007**, *1*, 24-38.
[http://dx.doi.org/10.1002/bbb.6]

[97] Rymowicz, W.; Rywińska, A.; Marcinkiewicz, M. High-yield production of erythritol from raw glycerol in fed-batch cultures of *Yarrowia lipolytica. Biotechnol. Lett.,* **2009**, *31*(3), 377-380.
[http://dx.doi.org/10.1007/s10529-008-9884-1] [PMID: 19037599]

[98] Rymowicz, W.; Rywińska, A.; Żarowska, B.; Juszczyk, P. Citric acid production from raw glycerol by acetate mutants of *Yarrowia lipolytica. Chem. Pap.,* **2006**, *60*, 391-394.
[http://dx.doi.org/10.2478/s11696-006-0071-3]

[99] Yang, L.B.; Zhan, X.B.; Zheng, Z.Y.; Wu, J.R.; Gao, M.J.; Lin, C.C. A novel osmotic pressure control fed-batch fermentation strategy for improvement of erythritol production by *Yarrowia lipolytica* from glycerol. *Bioresour. Technol.,* **2014**, *151*, 120-127.
[http://dx.doi.org/10.1016/j.biortech.2013.10.031] [PMID: 24215768]

[100] Lee, J.K.; Ha, S.J.; Kim, S.Y.; Oh, D.K. Increased erythritol production in *Torula sp.* with inositol and phytic acid. *Biotechnol. Lett.,* **2001**, *23*, 497-500.
[http://dx.doi.org/10.1023/A:1010386500326]

[101] Lee, J.K.; Jung, H.M.; Kim, S.Y. 1,8-dihydroxynaphthalene (DHN)-melanin biosynthesis inhibitors increase erythritol production in *Torula corallina*, and DHN-melanin inhibits erythrose reductase. *Appl. Environ. Microbiol.,* **2003**, *69*(6), 3427-3434.

[http://dx.doi.org/10.1128/AEM.69.6.3427-3434.2003] [PMID: 12788746]

[102] Stodola, F.H.; Koepsell, H.J.; Sharpe, E.S. A new disaccharide produced by *leuconostoc mesenteroides*. *J. Am. Chem. Soc.,* **1952**, *74*, 3202-3203. [http://dx.doi.org/10.1021/ja01132a540]

[103] Schiweck, H.; Munir, M.; Rapp, K.M.; Schneider, B.; Vogel, M. New developments in the use of sucrose as an industrial bulk chemical. *Zuckerindustrie,* **1990**, *115*, 555-565.

[104] Schiweck, H.; Munir, M.; Rapp, K.M.; Schneider, B.; Vogel, M. New developments in the use of sucrose as an industrial bulk chemical. In: *Carbohydrates as organic raw materials*; Lichtenthaler, F.W., Ed.; VCH: Weinheim, Germany, **1991**; pp. 57-94.

[105] Li, S.; Xu, H.; Yu, J.; Wang, Y.; Feng, X.; Ouyang, P. Enhancing isomaltulose production by recombinant *Escherichia coli* producing sucrose isomerase: culture medium optimization containing agricultural wastes and cell immobilization. *Bioprocess Biosyst. Eng.,* **2013**, *36*(10), 1395-1405. [http://dx.doi.org/10.1007/s00449-012-0877-z] [PMID: 23300051]

[106] Haghighatian, M.; Mofid, M.R.; Nekouei, M.K.; Yaghmaei, P.; Tafreshi, A.H. Isomalt production by cloning, purifying and expressing of the MDH gene from *Pseudomonas fluorescens* DSM 50106 in different strains of *E. coli. Pak. J. Biol. Sci.,* **2008**, *11*(16), 2001-2006. [http://dx.doi.org/10.3923/pjbs.2008.2001.2006] [PMID: 19266906]

[107] Börnke, F.; Hajirezaei, M.; Sonnewald, U. Potato tubers as bioreactors for palatinose production. *J. Biotechnol.,* **2002**, *96*(1), 119-124. [http://dx.doi.org/10.1016/S0168-1656(02)00042-1] [PMID: 12142148]

[108] Kunz, M.; Mattes, R.; Munir, M.; Vogel, M. *Transgenic plants which produce isomalt.,* **2001**. US20040064851A1

[109] Kato, K.; Moskowitz, A.H. Maltitol Naborrs L.O.B. Alternative sweeteners Marcel Dekker: New York, **2001**; pp. 283-295.

[110] Moon, H.J.; Jeya, M.; Kim, I.W.; Lee, J.K. Biotechnological production of erythritol and its applications. *Appl. Microbiol. Biotechnol.,* **2010**, *86*(4), 1017-1025. [http://dx.doi.org/10.1007/s00253-010-2496-4] [PMID: 20186409]

[111] Saha, B.C.; Racine, F.M. Biotechnological production of mannitol and its applications. *Appl. Microbiol. Biotechnol.,* **2011**, *89*(4), 879-891. [http://dx.doi.org/10.1007/s00253-010-2979-3] [PMID: 21063702]

[112] Nigam, P.; Singh, D. Processes of fermentative production of Xylitol — a sugar substitute. *Process Biochem.,* **1995**, *30*, 117-124.

[113] Yi, G.; Zhang, Y. One-pot selective conversion of hemicellulose (xylan) to xylitol under mild conditions. *ChemSusChem,* **2012**, *5*(8), 1383-1387. [http://dx.doi.org/10.1002/cssc.201200290] [PMID: 22764105]

[114] Rafiqul, I.S.; Sakinah, A.M. A perspective bioproduction of xylitol by enzyme technology and future prospects. *Int. Food Res. J.,* **2012**, *19*, 405-408.

[115] Communique, R. *Biotechnology and natural sweeteners Tahumatin*; Carolina: USA, **1987**.

[116] Edens, L.; Heslinga, L.; Klok, R.; Ledeboer, A.M.; Maat, J.; Toonen, M.Y.; Visser, C.; Verrips, C.T.

Cloning of cDNA encoding the sweet-tasting plant protein thaumatin and its expression in *Escherichia coli. Gene,* **1982**, *18*(1), 1-12.
[http://dx.doi.org/10.1016/0378-1119(82)90050-6] [PMID: 7049841]

[117] Lee, J.H.; Weickmann, J.L.; Koduri, R.K.; Ghosh-Dastidar, P.; Saito, K.; Blair, L.C.; Date, T.; Lai, J.S.; Hollenberg, S.M.; Kendall, R.L. Expression of synthetic thaumatin genes in yeast. *Biochemistry,* **1988**, *27*(14), 5101-5107.
[http://dx.doi.org/10.1021/bi00414a023] [PMID: 3167035]

[118] Faus, I.; del Moral, C.; Adroer, N.; del Río, J.L.; Patiño, C.; Sisniega, H.; Casas, C.; Bladé, J.; Rubio, V. Secretion of the sweet-tasting protein thaumatin by recombinant strains of Aspergillus niger var. awamori. *Appl. Microbiol. Biotechnol.,* **1998**, *49*(4), 393-398.
[http://dx.doi.org/10.1007/s002530051188] [PMID: 9615480]

[119] Faus, I.; Patino, C.; Del Rio, J.L.; Del Moral, C.; Barroso, H.S.; Blade, J.; Rubio, V. Expression of a synthetic gene encoding the sweet-tasting protein thaumatin in the filamentous fungus *Penicillium roquefortii. Biotechnol. Lett.,* **1997**, *19*, 1185-1191.
[http://dx.doi.org/10.1023/A:1018477602350]

[120] Witty, M. Preprothaumatin II is processed to biological activity in *Solanum tuberosum. Biotechnol. Lett.,* **1990**, *12*, 131-136.
[http://dx.doi.org/10.1007/BF01022429]

[121] Kondo, K.; Miura, Y.; Sone, H.; Kobayashi, K.; Iijima, H. High-level expression of a sweet protein, monellin, in the food yeast *Candida utilis. Nat. Biotechnol.,* **1997**, *15*(5), 453-457.
[http://dx.doi.org/10.1038/nbt0597-453] [PMID: 9131625]

[122] Penarrubia, L.; Kim, R.; Giovannoni, J.; Kim, S.H.; Fisher, R. Production of the sweet protein monellin in transgenic plants. *Biotechnology,* **1992**, *10*, 561-564.
[http://dx.doi.org/10.1038/nbt0592-561]

[123] Sun, H.J.; Kataoka, H.; Yano, M.; Ezura, H. Genetically stable expression of functional miraculin, a new type of alternative sweetener, in transgenic tomato plants. *Plant Biotechnol. J.,* **2007**, *5*(6), 768-777.
[http://dx.doi.org/10.1111/j.1467-7652.2007.00283.x] [PMID: 17692073]

[124] Xiong, L.; Sun, S. Molecular cloning and transgenic expression of the sweet protein mabinlin in potato tubers. *Plant Physiol.,* **1996**, *111*, 57.

[125] Tomes, D.T. *Methods and compositions for production of plant foodstuffs with enhanced sweet component flavour,* **1997**. WO97/42333

[126] Assadi-Porter, F.M.; Aceti, D.J.; Markley, J.L. Sweetness determinant sites of brazzein, a small, heat-stable, sweet-tasting protein. *Arch. Biochem. Biophys.,* **2000**, *376*(2), 259-265.
[http://dx.doi.org/10.1006/abbi.2000.1726] [PMID: 10775411]

[127] Abe, K.; Yamashita, H.; Arai, S.; Kurihara, Y. Molecular cloning of curculin, a novel taste-modifying protein with a sweet taste. *Biochim. Biophys. Acta,* **1992**, *1130*(2), 232-234.
[http://dx.doi.org/10.1016/0167-4781(92)90537-A] [PMID: 1562601]

[128] Masuda, Y.; Nirasawa, S.; Nakaya, K.; Kurihara, Y. Cloning and sequencing of a cDNA encoding a taste-modifying protein, miraculin. *Gene,* **1995**, *161*(2), 175-177.

[http://dx.doi.org/10.1016/0378-1119(95)00198-F] [PMID: 7665074]

[129] Kurihara, Y. Sweet proteins in general. In: *Thaumatin*; Witty, M.; Higginbotham, J.D., Eds.; CRC press: Boca Raton, **1994**; pp. 1-18.

[130] Rajasekaran, T.; Giridhar, P.; Ravishankar, G.A. Production of steviosides in *ex vitro* and *in vitro* grown *Stevia rebaudiana*.Bertoni. *J. Sci. Food Agric.*, **2007**, *87*, 420-424. [http://dx.doi.org/10.1002/jsfa.2713]

[131] Rajasekaran, T.; Ramakrishna, A.; Udaya Sankar, K.; Giridhar, P.; Ravishankar, G.A. Analysis of Predominant Steviosides in *Stevia rebaudiana* Bertoni.by Liquid Chromatography/Electrospray Ionization-Mass Spectrometry. *Food Biotechnol.*, **2008**, *22*, 179-188. [http://dx.doi.org/10.1080/08905430802043255]

[132] Giridhar, P.; Sowmya, K.S.; Ramakrishna, A.; Ravishankar, G.A. Rapid clonal propagation and stevioside profiles of *Stevia rebaudiana* Bertoni. *International Journal of Plant Developmental Biology*, **2010**, *4*, 47-52.

[133] Madan, S.; Ahmad, S.; Singh, G.N.; Kohli, K.; Kumar, Y.; Singh, R.; Garg, M. Stevia rebaudiana (Bert.) Bertoni- A review. Indian Journal of Natural products and Reports **2010**, *1*, 267-286.

[134] DuBois, G.E.; Prakash, I. Non-caloric sweeteners, sweetness modulators, and sweetener enhancers. *Annu. Rev. Food Sci. Technol.*, **2012**, *3*, 353-380. [http://dx.doi.org/10.1146/annurev-food-022811-101236] [PMID: 22224551]

[135] Hellfritsch, C.; Brockhoff, A.; Stähler, F.; Meyerhof, W.; Hofmann, T. Human psychometric and taste receptor responses to steviol glycosides. *J. Agric. Food Chem.*, **2012**, *60*(27), 6782-6793. [http://dx.doi.org/10.1021/jf301297n] [PMID: 22616809]

[136] Tang, Q.; Ma, X.; Mo, C.; Wilson, I.W.; Song, C.; Zhao, H.; Yang, Y.; Fu, W.; Qiu, D. An efficient approach to finding *Siraitia grosvenorii* triterpene biosynthetic genes by RNA-seq and digital gene expression analysis. *BMC Genomics*, **2011**, *12*, 343. [http://dx.doi.org/10.1186/1471-2164-12-343] [PMID: 21729270]

[137] Keasling, J.D. Manufacturing molecules through metabolic engineering. *Science*, **2010**, *330*(6009), 1355-1358. [http://dx.doi.org/10.1126/science.1193990] [PMID: 21127247]

[138] Leonard, E.; Ajikumar, P.K.; Thayer, K.; Xiao, W-H.; Mo, J.D.; Tidor, B.; Stephanopoulos, G.; Prather, K.L. Combining metabolic and protein engineering of a terpenoid biosynthetic pathway for overproduction and selectivity control. *Proc. Natl. Acad. Sci. USA*, **2010**, *107*(31), 13654-13659. [http://dx.doi.org/10.1073/pnas.1006138107] [PMID: 20643967]

[139] Nelson, G.; Hoon, M.A.; Chandrashekar, J.; Zhang, Y.; Ryba, N.J.; Zuker, C.S. Mammalian sweet taste receptors. *Cell*, **2001**, *106*(3), 381-390. [http://dx.doi.org/10.1016/S0092-8674(01)00451-2] [PMID: 11509186]

[140] Li, X.; Staszewski, L.; Xu, H.; Durick, K.; Zoller, M.; Adler, E. Human receptors for sweet and umami taste. *Proc. Natl. Acad. Sci. USA*, **2002**, *99*(7), 4692-4696. [http://dx.doi.org/10.1073/pnas.072090199] [PMID: 11917125]

[141] Brahmachari, G.; Mandal, L.C.; Roy, R.; Mondal, S.; Brahmachari, A.K. Stevioside and Related Compounds - Molecules of Pharmaceutical Promise: A Critical Overview. *Arch Pharm (Weinheim)*,

2011, *1*, 5-19.

[142] DuBois, G.E.; Crosby, G.A.; Lee, J.F.; Stephenson, R.A.; Wang, P.C. Dihydrochalcone sweeteners. Synthesis and sensory evaluation of a homoserine - dihydrochalcone conjugate with low aftertaste, sucrose-like organoleptic properties. *J. Agric. Food Chem.,* **1981**, *29*, 1269-1276. [http://dx.doi.org/10.1021/jf00108a042]

CHAPTER 3

Biotechnological Production of Amino Acids and Nucleotides

Volker F Wendisch*, Dorit Eberhardt, Marius Herbst and **Jaide V. K. Jensen**

Faculty of Biology & CeBiTec, Bielefeld University, D-33615 Bielefeld, Germany

Abstract: L-amino acids and nucleotides find various applications in food biotechnology. L-glutamic acid and its salts as well as 5'-nucleotides are used as flavor enhancers. Other L-amino acids are used as food or feed additives, in parenteral nutrition or as synthons for the chemical and pharmaceutical industries. L-amino acids and nucleotides are synthesized from precursors of central carbon metabolism. Based on the knowledge of the biochemical pathways microbial fermentation processes of food, feed and pharma amino acids and of nucleotides have been developed. Production strains of *Corynebacterium glutamicum,* which has been used safely for more than 50 years in food biotechnology, and *Escherichia coli* are constantly improved using metabolic engineering approaches. Research towards new processes is ongoing. Fermentative production of L-amino acids in the million-ton-scale has shaped modern biotechnology and its markets continue to grow steadily.

Keywords: *Corynebacterium glutamicum*, *Escherichia coli*, Essential amino acids, Feed additives, *Flavor enhancers*, Food additives, GMP, IMP, L-pyrrolysine fermentation, L-selenocysteine, Metabolic engineering, MSG, Nucleotides, Production strain development, Proteinogenic amino acids, XMP.

INTRODUCTION

As the building blocks of proteins, amino acids play an important role in human and animal nutrition. Essential amino acids cannot be synthesized by humans and animals and need to be taken up with the food or feed. Addition of essential amino

* **Address correspondence to Volker F. Wendisch:** Faculty of Biology & CeBiTec, Bielefeld University, Universit tsstr. 25, D-33615 Bielefeld, Germany; Email: volker.wendisch@uni-bielefeld.de.

acids improves feed quality and reduces manure. In the feed industry L-lysine is produced by fermentation in the million-ton scale while D,L-methionine is produced chemically. The amino acids L-glutamic acid and its salts finds application as flavor enhancer and annually about three million tons are obtained by fermentation. Since nucleotides provide the "umami" taste as glutamate, they are used as flavor enhancers and produced mainly by fermentation. This review focuses on microbial fermentation processes of food, feed and pharma amino acids and of nucleotides. Metabolic engineering of producing strains mainly *Corynebacterium glutamicum* and *Escherichia coli* is emphasized.

GLUTAMIC ACID, ITS SALTS AND 5'-NUCLEOTIDES AS FLAVOR ENHANCERS

Professor K. Ikeda discovered and isolated monosodium glutamate (MSG) in the early beginning of the 20th century in protein hydrolysates of "konbu", seaweed (*Laminaria japonica*) [1], while investigating the relationship between the chemical structure of a substance and its smell and taste [1]. "Konbu" is responsible for the dominant and distinct taste of the traditional Japanese soup base "dashi" [2]. Ikeda used methods and steps of classical chemistry and finally low-pressure evaporation resulted in the crystallization of a single substance, which could be identified as glutamic acid [1]. This substance, which elicits the specific taste of "umami", is the fifth taste quality besides salty, sweet, bitter and sour upon which the human taste is based [1]. Glutamate became more and more important after World War II, where especially protein-rich meat was very rare in Japan and MSG was used for seasoning tasteless food [2]. But for this purpose traditional production of L-glutamate by decomposition of soybean and wheat proteins followed by extraction in an industrial scale, like it was performed by Ajinomoto at this time [3], was too expensive and research started to aim at fermentative production of glutamic acid [4].

In the first approaches the synthesis of L-glutamate was achieved by fermentative production of 2-oxoglutaric acid with subsequent chemical conversion to glutamate. High yields for 2-oxoglutaric acid of about 50% on sugar basis were reported with the use of some bacteria of the genus *Pseudomonas*. Although several methods for the conversion of 2-oxoglutaric acid to L-glutamate had been developed, a one-step synthesis of L-glutamate from glucose and ammonia would have been superior [4]. In the 1950s, Japanese researchers of Kyowa Hakko Kogyo isolated a coryneform soil bacterium, which first was named *Micrococcus glutamicus* No. 534 [5] and later *Corynebacterium glutamicum* [6]. Studies showed that this bacterium was able to secrete glutamic acid under biotin limiting conditions, which finally led to a fermentation process with considerably reduced costs for the direct production of glutamate from cheap sugar and ammonia [6]. This was the starting point of the large-scale fermentative glutamate production with a steadily increasing market up to ~2.93 million tons in 2012 [7].

The umami taste can also be provided by the purine nucleotides disodium 5'-monoinosinate (IMP) and disodium 5'-monoguanylate (GMP) [8 - 10]. Shintaro Kodama isolated the nucleotide IMP from dried bonito tuna in 1913. GMP was isolated from shiitake broth in 1960 by Akira Kuninaka who also described synergy between glutamate, IMP, and GMP. The umami taste quality of IMP and GMP differs with IMP providing a strong initial taste enhancement and GMP proving a delayed, but about twofold stronger taste enhancement. Foods rich in IMP *e.g.* dried bonito tuna or in GMP *e.g.* shiitake broth are known, however, pure or enriched preparations of these nucleotides are available commercially. Their market volume increased steadily over decades and the global annual nucleotide production has reached ~34,000 tons in 2012 [7]. The 5′-nucleotides IMP and GMP are now produced by enzymatic degradation of ribonucleic acid or by chemical phosphorylation of the nucleosides inosine and guanine that are obtained by fermentation [11]. Since yeast is rich in ribonucleic acid (RNA), the first commercial yeast extract containing GMP was produced in 1974 and nowadays yeast extracts with various contents of IMP and GMP are available. Direct fermentation for the production of IMP and xanthosine 5'-monophosphate (XMP) has also been reported [12]. The highest titers for fermentative IMP and GMP production have been obtained with *Corynebacterium ammoniagenes* mutants (150 g/L; [13]).

While the worldwide glutamate consumption has increased dramatically over the recent decades, health and safety concerns have become more and more important [14, 15], especially since glutamate was associated with the so-called Chinese Restaurant Syndrome [16]. Moreover glutamate was also suspected to cause several other symptoms such as migraine, headaches, obesity or asthmatic bronchospasms [14, 17 - 20]. Up to now studies have failed or were insufficient to reproducibly demonstrate any relation between glutamate and certain symptoms. Therefore an Acceptable Daily Intake (ADI) has not been specified and the Joint Expert Committee on Food Additives of the United Nations Food and Agriculture Organization (JECFA) as well as the World Health Organization (WHO) and the European Community's Scientific Committee placed it in the safest category for food additives [21]. Due to FDA regulations it is required to include glutamate in the ingredients list of certain foods [22].

FOOD AND FEED USES OF AMINO ACIDS

Amino acids are used as additives to food and feed since they play important roles in human and animal nutrition as building blocks of proteins [1]. The uptake of amino acids – not only the essential ones – is primarily relevant for growth, but amino acids also play important roles in the regulation of gene expression, cell signaling, antioxidative responses, neurotransmission and immunity [23]. On this account a balanced amino acid ingestion is fundamental for growth, health, development and survival of animals and humans [23 - 26].

Feed use of amino acids

Livestock is increasing due to a growing world population and a higher demand of animal products. For that purpose it is desirable to optimize feeding in order to lower the production costs [27, 28]. In the feed industry the growth of livestock has been improved by the addition of amino acids to animal feed since 1960 [29]. Amino acids are important nutrients for animals, and especially the amino acids that cannot be synthesized by the animal (essential amino acids) are important to be supplied through the feed [30].

Julius von Liebig's "law of the minimum" can be applied to animal growth, where the benefit of protein in animal feeds is restricted by the scarcest resource. If one amino acid is deficient, the animal's body will not be able to use the other amino acids effectively and they will in the end be excreted. Taking into account that feedstuff from plant sources and other natural sources does not cover the whole for every amino acid [31], the efficiency of the plant-based feeds can be increased by adding the limiting amino acids to get the right nutrient balance [28, 32]. This is not only beneficial for animal growth, but also for the environment, not to mention the reduction of feed costs. For instance, the amount of L-lysine in lysine-rich soybean meal is 2.7-2.9%, while corn and wheat contain as little as 0.22 and 0.3% of L-lysine, respectively. The protein value of wheat can be improved by 25-50% by adding the first limiting amino acids L-lysine and L-threonine [29].

Furthermore, addition of the limiting amino acid reduces excretion of excess amino acids as nitrogen compounds. This excessive nitrogen release affects both soil and water quality and can moreover be released as N_2O in the atmosphere [30].

Common feedstuffs, such as corn, soybean, and wheat, contain all essential amino acids, but at differing concentrations resulting in deficient amounts of some amino acids [30]. Usually the plant material is deficient in L-lysine, L-methionine, L-threonine and L-tryptophan for mammal feed requirements, depending on the species and the growth stage of the animal [29, 31, 33].

Table 1. Limiting amino acids in various feed stuffs for growing pigs, dairy cows and poultry. According to [33] and [37]. The complete amino acid composition of different feed ingredients for animal diets is described by Li *et al.* [36].

	First limiting amino acids		
Feedstuff	**For growing pigs**	**For dairy cows**	**For poultry**
Cereals	L-lysine L-threonine		
Maize	L-lysine L-threonine L-tryptophan	L-lysine L-methionine	
Legumes	L-cystein L-methionine		
Barley	L-lysine L-threonine	L-lysine L-methionine	
Oilseed by-products	L-lysine L-methionine	L-lysine L-methionine L-leucine	L-cystein L-methionine L-lysine

On this account the essential amino acids L-lysine, L-threonine and L-tryptophan as well as the conditionally essential amino acids L-leucine, L-isoleucine, L-valine, L-phenylalanine, L-methionine, and L-histidine are of special interest to the feed industry (Table **1**) [1, 34, 35]. L-Arginine, glycine and L-proline can also be relevant since they are essential for many animals (*e.g.* avian species and fish) [33, 36 - 38]. For more information on amino acid essentiality see: [24, 33, 34, 39].

Supplementation of amino acids is a safe alternative to several protein sources. The safety requirements for feed supplements increased after the appearance of Bovine Spongiform Encephalopathy (BSE) and other prion-derived diseases [27, 31, 40]. To avoid the risk of infection, meat-based protein sources may be replaced by amino acids produced by microbial fermentation.

Food Use of Amino Acids

In human nutrition amino acids are predominantly added as flavor enhancers (L-aspartic acid, MSG, L-serine) or serve as precursors for the production of sweeteners like aspartame [1]. The addition of pure amino acids as a matter of dietary supplementation plays a minor role for healthy, adult humans. Since the amino acid supply usually is covered by the normal food uptake, the application of pure amino acids is restricted to changed nutritional requirements especially in the medical field, for example in parenteral and enteral nutrition [41 - 44]. Additionally, amino acids, especially branched chain species, are added in sports nutrition. Amino acids are also known to act as antioxidants and therefore facilitate preservative properties which is taken advantage of in the conservation of food and medical solutions [45 - 47]. The amino acid L-cysteine is used to increase dough proving and also as a flavor enhancer for vegetarian alternative products.

ECONOMIC IMPORTANCE AND COMMERCIAL RELEVANCE OF MICROBIAL AMINO ACIDS AND NUCLEOTIDES

Amino acids continue to be very important biotechnological compounds evident by the predicted rise of the global amino acids market to US$ 22.1 billion by 2018 [48]. The market size for amino acids is expected to keep on increasing partially due to a greater emphasis on health and therefore a consumer demand for healthy and nutritious foods, but mainly due to an increasing demand for meat for consumption, as the world population is growing and due to the increase in meat and dairy consumption in Asia and Central and South America [29]. The meat consumption is expected to rise to 52 kg per person per year by 2050 from 37.4 kg in 2000 [49]. As an example, the lysine market in China is rising, where it is the most consumed feed additive at around 610.000 tons in 2012, rising by 22% [50]. Although the demand for lysine keeps rising, the lysine market price for feed use has for the last decade remained more or less stable with an average price of around 1.5-2 US $/kg [7, 29]. The global market sizes of nucleotides and proteinogenic amino acids have been listed in Table **2**.

Table 2. Annual global production of amino acids and nucleotides with types of preferred production processes.

Substance	Predominant process(es)	World production (t/a)
L-Alanine	Enzymatic, extraction	500[a]
L-Arginine	Fermentation,	1,200[a]
L-Asparagine	Extraction	100-1,000[b]
L-Aspartic acid	Enzymatic	10,000[a]
L-Cysteine	Enzymatic, extraction	3,000[a]
L-Glutamine	Fermentation	1,300[a]
L-Glutamic acid	Fermentation	2,930,000[c]
Glycine	Chemical	22,000[a]
L-Histidine	Fermentation	400[a]
L-Isoleucine	Fermentation	400[a]
L-Leucine	Fermentation, extraction	500[a]
L-Lysine	Fermentation	1,950,000[c]
DL-Methionine	Chemical	750,000[d]
L-Phenylalanine	Fermentation, chemical	12,650[a]
L-Proline	Fermentation	350[a]
L-Serine	Fermentation	350[a]
L-Threonine	Fermentation	330,000[c]
L-Tryptophan	Enzymatic, fermentation	9,000[c]
L-Tyrosine	Fermentation, extraction	165[a]

(Table 2) contd.....

Substance	Predominant process(es)	World production (t/a)
L-Valine	Fermentation	500[a]
L-Selenocysteine	No process described for the 21st proteinogenic amino acid	
L-Pyrrolysine	No process described for the 22nd proteinogenic amino acid	
Nucleotides	Fermentation, extraction	34,000[c]

[a][51], [b][52], [c][7], [d][30].

Liebig's law is not only beneficial for animal growth; it also has a positive side effect. As an example, the production of 750,000 tons L-methionine would save around 15 million hectares of arable land [49]. Another example is the formula 50 kg soybean meal = 48.5 kg corn + 1.5 kg crystalline L-lysine; where the use of one ton of L-lysine can save around 33 tons of soybean meal [29]. This means that farmland can be saved by replacing soybean meal with corn and lysine, as the yield per hectare for corn is greater than for soybean [29]. It is no wonder that production of amino acids after 60 years of development still continues to be an interesting topic in which time, effort, and money are invested for further research and development of amino acid producing bacteria and optimization of the production and downstream processes.

The processes for amino acid production can roughly be divided into enzymatic, extraction, chemical and fermentative methods. The preferred production method depends on production costs, market demand, the availability of raw materials, *etc.* The methods used for production of the individual amino acids can be found in Table **2**. Most amino acids were initially obtained by extraction methods from protein rich resources. The main problem with this method can be product purity, satisfying the demand for the amino acid, the availability of the raw materials, and environmental concerns about the process as odor or waste treatment can be an issue. The chemical process is also not predominant, this is mainly due the high costs of the optical resolution step required for most chemical processes to deracemize the D,L-forms to the L-form. Glycine and methionine are mainly produced by this method because the optical resolution step is not necessary as glycine is achiral and for D,L-methionine, that is used in feed, the D-form is thought to be utilized by the animals by D-amino acids oxidase along with transaminase [53 - 55]. The use of biocatalysis is mainly limited by the cost of the substrates used, but it is the preferred method for the production of L-cysteine to replace previously common extraction of L-cysteine from human hair. Fermentation is the most applied method for commercial production of L-amino acids and in this chapter we mainly consider the microbial production of amino acids by fermentation.

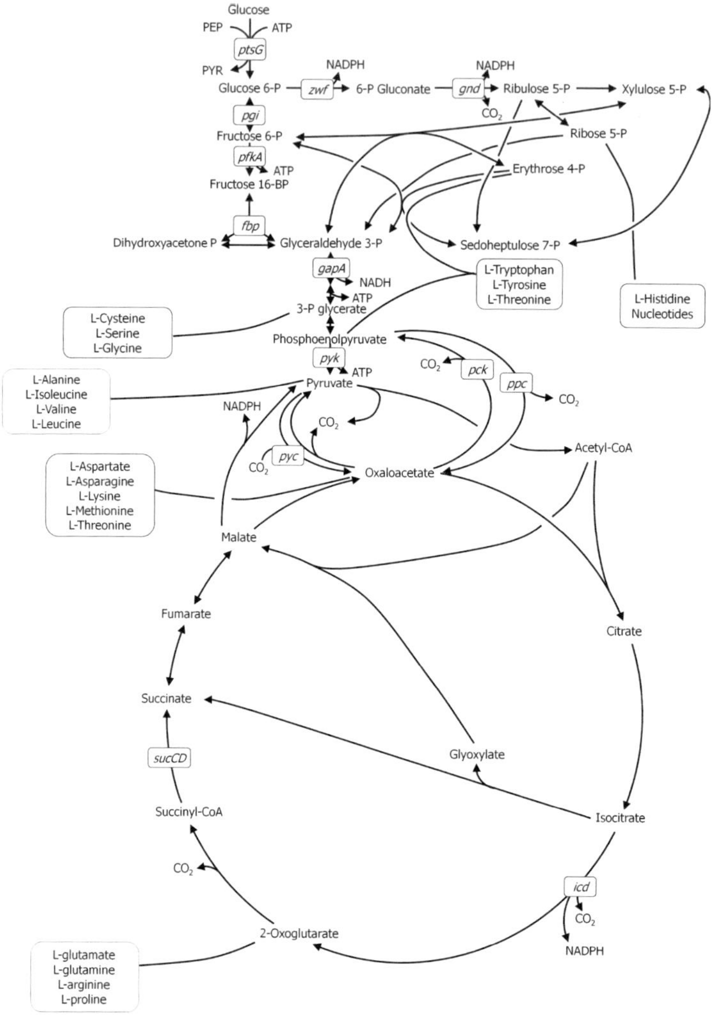

Fig. (1). Central carbon metabolism of *Corynebacterium glutamicum*. PEP, phosphoenolpyruvate; Pyr, pyruvate; *ptsG:*glucose-specific enzyme; *pgi:*: glucose-6-phosphate isomerase; *pfkA:*6-phosphofructokinase; *fbp:*fructose-1,6-bisphosphatase I; *zwf:*glucose-6-phosphate 1-dehydrogenase; *gnd:* 6-phosphogluconate dehydrogenase; *gapA:* glyceraldehyde-3-phosphate dehydrogenase A; *pyk:* pyruvate kinase; *pyc:* pyruvate carboxylase; *pck:* phosphoenolpyruvate carboxykinase; *ppc:* dehydrogenase A; *pyk:* pyruvate kinase; *pyc:* pyruvate carboxylase; *pck:* phosphoenolpyruvate carboxykinase; *ppc:* phosphoenolpyruvate carboxylase; *icd:* isocitrate dehydrogenase; *suc*CD: succinyl-CoA synthetase.

MICROBIAL PRODUCTION OF L-GLUTAMATE AND OTHER L-GLUTAMATE FAMILY AMINO ACIDS

L-Glutamate, L-glutamine, L-arginine and L-proline that belong to the glutamate family of amino acids are produced by fermentation. In the following we will discuss metabolic engineering of production strains of these amino acids. L-Glutamate is derived from 2-oxoglutarate (Fig. **1**) and the biosynthetic pathways of the L-glutamate-derived amino acids can be found in Fig. (**2**).

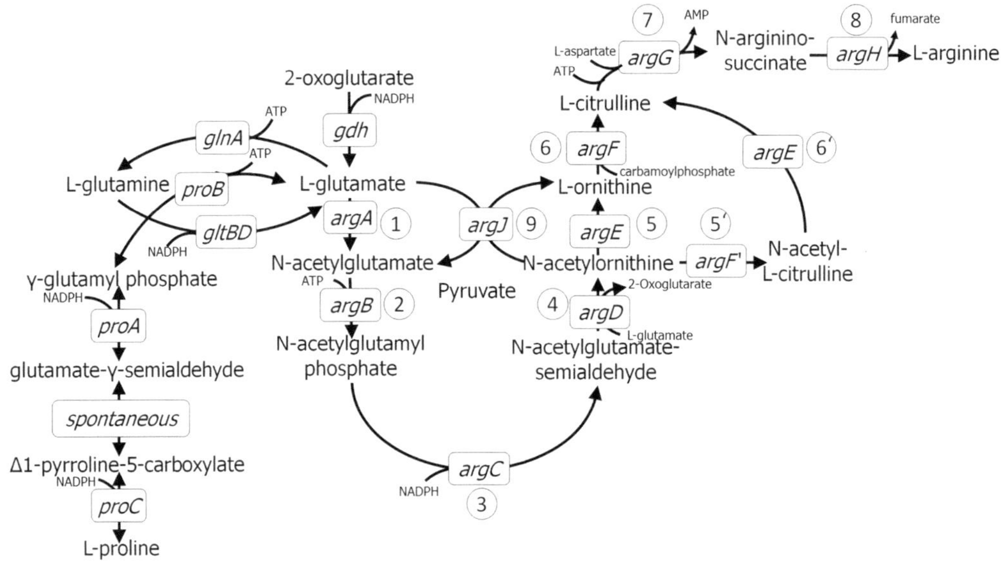

Fig. (2). Biosynthetic pathways of L-glutamate and the L-glutamate derived amino acids L-arginine, L-glutamine and L-proline. L-arginine biosynthesis according to [80]. Two pathways exist, the linear (as occuring in *E. coli*) is characterized by N-acetylglutamate synthase *(argA)*and acetylornithinase *(argE),* and the cyclic pathway (occuring in *C. glutamicum)* where ornithine N-acetyltransferase *(argJ),* bypasses the reactions of N-acetylglutamate synthase and acetylornithinase by transfering the acetyl group of N-acetyl-L-ornithine to L-glutamate. For an overview of the enzymes of the L-arginine pathway and their catalysed reactions see (Table **4**). *gdh:*glutamate dehydrogenase; glnA: glutamine synthetase; *gltBD:* glutamate synthase; *proB:* gamma-glutamyl kinase; *proA:* gamma-glutamyl phosphate reductase; *proC:*pyrroline---carboxylate reductase; *argA:* N-acetylglutamate synthase; *argB:* N-acetylglutamate kinase; *argC*: N-acety--gamma-glutamyl-phosphate reductase; *argD*: acetylornithine aminotransferase; *argJ:* ornithine acetyltransferase / N-acetylglutamate synthase; *argE:* acetylornithine deacetylase; *aargF:* ornithine carbamoyltransferase; *argG:* argininosuccinate synthetase; *argH:* argininosuccinate lyase. Oxoglutarate is an intermediate of central carbon metabolism (see Fig. **1**).

Glutamate Production

Due to its natural ability to synthesize high amounts of L-glutamate, *Corynebacterium glutamicum* and its close relatives are almost exclusively used for industrial L-glutamate production. Besides that, also engineered strains of *Brevibacterium lactofermentum*, *Brevibacterium flavum* and *Escherichia coli* are known [56].

Eliciting Glutamate Production by C. glutamicum

L-Glutamate production by *C. glutamicum* must be elicited. Since its discovery it is known that *C. glutamicum* produces glutamate under biotin starvation conditions. As *C. glutamicum* is auxotrophic for biotin this vitamin has to be supplied with the medium. A number of other elicitors of glutamate production are known (a selection is shown in Table **3**). They have in common that they affect the cell wall and/or the cytoplasmic membrane. Ethambutol, a chemotherapeutic used against *Mycobacterium tuberculosis* inhibits an arabinosyltransferase, active in cell wall arabinogalactan biosynthesis, and elicits L-glutamate production [57]. Penicillin G inhibits transpeptidase and, thus, cell wall biosynthesis [58] while Tween 40 and other surfactants affect membrane tension [59 - 61]. A temperature upshock has been used to elicit glutamate production in certain *C. glutamicum* strains [64]. However, this trait is not stable [65]. In addition to these cell wall/cytoplasma membrane effects, the metabolite flux at the branch point 2-oxoglutarate is channeled away from the TCA cycle to glutamate biosynthesis [66] and carriers export glutamate out of the cell [67 - 73].

Table 3. Elicitors and titers of L-glutamate production by *C. glutamicum* wild type.

Elicitor of glutamate production	Titer	Source
Biotin limitation	12 g/L from 80 g/L glucose	[62]
Ethambutol (30 mg/L)	8.2 g/L from 40 g/L glucose	[57]
Penicillin G (10 μM)	20 g/L from 80 g/L glucose	[63]
Tween 40 (4 g/L)	18 g/L from 80 g/L glucose	[58]
thermal upshock (33 -> 39 °C)	85 g/L in fed-batch	[64]

De-Bottlenecking Glutamate Biosynthesis

Glutamate dehydrogenase (GDH), which catalyzes the reversible reductive amination of 2-oxoglutarate, is the key enzyme of L-glutamate formation under high ammonium concentrations in many bacteria [74]. Both NADH- and NADPH-dependent enzymes are known. In *C. glutamicum* the *gdh* gene encodes NADPH-dependent glutamate dehydrogenase. This enzyme has a low ammonium affinity but high catalytic capacity and thus is relevant only under high ammonium concentrations [74, 75]. Interestingly, deletion of *gdh* did not abrogate glutamate production by *C. glutamicum* [76], since the glutamine synthetase (GS) – glutamine-oxoglutarate aminotransferase (GOGAT) can substitute for GDH [77, 78]. An enhancement of GDH activity alone did not change the flux distribution at the 2-oxoglutarate branch point and, thus, L-glutamate yields remained unchanged [79]. The 2-oxoglutarate dehydrogenase complex (ODHC) competes with GDH for the substrate 2-oxoglutarate and catalyzes the oxidative decarboxylation of 2-oxoglutarate to succinyl-CoA. The enzyme consists of three subunits (E1o, E2o and LPD, encoded by *odhA*, *sucB* and *lpd*) [81]. A decrease in ODHC activity is observed when specific L-glutamate production conditions are adjusted (*e.g.* biotin limitation, Tween 40, Penicillin G) [58, 63, 79].

A disruption of *odhA* was first reported to result in an increased L-glutamate excretion [82] but it was later on observed that constitutive overproduction of L-glutamate was caused by a mutation of the NCgl1221 gene (see section Glutamate export) and disruption of *odhA* alone did not result in L-glutamate overproduction [68].

ODHC activity is exquisitely regulated in *C. glutamicum* involving a specific inhibitory protein named OdhI [83]. OdhI in turn is regulated by protein phosphorylation and can be deactivated by phosphorylation of its threonine residues (T14, T15) *via* serine/threonine protein kinases PknG, PknA, PknB and PknL [83 - 85]. In its phosphorylated form OdhI cannot bind to the OdhA subunit and ODHC remains active [83, 85, 86]. Exchanging the threonine residues in OdhI (T14A; T15A) resulted in phosphorylation independent inhibition of ODHC [83]. Deletion of *pknG* was shown to be beneficial in the presence of ethambutol or under strong biotin limitation [87]. Nevertheless, the OdhI phosphorylation status was not affected by *pknG* deletion under Tween 40, normal biotin limitation, and temperature shift conditions since serine/threonine protein kinases are responsible for phosphorylation under these conditions [88]. In Penicillin G treated cells OdhI levels are increased and homologous overexpression of *odhI* led to higher L-glutamate yields [63].

Precursor Supply for Glutamate Production

Withdrawal of 2-oxoglutarate from the TCA cycle for glutamate biosynthesis requires replenishment of the TCA cycle by anaplerotic reactions. In *C. glutamicum*, anaplerosis may occur *via* the glyoxylate cycle, by PEP carboxylase (PEPCx encoded by *ppc*) or by pyruvate carboxylase (PCx encoded by *pyc*) [89, 90]. PEPCx was found to be dispensable for growth and L-glutamate and L-lysine formation [91], while *pyc* overexpression improved L-glutamate production [92]. However, since PCx is a biotin-dependent enzyme [90], glutamate production under biotin-limiting conditions requires PEPCx.

Cofactor Supply for Glutamate Production

The formation of L-glutamate in *C. glutamicum* is NADPH neutral because NADPH required in the GDH reaction is provided by the reaction of $NADP^+$ dependent isocitrate dehydrogenase (ICDH), which oxidizes isocitrate to 2-oxoglutarate and generates NADPH [93]. Therefore, NADPH provision by the pentose phosphate pathway (PPP) is low under glutamate producing conditions [94]. Replacing the endogenous $NADP^+$-dependent GDH by a heterologous NAD^+-dependent GDH from glutamate degrading *Peptostreptococcus assacharolyticus* was attempted in a L-lysine producing strain. However, the *C. glutamicum* metabolism reacted flexible by reducing the NADPH provided by the PPP [95]. Therefore, this strategy might not be beneficial for L-glutamate production.

Glutamate Export

As the majority of L-glutamate inducing culture conditions can be associated with the cell envelope structure of *C. glutamicum* (*e.g.* biotin limitation, fatty acid ester surfactant or penicillin treatment) it was initially proposed that L-glutamate overproduction was only caused by alterations in the envelope structure (leaky model). This hypothesis was discarded [68] as soon as experimental evidence for glutamate efflux being mediated by a transport system became available.

The export of L-glutamate is carried out by the mechanosensitive channel protein NCgl1221 (also called YggB) [68, 96]. Mutations in the *NCgl1221* gene which lead to a C-terminal truncated protein form (423 amino acids instead of 533; V419::IS1207) or other mutations (*e.g.* A100T; A111V/T; P424L; W15CSLW) increased the L-glutamate yield under biotin excess accompanied with decreased biomass formation (up to 92.3 mM instead of 2.4 mM L-glutamate in the wild type ATCC 13869; OD_{620} = 41.0 instead of 63.2) [68 - 73]. There is evidence that the channel domain of NCgl1221 mediates glutamate efflux and that a C-terminal extension specific to NCgl1221, but absent from related mechanosensitive channel proteins, has a significant impact for function and/or activation of MscCG [72].

Avoidance of Glutamate Re-import and Degradation

C. glutamicum is able to catabolize L-glutamate as an energy and nitrogen source [97]. Nevertheless this does not occur until glucose has been exhausted, since L-glutamate import does not start until glucose is utilized [98]. Therefore, it is not necessary to interfere with the L-glutamate catabolism, but stop the fermentation process before the carbon source has been utilized.

There are two import systems known for L-glutamate in *C. glutamicum*, namely GluABCD and GltS [99, 100]. The ABC transporter system GluABCD is responsible for the primary active transport of L-glutamate when other carbon sources than L-glutamate are absent [100]. The GluABCD import system is regulated by catabolite repression and is therefore inactive during the production of L-glutamate from carbon sources such as glucose or sucrose [100 - 101]. A deletion of the *gluABCD* operon was shown to have no impact on L-glutamate production from glucose [100]. Nevertheless a detectable L-glutamate import activity could be observed when *C. glutamicum* was grown on acetate as a sole carbon source [98]. The secondary passive uptake system GltS is sodium dependent. The L-glutamate excretion could be increased by about 10 % when the *gltS* gene was deleted [99].

L-glutamate plays an important role as an ammonium donor in many transamination reactions for example in the amination of 2-oxo acids to form amino acids and other aminated products (histidinol-phosphate transaminase, branched-chain-amino acid transferase, aromatic amino acid transaminase, valine-pyruvate transaminase, glutamate-prephenate aminotransferase, succinyl-amino oxopimelate transaminase, aspartate transaminase). However, the interference with glutamate production by its role as amino donor in transaminase reactions has not been described.

L-Glutamine Production

L-glutamine has been produced by microbial fermentation since the late 1960s [102]. L-glutamine can be formed from L-glutamate in a single enzymatic step by glutamine synthetase (GS encoded by *glnA*). For this purpose it is important to supply sufficient amounts of the precursor L-glutamate (see section L-glutamate). As producing organisms coryneform bacteria, predominantly *C. glutamicum*, are preferred due to their ability to form the precursor L-glutamate in adequate amounts [103, 104].

De-bottlenecking Glutamine Biosynthesis

In *C. glutamicum* the enzyme activity of GS was increased by overexpressing the *glnA* gene [103] or its variants [104]. When *glnA* was overexpressed in *B. flavum* ~12 % more L-glutamine (~45 g/L instead of ~38 g/L) and less by-product L-glutamate was formed [103]. GS has a typical adenylylation site (Tyr405), which is responsible for the regulation of its activity [105]. While deadenylylated GS is active, the adenylylated version of GS is inactive. The adenylylation is maintained by the adenylyltransferase GlnE and is dependent on the nitrogen level in the cell [105 - 107]. By replacing the tyrosine residue with phenylalanine (Y405F) the enzyme was cured of its nitrogen dependent activity most likely due to the lack of adenylation [105].

The expression of the *glnA* gene is repressed under nitrogen surplus by AmtR [106]. Derepression is observed when ammonium is limited. In that case a signal transduction cascade is activated and the protein GlnK (P_{II}-type protein, previosly GlnB) is adenylylated by the adenylyltransferase GlnD [108 - 110]. In its adenylylated form GlnK probably binds to AmtR and therefore prevents transcription attenuation as AmtR releases the target DNA [97, 106, 108, 109, 111 - 113]. Deletion of *amtR* in *C. glutamicum* leads to increased amounts of L-glutamine and also L-Lysine [114]. How the cascade is initiated is unkown, but ammonium could be involved as a cellular nitrogen status marker [115].

When the *glnE* gene, coding for adenylyltransferase, is deleted, strains were able to produce 5 times more L-glutamine than the wild type (30±4 mM *vs.* 153±36 mM) and the cells showed a nitrogen starvation response even at high ammonium concentrations [116]. This is of high interest for the industrial production of amino acids by *C. glutamicum* because of pH regulation by ammonium hydroxide [115].

Avoidance of Glutamine Degradation

Since glutamine can be used to aminate 2-oxoglutarate in the reaction of glutamate synthase (GOGAT), decreasing GOGAT activity led to decreased L-glutamine degradation [103]. Deletion of the two subunits *gltB* and *gltD* encoding GOGAT led to a slight decrease of the growth rate [117], but whether it is beneficial for L-glutamine production needs to be elucidated. L-glutamine is also an ammonium donor or precursor for many other enzymes (carbamoylphosphate synthase, glucosamine-fructose-6-phosphate aminotransferase, amidophosphor-ibosyltransferase). Deletion or reduced expression of their respective genes (*carA* and *carB*, *glmS*, *purF*) is also a potential strategy for strain engineering.

Precursor Supply for Glutamine Production and Avoiding by-Products

For attenuation or elimination of the competing reactions of isocitrate lyase (*aceA*) and 2-oxoglutarate dehydrogenase (ODHC) see sub-chapter L-glutamate [103]. Increasing the formation of the precursor L-glutamate by overexpression of glutamate dehydrogenase (encoded by *gdh*, see section L-glutamate) and simultaneous enhancement of GS by *glnA* overexpression, shortened the cultivation process and increased L-glutamine formation coupled with the reduced formation of the by-product L-glutamate [104]. For a detailed description of improving L-glutamate supply see the sub-chapter on L-glutamate production.

Export and Avoidance of Re-import

The export system for L-glutamine has not been characterized in industrially relevant bacteria although specific exporters exist [118]. Uptake of L-glutamine in *E. coli* is mediated by L-glutamine permease, which is encoded by *glnHPQ* [119]. A gene predicted to encode a L-glutamine importer in *C. glutamicum* was proposed based on a bioinformatic analysis [118, 120]. Possibly, deletion of the importer genes may improve L-glutamine production.

L-Arginine Production

Since L-arginine is synthesized from L-glutamate it is attractive to engineer the L-arginine pathway in *C. glutamicum*. Besides *C. glutamicum* also *C. acetoacidophilum*, *C. flavum*, *Bacillus subtilis* and *E. coli* have been used for L-arginine production [56, 121 - 125]. As some yeasts such as *S. cerevisiae* are known to exhibit high intracellular L-arginine concentrations they are exploited in extraction processes [126]. Biosynthesis of L-arginine in bacteria is summarized in Fig. (**2**) and Table **4** lists the involved enzymes, their abbreviations and gene names.

Table 4. Genes and enzymes of the L-arginine biosynthesic pathway in microorganisms according to [80, 129, 133].

Gene	Enzyme name	Short	Reaction[a]
argA	N-acetylglutamate synthase	NAGS	1
arg(A)	acetyl-CoA domain of classical NAGS	S-NAG	1
argJ	ornithine N-acetyltransferase	OAT	9
argJ	bifunctional enzymes that carry OAT and NAGS activity	OAT^{bi}	9 + 1
argB	acetylglutamate kinase	NAGK	2
argC	N-acetyl-gamma-glutamyl-phosphate reductase		3
argD	acetylornithine/succinyl diaminopimelate aminotransferase		4
argE	acetylornithinase	AO	5 + 6’
argF	ornithine carbamoyltransferase	OTC	6
argI	paralogous carbamoyltransferase gene in *E. coli* K12	OTC	6
argF’	acetylornithine carbamoyltransferase		5’

(Table 4) contd.....

Gene	Enzyme name	Short	Reaction[a]
argG	argininosuccinate synthase		7
argH	argininosuccinase (argininosuccinate lyase)		8

[a]reactions according to Fig. (**2**).

De-bottlenecking L-Arginine Biosynthesis

In many Gram-positive and Gram-negative bacteria including *E. coli* and *C. glutamicum* transcription of the L-arginine biosynthesis operon is regulated by the arginine repressor ArgR [127, 128]. ArgR binds to the so called ARG boxes which are located in the promoter regions of operons encoding enzymes of the L-arginine biosynthetic pathway [127, 134 - 136]. To derepress transcription of *argCJBDFRGH* in *C. glutamicum*, the *argR* gene was deleted [137]. Whereas the deletion of *argR* alone did not result in a high L-arginine yield, the simultaneous overexpression of feedback resistant *argB* gene variants ($argB^{FBR}$) increased the L-arginine yield significantly [137].

In *C. glutamicum* the acetyl-group is recycled in the OAT reaction and therefore ArgB is feedback regulated to maintain the regulation of L-arginine synthesis [129], while in *E. coli* and other microorganisms a linear pathway, in which ArgA and ArgE replace OAT activity, ArgA is feedback inhibited as it catalyzes the initial step of L-arginine biosynthesis [138].

Several enzyme variants of ArgA which are feedback insensitive are known [139 - 140]. Overexpression of these genes with L-glutamate or L-glutamine supplementation together with *argI* and *carAB* increased the L-arginine production up to 15 times in comparison to the overexpression of the wild type *argA* [140]. The necessity of L-glutamate supplementation, however, strongly indicates that the rate limitating reaction is catalyzed by GDH (see also below).

Homologous overexpression of *argJ* in *C. crenatum* increased the L-arginine production by about 17 % [141]. When the whole *argCJBDFRGH* operon was overexpressed in the *C. crenatum* strain SYPA 5-5, which is insensitive to ArgR regulation, the L-arginine production could be increased by about 25 % [142].

Avoidance of By-products of L-Arginine Production

During L-arginine production, L-citrulline, L-ornithine, L-proline and L-glutamate may be side-products. L-glutamate formation can be prevented by avoiding biotin limitation which is known to cause L-glutamate production (See section on L-glutamate). L-proline formation can be avoided by the deletion of *proB*as shown for L-ornithine production [143, 144]. Based on these results it is likely that Δ*proB* is also conducive to L-arginine formation. It is likely that L-ornithine and L-citrulline formation can be reduced by overproduction and deregulation of the L-arginine biosynthesis enzymes converting L-ornithine and L-citrulline to L-arginine.

Precursor Supply for L-Arginine Production

As L-arginine is formed from L-glutamate, an increased L-glutamate supply is beneficial for L-arginine biosynthesis. Besides L-glutamate, also carbamoyl phosphate is needed for the carbamylation of L-ornithine (catalyzed by the *argF* gene product) [129]. In *E. coli* it could be shown that *carAB* overexpression with concomitant *argI* overexpression (carbamoylphosphate synthase) did not result in higher L-arginine yields, as long as no L-ornithine was added to the medium [145]. This clearly indicates that in *E. coli* the rate limiting step is not attributed to carbamoyl phosphate availability. In L-arginine production by *C. glutamicum* L-citrulline is a by-product; therefore the availability of carbamoylphosphate does also not seem to be a limiting factor [146]. Plasmid-borne heterologous overexpression of *carAB* in *C. glutamicum* led only to a small increase of the L-arginine titer in the fermentation supernatant [125].

During biosynthesis of L-arginine from L-glutamate three nitrogen atoms are added. The nitrogen donors in these reactions, L-glutamate, L-glutamine and L-aspartate need to be regenerated at the expense of ATP and NADPH (see Fig. **2** and Fig. **3**). Thus, L-arginine biosynthesis is energetically demanding (see also section Cofactor supply for arginine production).

Overexpression of *cg1722* (*NCgl1469*) (coding for an enzyme that exhibits acetylglutamate synthase and acetylornithinase activity) in *C. glutamicum* resulted in an increase in formation of L-ornithine (about 0.147 instead of about 0.138

g/L), which is a precursor of L-arginine [147]. By adding L-glutamate to the medium, the positive effect was even higher, indicating that the supply of the precursor L-glutamate could still be a bottleneck in the L-arginine pathway [147]. Increasing L-ornithine production in the L-arginine producing strain Δ*argR* NCgl1469 was only beneficial for L-arginine production when L-glutamate was supplemented [147]. Nevertheless other sources didn't confirm these results and showed that additional L-glutamate was neither taken up by the cell nor converted to L-ornithine [143]. It has to be noted that glutamate uptake is repressed by glucose [98].

Cofactor Supply for L-Arginine Production

The biosynthesis of one L-arginine molecule requires 3 molecules of ATP and 4 molecules of NADPH. In *C. crenatum* a NAD kinase (*ppnK*) was overexpressed in an L-arginine producing strain. It was shown that the arginine concentration could be improved from about 24 g/L to about 26 g/L [148]. PpnK phosphorylates the oxidized form of NAD^+, thus, only increasing the NADP concentration in the cell, but not the NADH/NADPH ratio [149]. For a detailed description of targets for increasing the NADPH supply, see the sub-chapter L-lysine. How the increase of ATP can influence the L-arginine yield remains uninvestigated.

L-Arginine Export

In *C. glutamicum* surplus L-arginine is exported by the *lysE* gene product as it cannot catabolize L-arginine [150 - 152]. *lysE* expression is activated by the transcriptional regulator LysG [151]. The analogous gene *argO* (*yggA*) in *E. coli* has a very similar primary structure and has also been attributed to L-lysine and L-arginine export [121]. Transcription of *argO* is regulated by Lrp and ArgP [153]. Heterologous as well as homologous overexpression of *lysE* and *argO* in *C. crenatum* had a positive impact on L-arginine accumulation in the medium by about 14 % [154].

Avoidance of L-Arginine Re-import and Degradation

L-arginine uptake is characterized in *E. coli* by the two ABC transporters ArgTHisPQM and ArtIMPQ [155, 156]. Until now there are no engineering efforts reported in this field. By contrast, L-arginine import in *C. glutamicum* seems to be absent [118]. Gene cg3045, however, was "proposed to function as a permease of an uptake system for L-glutamine or L-arginine" [118].

Industrial Production Strains and Processes, and Improvement of the Fermentation Quality of Strains

L-arginine is produced by *C. glutamicum* from sugar syrup or hydrolysed starch at an industrial scale [122]. Of high importance is also the high concentration of a nitrogen source because the generation of each L-arginine molecule requires three ammonium molecules [122].

A characteristic of L-arginine biosynthesis is the high demand for bicarbonate, which is assimilated during the formation of carbamoyl phosphate [157]. Therefore, sparging with carbon dioxide is important and might be advantageous for L-arginine production [122]. For an improved temperature resistance strain engineering strategies, concerning an increase of productivity under higher, suboptimal temperatures (*e.g.* 38 °C) to reduce the cooling requirements of the process, have been exploited [146]. Also *C. crenatum* strains, which have improved oxygen uptake abilities have been investigated [158].

C. glutamicum strains obtained by undirected breeding methods were reported to produce about 25 g/L L-arginine [122]. The US patent 6897048 B2 reported an L-arginine concentration of 14 g/L with the *E. coli* strain K12 P4XB2 [123]. For L-arginine fermentations at an industrial scale approximately 100 g/L were reached in 2010 [159].

L-Proline Production

L-Proline, the only proteinogenic amino acid with a secondary amine, is commonly used by the chemical industry as an organocatalyst [160]. It serves as a feed additive and as a precursor in the pharmaceutical and cosmetic industry [161, 162]. Initially, L-proline was extracted from protein hydrolysates, but has now been produced by fermentation for five decades [163, 164]. Proline, besides in protein synthesis, has been associated to function as carbon, nitrogen, and energy source, as a virulence factor of some pathogenic bacteria [165 - 167], and as a protectant against osmotic stress [168 - 169].

L-Proline is derived from L-glutamate and its biosynthetic pathway has been studied in a range of microorganisms (Fig. **2**) [170, 171]. L-Proline can also be produced from L-ornithine in plants, animals, and some bacteria [172]. First L-glutamate is converted into the labile intermediate γ-glutamyl phosphate at the cost of one ATP by γ-glutamyl kinase (GK) encoded by *proB*. Then a reduction to glutamate-γ-semialdehyde occurs, consuming NADPH by the *proA* gene product, γ-glutamyl phosphate reductase (GP). The enzymes catalyzing the two first steps have been proposed to function in a complex, owing the liability of the intermediate [173 - 174]. Then glutamate-γ-semialdehyde cyclizes into Δ1-pyrroline-5-carboxylate by a spontaneous reaction, and finally proline is reduced by pyrroline-5- carboxylate reductase, the *proC* gene product [174].

L-Proline Producing Strains

Most published or patented proline producers have been obtained by classical strain development by means of mutagenesis and screening. The two most common traits among proline producers are an inactivated proline degradation system and/or feedback desensitized GK [164, 169, 175, 176]. It has been demonstrated several times that feedback inhibition of γ-glutamyl kinase by proline seems to be the primary mechanism for the control of proline synthesis [177]. An *E. coli* strain with feedback resistant GK was shown to have the amino residue change D107N [178] and for *C. glutamicum* the change G149D was described [179]. Common for the mutations in GK reported for different organisms that led to proline overproduction, is that they occur in the same region, that appears to be the allosteric binding site.

B. flavum was the first L-proline producing strain reported, with a yield of 0.12 g proline / g glucose [163]. In *Listeria monocytogenesis* random mutagenesis was performed on the *proBA* operon and recombinants were screened for resistance to the L-proline analog L-azetidine-2 carboxylic acid (AZ). The three mutations that led to an altered GK and to AZ resistance of an *L. monocytogenes proBA* mutant were V121I, A144V, and E146K. When tested in an *E. coli* Δ*proBA* strain, the mutant GKs led to L-proline accumulation in the mg/L range [169]. A *Thermus* *thermophiles* mutant resistant to 3,4-dehydroproline was constructed by site-directed mutagenesis of *proB*. To disrupt proline degradation the strain was mutated by UV irradiation and one of the isolated mutants unable to utilize proline for growth produced 2 mg/L in 12h [180]. Recently, a *C. glutamicum* strain overproducing L-proline through the L-ornithine pathway was published and L-proline concentrations of 13 g/L were reached [181]. The strain with a Δ*argFR* background (for further information see the L-arginine section) was harboring a plasmid encoding ornithine cyclodeaminase from *Pseudomonas putida*, converting L-ornithine to L-proline and ammonia. This strain had a yield of 0.31 g L-proline/ g glucose and could be further increased to 0.36 g/g by overexpressing feedback desensitized NAGK [181].

C. glutamicum can utilize and co-utilize a wide range of carbon sources including a large variety of aromatic compounds [182 - 184]. The TCA cycle utilizes phenol intermediates as energy sources. Lee *et al.* [185] demonstrated the use of 8.5 mM phenol along with glucose and $FeSO_4$, to increase L-proline production by *C. glutamicum*. It was shown that the increase in L-glutamate and L-proline was caused by iron decreasing the DNA-binding affinity for the repressor ArgR [185].

MICROBIAL PRODUCTION OF L-LYSINE AND OTHER ASPARTATE FAMILY AMINO ACIDS

Production of the aspartate family amino acids L-lysine and L-threonine will be described here. Production of L-methionine is described together with cysteine in a separate sub-chapter. Production of L-aspartic acid and L-asparagine is described together with L-alanine in a separate sub-chapter. The biosynthetic pathways of the amino acids of the aspartate family are depicted in Fig. (**3**).

L-Lysine Production

Among the proteinogenic amino acids addressed in this sub-chapter L-lysine is one of the most important due to its application in the animal feed industry with an estimated annual market volume of 1.95 million tons in 2013 [7]. This high demand is on one hand attributed to the fact that the addition of L-lysine to typical L-lysine-poor animal feeds, like wheat, corn or barley, was shown to be very effective. On the other hand the continuously increasing consumption of white meat in North America and Europe as well as the increasing demand for affordable meat in developing countries, are driving forces for a strongly growing L-lysine market with an increase of ~8 % per year [186, 187].

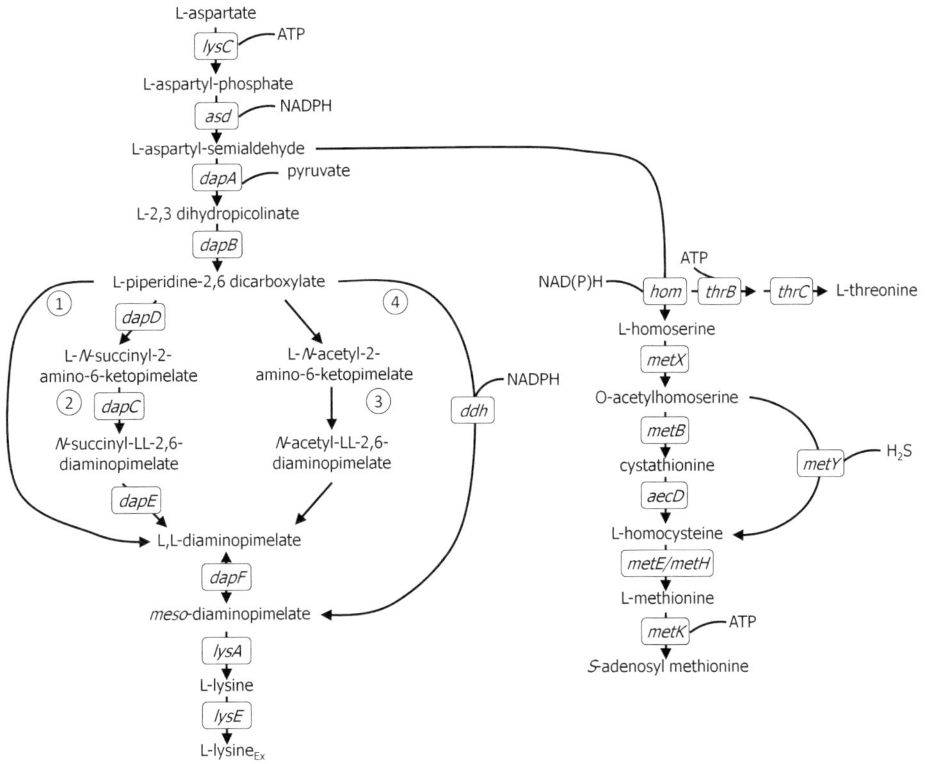

Fig. (3). L-Lysine biosynthetic pathways in bacteria and L-methionine and L-threonine biosynthetic pathways of *C. glutamicum:* (**1**)aminotransferase pathway; (**2**) succinylase pathway (present in *C. glutamicum*); (**3**) acetylase pathway (present in *E. coli*); (**4**) dehydrogenase pathway (present in *C. glutamicum*); *lysC:* aspartokinase gene; *asd:* aspartate semialdehyde dehydrogenase gene; *dapA:* dihydrodipicolinate synthase gene *; dapB:*dihydrodipicolinate reductase gene; *dapD:* tetrahydrodipicolinate succinylase gene; *dapC:* N-succinyldiaminopimelate aminotransferase gene; *dapE:* succinyl-diaminopimelate desuccinylase gene; *dapF:* diaminopimelate epimerase gene; *ddh:* meso-diaminopimelate dehydrogenase gene *lysA:* diaminopimelate decarboxylase gene; *lysE.* L-lysine exporter; *hom:* homoserine dehydrogenase gene; *thrB:* homoserine kinase gene; *thrC:* threonine synthase gene; *metX:* homoserine O-acetyltransferase gene; *metB:* cystathionine gamma-synthase gene; *aecD:* cystathionine beta-lyase gene; *metE/metH:* B_{12} -independent/dependent synthase gene; *metK:* Sadenosylmethionine synthetase. L-aspartate is derived from oxaloacetate, an intermediate of central carbon metabolism (see Fig. **1**).

Since 1958 *C. glutamicum*, which was discovered in 1956 due to its capability of secreting L-glutamic acid, has been the most important organism used for large-scale production of L-lysine. The first plant was built in Japan by Kyowa Hakko, which is still one of the main players in the business [5 - 6, 187]. To cover the steadily increasing demand, strain improvement and process engineering have been performed continuously and has led to efficient manufacturing of L-lysine by renewable resources [188, 189].

L-lysine Biosynthesis

Two different routes for the biosynthesis of L-lysine are known in microorganisms. The first one is the α-aminoadipate route (AA), which starts from 2-oxoglutarate and acetyl-CoA and the second one is the diaminopimelate route (DAP) starting from L-aspartate (see Fig. **3**) for variants of the DAP route). Moreover, two variants of the AA pathway and three variants of the DAP pathway exist [190]. Two of the three variants of the DAP pathway can be distinguished by either succinylated or acetylated intermediates, while in the third variant D,L-diaminopimelate is directly formed from tetrahydropicolinate [187]. In most microorganisms only one of the variants is present, but in a few organisms two pathways operate together [191 - 194]. This is known for different species of the genera *Corynebacterium* including *C. glutamicum* where the DAP dehydrogenase and succinyl variants co-exist and increase its flexibility to react to different ammonia levels in the environment [187, 194]. A key feature of all variants is feedback inhibition of the first enzyme aspartate kinase [187].

Strain Development for Microbial Production of L-lysine

To match the demands of steadily decreasing market prices the main goals of strain development are increased conversion of the carbon source and increased volumetric productivity [186]. In former times strains were generated by random mutagenesis and selection. With this procedure production strains could be identified with for example feedback resistance to a mixture of the L-lysine analogue S-(2-aminoethyl) cysteine plus L-threonine, where a feedback resistant aspartate kinase no longer was rate limiting for L-lysine production [186]. On one hand the strains showed a sugar conversion of 50% and L-lysine titers of 100 g/L, but on the other hand those strains accumulated a substantial number of unwanted mutations as well as auxotrophies. That was however detrimental for cheap and stable processes due to the additional nutrient requirements, higher sensitivity to unfavorable temperature or pH as well as increased affection by certain limitations [1, 56, 195]. The development of several tools for genetic engineering in the last 20 years and the availability of the whole genome sequence of *C. glutamicum* led to more site-directed and rational strain design [186, 196].

De-bottlenecking and L-lysine Export

As the key enzyme and major bottleneck for L-lysine production *lysC* which is organized in an operon together with *asd* was considered quite early [197]. Several mutations in the coding region of the regulatory β-subunit of the *lysC* gene led to L-lysine or L-threonine feedback resistant LysC [187]. This approach, as well as the overexpression of the aspartate kinase, enhanced L-lysine formation significantly, which is also reflected by several patents describing feedback resistant aspartate kinase mutants [197 - 201]. Furthermore Ohnishi *et al.* [195] demonstrated that two single amino acid substitutions in the pyruvate carboxylase (Pyc^{P458S}) and homoserine dehydrogenase (Hom^{V59A}) increased L-lysine production from 50 to 80 g/L with a productivity of 3.0 g/L h^{-1}.

A further promising target for increasing L-lysine formation is the *dapA* gene coding for the dihydropicolinate synthase. An increased copy number was shown to be beneficial and several nucleotide exchanges within the extended -10 region of the *dapA* promoter resulted in an enhanced enzyme activity up to 2.5-fold compared to the wild type [188].

Identification and subsequent overexpression of the L-lysine exporter coding gene *lysE* led to a 5-fold increased lysine excretion rate compared to the wild-type [150 - 151, 202]. Further work also showed that overexpression of *dapF* and *dapC* as well as increasing the copy number of further genes involved in L-lysine biosynthesis like *dapB*, *ddh* and *lysA* or the exchange of the native promoter with the *sod* promoter of certain genes (*e.g. dapB*, *lysC*, *pyc*) improves L-lysine production additionally [186, 203, 204].

Reduction of By-products and Competing Pathways of L-lysine Production

Trehalose is the dominating by-product in L-lysine production by *C. glutamicum* when grown on glucose. Overexpression of *zwf* as well as mutated *zwf* under the control of the *sod* promoter led to a decrease in trehalose formation by 53 and 67%, respectively, as a positive side effect of an increased flux through the PPP [205]. The use of sucrose can also prevent the formation of certain by-products at a L-lysine yield comparable to that on glucose, in contrast to fructose where the formation of lactate reduces L-lysine production by 30 % [206].

Precursor Supply for L-lysine Production and Redirection of Carbon Metabolism

Phosphoenolpyruvate carboxylase (Ppc) and pyruvate carboxylase (Pyc) are the anaplerotic enzymes of *C. glutamicum* responsible for the formation of oxaloacetate (Fig. **1**), a direct precursor of the aspartate-derived amino acids [89, 207]. Experiments showed that the overexpression of *pyc* improved L-lysine production [92]. Moreover, this improvement additionally could be triggered by the mutation C1372T [195]. Further investigations showed the presence of enzymes decarboxylating C_4 metabolites of the TCA cycle into C_3 metabolites of glycolysis [208]. Deletion of the genes coding for phosphoenolpyruvate carboxykinase (Pck) and glucose 6-phosphate dehydrogenase (Zwf) were thought to enhance L-lysine by increasing anaplerotic net flux, but only the deletion of *pck* led to a significant improvement while neither the deletion nor the overexpression of *zwf* had an effect [209].

Another approach for redirecting the carbon flux was to couple the tricarboxylic acid (TCA) cycle and the succinylase branch of the L-lysine biosynthetic pathway. This was done by targeted interruption of succinyl-CoA-synthetase activity by deleting *sucCD*. The deletion resulted in a vital strain that only exhibited a slightly reduced specific growth rate and a 60% increase in L-lysine production [210].

Inhibition of the dominant TCA cycle enzyme, isocitrate dehydrogenase (ICD) on the translational level, by exchanging the start codon was employed for shifting the TCA cycle flux to the anaplerotic carboxylation, which led to a more than 40% increase L-lysine production [211].

NADPH Supply for L-lysine Production

NADPH formation and consumption are dependent on the carbon source [212 - 214], the genetic background [94] and the particular physiological growth state, which has been shown by ^{13}C metabolic flux studies [94, 215]. These studies also showed that NADPH supply is quite important for L-lysine production where NADPH is consumed in four steps through the assimilation of ammonium and that NADPH supply by the PPP correlates with the L-lysine yield [94, 195, 205, 216, 217]. Moreover, Kiefer *et al.* [218] observed an apparent NADPH limitation during phases of maximal L-lysine production in fed-batch cultivation. Also the production on fructose, where only 14.4 % of the carbon in comparison to 62.3 % in case of glucose are channeled through the PPP, lead to NADPH limitation [218]. This limitation becomes obvious by the fact that four NADPH are required for the synthesis of one L-lysine. In consideration of these insights several approaches aimed at increasing L-lysine production, and for example L-valine production too, by redirecting the carbon fluxes through the oxidative PPP [205, 219].

One approach was to enhance the expression of the fructose 1,6-bisphosphatase gene *fbp* increasing the PPP flux by 10% [214, 215]. In a strain carrying the feedback resistant allele of the *lysC* gene, overexpression of *fbp* increased L-lysine production on sucrose, glucose and fructose up to 40% [215]. Overexpression of *fbp* also has another effect, as shown by Georgi *et al.* [216], who described an increased L-lysine production on sucrose by a minimally mutated production strain due to reduced intracellular concentrations of fructose 1,6-bisphosphate, an inhibitor of glucose 6-phosphate dehydrogenase and 6-phosphogluconate dehydrogenase, which are two of the three NADPH generating enzymes involved. Also single mutations derived from classical derived production strains in the genes *fbp* and *lysC* leading to base substitutions were described as beneficial for L-lysine production most likely due to improved enzyme kinetics [216, 220, 221]. Another approach for increased L-lysine production on glucose is the deletion of the phosphoglucose isomerase gene *pgi* to force the entire carbon flux through the PPP [222].

By replacing endogenous NAD-dependent glyceraldehyde 3-phosphate dehydrogenase (GAPDH) of *C. glutamicum* with nonphosphorylating NADP-dependent glyceraldehyde 3-phosphate dehydrogenase (GapN) of *Streptococcus mutans* Takeno *et al.* [223] reconstructed a functional NADPH generating glycolytic pathway that, however, was only functional in an uncharacterized suppressor mutant with enhanced sugar uptake ability and increased L-lysine production on glucose, fructose and sucrose.

The membrane-integral nicotinamide nucleotide transhydrogenase PntAB of *E. coli*, which can use the electrochemical proton gradient across the cytoplasmic membrane to drive the reduction of $NADP^+$*via* the oxidation of NADH, was transferred into a *C. glutamicum* L-lysine producer. This approach to increase the NADPH pool, resulted in an increased biomass formation as well as increased L-lysine production of up to ~10% on glucose, 70% on fructose and 300% on sucrose, whereas the specific growth rates were reduced [224].

Strains and Industrial Processes of L-lysine Production

In 1989 Hirao *et al.* [225] reported a L-lysine producing *C. glutamicum* strain generated *via* random mutagenesis and selection with so far unreported high titer and productivity of 100 g/L and 2.1 g/L h^{-1}, respectively, in fed-batch fermentations. Combining several of the targets described in the previous sub sections genetic engineering resulted in a strain with a productivity of 4.0 g/L h^{-1} and a titer of 120 g/L in a fed-batch process [211].

For the industrial production of L-lysine, sucrose obtained from sugarcane or beet molasses or starch hydrolysates of different qualities is mainly used as a carbon source, but also defined carbon sources such as glucose are utilized [186, 226]. For fermentation processes using auxotrophic production strains the medium can be supplemented with peptones, corn steep liquors (CSL) or soybean meal hydrolysates [186]. Ammonia, as nitrogen source, can be added pure or as salts. Further media components are vitamins in particular biotin as well as salts and trace elements, depending on the requirement of the used strain [186].

Cultivation strategies for industrial processes usually include multiple-step seed train cultivations with vessel volumes of 1-5 m^3 up to 10-50 m^3 prior to the main cultivation. Two or more steps of seed train cultivations are necessary because *C.*

glutamicum typically exhibits a long lag-phase at biomass concentrations below 0.1 g/L [186, 226]. Fermenter conditions like temperature, pH and oxygen supply are controlled to ensure optimal conditions for exponential growth and production in the fermenters. Growth can be observed offline by measuring biomass and carbon source concentrations and monitored online *via* heat evolution and oxygen uptake rate (OUR), which reduces the analytical load in industrial practice dramatically [186].

The demand for cost reduction has driven industrial fermenter sizes from 50 up to 750 m^3 in the last 30 years. High oxygen transfer rates (OTR) in these stirred-tank reactors require high specific power intake. High cell-specific production rates, resistance to high osmotic pressures and resistance to high shear stress are the main characteristics of successful strains for industrial processes.

The uptake rate of the carbon and nitrogen sources is controlled and adapted to the capability of the reactor considering heat evolution and OTR by closed- or open-loop control [186, 227 - 229]. Although *C. glutamicum* can tolerate oxygen limitation, in the case of L-lysine production, this would lead to the undesired formation of by-products such as acetate, lactate and L-alanine [186, 230, 231].

Process strategies for commercial scale are fed-batch or repeated fed-batch and continuous cultures for increased productivity [186, 226]. Fed-batch processes are advantageous especially for L-lysine production due to the amino acid auxotrophies of L-lysine producer strains, which allow for feeding profiles clearly separating growth and production phases [186]. A further advantage of the fed-batch process for L-lysine production is the fact that the solubility does not limit the concentration that can be achieved, which is also beneficial for downstream processing, where the crystalline L-lysine-HCl is the dominant product form compared to $(L\text{-lysine})_2SO_4$ [186].

Besides crystalline L-lysine preparations, liquid L-lysine (50% purity) and granulated L-ysine sulfate are in the market. Evonik Degussa GmbH developed a downstream process with a reduced number of unit operations and reduced waste emission with a yield of 95 % compared to 90 % for conventional processes [186]. The product of the process is sold as Biolys® and it contains granulated L-lysine sulfate as well as the entire fermentation broth. This is possible due to the GRAS (Generally Recommended As Safe) status of *C. glutamicum*. In that way the organism itself can be used as valuable feed additive [186].

L-Threonine Production

Owing to its use in the feed industry L-threonine is one of the major amino acids produced by fermentation [232] and as for many other amino acids the demand for L-threonine has steadily increased over the past years [233]. L-Threonine is industrially mainly produced by mutant strains of *E. coli* obtained by classical mutagenesis and screening procedures [226, 234, 235], but also L-threonine producing *Serratia marcescens, B. flavum,* and *C. glutamicum* strains exist [236].

L-Threonine Biosynthesis

From L-aspartate, L-threonine is synthesized in five enzymatic steps whereby the first step is catalyzed by aspartate kinase (AK). In *E. coli* there are three AK isoenzymes encoded by *thrA*, *metL*, and *lysC*. ThrA activity is inhibited by L-threonine and its synthesis is inhibited by L-threonine and L-methionine, MetL synthesis is repressed by L-methionine, and LysC activity and its synthesis are inhibited by L-lysine. After the first two steps aspartate semialdehyde (ASA) is formed, which is the last common precursor of the L-aspartate derived amino acids. From that step the activities encoded by *hom*, *thrB*, and *thrC* are required for L-threonine biosynthesis. In *C. glutamicum,* the carbon flux at the ASA branch point is greater towards L-threonine than to L-lysine synthesis [237]. Homoserine dehydrogenase (HD) and homoserine kinase are both feedback-inhibited by L-threonine and L-methionine. HD mutants resistant to L-threonine have been reported [238 - 240].

Strain Construction Strategies for L-threonine Production

Previous efforts to construct L-threonine producing strains include feedback alleviation of AK I (Thr^{S345F}) and III ($LysC^{T342I}$) [241 - 243], removing transcriptional attenuation by replacing the promoter of the *thrABC* operon [241], diminishing the effect of competing pathways by deleting the genes *lysA*, *metA*, and *tdh* [241], decreasing the activity of threonine dehydratase, catalyzing the first step towards L-isoleucine [244], reducing the activity of IlvA [243, 245] and increasing the copy number of *hom-thrB* [246]. Plasmid-based expression of *thrABC*, where *thrA* had the S345F amino acid substitution to increase the flux from aspartate to threonine [241], replacing the native promoter of *ppc* on the chromosome with the trc promoter to increase the PPC flux [241], increasing the flux through the glyoxylate shunt by deleting *iclR* encoding a regulator affecting the expression of *aceBA* [241, 247] and heterologous expression of the *E. coli* threonine operon in *C. glutamicum* resulted in a 30% increase in L-threonine production [248].

Transport Engineering for L-threonine Production

The overexpression of the transporters *rhtA* and *rhtB*, involved in export of L-threonine and homoserine increased L-threonine production [249]. Also the overexpression of *rhtC* encoding a threonine exporter in an L-threonine producing *E. coli* strain was beneficial for L-threonine accumulation [250]. The overproduction of L-threonine by *C. glutamicum* was associated with an intracellular increase in L-threonine due to a limitation of L-threonine export by diffusion [251] and *via* ThrE [252]. Consequently, the overexpression of *thrE* resulted in an increase in extracellular L-threonine but L-lysine and glycine also accumulated [252]. The use of the RhtA, RhtB, and RhtC transporters has also been demonstrated in *C. glutamicum*, where RhtA and RhtC were most effective, however, RhtA also entailed excretion of homoserine [253]. L-Threonine export remains to be a challenge in L-threonine production by *C. glutamicum* [254]. To avoid re-uptake of the product L-threonine, the deletion of the L-threonine uptake system *tdcC* [255] has been employed [241].

Alternative Substrates for L-threonine Production

E. coli K-12 that does not catabolize sucrose as sole carbon source was engineered to utilize sucrose by the heterologous expression of *sacC* encoding 6-phosphate hydrolase from *Mannheimia succiniciproducens*, where sucrose was hydrolyzed into fructose and glucose in the extracellular space. An L-threonine producing *E. coli* strain was in this way able to produce 51 g/L L-threonine in fed-batch with a yield of 0.284 g threonine/g sucrose [256].

Enzyme Proximity for L-threonine Production

The spatial organization of enzymes using a scaffold system can increase the efficiency of the biosynthetic pathway, reduce production time, and prevent accumulation of toxic intermediates [257 - 259]. DNA scaffold systems have been demonstrated to be useful for the production of resveratrol, 1,2-propandiol, mevalonate, and L-threonine [260, 261]. The accumulation of the toxic intermediates homoserine and phosphohomoserine that during fermentation might inhibit cell growth [262, 263] were likely diminished through the scaffold strategy [260]. For L-threonine production in *E. coli* the assembly of the artificial enzymatic cascade on the DNA scaffold resulted in a more than 50% reduction of the production time [260].

Fermentation Processes for L-threonine Production

Different feeding strategies and methods have been reported for the increased production of L-threonine in fed-batch fermentations, for instance DO-control pulse [264, 265]. Optimization of a fed-batch process for the production of L-threonine by *E. coli* included assessing the effect of different carbon sources, their concentrations, feeding methods, and the C/N ratio. The optimized process resulted in a yield of 118 g/L and a productivity of 3.1 g/L h^{-1}, which was reached within 38 h with a conversion ratio of 46% from glucose [264]. In fermentations of *E. coli* acetic acid accumulation has been reported to be an issue as it retards growth and reduces the glucose uptake rate [241]. This issue was addressed by amplification of *acs* encoding acetyl-CoA synthetase in a L-threonine producing strain [241].

MICROBIAL PRODUCTION OF BRANCHED-CHAIN AMINO ACIDS

The industrial production of the branched chain amino acids (BCAA) L-valine, L-leucine, and L-isoleucine is mainly performed with strains of *C. glutamicum,B. lactofermentum* and *E. coli* [175, 266 - 271]. Due to their similarities regarding precursors and enzymes the three amino acids are summarized together. The enzymatic steps the pathways share are: the acetohydroxyacid synthase (AHAS) reaction, the acetohydroxyacid isomeroreductase (AHAIR) reaction, the dihydroxyacid dehydratase (DHAD) reaction and the transaminase (TA) reaction

(Fig. **4** and Table **5**).

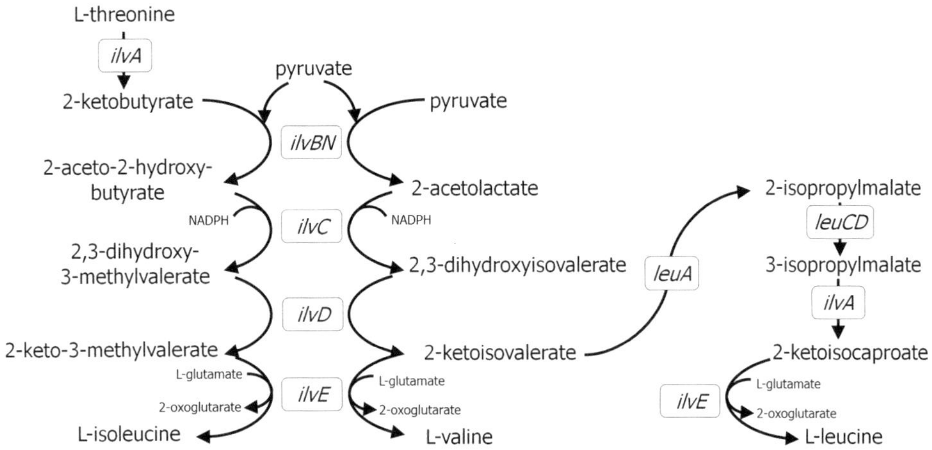

Fig. (4). Biosynthesis pathway of BCAAs (adapted from Patek, 2007). For a description of gene abbreviations and their encoded enzymes see Table **5**. L-threonine is derived from L-aspartate (see Fig. **3**) and pyruvate is an intermediate of central carbon metabolism (see Fig. **1**).

Precursor Supply for Branched-chain Amino Acids

The formation of L-isoleucine starts with the condensation of 2-oxobutyrate and pyruvate by AHAS. One L-threonine molecule is needed, as substrate, for its conversion to 2-oxobutyrate by threonine dehydratase (TD). For the optimization of the L-threonine supply see the corresponding sub-chapter. The increased supply of pyruvate, which is needed for the synthesis of all BCAAs is discussed in the sub-chapter L-lysine. One approach to increase pyruvate availability is to delete genes of the pyruvate dehydrogenase complex (*aceE* in *C. glutamicum* or *aceF* in *E. coli*). The pyruvate dehydrogenase complex consumes pyruvate and, thus, competes with use of pyruvate in branched-chain amino acid biosynthesis. Accordingly, deletion of genes of the pyruvate dehydrogenase complex proved beneficial for L-valine production [268, 272]. Also the deletion of the *pqo* gene (encoding for pyruvate:quinone oxidoreductase) in *C. glutamicum* was performed to increase L-valine production by preventing the decarboxylation of pyruvate to acetate [268].

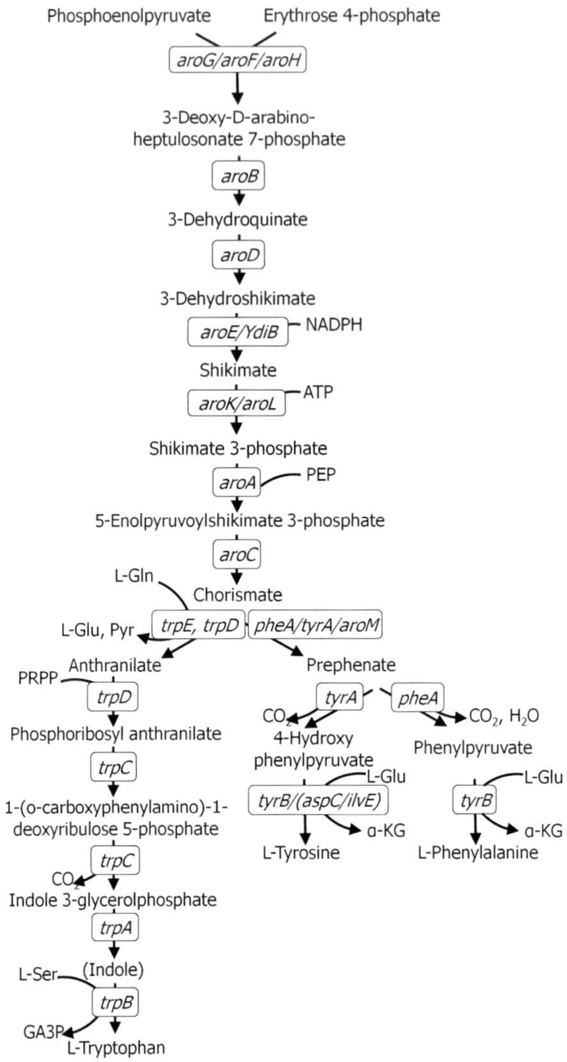

Fig. (5). Aromatic amino acid biosynthetic pathways of *E. coli*. PEP, phosphoenolpyruvate; PRPP, 5-phosphoribosyl-α-pyrophosphate; L-Ser, L-serine; GA3P, glyceraldehyde 3-phosphate; L-Gln, L-glutamine L-Glu, L-glutamate; Pyr, pyruvate; α-KG, 2-oxoglutarate; *aroG/aroF/aroH:* 3-deoxy-D-arob-no-heptulosonate 7-phosphate synthase; *aroB:* 3-dehydroquinate synthase; *aroD* 3-dehydroquinate dehydrathase; *aroE*: shikimate dehydrogenase; *aroK/aroL*: shikimate kinase I / II; *aroA*: 5-enolpyruvoylshikimate 3-phosphate synthase; *aroC*: chorismate synthase; *trpE*: anthranilate synthase; *trpD*: anthranilate PRPP transferase; *trpC*: phosphoribosyl anthranilate isomerase / indoleglycerol phosphate synthase; *trpA/trpB*: tryptophan synthase; *tyrA/pheA*: chorismate mutase / prephenate dehydratase; *tyrB*: tyrosine aminotransferase; *aspC*: aspartate aminotransferase *ilvE*: branched chain amino acid transferase. Phosphoenolpyruvate and erythrose-4-phosphate are intermediates of central carbon metabolism (see Fig. **1**).

L-glutamate as amino donor for the transamination reaction catalysed by TA [273], may be considered a precursor. While in *E. coli* an alternative TA encoded by *avtA* uses L-alanine as an amino group donor in L-valine synthesis [274], metabolic engineering of BCAA production involving transaminases has not been followed often.

Avoidance of By-products

Pantothenate can be considered a by-product of the BCAAs L-valine and L-leucine as it is derived from 2-oxoisovalerate. Deletion of the genes *panBC* encoding the enzyme converting 2-oxoisovalerate to pantothenate had a beneficial effect on L-valine production by *C. glutamicum* [266, 267]. Since all BCAAs are derived from the same precursors, the formation of the other two BCAAs are undesired. Therefore, the avoidance of *e.g.* the synthesis of L-isoleucine by deletion of *ilvA* is advantageous for the production of L-valine and L-leucine, but necessitates supplementation of the medium with isoleucine [266, 267]. Since the BCAAs in *C. glutamicum* are imported by BrnQ high concentrations of one BCAA can competitively inhibit uptake of the BCAA required as supplement [289]. 2-Oxoisovalerate is the branch point of L-valine and L-leucine biosynthesis and IPMS, encoded by *leuA* and TA, encoded by *ilvE*, compete for the intermediate. The downregulation of the expression of *leuA* as well as *ilvA* in *C. glutamicum* by exchanging the native promoter resulted, in L-isoleucine and L-leucine bradytrophy and a moderate increase in L-valine production [290]. In *E. coli* the production of L-leucine was increased by obtaining BCAA auxotrophic mutants by N-methyl-N'-nitro-N-nitrosoguanidine treatment. The auxotrophy was caused by a disruption of the *ilvE* gene [291]. Concomitant overexpression of *tyrE* restored L-leucine production and yielded in a titer of 2.7 g/L.

Cofactor Supply

In the synthesis of the BCAAs two NADPH molecules are needed. The first one is oxidized during the synthesis of 2,3-dihydroxyisovalerate by AHAIR. The second NADPH is required for to the regeneration of L-glutamate, which is consumed by transamination of TA. One achievement in this field is the change of the cofactors in *C. glutamicum* for L-valine production. The NADPH-dependent AHAIR was exchanged by protein engineering of the NADP binding region to an NADH-dependent AHAIR* version. Additionally an NADH-dependent leucine dehydrogenase (LeuDH) from *Lysinibacillus sphaericus* was introduced in place of the endogenous TA. This enzyme catalyzes the reversible deamination of the BCAAs using NAD^+ as a cofactor. The resulting strain accumulated 1470 mM L-valine under oxygen-deprivation conditions after 24 h [292]. Other approaches for an increased L-valine production are the deletion of *pgi* in *C. glutamicum* [268] or the *in silico* deletion of *pfkA* in *E. coli* [293]. For a detailed description of the cofactor engineering for amino acid production, see sub-chapter L-lysine.

De-bottlenecking BCAA Biosynthesis

The biosynthetic pathways of the BCAAs are intensively regulated by their end products L-valine, L-leucine and L-isoleucine, both at transcriptional and post translational level Table **5**. As many key enzymes are feedback regulated by their end products (Table **5**), feedback resistant variants have been overexpressed in many BCAA producing strains in *E. coli* as well as in *C. glutamicum*. In *E. coli* the overexpression of $AHAS^{fbr}$ increased the L-valine yield by around 20 % [294]. Introduction of TD^{fbr} into L-isoleucine producing strains of *C. glutamicum* and *E. coli* was performed to increase the carbon flux to L-isoleucine [269, 295 - 296]. $IPMS^{fbr}$, which is resistant to L-leucine, was introduced in *E. coli* to obtain an L-leucine producing strain [270].

Table 5. Genes and enzymes of the L-arginine biosynthesic pathway in microorganisms according to [80, 129 - 133].

Enzyme	Gene	L-isoleucine	L-valine	L-leucine	Regulation	Reference
TD, threonine dehydratase/ threonine deaminase	*ilvA*	+	Δ	Δ	fb (thr); a (val)	[245, 275 - 278]
AHAS, acetohydroxyacid synthase	*ilvBN** *ilvGMEc* *ilvIH Ec*	+	+	+	*ilvN*: tr (val, leu); fb (val) *ilvM*: tr (val, ile, leu)Ec *ilvH*: tr (leu)Ec; fb (val, ile, leu)Ec	[279 - 288]

(Table 5) contd.....

Enzyme	Gene	L-isoleucine	L-valine	L-leucine	Regulation	Reference
AHAIR, acetohydroxyacid isomeroreductase	*ilvC*	+	+	+	fb (val, leu)Cg	[285]
DHAD, dihydroxyacid dehydratase	*ilvD*	+	+	+	fb (val, leu)Cg	[285]
TA, transaminase	*ilvE* *(avtA)* *(tyrBEc)*	+	+	+	tr (val, leu, ile)Ec	[273 - 274, 285]
IPMS, isopropylmalate synthase	*leuA*	–	Δ	+	fb (leu)Cg tr (leu)	[287, 288]
IPMI, isopropylmalate dehydratase	*leuCD*	–	–	+	tr (leu)	[287, 288]
IPMD, isopropylmalate dehydrogenase	*leuB*	–	–	+	tr (leu)	[287, 288]

*For amino acid production the activity of the enzyme should be increased (+), decreased or abrogated (Δ) or has not been reported to as target (–). Gene names written in brackets are present in *E. coli*, but not in *C. glutamicum*. Transcriptional regulation (tr) and allosteric enzyme regulation by feedback inhibition (fb) or activation (a) as well as the effector molecule (in brackets) are indicated; abbreviations Ec for *E. coli* and Cg for *C. glutamicum* are used if regulation has only been reported for one of the two organisms.

For *C. glutamicum* it could be shown that overexpression of *ilvBNCE* was beneficial for the production of L-valine [266, 268, 272].

Export of Branched-chain Amino Acids

In *C. glutamicum* a BCAA exporter encoded by *brnFE* was identified [297]. Due to sequence homology the exporter *ygaZH* was identified for L-valine export in *E. coli* [293]. The homologous overexpression of *brnFE* in *C. glutamicum* resulted in an increase in BCAA export [297]. *In silico* metabolic engineering of *E. coli* predicted that overexpression of *ygaZH* would result in an increased L-valine production [293]. Recently, an L-leucine exporter (encoded by *leuE*; former: *yeaS*) has been identified in *E. coli* [298].

Avoidance of Re-import of Branched-chain Amino Acids

E. coli possesses an ATP dependent uptake system for BCAAs that is a member of the ABC (ATP binding cassette) systems named LivFGHJM, and is known to import all three BCAAs [118]. In *C. glutamicum* the unrelated uptake system for BCAAs BrnQ functions as a Na^+ symport system and is also present in *E. coli* [299 - 301]. The effect of deleting genes for BCAA uptake systems for BCAA production by *C. glutamicum* or *E. coli* has not yet been reported. As most import systems are relevant for all BCAAs it is difficult to delete the systems completely, due to the fact that many production strains are auxotrophic for other BCAAs, which are supplemented to the medium (see sub-section Avoidance of by-products).

Branched-chain Amino Acid Production Titers

Depending on the growth conditions, L-valine titers from ~7-11 g/L can be obtained in batch cultivations of *C. glutamicum* production strains (shaking flask scale) [289]. In fed-batch experiments a total volumetric productivity of about 0.7 g L-valine per L per hour was obtained [302]. Using a fed-batch process, L-valine titers of about 25 g/L [272] to 48 g/L [268] and up to 105 g/L were achieved [159]. L-isoleucine titers of *C. glutamicum* and *E. coli* strains reached about 10 to 20 g/L [269, 296, 303, 304]. A fed-batch process was established for *C. glutamicum* with a maximum L-isoleucine concentration of about 21 g/L [305] and about 40 g/L have been reported elsewhere [159]. The titers described for L-leucine production in *E. coli* strains are around 5 to 10 g/L [270, 306].

MICROBIAL PRODUCTION OF AROMATIC AMINO ACIDS

L-phenylalanine, L-tryptophan, and L-tyrosine constitute the aromatic amino acids (AAA) and are of interest due to their wide uses in the pharmaceutical, chemical, and food & feed industries [307]. L-phenylalanine and L-tryptophan are produced by chemo-enzymatic processes and by fermentation, whereas L-tyrosine is mainly obtained by extraction from protein hydrolysates [1, 56]. The aromatic amino acid biosynthetic pathways are depicted in Fig. (**5**).

Biosynthesis of Aromatic Amino Acids

The common AAA pathway is also known as the shikimate pathway. It starts with the condensation of phosphoenolpyruvate (PEP) and erythrose 4-phosphate (E4P) by 3-deoxy-D-arabino-heptulosonate-7-phosphate (DAHP) synthase to form DAHP. This step controls the carbon flux into the pathway [308]. In *E. coli,* three isoenzymes of DAHP synthase, encoded by *aroF*, *aroG*, and *aroH* exist, that are feedback inhibited by L-tyrosine, L-phenylalanine, and L-tryptophan, respectively [309]. In *C. glutamicum,* there are two DAHP isoenzymes, encoded by *aroG* and *aroF*, both isoenzymes are inhibited by L-tyrosine and the *aroG* encoded isoenzyme is furthermore inhibited by chorismate, L-phenylalanine, and prephenate [310, 311]. AroF of *C. glutamicum* forms a complex with chorismate mutase, converting chorismate to prephenate [312]. Shikimate is formed *via* three additional steps, and in *E. coli* the isoenzymes AroE and YdiB catalyze the shikimate forming reaction [313, 314]. Three more steps lead to chorismate. Phosphorylation of shikimate to shikimate-3-phosphate is catalyzed by the isoenzymes AroK and AroL, where AroL has high affinity to its substrates [315].

Chorismate is a branch point for the three AAAs as well as for ubiquinone, menaquinone, siderophores, folate, and enterochelin [316]. The transcriptional regulators TyrR and TrpR regulate the expression of genes of AAA biosynthesis and transport in *E. coli* [309] whereas *C. glutamicum* does not appear to possess such regulators [317].

L-Tryptophan Biosynthesis

In the first step of L-tryptophan biosynthesis chorismate is converted to anthranilate. Anthranilate synthase (ANS) and anthranilate phosphoribosyl-transferase (PRT) are inhibited by L-tryptophan in corynebacteria and *E. coli* [318 - 321]. Tryptophan synthase (TS) is inhibited by L-tryptophan [322]. L-Serine (see sub-chapter on L-serine and glycine) and phosphoribosylpyrophosphate (PRPP) are required for L-tryptophan biosynthesis. Transcriptional repression of the L-tryptophan biosynthetic genes is exerted by L-tryptophan [318, 320, 323].

L-Phenylalanine and L-Tyrosine Biosynthesis

The L-henylalanine and L-tyrosine biosynthetic pathways branch at prephenate. Conversion of chorismate to prephenate is catalyzed by chorismate mutase, a moiety of bifunctional TyrA and PheA, both of which also catalyze the second step of the respective pathways [307]. The first committed step of L-phenylalanine is catalyzed by the bifunctional chorismate mutase/prephenate dehydratase, encoded by *pheA* [324]. In *E. coli* CM-PDT is feedback inhibited by L-phenylalanine and L-tryptophan, and in *C. glutamicum* CM is monofunctional and encoded by *csm* and PD is encoded by *pheA* and is feedback inhibited by L-phenylalanine [310, 325, 326]. The types of regulation exerted on L-phenylalanine biosynthesis are feedback inhibition of rate-limiting enzymes, transcriptional repression, and attenuation at the transcriptional/translational level [309].

Strain Construction Strategies for Aromatic Amino Acid Production

Classical strain development usually resulted in auxotrophic or regulatory mutants [56]. For L-phenylalanine production L-tyrosine auxotrophic mutants were of interest as their synthesis diverges at prephenate, and as L-tyrosine production is a waste of carbon and energy, and, thus, is an unwanted by-product [327 - 329].

The typical approach for rational strain engineering for the production of AAAs is the alleviation of regulatory circuits by constructing feedback resistant mutants, using auxotrophic strains to reduce or eliminate by-product formation, overexpression and increase of activity of rate-limiting enzymes, and increasing the availability of precursors for the shikimate pathway [307, 312, 330 - 333].

Engineering the Shikimate Pathway and Production of Its Precursors

For the generation of aromatic metabolites in both *E. coli* and *C. glutamicum* PEP and E4P are needed for the committing and first step of the shikimate pathway. Therefore, efforts were made to increase the precursor supply (see reviews [56, 307, 324]).

The strategies employed to increase the PEP supply included overexpression of *ppsA* [334] and recycling PEP from oxaloacetate [335]. The main glucose transport system, phosphoenolpyruvate: carbohydrate phosphotransferase system (PTS) was inactivated since it competes with the shikimate pathway for PEP [336]. For the same reason, the genes encoding PEP carboxylase or pyruvate kinases were deleted in L-phenylalanine producing *E. coli* strains [331, 334, 337 - 339]. Also mutation of the gene encoding PEP carboxylase was combined with overproduction of gluconeogenic PEP carboxykinase [340] or inactivation of PEP carboxylase [341] and pyruvate kinases [339] were combined. PEP synthase was overproduced to recycle pyruvate formed by the PTS or pyruvate kinases back to PEP [342] and inactivation of the PTS was combined with increasing glucokinase activity for phosphorylation of free glucose [343]. To avoid anaplerosis by PEP carboxylase under glycolytic conditions, the glyoxylate cycle genes were induced [344]. The regulator CsrA was inactivated as it positively regulates *pykF* and negatively regulates *pckA* and *pps* [345]. Alternatively overexpression of *crsB*, where CsrB inhibits CsrA activity, was shown to increase L-phenylalanine production [346].

For an increase of E4P the overexpression of *tktA* [340] and *tal* [347] which encode pentose phosphate pathway enzymes transketolase and transaldolase, respectively, was shown to be beneficial. The overexpression of *tktA* alone was shown to be more efficient than the overexpression of *tal*, and overexpression of both did not result in further increase in production of aromatics as overexpression of *tktA* alone seemed to satisfy the E4P supply [347]. Further strategies include the overexpression of the genes encoding enzymes of the non-oxidative PPP [343, 347, 348] and the inactivation of phosphoglucose isomerase [349].

A balanced supply of the precursors is required for significant yield improvements, demonstrated for DAHP production by *E. coli* where the maximum theoretical yield was almost achieved by simultaneous overexpression of transketolase, PEP synthase and DAHP synthase [338]. Another example of this is the plasmid-based overexpression of *ppsA* and/or *tktA* in a L-tryptophan producing *E. coli* strain. While overexpression of *ppsA* and *tktA* alone hardly increased production (36 g/L and 38 g/L L-tryptophan as compared to 36 g/L by the control), combined overexpression of both genes resulted in 40 g/L L-tryptophan [350]. The overexpression of *ppsA* and *tktA* has previously been shown to impair growth, result in increased glucose consumption, and excretion of pyruvate and acetate in nutrient-rich and glucose containing media [338]. This effect could however almost completely be circumvented by lowering the gene dosage and by changing the fermentation process [338, 350]. These strategies were additionally shown to reduce plasmid loss that along with retarded growth has been connected with *tktA* overexpression [350]. Moreover beneficial engineering of PEP carboxylase and transketolase for L-phenylalanine and L-tyrosine producing *C. glutamicum* strains was demonstrated [351, 352]. Other common targets for engineering have been shikimate kinase (AroK or AroL) and quinate/shikimate dehydrogenase (YdiB), as they are rate-limiting enzymes [353 - 355]. Recently, the construction of a synthetic operon containing the genes *tktA* and *zwf* of the PPP and *aroB*, $aroG^{fbr}$, *aroE*, and *aroD* of the shikimate pathway for the increased production of shikimate in *E. coli* was reported [356].

Export of Aromatic Amino Acids

Removing the regulatory control of amino acid pathways completely can be difficult, especially when regulation is exerted at multiple levels. One way to overcome this is by maintaining a low intracellular level of the amino acid to be produced; this can be achieved by transport engineering. Although AAAs can efficiently penetrate the bacterial cell membrane due to their hydrophobic properties, proteins involved in AAA transport have been identified [357]. In *E. coli* the AAA transport systems TyrP for L-tyrosine, L-PheP for phenylalanine, Mtr and TnaB for L-tryptophan, and AroP for all AAAs exist [309]. Overproduction of the inner membrane protein YddG in L-phenylalanine, L-tyrosine, and L-tryptophan overproducing *E. coli* strains resulted in an increase in production of three, three, and 1.5 times, respectively [358]. A deletion of *yddG* correspondingly decreased the accumulation of the AAAs [357]. However, YddG does not seem to have a high affinity for AAAs and the export thereof might not be its native function [357].

L-Phenylalanine Production

L-phenylalanine is one of the commercially most important amino acids, mainly due to its use as a precursor for the production of the sweetener aspartame [359]. Its other uses are as a feed and food additive, in infusion fluids, as a flavor enhancer, and for the use in the synthesis of active compounds [307]. The bacteria *Bacillus subtilis* [360], *C. glutamicum* [361 - 363], and *E. coli* [364, 365] have been employed as L-phenylalanine producers [56, 307, 312, 324]. L-phenylalanine is also produced chemically [307, 366] and enzymatically [324]. The chemical synthesis is, however, being replaced due to issues concerning the synthesis, the demand for the L-form, and the consumers' preference [367].

Strain Construction Strategies for L-Phenylalanine Production

Although some of the most successful microbial producers of biotechnological compounds were obtained by classical strain improvements, this was not the case after decades of classical strain improvement for L-phenylalanine production; the strains obtained did not have sufficient titers for industrial production [324].

Strategies employed for the rational engineering of L-phenylalanine producers include: construction of feedback resistant PheA mutants [326, 368] and expression of genes for phenylalanine dehydrogenases that catalyze NAD(P)H-dependent reductive amination of phenylpyruvate to L-phenylalanine [369]. For this purpose *phedh* from *Acinetobacter lwoffii* and *Bacillus lentus* have been heterologously expressed in *E. coli* [370, 371]. Using the bifunctional enzyme CM-PD, encoded by *pheA*, from *E. coli* that is resistant to inhibition of L-phenylalanine in *C. glutamicum* resulted in a strain producing 23 g/L L-phenylalanine in fed-batch without the accumulation of L-tyrosine [359]. Zhang *et al.* [372] demonstrated an example of the tight regulation of L-phenylalanine synthesis in *C. glutamicum* as overexpression of *aroF*, *aroG*, or *pheA* only resulted in a small increase in L-phenylalanine production. The wild-type *aroH* of *E. coli* was cloned into *C. glutamicum*, as *C. glutamicum* does not contain this gene, to improve the flux through the shikimate pathway for L-phenylalanine production [372]. The co-overexpression of *aroF* and feedback resistant *pheA* from *E. coli* resulted in a titer of 4.5 g/L [372]. In L-phenylalanine producing *E. coli*, the *pheL* attenuator region, and the gene for TyrR that represses *pheA* were deleted. An operon encoding AroF and PheA feedback resistant to L-tyrosine and L-phenylalanine, respectively, was introduced into the strain on a medium copy number plasmid under the control of the IPTG inducible promoter P_{tac}. Although the strain produced L-phenylalanine, the by-products shikimic and 3-dehydroshikimic acids accumulated. Therefore *aroL*, encoding the major shikimate kinase, was cloned into the plasmid carrying the *aroF*^{fbr}-*pheA*^{fbr} operon. This reduced the by-product accumulation but L-phenylalanine production was not increased as DAHP and DAH were accumulating, hence *aroB* was added to the operon. This abolished by-product formation but still no increase in L-phenylalanine production was observed [324]. The issue appeared to be instability of $AroF^{fbr}$ [329]. Enzyme activity assays showed that although $AroF^{fbr}$ had a lower sensitivity towards L-tyrosine than the wild-type AroF and the enzyme activities were similar, a mutant having $AroF^{N8K}$ lost 50% activity in 20 min and had a decreased thermostability [373]. In the end the use of the wild-type AroF in the operon *aroF-pheA^{fbr}-aroL* and restricted L-tyrosine feeding in the fed-batch fermentation process turned out to be optimal for L-phenylalanine production. The resulting titer was 34 g/L as compared to the *aroF^{fbr}* containing operon that resulted in a titer of 28 g/L L-phenylalanine [329]. Introducing *aroB* into the operon entailed a titer of 38 g/L [374].

L-phenylalanine production with an *E. coli* strain was frequently abolished by bacteriophage BP-1 infestation. Mutagenesis with *N*-methyl-*N*'-nitro-*N*-nitrosoguanidine and selection for resistance to the phage resulted in the isolation of a strain with high tolerance and combined with an improved feeding strategy the strain was able to accumulate ~58 g/L [375]. An *E. coli* strain was engineered to produce L-phenylalanine with glycerol as carbon source. To increase L-phenylalanine accumulation one of the measures taken was the increased expression of *glpK*, encoding a glycerol facilitator [371]. Also different glucose uptake systems have been exploited [376].

L-Tryptophan Production

L-Tryptophan is mainly produced for the feed industry and besides its use as a food additive it is also used in infusion liquids, for pellagra treatment, for sleep induction and nutritional therapy [307, 312, 377, 378]. The high production cost of L-tryptophan however, limits its market [307]. The two organisms mainly used for L-tryptophan production are *E. coli* and *C. glutamicum* [56, 324].

Strain Construction Strategies for L-Tryptophan Production

Strategies to increase L-tryptophan production include promoter swapping [379], deleting *trpR* encoding the tryptophan transcriptional repressor [377, 380], construction of feedback resistant mutants of TrpE and AroG [381 - 382], deleting *tnaA* encoding tryptophanase, which catalyzes the conversion of L-tryptophan to indole [358] and amplification of the tryptophan operon and the introduction of multiple copies of the tryptophan operon in the chromosome [377, 383 - 385]. Overexpression of *serA* [386], *prs*, and *ywlF* as L-serine and PRPP are needed for L-tryptophan production [387]. For L-tryptophan production, an *E. coli* strain was engineered with inactivated *trpR*, *tnaA*, and *ptsG*, along with plasmid-based overexpression of $argG^{Fbr}$, $trpE^{Fbr}$, and *tktA*. Moreover replacing the *trp* promoter of the tryptophan operon inactivated the tryptophan attenuator [379].

By-product Elimination During L-Tryptophan Production

In *E. coli* cultures the primary inhibitory metabolite is acetate [388]. Acetate can be a by-product in L-tryptophan production, which can reduce the substrate conversion ratio. Therefore, Gu *et al.* [389] introduced the polyhydroxybutyrate (PHB) biosynthetic pathway from *Ralstoniaeutropha* into an L-tryptophan producing *E. coli* strain in order to divert acetyl-CoA into PHB biosynthesis instead of acetate biosynthesis, thereby reducing acetate secretion. While the strain was slightly growth impaired and consumed glucose slower compared to the control, L-tryptophan production was increased by 11.6%.

In *E. coli* acetate assimilation is accomplished by Pta, AckA and acetyl-CoA synthetase (Acs) [390]. The deletion of *pta* that has been shown to be effective in decreasing acetate accumulation [391]. Overexpression of *acs* has been shown to increase acetate assimilation and to decrease its accumulation [392], while the deletion of *pta* decreased *acs* expression and activity [393]. An *E. coli* strain with a Δ*pta-mtr* background and overproduction of the transport protein YddG resulted in a yield of about 49 g/L, although deletions of *pta* and/or *mtr* entailed a decreased growth rate [358, 393].

L-Tryptophan Transport Engineering

TnaB was identified to be the main transporter for L-tryptophan uptake [394] and mutations/deletion of *mtr* encoding the L-tryptophan permease Mtr entailed a loss of L-tryptophan specific uptake thereby increasing the tryptophan concentration [397]. An *E. coli* strain with the background Δ*trpR*Δ*tnaA*Δ*mtr* overexpressing *yddG*, *tktA*, and *ppsA* exhibited a decreased intracellular and increased extracellular concentration of L-tryptophan [358]. Deleting *mtr* and *tnaB* in *E. coli* increased L-tryptophan production significantly, an additional deletion of *aroP* however retarded growth, which was a consequence of a low intracellular concentration of the AAAs [394 - 396].

L-Tryptophan Fermentation Processes

In L-tryptophan production, feeding strategies have been shown to affect production positively or negatively. One strategy to increase tryptophan production is to decrease the reuptake of the amino acid by the use of low glucose concentrations in the stationary phase [358, 398]. DO-stat feeding [350] and a pH-feedback feeding method, using a pH range where the fermentation process was divided into three stages, have also been reported [399]. A steady state glucose concentration was maintained and a titer of about 44 g/L was obtained [399].

L-Tyrosine Production

L-Tyrosine is used as a dietary supplement and in the production of 3,4-dihydroxyphenyl-L-alanine (L-DOPA), an anti-Parkinson's drug, [307], and is a precursor for melanin-based UV absorbers, cation exchangers, drug carriers, and amorphous semiconductors [400]. Lastly, it is a precursor of *p*-hydroxycinnamic acid, a monomer from which liquid crystal polymers, used for electronic applications, can be produced [401]. *C. glutamicum*, *C. glutamicum* ssp. *lactofermentum*, *Arthrobacter globiformis* [325, 330, 351, 402 - 404], and *E. coli* [405] have been engineered for L-tyrosine production.

Strain Construction Strategies for L-Tyrosine Production

E. coli was engineered for overproduction of L-tyrosine by deleting the gene encoding the repressor *tyrR* and by overexpressing *aroG*, and *tyrA* encoding chorismate mutase/prephenate dehydrogenase, where both AroG and TyrA were feedback resistant [405]. The plasmid-based expression of $aroG^{fbr}$ and $tyrA^{fbr}$ was achieved with the constitutive $P_{Ltet\text{-}O1}$ promoter. Furthermore, overexpressing *ppsA* and *tktA* increased the availability of the precursors PEP and E4P. A fed-batch fermentation of this strain resulted in a titer of ~10 g/L L-tyrosine [405]. Another approach for increasing L-tyrosine production in *E. coli* was by the use of global transcription machinery engineering (gTME) [406]. This was performed in an *E. coli* Δ*tyrR* strain with a deletion of *pheA* to remove the competing pathway of L-phenylalanine, and chromosomal expression of *aroG*^{fbr}and *tyrA*^{fbr} using synthetic promoters and two copies to achieve production comparable to plasmid-based overexpression. This strain was used to construct gTME-derived libraries, and screened using a high throughput assay for L-tyrosine detection, based on the L-tyrosine-mediated conversion of L-tyrosine to the pigment melanin [407]. A different approach towards increased L-tyrosine production was the construction of two modules containing all genes of the shikimate and L-tyrosine pathway. The first module consisted of the first six genes and the second of the last five genes for production of L-tyrosine from shikimate. YdiB was used as dehydrogenase, and AroK as shikimate kinase. The genes of the modules were arranged in a way that the last gene of the pathway was placed closest to the promoter. YdiB and AroB exhibited low levels, wherefore the N- terminal of *aroB* was codon optimized [408]. The CM domain of PheA from *E. coli* and TyrC encoding cyclohexadienyl dehydrogenase from *Zymomonas mobilis* are both feedback resistant to L-tyrosine and their overexpression resulted in an increase in L-tyrosine production [409]. Lastly, the optimization of a fermentation process for L-tyrosine production by *E. coli* was described where factors such as feeding profile, media components, foaming, crystallization and strain stability were considered [410].

MICROBIAL PRODUCTION OF L-CYSTEINE AND L-METHIONINE

L-Cysteine

The sulfur-containing proteinogenic amino acid L-cysteine possesses high importance as it is the sole metabolic entrance for reduced sulfur in several organisms and has a commercial application with an annual market volume of 4000 tons in 2004 [411]. It has its main use in the pharmaceutical, cosmetic and food industries, for example as processing aid for baking or for the production of flavors [411].

In general, L-cysteine can be produced chemically *via* Asinger-thiazolin--synthesis as racemic mixture, by acidic hydrolysis of keratin or biotechnologically [411, 412]. Primary sources of keratin are human hair and poultry feathers, which is not compatible with consumer demands of, for example, kosher food [413]. Therefore, enzymatic as well as fermentative processes have been developed [411, 414].

L-Cysteine Biosynthesis

In most bacteria L-serine is the precursor of L-cysteine (Fig. **6**) and 3-phosphoglycerate is the glycolytic intermediate from which L-serine can be synthesized by a three-step pathway in bacteria like *E. coli* and *C. glutamicum* [411, 415].

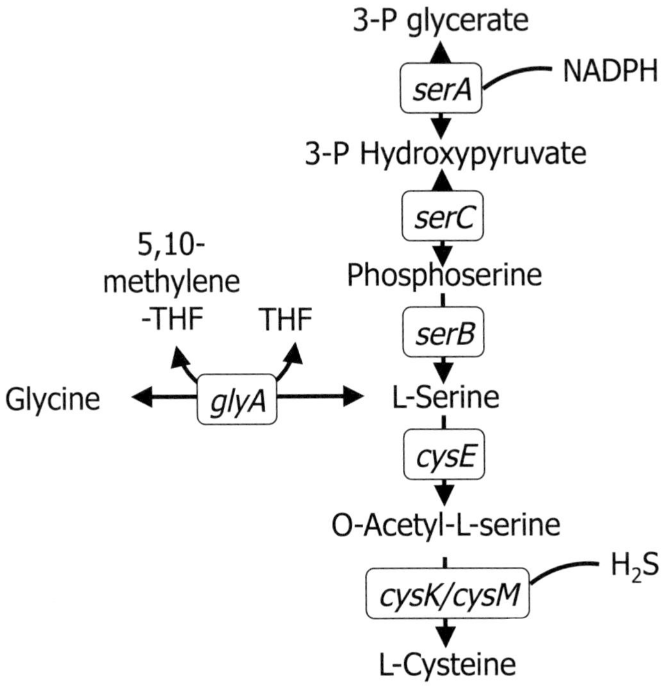

Fig. (6). L-serine, L-glycine and L-cysteine biosynthetic pathways of C. glutamicum. *serA:* 3-phosphoglycerate dehydrogenase gene; *serC*: phosphoserine aminotransferase gene; *serB*: phosphoserine phosphatase gene; *cysE*: serine O-acetyltransferase gene; *cysK/cysM*: cysteine synthase A/B gene; *glyA*: L-serine hydroxymethyltransferase phosphatase gene. 3-P-glycerate is an intermediate of central carbon metabolism (see Fig. **1**).

In the first step of L-cysteine biosynthesis, the *cysE* encoded L-serine O-acetyltransferase (CysE) catalyzes the acetylation of the β-hydroxyl group of L-serine with acetyl-CoA. In the second step the acetyl group of O-acetyl L-serine is eliminated with concomittant addition of H_2S by *cysK* encoded O-acetyl L-serine sulfhydrylase (CysK) to yield L-cysteine [411, 416]. For *E. coli* as well as for *C. glutamicum* it is known that CysE is feedback-inhibited by L-cysteine and, thus, is the rate limiting step [411, 417, 418]. However, the biosynthesis of L-cysteine in *E. coli* is not only regulated on the enzymatic level by feedback inhibition but also on the level of gene expression by the transcriptional activator CysB which is induced by N-acetyl L-serine and sulfur limitation [411, 417].

De-bottlenecking L-cysteine Production

Biotechnological production of L-cysteine is carried out enzymatically or *via* direct fermentation from glucose [1, 411]. An industrial enzymatic process uses DL-2-amino-D2-thiazolin-4-carbonic acid as substrate (DL-ATCA). For the bioconversion to L-cysteine three enzymes ATCA racemase, L-ATCA-hydrolase, and S-carbamoyl-L-cysteine hydrolase from *Pseudomonas* sp. are involved [1].

Catalyzing the initial step of the L-cysteine biosynthetic pathway, CysE is the rate limitig step. Thus, approaches aimed at desensitizing CysE from *E. coli* to feedback inhibition, both by site-directed or random mutagenesis as well as by employing naturally feedback-resistant CysE from higher plants [419, 420].

Nakamori *et al.* [421] obtained a feedback resistant *E. coli* CysE variant with the methionine residue 256 exchanged with alanine *via* site-directed mutagenesis, which led to production of 0.79 g/L L-cysteine. This could be improved to 0.99 g/L by a further amino acid exchange (M201R) [419] and finally 1.7 g/L were obtained with *E. coli* JM39-8 as host strain [422].

L-Cysteine Degradation

In L-cysteine production, its degradation remains to be an issue. In *E. coli* five cysteine desulfhydrases (CysK, CysM, MetC, MalY and tryptophanase (TnaA)) have been characterized [423, 424]. Inactivation of these genes doubled L-cysteine production. In *C. glutamicum* only one enzyme (encoded by *aecD*) with cysteine desulfhydrase side activity has been identified so far [425]. As well as in *E. coli*, disruption of this gene increased L-cysteine production by *C. glutamicum* [425].

L-Cysteine Export

A further target for efficient L-cysteine production is the L-cysteine export [411]. Two L-cysteine transport proteins, YdeD and YfiK, which are membrane-integral, have been reported for *E. coli* [414, 426]. Overexpression of the corresponding genes in a strain with feedback resistant CysE in the presence of thiosulfate, led to the production of up to 120 mg/L L-cysteine [414, 426]. Furthermore, the ATP-binding cassette transporter CydCD of *E. coli* was shown to transport L-cysteine from cytoplasm to periplasm [427].

Strains, Processes, and Conditions for L-cysteine Production

Maier *et al.* [428] described a fermentation process where up to 3.8 g/L L-cysteine was produced from glucose and thiosulfate with a strain carrying a plasmid with modified *cysE* gene (for CysE-T167A-G245S) [428]. Due to the broad substrate specificity of CysK and CysM this process is also capable for production of unnatural amino acids.

L-Methionine

The essential sulfur-containing amino acid L-methionine, which is part of the aspartate-family, is of high industrial importance with a worldwide annual production of 750.000 tons [30]. Especially in poultry feed methionine is the main limiting amino acid [429]. It is almost exclusively produced as a racemic mixture *via* chemical synthesis [430]. Although animals can metabolize both stereoisomers in principal, L-methionine is the better sulfur amino acid source [431]. Furthermore, chemical synthesis of the racemic mixture uses hazardous chemicals and produces substantial waste streams [432]. This illustrates why research aimed at the development of a sustainable biotechnological process based on renewable sources [433]. That extensive research has been carried out recently can also be attributed to the number of patents published in the last three years *e.g.* [220, 434 - 439].

L-Methionine Biosynthesis

E. coli as well as *C. glutamicum* are the most important L-methionine producing organisms [440]. They share the same pathway starting from L-aspartate *via* L-homoserine towards L-methionine consisting of seven reactions catalyzed by diverse organism specific enzymes (Fig. **3**) [440] and references therein. In *C. glutamicum* the fifth and sixth reaction can be by-passed *via* MetY-mediated (O-acetylhomoserine sulfhydrylase) incorporation of sulfur from hydrogen sulfide [440]. While *C. glutamicum* possesses only one homoserine dehydrogenase encoding gene (*hom*), *E. coli* harbours two (*thrA* and *metL*) [440]. The transcription of *hom* is repressed by L-methionine and the gene product is inhibited by L-threonine [441]. The *E. coli* genes *lysC*, *metL* and *thrA* are regulated on the transcriptional level and LysC as well as ThrA are also feedback-inhibited by L-lysine and L-threonine [440]. In an eighth step L-methionine is converted to *S*-adenosylmethionine by the *metK* encoded *S*-adenosylmethionine synthetase, that is the major methyl donor of methylation reactions in *C. glutamicum* as well as in *E. coli* [440].

De-bottlenecking L-methionine Biosynthesis, Export, and Co-factor Supply

Methionine-producing mutants were selected by their (multiple) resistance(s) to methionine analogues, which is an effective method to construct L-methionine producing strains with deregulated genes and feedback-resistant enzymes involved in L-methionine biosynthesis for more information see review by Kumar and Gomes (2005). The occurring resistances can be originated from mutations in operators causing de-repression, mutations in structural genes leading to feedback resistant enzymes, mutations in a structural gene of an amino acyl-tRNA synthase, provoking discrimination between the amino acid and the analogue, and mutations in a permease gene reducing the ability to take up the analogue [442].

Nowadays genetic engineering is a powerful tool for enhancing L-methionine production, eliminating the need for exhausting screening procedures [442]. First approaches in the field of rational strain design for L-methionine production in *E. coli* were performed by Nakamori *et al.* [443], who identified a single amino acid substitution (S54N) of the *met* repressor encoded by the gene *metJ*, leading to derepression of the L-methionine biosynthetic enzymes. A strain harboring different feedback resistant *metA* alleles and a *metJ* deletion produced 3.4 g/L L-methionine when *metL* was overexpressed [444]. Furthermore, Maier *et al.* [445] could show that overexpression of the putative L-methionine transporter YjeH can enhance L-methionine production in a *metJ* deletion strain.

Park *et al.* [446] constructed a L-methionine producing *C. glutamicum* strain derived from the L-lysine-producer MH20-22B (leu$^-$, *lysC*^{FBR}) by replacing *hom* with *hom*^{FBR} and disrupting *thrB*. This strain produced up to 2.9 g/L L-methionine in a fed-batch culture, which is in the range of the production of mutants obtained by classical strain construction [442]. By another approach, a L-methionine titer of 16 g/L could achieved by homologous overexpression of *metX* and *metY* in an aminoethylcysteine resistant *C. glutamicum* L-lysine producer [447].

Comparative analysis of theoretical L-methionine yields showed that the use of the reduced sulphur compound methyl mercaptan as a sulphur source has the highest potential to increase the L-methionine production by making L-methionine synthesis independent of sulphur reduction and one-carbon metabolism, due to saving 8 NADPH normally required for reduction of sulphate to sulphite [433]. Fermentative L-methionine production processes have only very recently been reported by the companies CJ in South Korea and Metabolic Explorer in France.

L-Methionine Processes and Conditions

L-methionine can be produced by resolution of the chemically synthesized racemic mixture using continuous flow immobilized enzyme bioreactors containing fungal aminoacylases [448]. Further approaches for the production of biologically active L-methionine employ bioconversion of (expensive) precursors or submerged fermentation [442].

MICROBIAL PRODUCTION OF L-HISTIDINE

L-histidine is used in infusions, special diets, and as raw material for pharmaceutical derivatives [449 - 450]. Industrially, L-histidine is mainly produced by fermentation with *C. glutamicum* and *B. flavum* [451, 452]. Most producers have been obtained by classical strain improvement [453 - 455], but recent publications describe the rational engineering of strains [449, 452].

Histidine Biosynthesis

The biosynthesis of L-histidine occurs through 10 enzymatic steps starting from ribose-5-phosphate (Fig. **1**). The first step of the pathway is catalyzed by ATP-phosphoribosyl transferase (PRT), encoded by *hisG*, and is feedback inhibited by L-histidine [455 - 456]. The transcription of the *C. glutamicum hisDCB-orf1-or-2-hisHA-impA-hisFI* is regulated by attenuation [449]. Readers interested in further details in L-histidine biosynthesis by *C. glutamicum* are referred to a thorough review by Kulis-Horn [457].

Strain Engineering and Fermentation Processes for L-histidine Production

Strategies to construct L-histidine producers include: construction and amplification of feedback resistant PRT overproduced in *C. glutamicum* [452, 458], replacing the *hisD* promoter to alleviate repression control [449], plasmid-based expression of *hisE* and *hisG* [449], amplification of *prsA* encoding mutant PRPP synthetase feedback resistant to purine nucleotides in *E. coli* [459], and increased expression of *talB* in *E. coli* [460]. Since ATP is utilized in the first step of histidine biosynthesis, histidine overproduction can entail a depletion of the adenine pool. The recycling of 5-phosphoribosyl-4-carboxamide-5-aminoim-dazo-le (AICAR), an intermediate of L-histidine biosynthesis, by the purine biosynthesis enzyme AICAR transformylase is important for histidine production [461]. An example of this was an observed increase in production entailed by the amplification of *purH* encoding AICAR transformylase [461]. Optimization of L-histidine production has not only been described based on strain improvement but also by means of the fermentation conditions and processes. The production of L-histidine by *B. flavum* in fed-batch was optimized by boundary control of the specific growth rate [462] and by controlling the concentrations of the carbon sources glucose and acetate [451].

PRODUCTION OF L-SERINE AND GLYCINE

Glycine is the simplest among the proteinogenic amino acids because it does not have a side chain and has no stereo center. Since it is the only proteinogenic amino acid without optical activity, chemical production *via* classical Strecker-synthesis is the exclusive production route, even though microbial production could be demonstrated [463].

L-Serine can be obtained by extraction, by enzymatic or cellular conversion of glycine and a one-carbon compound or by fermentation using glucose [416]. The annual production by fermentation is about 300 tons [51]. It has its main applications in the feed and cosmetics industries [464].

Biosynthesis of L-serine and Glycine

Microbial synthesis of L-serine is closely linked to the synthesis of glycine. As shown in Fig. (**6**), L-serine can be synthesized from 3-phosphoglycerate in a three step pathway as known for example for *C. glutamicum* [416] or L-serine can be synthesized from glycine *via* incorporation of one-carbon compounds [464].

De-bottlenecking and Degradation for L-serine Production

Peters-Wendisch *et al.* [416, 465] investigated L-serine biosynthesis in *C. glutamicum* and could show that the deletion of the 197 carboxy-terminal amino acids of the 3-phosphoglycerate dehydrogenase encoded by *serA*, leads to a feedback-resistant enzyme. However, overexpression of $serA^{\Delta 197}$, *serB* and *serC* was not sufficient for L-serine production most likely due to the high L-serine degradation capacity of *C. glutamicum* [416]. By additionally disrupting the genes *glyA* and *sdaA* coding for serine hydroxymethyltransferase and serine dehydratase, L-serine degradation was avoided and a L-serine production of up to 9 g/L entailed [416]. This principle was later applied to *E. coli* [466].

L-Serine Strains and Processes

Stolz *et al.* [467] created a genetically very stable *C. glutamicum* strain (with intact *glyA*) suitable for the conversion of glucose to L-serine by deleting the serine dehydratase gene *sdaA* and *pabABC* encoding aminodeoxychorismate synthase and lyase, which are responsible for 5,6,7,8-tetrahydrofolate formation. As a consequence, the folate-auxotrophic strain could be controlled for growth-decoupled L-serine production by folate supplementation. Furthermore this strain carried a plasmid for homologous overexpression of *serACB* [467]. This strain was tested under industrial conditions in a 20-L reactor with corn steep liquor (CSL) medium containing 35 g/L CSL, 15 g/L glucose, and 15 g/L fructose [467].

Early L-serine formation during fermentation indicates that CSL delivers a suitable folate supply. Moreover, adjusting the folate concentration can serve as an effective tool for inhibition of L-serine catabolism [467].

PRODUCTION OF L-ALANINE, L-ASPARAGINE, AND L-ASPARTATE

The three amino acids L-asparagine, L-aspartate and L-alanine are not produced by fermentation but by biocatalysis or bioconversion. L-Asparagine is primarily produced by extraction [52].

L-Aspartate can be produced by the enzymatic amination of fumaric acid by aspartase. Radiated *E. coli* K-12 mutants are known to convert fumarate and ammonium efficiently into L-aspartate, a reaction catalysed by aspartate ammonia lyase [468]. The cells can be immobilized by polyurethane prepolymers [469 - 470]. Also *B. stearothermophilus* cultures are reported to produce L-aspartate at high temperatures [471].

L-Alanine is produced by subsequent conversion of L-aspartate, which is catalysed by L-aspartic-β-decarboxylase from *Pseudomonas dacunhae* [472 - 473]. One approach is to use a mixture of immobilized *E. coli* for conversion of fumarate to L-asparate and *P. dacunhae* cells for production of L-aspartic-β-decarboxylase, resulting in a combined conversion of fumarate and ammonium to carbon dioxide and L-alanine [474]. Side reactions (catalysed by fumarase and alanine racemase) can be avoided by heating or pH shifts due to the low thermo- or pH-stability of these enzymes [474]. The same holds true for L-aspartate production by *E. coli* [469, 475]. More recently, a *C. glutamicum* process under oxygen-deprivation conditions for conversion of glucose to L-alanine has been described [476]. In this strain the genes encoding alanine racemase, lactate dehydrogenase and PEP carboxylase (*ΔldhAΔppcΔalr*) were deleted and the alanine dehydrogenase gene (*alaD*) from *A. oxysans* or *L. sphaericus* was heterologously expressed. Additionally endogenous *gapA* which encodes glyceraldehyde 3-phosphate dehydrogenase was overexpressed homologously [476].

BIOSYNTHESIS OF L-SELENOCYSTEINE AND L-PYRROLYSINE

Amino acids occur widespread in nature. They are simple organic compounds containing at least one amino and one carboxylic function. D-amino acids, *e.g.* D-alanine in bacterial cell walls, and β- and γ-amino acids like β-alanine as component of vitamin B5 and γ-aminobutyrate as neurotransmitter have important functions in nature. However, the L-a-amino acids and glycine stand out, as they constitute the biopolymer proteins in all organisms. Mature proteins may contain a variety of different amino acids, whereas 20 amino acids are typically genetically encoded and incorporated into proteins by the ribosome during translation. Two exceptions exist since a translational stop codon can be decoded as sense codon for either L-selenocysteine or L-pyrrolysine. Thus, L-selenocysteine or L-pyrrolysine are also genetically encoded L-α-amino acids with specific amino acyl-tRNAs, but different strategies of genetic code expansion are used. They were not discovered until five and six decades after the deciphering of the genetic code [477 - 480]. Based on bioinformatic analysis of tRNAs encoded in about 150 bacterial and archaeal genomes it appears unlikely that further widely distributed and genetically encoded amino acids exist [481].

Since L-selenocysteine and L-pyrrolysine are synthesized when bound to tRNA, no fermentative production process has been developed yet.

Biosynthesis of L-selenocysteine and its TRNA

The physiological role of L-selenocysteine is not fuly understood since this amino acid can be substituted by L-cysteine and homologs of L-selenoprotein having L-cysteine instead of L-selenocysteine exist. L-selenocysteine differs from L-cysteine *e.g.* by higher electrophilicity, lower protonation of the thiol group and higher reactivity at physiological pH and lower redox potential. L-selenocysteine may be an ancient trait lost unevenly during evolution due to the low availability of selenium in nature. In bacteria such as *E. coli* [482 - 484] the open reading frame for selenoproteins is disrupted by an UGA translational stop codon which is embedded in a secondary structure element of its mRNA, the SECIS element. The stop codon in the SECIS element is decoded with the help of selenocysteyl-tRNA. Selenocysteine-specific tRNA (encoded by *selC*) is charged with L-serine by seryl-tRNA synthetase (Fig. 7). Subsequently, selenocysteine synthase (encoded by *selA*) converts the seryl moiety of this tRNA to a selenocysteyl-moiety using selenomonophosphate as selenium donor. Selenomonophosphate is in turn generated by selenophosphate synthetase (encoded by *selD*). For delivery to the ribosome selenocysteyl-tRNA needs to be bound by a specialized translation elongation factor (SelB) and GTP (Fig. 7). Finally, a conformational change in the quaternary selenocysteyl-tRNA-SelB-GTP-SECIS complex allows for insertion of the charged tRNA into the ribosomal A site.

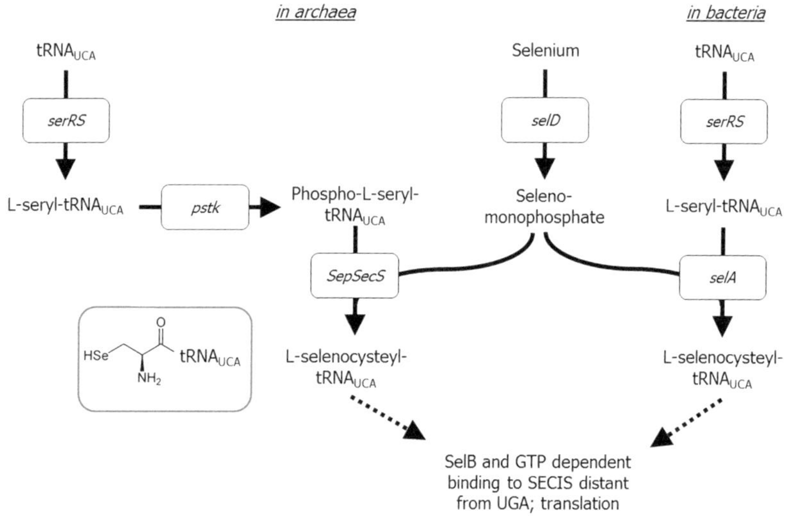

Fig. (7). UGA stop codon decoding and biosynthesis of selenocysteyl-tRNA in Archaea and Bacteria. *pstk*, O-phosphoseryl-tRNA kinase; *selB*, bacterial EF-TU-like translation elongation factor; *SepSecS*, O-phosphoseryl-tRNA:selenocysteine synthase; *serRS*, seryl-tRNA synthetase; *selD*, selenophosphate synthetase; *selA*, selenocysteine synthase. Both, Sec synthesis and Sec insertion are different in Archaea and Eukarya [484].

The archaeal seryl-tRNA is first phosphorylated by O-phosphoseryl-tRNA kinase which is converted to selenocysteyl-tRNA using selenomonophosphate by O-phosphoseryl-tRNA:selenocysteine synthase. The SECIS elements are found in the nontranslated regions of selenoprotein mRNAs and are bound by SECIS-binding protein 2 and the ribosomal protein L30.

Biosynthesis of L-pyrrolysine and its tRNA

The major physiological role of L-pyrrolysine appears to be methylamine methyltransfer in methanogenic archaea since loss of function of the *pylTSBCD* gene cluster results in the inability to grow with methylamines. The *pylTSBCD* gene cluster in the archaeon *Methanosarcina acetivorans* is sufficient for biosynthesis and genetic encoding of L-pyrrolysine [485]. L-pyrrolysine synthesis involves the PylB, PylC and PylD proteins [486 - 487]. L-lysine is methylated by SAM-dependent PylB to 3-methyl-D-ornithine (Fig. **8**). Amino acid ligase PylC forms the dipeptide 3-methyl-D-ornithinyl-N6-L-lysine by ligating another molecule of L-lysine with 3-methyl-D-ornithine. Oxidative deamination of this dipeptide by PylD, dehydration and ring formation terminates L-pyrrolysine biosynthesis (Fig. **8**). Free L-pyrrolysine is ligated to the cognate tRNA^{pylT} by class II aminoacyl-tRNA synthetase PylS, which consists of two subunits (PylSn and PylSc) in bacteria [485]. Although L-pyrrolysine is synthesized as a free amino acid (in contrast to L-selenocysteine which only occurs bound to tRNA) a process for its microbial production has not been described. Interestingly, the orthologous pair, PylS and tRNA^{pylT} have been used in synthetic biology for incorporation of modifiable pyrrolysine analogs into recombinant proteins [488].

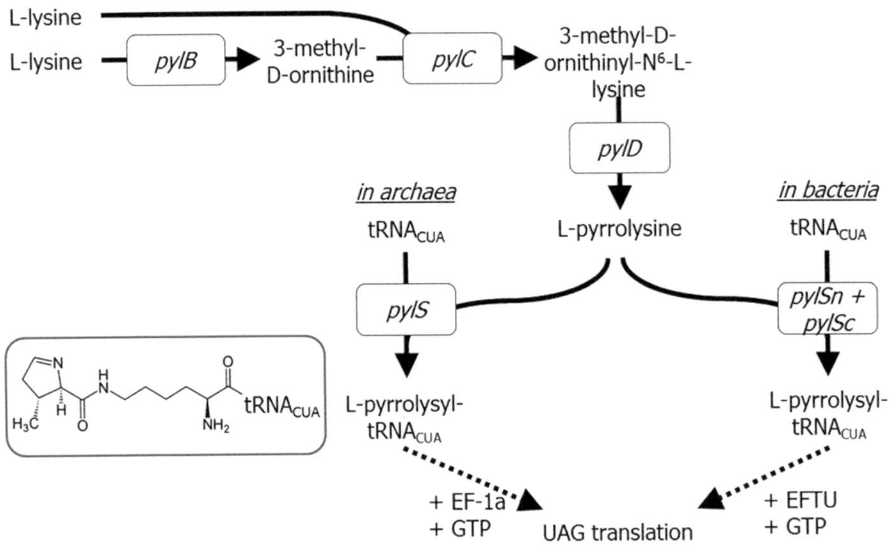

Fig. (8). UAG stop codon decoding and proposed pathway of L-pyrrolysine biosynthesis in Archaea and Bacteria. EF-TU: bacterial translation elongation factor; EF-1a, archaeal translation elongation factor; PylB, PylC and PylD, pyrrolysine biosynthesis proteins; PylS, pyrrolysyl-tRNA synthetase; PylSn, N-terminal part of pyrrolysyl-tRNA synthetase; PylSc, C-terminal part of pyrrolysyl-tRNA synthetase. L-lysine is an L-aspartate family amino acid (see Fig. **3**).

MICROBIAL PRODUCTION OF NUCLEOTIDES

Besides L-glutamate, the umami taste is also provided by the purine nucleotides disodium 5'-monoinosinate (IMP) and disodium 5'-monoguanylate (GMP) [8 - 10]. Shintaro Kodama isolated the nucleotide IMP from dried bonito tuna in 1913. GMP was isolated from shiitake broth in 1960 by Akira Kuninaka who also described synergy between glutamate, IMP, and GMP. The umami taste quality of IMP and GMP differ, with IMP providing a strong initial taste enhancement and with GMP proving a delayed, but about twofold stronger taste enhancement. Foods rich in IMP, *e.g.* dried bonito tuna, or in GMP, *e.g.* shiitake broth, are known. Additionally, pure or enriched preparations of these nucleotides are available commercially. Their market volume increased steadily over decades and the global annual nucleotide production has reached ~34,000 tons in 2012 [7]. The 5′-nucleotides IMP and GMP are now produced by enzymatic degradation of ribonucleic acid or by chemical phosphorylation of the nucleosides inosine and guanine that are obtained by fermentation [11]. Since yeast is rich in ribonucleic acid (RNA), the first commercial Yeast Extract containing GMP was produced in 1974 and nowadays Yeast Extracts with various contents of IMP and GMP are available. Direct fermentation of IMP and XMP has also been reported [12].

RNA Degradation and Nucleotide Extraction

Since yeasts such as *Saccharomyces, Pichia, Candida* and *Hansenula* are rich in RNA, they are used for obtaining this nucleic acid. Heat-based autolysis followed by ethanol precipitation yields RNA, which subsequently is hydrolyzed chemically, using calcium hydroxide, or enzymatically. For enzymatic hydrolysis nucleases, which may be immobilized for increased stability, such as nuclease P1 from *Penicillium* species are employed [489]. It has to be noted that these preparations contain the flavorless purine nucleotide AMP and the flavorless pyrimidine nucleotides UMP and CMP as significant by-products. Therefore, besides the purified flavor enhancers IMP and GMP often yeast extracts enriched in IMP and GMP are commercially available.

Fermentation of nucleosides and nucleotides

Fermentation leads to inosine, guanosine, 5'-inosinic acid, and 5'-xanthylic acid, which are phosphorylated to the respective nucleotides. Chemical phosphorylation using phosphoryl chloride ($POCl_3$) [490] and enzymatic phosphorylation by mutants of acid inosine kinase [491] or phosphatase/pyrophosphate phosphorylase *e.g.* from *Morganella morganii* [492] are used.

IMP, XMP and GMP are produced *de novo* from PRPP and L-glutamine in the purine biosynthetic pathway. Glutamine and PRPP are provided by the central metabolism with PRPP derived from ribose 5-phosphate. Alternatively, free purine nucleobases can be phosphorylated to the respective monophosphates in the nucleotide salvage pathways. Free nucleobases are generated in nucleic acid breakdown by nucleases and 5' nucleotidases [493, 494].

Hosts

Nucleotides are produced almost exclusively using bacteria, primarily *C. ammoniagenes* and also *C. glutamicum, B. subtilis*, and to lesser extent *E. coli* [495]. The highest titers for fermentative inosine production have been obtained with *C. ammoniagenes* mutants (150 g/L; [496]). Classical mutagenesis was used extensively to screen and select for better producing strains. Adenine or guanine auxotrophs exhibiting high 5'-nucleotidase activity were isolated and used for inosine production [495]. Direct production of IMP has been reported in adenine auxotrophs with low 5'-nucleotidase activity and/or increased cytoplasma membrane permeability [11, 12]. Strains secreting GMP have not been isolated, but guanine-adenine double auxotrophs with low 5'-nucleotidase activity and/or increased cytoplasma membrane permeability secrete XMP [11]. XMP is then converted in a whole cell biotransformation by recombinant *E. coli* overproducing and secreting XMP aminase [497].

Metabolic Engineering of Production Hosts

For de-bottlenecking of *de novo* purine biosynthesis, its rate-limiting step catalyzed by PRPP amidotransferase was targeted and feedback-deregulated variants of the enzyme were used [12, 498].

To avoid inosine depletion, conversion of inosine to adenosine, adenylosuccinate and hypoxanthine, was targeted by deletion of genes for adenosine deaminase, adenylosuccinate synthase and purine nucleoside phosphorylase, respectively [499, 500].

To increase precursor supply, glycolysis and in *E. coli* also the Entner-Doudoroff pathway were disrupted by deletion of the genes encoding the initial enzymes glucose 6-phosphate isomerase and 6-phosphogluconate dehydratase, respectively [498, 501]. Disruption of the transketolase gene in *C. ammoniagenes* resulted in shikimate auxotrophy, but improved inosine production [502].

To improve secretion of purine nucleosides the *E. coli* gene *yeaS/leuE* encoding a neutral amino acid exporter [503] or the *B. subtilis* gene *pbuE/ydhL* encoding the purine base efflux pump was overexpressed [503 - 505].

CONFLICT OF INTEREST

The authors confirm that they have no conflict of interest to declare for this publication.

ACKNOWLEDGEMENTS

Work in the lab of VFW is supported in part by SynMet, a 09-EuroSYNBIO-FP-023 project, funded in part by DFG through grant no. WE 2320/2-1, by PROMYSE (EU, FP7 project 289540), by a ZIM project (KF2969003SB2), and by BMBF projects SysEnCor (0315598E) and Genome reduction (0316017A).

REFERENCES

[1] Leuchtenberger, W.; Huthmacher, K.; Drauz, K. Biotechnological production of amino acids and derivatives: current status and prospects. *Appl. Microbiol. Biotechnol.*, **2005**, *69*(1), 1-8. [http://dx.doi.org/10.1007/s00253-005-0155-y] [PMID: 16195792]

[2] Lindemann, B.; Ogiwara, Y.; Ninomiya, Y. The discovery of umami. *Chem. Senses*, **2002**, *27*(9), 843-844. [http://dx.doi.org/10.1093/chemse/27.9.843] [PMID: 12438211]

[3] Kinoshita, S. A Short History of the Birth of the Amino Acid Industry in Japan. In: *Handbook of Corynebacterium glutamicum L*; Eggeling, M.B., Ed.; CRC Press, **2010**; pp. 3-5.

[4] Udaka, S. The discovery of *Corynebacterium glutamicum* and Birth of Amino Acid Fermentation Industry in Japan. In: *Corynebacteria: Genomics and Molecular Biology*; Burkovski, A., Ed.; Caister Academic Press: Norwich, **2008**; pp. 1-6.

[5] Kinoshita, S.; Udaka, S.; Shimono, M. Studies on the amino acid fermentation. Production of L-

glutamic acid by various microorganisms. *J. Gen. Appl. Microbiol.,* **1957**, *3*, 193-205. [http://dx.doi.org/10.2323/jgam.3.193] [PMID: 15965888]

[6] Udaka, S. Screening method for microorganisms accumulating metabolites and its use in the isolation of *Micrococcus glutamicus. J. Bacteriol.,* **1960**, *79*, 754-755. [PMID: 13840150]

[7] *Ajinomoto www.ajinomoto.com/en/ir/pdf/FY13Q1_data_E.pdf,* [18.09.2013];

[8] Fagerson, S.E. Flavours of amino acids and peptides. In: *International Flavours and Food Additives*; , **1976**; 7, .

[9] Ninomiya, K. Natural occurrence. *Food Rev. Int.,* **1998**, *14*(2-3), 177-211. [http://dx.doi.org/10.1080/87559129809541157]

[10] Solms, J. Nonvolatile compounds and flavor. In: *Gustation and Olfaction, G*; Ohlhoff, A.F., Ed.; Academic Press: London, **1971**; pp. 92-110.

[11] Kuninaka, A. *Recent Studies of 5'-Nucleotides as New Flavor Enhancers*; Flavor Chemistry, **1969**, pp. 261-274.

[12] Shimaoka, M.; Takenaka, Y.; Kurahashi, O.; Kawasaki, H.; Matsui, H. Effect of amplification of desensitized *purF* and *prs* on inosine accumulation in *Escherichia coli. J. Biosci. Bioeng.,* **2007**, *103*(3), 255-261. [http://dx.doi.org/10.1263/jbb.103.255] [PMID: 17434429]

[13] Y., K. Inosine accumulation by mutants of *Brevibacterium ammoniagenes* - strain improvement and culture conditions. *Agric. Biol. Chem.,* **1978**, *42*, 399-405. [http://dx.doi.org/10.1271/bbb1961.42.399]

[14] He, K.; Zhao, L.; Daviglus, M.L.; Dyer, A.R.; Van Horn, L.; Garside, D.; Zhu, L.; Guo, D.; Wu, Y.; Zhou, B.; Stamler, J. Association of monosodium glutamate intake with overweight in Chinese adults: the INTERMAP Study. *Obesity (Silver Spring),* **2008**, *16*(8), 1875-1880. [http://dx.doi.org/10.1038/oby.2008.274] [PMID: 18497735]

[15] Hermanussen, M.; García, A.P.; Sunder, M.; Voigt, M.; Salazar, V.; Tresguerres, J.A. Obesity, voracity, and short stature: the impact of glutamate on the regulation of appetite. *Eur. J. Clin. Nutr.,* **2006**, *60*(1), 25-31. [http://dx.doi.org/10.1038/sj.ejcn.1602263] [PMID: 16132059]

[16] Kwok, R.H. Chinese-restaurant syndrome. *N. Engl. J. Med.,* **1968**, *278*(14), 796. [http://dx.doi.org/10.1056/NEJM196804042781419] [PMID: 25276867]

[17] Freeman, M. Reconsidering the effects of monosodium glutamate: a literature review. *J. Am. Acad. Nurse Pract.,* **2006**, *18*(10), 482-486. [http://dx.doi.org/10.1111/j.1745-7599.2006.00160.x] [PMID: 16999713]

[18] Spergel, J.M.; Fiedler, J. Food allergy and additives: triggers in asthma. *Immunol. Allergy Clin. North Am.,* **2005**, *25*(1), 149-167. [http://dx.doi.org/10.1016/j.iac.2004.09.012] [PMID: 15579369]

[19] Woods, R.K.; Weiner, J.; Abramson, M.; Thien, F.; Walters, E.H. Patients' perceptions of food-induced asthma. *Aust. N. Z. J. Med.,* **1996**, *26*(4), 504-512. [http://dx.doi.org/10.1111/j.1445-5994.1996.tb00596.x] [PMID: 8873933]

[20] Radnitz, C. Food-triggered migraine. A critical review. *Ann. Behav. Med.,* **1990**, *12*, 51-65. [http://dx.doi.org/10.1207/s15324796abm1202_1]

[21] Jinap, S.; Hajeb, P. Glutamate. Its applications in food and contribution to health. *Appetite,* **2010**, *55*(1), 1-10. [http://dx.doi.org/10.1016/j.appet.2010.05.002] [PMID: 20470841]

[22] FDA. Questions and Answers on Monosodium glutamate (MSG). http://www.fda.gov/Food/IngredientsPackagingLabeling/FoodAdditivesIngredients/ucm328728.htm [29.07.2013];

[23] Wu, G.; Wu, Z.; Dai, Z.; Yang, Y.; Wang, W.; Liu, C.; Wang, B.; Wang, J.; Yin, Y. Dietary requirements of "nutritionally non-essential amino acids" by animals and humans. *Amino Acids,* **2013**, *44*(4), 1107-1113. [http://dx.doi.org/10.1007/s00726-012-1444-2] [PMID: 23247926]

[24] Wu, G. Amino acids: metabolism, functions, and nutrition. *Amino Acids,* **2009**, *37*(1), 1-17. [http://dx.doi.org/10.1007/s00726-009-0269-0] [PMID: 19301095]

[25] Ren, W.; Yin, Y.; Liu, G.; Yu, X.; Li, Y.; Yang, G.; Li, T.; Wu, G. Effect of dietary arginine supplementation on reproductive performance of mice with porcine circovirus type 2 infection. *Amino Acids,* **2012**, *42*(6), 2089-2094. [http://dx.doi.org/10.1007/s00726-011-0942-y] [PMID: 21617969]

[26] Wu, G. Functional amino acids in nutrition and health. *Amino Acids,* **2013**, *45*(3), 407-411. [http://dx.doi.org/10.1007/s00726-013-1500-6] [PMID: 23595206]

[27] Bonneau, M.; Laarveld, B. Biotechnology in animal nutrition, physiology and health. *Livest. Prod. Sci.,* **1999**, *59*(2-3), 223-241. [http://dx.doi.org/10.1016/S0301-6226(99)00029-9]

[28] Toride, Y. Lysine and other amino acids for feed: production and contribution to protein utilization in animal feeding. *Protein sources for the animal feed industry,* **2002**, 161-165.

[29] Ajinomoto. http://www.ajinomoto.com/ir/pdf/Feed-useAA-Oct2011.pdf [29.07.2013];

[30] *Ajinomoto,* **2010**.

[31] FAO Protein sources for the animal feed industry. *Expert consultation and Workshop. of the Food and Agriculture Organization of the United Nations, Executive Summary,* **2004**.

[32] Miller, E. L. Nutrition Laboratory Department of Clinical Veterinary Medicine University of Cambridge, **2004**. UK

[33] Boisen, S.; Hvelplund, T.; Weisbjerg, M.R. Ideal amino acid profiles as a basis for feed protein evaluation. *Livest. Prod. Sci.,* **2000**, *64*(2-3), 239-251. [http://dx.doi.org/10.1016/S0301-6226(99)00146-3]

[34] Reeds, P.J. Dispensable and indispensable amino acids for humans. *J. Nutr.,* **2000**, *130*(7), 1835S-1840S. [PMID: 10867060]

[35] WHO Protein and amino acid requirements in human nutrition Technical Report Series, 935.

[36] Li, X.; Rezaei, R.; Li, P.; Wu, G. Composition of amino acids in feed ingredients for animal diets.

Amino Acids, **2011**, *40*(4), 1159-1168.
[http://dx.doi.org/10.1007/s00726-010-0740-y] [PMID: 20842395]

[37] Baker, D.H. Advances in protein-amino acid nutrition of poultry. *Amino Acids,* **2009**, *37*(1), 29-41.
[http://dx.doi.org/10.1007/s00726-008-0198-3] [PMID: 19009229]

[38] Li, P.; Mai, K.; Trushenski, J.; Wu, G. New developments in fish amino acid nutrition: towards functional and environmentally oriented aquafeeds. *Amino Acids,* **2009**, *37*(1), 43-53.
[http://dx.doi.org/10.1007/s00726-008-0171-1] [PMID: 18751871]

[39] Fürst, P.; Stehle, P. What are the essential elements needed for the determination of amino acid requirements in humans? *J. Nutr.,* **2004**, *134*(6) Suppl., 1558S-1565S.
[PMID: 15173430]

[40] Fuller, D.B. Industrial amino acids in nonruminant animal nutrition. In: *Biotechnology in Animal Feeds and Animal Feeding*; Chesson, R.J., Ed.; VHC: Weinheim, **1995**; pp. 93-113.

[41] Naylor, C.D.; O'Rourke, K.; Detsky, A.S.; Baker, J.P. Parenteral nutrition with branched-chain amino acids in hepatic encephalopathy. A meta-analysis. *Gastroenterology,* **1989**, *97*(4), 1033-1042.
[PMID: 2506095]

[42] Pearlstone, D.B.; Lee, J.I.; Alexander, R.H.; Chang, T.H.; Brennan, M.F.; Burt, M. Effect of enteral and parenteral nutrition on amino acid levels in cancer patients. *JPEN J. Parenter. Enteral Nutr.,* **1995**, *19*(3), 204-208.
[http://dx.doi.org/10.1177/0148607195019003204] [PMID: 8551648]

[43] Verbruggen, S.; Sy, J.; Arrivillaga, A.; Joosten, K.; van Goudoever, J.; Castillo, L. Parenteral amino acid intakes in critically ill children: a matter of convenience. *JPEN J. Parenter. Enteral Nutr.,* **2010**, *34*(3), 329-340.
[http://dx.doi.org/10.1177/0148607109347897] [PMID: 20093426]

[44] Stein, J.; Boehles, H.J.; Blumenstein, I.; Goeters, C.; Schulz, R. Amino acids - Guidelines on Parenteral Nutrition, Chapter 4. *Ger. Med. Sci.,* **2009**, *7*, Doc24.
[PMID: 20049071]

[45] Oshimura, E.; Tobita, K. *Aqueous preservative solution with high amino acid content US20080260672 A1*; , **2008**.

[46] Shibasaki, I. Food preservation with nontraditional antimicrobial agents. *J. Food Saf.,* **1982**, *4*, 35-38.
[http://dx.doi.org/10.1111/j.1745-4565.1982.tb00432.x]

[47] Smith. *F. X. U. L-HISTIDINE IN OPHTHALMIC SOLUTIONS*; , **2003**.

[48] Inc.. *G. I. A.http://www.strategyr.com/Amino_Acids_Market_Report.asp,* [23.09.2013];

[49] EVONIK.
http://corporate.evonik.com/sites/dc/Downloadcenter/Evonik/Corporate/en/Company/Responsibility/evonik-industries-amino-acids.pdf [24.07.2013];

[50] Reportlinker. http://www.reportlinker.com/p0481352/China-Feed-Amino-Acid-Industry-Report.html [24.07.2013];

[51] Demain, A.L. REVIEWS: The business of biotechnology. *Ind. Biotechnol. (New Rochelle N.Y.),* **2007**, *3*(3), 269-283.

[http://dx.doi.org/10.1089/ind.2007.3.269]

[52] Leuchtenberger, W. Amino acids - technical production and use Biotech *Products of primary metabolism,* **1996**, *6*, 465-502.

[53] Eggeling, L.; Pfefferle, W.; Sahm, H. Amino acids In: *In Basic biotechnology*; Cambridge university press New York, **2006**. US

[54] Dibner, J.J.; Knight, C.D. Conversion of 2-hydroxy-4-(methylthio)butanoic acid to L-methionine in the chick: a stereospecific pathway. *J. Nutr.,* **1984**, *114*(9), 1716-1723. [PMID: 6381670]

[55] Friedman, M. Chemistry, nutrition, and microbiology of D-amino acids. *J. Agric. Food Chem.,* **1999**, *47*(9), 3457-3479. [http://dx.doi.org/10.1021/jf990080u] [PMID: 10552672]

[56] Ikeda, M. Amino acid production processes. *Adv. Biochem. Eng. Biotechnol.,* **2003**, *79*, 1-35. [http://dx.doi.org/10.1007/3-540-45989-8_1] [PMID: 12523387]

[57] Radmacher, E.; Stansen, K.C.; Besra, G.S.; Alderwick, L.J.; Maughan, W.N.; Hollweg, G.; Sahm, H.; Wendisch, V.F.; Eggeling, L. Ethambutol, a cell wall inhibitor of *Mycobacterium tuberculosis*, elicits L-glutamate efflux of *Corynebacterium glutamicum. Microbiology,* **2005**, *151*(Pt 5), 1359-1368. [http://dx.doi.org/10.1099/mic.0.27804-0] [PMID: 15870446]

[58] Kim, J.; Hirasawa, T.; Sato, Y.; Nagahisa, K.; Furusawa, C.; Shimizu, H. Effect of *odhA* overexpression and *odhA* antisense RNA expression on Tween-40-triggered glutamate production by *Corynebacterium glutamicum. Appl. Microbiol. Biotechnol.,* **2009**, *81*(6), 1097-1106. [http://dx.doi.org/10.1007/s00253-008-1743-4] [PMID: 18923827]

[59] Nampoothiri, K.M.; Hoischen, C.; Bathe, B.; Möckel, B.; Pfefferle, W.; Krumbach, K.; Sahm, H.; Eggeling, L. Expression of genes of lipid synthesis and altered lipid composition modulates L-glutamate efflux of *Corynebacterium glutamicum. Appl. Microbiol. Biotechnol.,* **2002**, *58*(1), 89-96. [http://dx.doi.org/10.1007/s00253-001-0861-z] [PMID: 11831479]

[60] Duperray, F.; Jezequel, D.; Ghazi, A.; Letellier, L.; Shechter, E. Excretion of glutamate from *Corynebacterium glutamicum* triggered by amine surfactants. *Biochim. Biophys. Acta,* **1992**, *1103*(2), 250-258. [http://dx.doi.org/10.1016/0005-2736(92)90094-3] [PMID: 1543710]

[61] Takinami, K.; Yoshii, H.; Tsuri, H.; Okada, H. Biochemical Effects of Fatty Acid and Its Derivatives on L-Glutamic Acid Fermentation. 3. Biotin-Tween 60 Relationship in Accumulation of L-Glutamic Acid and Growth of Brevibacterium Lactofermentum. *Agric. Biol. Chem.,* **1965**, *29*(4), 351. [http://dx.doi.org/10.1271/bbb1961.29.351]

[62] Sato, H.; Orishimo, K.; Shirai, T.; Hirasawa, T.; Nagahisa, K.; Shimizu, H.; Wachi, M. Distinct roles of two anaplerotic pathways in glutamate production induced by biotin limitation in *Corynebacterium glutamicum. J. Biosci. Bioeng.,* **2008**, *106*(1), 51-58. [http://dx.doi.org/10.1263/jbb.106.51] [PMID: 18691531]

[63] Kim, J.; Fukuda, H.; Hirasawa, T.; Nagahisa, K.; Nagai, K.; Wachi, M.; Shimizu, H. Requirement of de novo synthesis of the OdhI protein in penicillin-induced glutamate production by *Corynebacterium glutamicum. Appl. Microbiol. Biotechnol.,* **2010**, *86*(3), 911-920.

[http://dx.doi.org/10.1007/s00253-009-2360-6] [PMID: 19956942]

[64] Delaunay, S.; Gourdon, P.; Lapujade, P.; Mailly, E.; Oriol, E.; Engasser, J.M.; Lindley, N.L.; Goergen, J.L. An improved temperature triggered process for glutamate production with *Corynebacterium glutamicum. Enz. Microb. Biotechnol.,* **1999**, *25*, 762-768. [http://dx.doi.org/10.1016/S0141-0229(99)00120-9]

[65] Stansen, C.; Uy, D.; Delaunay, S.; Eggeling, L.; Goergen, J.L.; Wendisch, V.F. Characterization of a *Corynebacterium glutamicum* lactate utilization operon induced during temperature-triggered glutamate production. *Appl. Environ. Microbiol.,* **2005**, *71*(10), 5920-5928. [http://dx.doi.org/10.1128/AEM.71.10.5920-5928.2005] [PMID: 16204505]

[66] Shimizu, H.; Hirasawa, T. Production of Glutamate and Glutamate-Related Amino Acids: Molecular Mechanism Analysis and Metabolic Engineering. In: *Amino Acid Biosynthesis – Pathways, Regulation and Metabolic Engineering*; Wendisch, V.F., Ed.; Springer: Heidelberg, Germany, **2007**. [http://dx.doi.org/10.1007/7171_2006_064]

[67] Gutmann, M.; Hoischen, C.; Krämer, R. Carrier-mediated glutamate secretion by *Corynebacterium glutamicum* under biotin limitation. *Biochim. Biophys. Acta,* **1992**, *1112*(1), 115-123. [http://dx.doi.org/10.1016/0005-2736(92)90261-J] [PMID: 1358200]

[68] Nakamura, J.; Hirano, S.; Ito, H.; Wachi, M. Mutations of the *Corynebacterium glutamicum NCgl1221* gene, encoding a mechanosensitive channel homolog, induce L-glutamic acid production. *Appl. Environ. Microbiol.,* **2007**, *73*(14), 4491-4498. [http://dx.doi.org/10.1128/AEM.02446-06] [PMID: 17513583]

[69] Hashimoto, K.; Nakamura, K.; Kuroda, T.; Yabe, I.; Nakamatsu, T.; Kawasaki, H. The protein encoded by *NCgl1221* in *Corynebacterium glutamicum* functions as a mechanosensitive channel. *Biosci. Biotechnol. Biochem.,* **2010**, *74*(12), 2546-2549. [http://dx.doi.org/10.1271/bbb.100636] [PMID: 21150093]

[70] Nakayama, Y.; Yoshimura, K.; Iida, H. A gain-of-function mutation in gating of Corynebacterium glutamicum NCgl1221 causes constitutive glutamate secretion. *Appl. Environ. Microbiol.,* **2012**, *78*(15), 5432-5434. [http://dx.doi.org/10.1128/AEM.01310-12] [PMID: 22610427]

[71] Hashimoto, K.; Murata, J.; Konishi, T.; Yabe, I.; Nakamatsu, T.; Kawasaki, H. Glutamate is excreted across the cytoplasmic membrane through the NCgl1221 channel of Corynebacterium glutamicum by passive diffusion. *Biosci. Biotechnol. Biochem.,* **2012**, *76*(7), 1422-1424. [http://dx.doi.org/10.1271/bbb.120366] [PMID: 22785475]

[72] Becker, M.; Börngen, K.; Nomura, T.; Battle, A.R.; Marin, K.; Martinac, B.; Krämer, R. Glutamate efflux mediated by *Corynebacterium glutamicum* MscCG, *Escherichia coli* MscS, and their derivatives. *Biochim. Biophys. Acta,* **2013**, *1828*(4), 1230-1240. [http://dx.doi.org/10.1016/j.bbamem.2013.01.001] [PMID: 23313454]

[73] Nakayama, Y.; Yoshimura, K.; Iida, H. Electrophysiological characterization of the mechanosensitive channel MscCG in Corynebacterium glutamicum. *Biophys. J.,* **2013**, *105*(6), 1366-1375. [http://dx.doi.org/10.1016/j.bpj.2013.06.054] [PMID: 24047987]

[74] Börmann, E.R.; Eikmanns, B.J.; Sahm, H. Molecular analysis of the *Corynebacterium glutamicum gdh* gene encoding glutamate dehydrogenase. *Mol. Microbiol.,* **1992**, *6*(3), 317-326.

[http://dx.doi.org/10.1111/j.1365-2958.1992.tb01474.x] [PMID: 1552846]

[75] Tesch, M.; Sahm, H.; Sahm, H. *In vivo* fluxes in the ammonium-assimilatory pathways in *corynebacterium glutamicum* studied by 15N nuclear magnetic resonance. *Appl. Environ. Microbiol.,* **1999**, *65*(3), 1099-1109.
[PMID: 10049869]

[76] Kholy, E.R.; Eikmanns, B.J.; Gutmann, M.; Sahm, H. Glutamate Dehydrogenase Is Not Essential for Glutamate Formation by Corynebacterium glutamicum. *Appl. Environ. Microbiol.,* **1993**, *59*(7), 2329-2331.
[PMID: 16349003]

[77] Jakoby, M.; Tesch, M.; Sahm, H.; Krämer, R.; Burkovski, A. Isolation of the *Corynebacterium glutamicum glnA* gene encoding glutamine synthetase I. *FEMS Microbiol. Lett.,* **1997**, *154*(1), 81-88.
[http://dx.doi.org/10.1111/j.1574-6968.1997.tb12627.x] [PMID: 9297824]

[78] Tesch, M.; Eikmanns, B.J.; de Graaf, A.A.; Sahm, H. Ammonia assimilation in *Corynebacterium glutamicum* and a glutamate dehydrogenase-deficient mutant. *Biotechnol. Lett.,* **1998**, *20*(10), 953-957.
[http://dx.doi.org/10.1023/A:1005442618575]

[79] Shimizu, H.; Tanaka, H.; Nakato, A.; Nagahisa, K.; Kimura, E.; Shioya, S. Effects of the changes in enzyme activities on metabolic flux redistribution around the 2-oxoglutarate branch in glutamate production by *Corynebacterium glutamicum. Bioprocess Biosyst. Eng.,* **2003**, *25*(5), 291-298.
[http://dx.doi.org/10.1007/s00449-002-0307-8] [PMID: 14505173]

[80] Xu, Y.; Labedan, B.; Glansdorff, N. Surprising arginine biosynthesis: a reappraisal of the enzymology and evolution of the pathway in microorganisms. *Microbiol. Mol. Biol. Rev.,* **2007**, *71*(1), 36-47.
[http://dx.doi.org/10.1128/MMBR.00032-06] [PMID: 17347518]

[81] Eikmanns, B.J. Central metabolism: Tricarboxylic acid cycle and anaplerotic reactions. In: *Handbook on Corynebacterium glutamicum*; Eggeling, L.; Bott, M., Eds.; CRC Press: Boca Raton, USA, **2005**; pp. 241-276.
[http://dx.doi.org/10.1201/9781420039696.ch11]

[82] Asakura, Y.; Kimura, E.; Usuda, Y.; Kawahara, Y.; Matsui, K.; Osumi, T.; Nakamatsu, T. Altered metabolic flux due to deletion of *odhA* causes L-glutamate overproduction in *Corynebacterium glutamicum. Appl. Environ. Microbiol.,* **2007**, *73*(4), 1308-1319.
[http://dx.doi.org/10.1128/AEM.01867-06] [PMID: 17158630]

[83] Niebisch, A.; Kabus, A.; Schultz, C.; Weil, B.; Bott, M. Corynebacterial protein kinase G controls 2-oxoglutarate dehydrogenase activity *via* the phosphorylation status of the OdhI protein. *J. Biol. Chem.,* **2006**, *281*(18), 12300-12307.
[http://dx.doi.org/10.1074/jbc.M512515200] [PMID: 16522631]

[84] Barthe, P.; Roumestand, C.; Canova, M.J.; Kremer, L.; Hurard, C.; Molle, V.; Cohen-Gonsaud, M. Dynamic and structural characterization of a bacterial FHA protein reveals a new autoinhibition mechanism. *Structure,* **2009**, *17*(4), 568-578.
[http://dx.doi.org/10.1016/j.str.2009.02.012] [PMID: 19368890]

[85] Schultz, C.; Niebisch, A.; Schwaiger, A.; Viets, U.; Metzger, S.; Bramkamp, M.; Bott, M. Genetic and biochemical analysis of the serine/threonine protein kinases PknA, PknB, PknG and PknL of

Corynebacterium glutamicum: evidence for non-essentiality and for phosphorylation of OdhI and FtsZ by multiple kinases. *Mol. Microbiol.,* **2009**, *74*(3), 724-741. [http://dx.doi.org/10.1111/j.1365-2958.2009.06897.x] [PMID: 19788543]

[86] Krawczyk, S.; Raasch, K.; Schultz, C.; Hoffelder, M.; Eggeling, L.; Bott, M. The FHA domain of OdhI interacts with the carboxyterminal 2 oxoglutarate dehydrogenase domain of OdhA in *Corynebacterium glutamicum. FEBS Lett.,* **2010**, *584*(8), 1463-1468. [http://dx.doi.org/10.1016/j.febslet.2010.03.028] [PMID: 20303957]

[87] Schultz, C.; Niebisch, A.; Gebel, L.; Bott, M. Glutamate production by *Corynebacterium glutamicum*: dependence on the oxoglutarate dehydrogenase inhibitor protein OdhI and protein kinase PknG. *Appl. Microbiol. Biotechnol.,* **2007**, *76*(3), 691-700. [http://dx.doi.org/10.1007/s00253-007-0933-9] [PMID: 17437098]

[88] Boulahya, K.A.; Guedon, E.; Delaunay, S.; Schultz, C.; Boudrant, J.; Bott, M.; Goergen, J.L. *OdhI* dephosphorylation kinetics during different glutamate production processes involving *Corynebacterium glutamicum. Appl. Microbiol. Biotechnol.,* **2010**, *87*(5), 1867-1874. [http://dx.doi.org/10.1007/s00253-010-2599-y] [PMID: 20449744]

[89] Eikmanns, B.J.; Follettie, M.T.; Griot, M.U.; Sinskey, A.J. The phosphoenolpyruvate carboxylase gene of *Corynebacterium glutamicum*: molecular cloning, nucleotide sequence, and expression. *Mol. Gen. Genet.,* **1989**, *218*(2), 330-339. [http://dx.doi.org/10.1007/BF00331286] [PMID: 2779518]

[90] Peters-Wendisch, P.G. V.F.; Paul, S.; Eikmanns, B.J.; Sahm, H. Pyruvate carboxylase as an anaplerotic enzyme in Corynebacterium glutamicum. *Microbiology-SGM,* **1997**, *143*, 1095-1103. [http://dx.doi.org/10.1099/00221287-143-4-1095]

[91] Peters-Wendisch, P.G.; Eikmanns, B.J.; Thierbach, G.; Bachmann, B.; Sahm, H. Phosphoenolpyruvate carboxylase in *Corynebacterium glutamicum* is dispensable for gwoth and lysine production. *FEMS Microbiol. Lett.,* **1993**, *112*(3), 269-274. [http://dx.doi.org/10.1111/j.1574-6968.1993.tb06461.x]

[92] Peters-Wendisch, P.G.; Schiel, B.; Wendisch, V.F.; Katsoulidis, E.; Möckel, B.; Sahm, H.; Eikmanns, B.J. Pyruvate carboxylase is a major bottleneck for glutamate and lysine production by *Corynebacterium glutamicum. J. Mol. Microbiol. Biotechnol.,* **2001**, *3*(2), 295-300. [PMID: 11321586]

[93] Eikmanns, B.J.; Rittmann, D.; Sahm, H. Cloning, sequence analysis, expression, and inactivation of the *Corynebacterium glutamicum icd* gene encoding isocitrate dehydrogenase and biochemical characterization of the enzyme. *J. Bacteriol.,* **1995**, *177*(3), 774-782. [PMID: 7836312]

[94] Marx, A.; Striegel, K.; de Graaf, A.A.; Sahm, H.; Eggeling, L. Response of the central metabolism of *Corynebacterium glutamicum* to different flux burdens. *Biotechnol. Bioeng.,* **1997**, *56*(2), 168-180. [http://dx.doi.org/10.1002/(SICI)1097-0290(19971020)56:2<168::AID-BIT6>3.0.CO;2-N] [PMID: 18636622]

[95] Marx, A.; Eikmanns, B.J.; Sahm, H.; de Graaf, A.A.; Eggeling, L. Response of the central metabolism in *Corynebacterium glutamicum* to the use of an NADH-dependent glutamate dehydrogenase. *Metab. Eng.,* **1999**, *1*(1), 35-48.

[http://dx.doi.org/10.1006/mben.1998.0106] [PMID: 10935753]

[96] Yao, W.; Deng, X.; Liu, M.; Zheng, P.; Sun, Z.; Zhang, Y. Expression and localization of the *Corynebacterium glutamicum NCgl1221* protein encoding an L-glutamic acid exporter. *Microbiol. Res.,* **2009**, *164*(6), 680-687.
[http://dx.doi.org/10.1016/j.micres.2009.01.001] [PMID: 19233628]

[97] Burkovski, A. Ammonium assimilation and nitrogen control in Corynebacterium glutamicum and its relatives: an example for new regulatory mechanisms in actinomycetes. *FEMS Microbiol. Rev.,* **2003**, *27*(5), 617-628.
[http://dx.doi.org/10.1016/S0168-6445(03)00067-6] [PMID: 14638415]

[98] Krämer, R.; Lambert, C.; Hoischen, C.; Ebbighausen, H. Uptake of glutamate in Corynebacterium glutamicum. 1. Kinetic properties and regulation by internal pH and potassium. *Eur. J. Biochem.,* **1990**, *194*(3), 929-935.
[http://dx.doi.org/10.1111/j.1432-1033.1990.tb19488.x] [PMID: 1980106]

[99] Trötschel, C.; Kandirali, S.; Diaz-Achirica, P.; Meinhardt, A.; Morbach, S.; Krämer, R.; Burkovski, A. *GltS*, the sodium-coupled L-glutamate uptake system of *Corynebacterium glutamicum*: identification of the corresponding gene and impact on L-glutamate production. *Appl. Microbiol. Biotechnol.,* **2003**, *60*(6), 738-742.
[http://dx.doi.org/10.1007/s00253-002-1170-x] [PMID: 12664155]

[100] Kronemeyer, W.; Peekhaus, N.; Krämer, R.; Sahm, H.; Eggeling, L. Structure of the *gluABCD* cluster encoding the glutamate uptake system of *Corynebacterium glutamicum. J. Bacteriol.,* **1995**, *177*(5), 1152-1158.
[PMID: 7868586]

[101] Krämer, R.; Lambert, C. Uptake of glutamate in *Corynebacterium glutamicum*. 2. Evidence for a primary active transport system. *Eur. J. Biochem.,* **1990**, *194*(3), 937-944.
[http://dx.doi.org/10.1111/j.1432-1033.1990.tb19489.x] [PMID: 1980107]

[102] Kusumoto, I. Industrial production of L-glutamine. *J. Nutr.,* **2001**, *131*(9) Suppl., 2552S-2555S.
[PMID: 11533312]

[103] Nakamura, J.; Izui, H.; Moriguchi, K.; Kawashima, H.; Nakamatsu, T. Method for producing L-glutamine by fermentation and L-glutamine producing bacterium. *E,* **2002**, *B1*, 1229121.

[104] Mikiro, H.a. Method for production of l-glutamine. *E,* **2005**, *A1*, 1978094.

[105] Jakoby, M.; Krämer, R.; Burkovski, A. Nitrogen regulation in Corynebacterium glutamicum: isolation of genes involved and biochemical characterization of corresponding proteins. *FEMS Microbiol. Lett.,* **1999**, *173*(2), 303-310.
[http://dx.doi.org/10.1111/j.1574-6968.1999.tb13518.x] [PMID: 10227160]

[106] Nolden, L.; Farwick, M.; Krämer, R.; Burkovski, A. Glutamine synthetases of *Corynebacterium glutamicum*: transcriptional control and regulation of activity. *FEMS Microbiol. Lett.,* **2001**, *201*(1), 91-98.
[http://dx.doi.org/10.1111/j.1574-6968.2001.tb10738.x] [PMID: 11445173]

[107] Burkovski, A. Nitrogen control in *Corynebacterium glutamicum*: proteins, mechanisms, signals. *J. Microbiol. Biotechnol.,* **2007**, *17*(2), 187-194.

[PMID: 18051748]

[108] Nolden, L.; Ngouoto-Nkili, C.E.; Bendt, A.K.; Krämer, R.; Burkovski, A. Sensing nitrogen limitation in Corynebacterium glutamicum: the role of glnK and glnD. *Mol. Microbiol.,* **2001**, *42*(5), 1281-1295. [http://dx.doi.org/10.1046/j.1365-2958.2001.02694.x] [PMID: 11886559]

[109] Strösser, J.; Lüdke, A.; Schaffer, S.; Krämer, R.; Burkovski, A. Regulation of GlnK activity: modification, membrane sequestration and proteolysis as regulatory principles in the network of nitrogen control in Corynebacterium glutamicum. *Mol. Microbiol.,* **2004**, *54*(1), 132-147. [http://dx.doi.org/10.1111/j.1365-2958.2004.04247.x] [PMID: 15458411]

[110] Beckers, G.; Strösser, J.; Hildebrandt, U.; Kalinowski, J.; Farwick, M.; Krämer, R.; Burkovski, A. Regulation of AmtR-controlled gene expression in *Corynebacterium glutamicum*: mechanism and characterization of the *AmtR* regulon. *Mol. Microbiol.,* **2005**, *58*(2), 580-595. [http://dx.doi.org/10.1111/j.1365-2958.2005.04855.x] [PMID: 16194241]

[111] Jakoby, M.; Nolden, L.; Meier-Wagner, J.; Krämer, R.; Burkovski, A. AmtR, a global repressor in the nitrogen regulation system of Corynebacterium glutamicum. *Mol. Microbiol.,* **2000**, *37*(4), 964-977. [http://dx.doi.org/10.1046/j.1365-2958.2000.02073.x] [PMID: 10972815]

[112] Burkovski, A. I do it my way: Regulation of ammonium uptake and ammonium assimilation in *Corynebacterium glutamicum. Arch. Microbiol.,* **2003**, *179*(2), 83-88. [PMID: 12560985]

[113] Wendisch, V.F. Genetic regulation of *Corynebacterium glutamicum* metabolism. *J. Microbiol. Biotechnol.,* **2006**, *16*(7), 999.

[114] Buchinger, S.; Strösser, J.; Rehm, N.; Hänssler, E.; Hans, S.; Bathe, B.; Schomburg, D.; Krämer, R.; Burkovski, A. A combination of metabolome and transcriptome analyses reveals new targets of the Corynebacterium glutamicum nitrogen regulator AmtR. *J. Biotechnol.,* **2009**, *140*(1-2), 68-74. [http://dx.doi.org/10.1016/j.jbiotec.2008.10.009] [PMID: 19041910]

[115] Rehm, N.; Burkovski, A. Engineering of nitrogen metabolism and its regulation in Corynebacterium glutamicum: influence on amino acid pools and production. *Appl. Microbiol. Biotechnol.,* **2011**, *89*(2), 239-248. [http://dx.doi.org/10.1007/s00253-010-2922-7] [PMID: 20922371]

[116] Rehm, N.; Buchinger, S.; Strösser, J.; Dotzauer, A.; Walter, B.; Hans, S.; Bathe, B.; Schomburg, D.; Krämer, R.; Burkovski, A. Impact of adenylyltransferase GlnE on nitrogen starvation response in Corynebacterium glutamicum. *J. Biotechnol.,* **2010**, *145*(3), 244-252. [http://dx.doi.org/10.1016/j.jbiotec.2009.11.024] [PMID: 19963018]

[117] Beckers, G.; Nolden, L.; Burkovski, A. Glutamate synthase of Corynebacterium glutamicum is not essential for glutamate synthesis and is regulated by the nitrogen status. *Microbiology,* **2001**, *147*(Pt 11), 2961-2970. [http://dx.doi.org/10.1099/00221287-147-11-2961] [PMID: 11700347]

[118] Marin, K.K. R., Amino Acid Transport Systems in Biotechnologically Relevant Bacteria. In: *Amino Acid Biosynthesis*; Wendisch, V.F., Ed.; Microbiol. Monogr, **2007**; Vol. 5, . [http://dx.doi.org/10.1007/7171_2006_069]

[119] Nohno, T.; Saito, T.; Hong, J.S. Cloning and complete nucleotide sequence of the *Escherichia coli*

glutamine permease operon (*glnHPQ*). *Mol. Gen. Genet.,* **1986**, *205*(2), 260-269. [http://dx.doi.org/10.1007/BF00430437] [PMID: 3027504]

[120] Ren, Q.; Paulsen, I.T. Comparative analyses of fundamental differences in membrane transport capabilities in prokaryotes and eukaryotes. *PLOS Comput. Biol.,* **2005**, *1*(3), e27. [http://dx.doi.org/10.1371/journal.pcbi.0010027] [PMID: 16118665]

[121] Jayaraman, G. M. *A microbial process for arginine production,* **2004**. EP 1 574 566 B1,

[122] Utagawa, T. T. Production of arginine by fermentation *J. Nutr.,* **2004**, *134* 10, 2854S-2857S. discussion 2895S.

[123] Sakanyan, V.; Marc, M.; Hovsepyan, A.; Lecocq, M. *Microorganisms and method for L-arginine production by fermentation,* **2005**. US 6,897,048 B2

[124] Kim, H. W. C. **2008**.

[125] Kuwabara, Y. H., K.; Nakamatsu, T.; Kurahashi, O.; Mori, Y.; Ito, H. Carbamoyl-phosphate synthetase gene of coryneform bacteria and method for producing L-arginine, **2003**. US 2003/0124685 A1

[126] Fujita, Y.; Ogasawara, M.; Kawasaki, H.; Miyauchi, D.; Takano, R.; Sasaki, E. Arginine-rich yeast extract and process for production thereof, **2010**. WO2011074359 A1

[127] Maas, W.K. The arginine repressor of *Escherichia coli. Microbiol. Rev.,* **1994**, *58*(4), 631-640. [PMID: 7854250]

[128] Yim, S.H.; Jung, S.; Lee, S.K.; Cheon, C.I.; Song, E.; Lee, S.S.; Shin, J.; Lee, M.S. Purification and characterization of an arginine regulatory protein, ArgR, in *Corynebacterium glutamicum. J. Ind. Microbiol. Biotechnol.,* **2011**, *38*(12), 1911-1920. [http://dx.doi.org/10.1007/s10295-011-0977-9] [PMID: 21559975]

[129] Sakanyan, V.; Petrosyan, P.; Lecocq, M.; Boyen, A.; Legrain, C.; Demarez, M.; Hallet, J.N.; Glansdorff, N. Genes and enzymes of the acetyl cycle of arginine biosynthesis in *Corynebacterium glutamicum*: enzyme evolution in the early steps of the arginine pathway. *Microbiology,* **1996**, *142*(Pt 1), 99-108. [http://dx.doi.org/10.1099/13500872-142-1-99] [PMID: 8581175]

[130] Marc, F.; Weigel, P.; Legrain, C.; Almeras, Y.; Santrot, M.; Glansdorff, N.; Sakanyan, V. Characterization and kinetic mechanism of mono- and bifunctional ornithine acetyltransferases from thermophilic microorganisms. *Eur. J. Biochem.,* **2000**, *267*(16), 5217-5226. [http://dx.doi.org/10.1046/j.1432-1327.2000.01593.x] [PMID: 10931207]

[131] Caldovic, L.; Tuchman, M. N-acetylglutamate and its changing role through evolution. *Biochem. J.,* **2003**, *372*(Pt 2), 279-290. [http://dx.doi.org/10.1042/bj20030002] [PMID: 12633501]

[132] Glansdorff, N. Microbial Arginine Biosynthesis: Pathway, Regulation and Industrial Production. In: *Amino Acid Biosynthesis*; Wendisch, V.F., Ed.; Microbiol. Monographs, **2007**. [http://dx.doi.org/10.1007/7171_2006_061]

[133] Lu, C.D. Pathways and regulation of bacterial arginine metabolism and perspectives for obtaining arginine overproducing strains. *Appl. Microbiol. Biotechnol.,* **2006**, *70*(3), 261-272. [http://dx.doi.org/10.1007/s00253-005-0308-z] [PMID: 16432742]

[134] Tian, G.; Lim, D.; Carey, J.; Maas, W.K. Binding of the arginine repressor of *Escherichia coli* K12 to its operator sites. *J. Mol. Biol.,* **1992**, *226*(2), 387-397.
[http://dx.doi.org/10.1016/0022-2836(92)90954-I] [PMID: 1640457]

[135] Grandori, R.; Lavoie, T.A.; Pflumm, M.; Tian, G.; Niersbach, H.; Maas, W.K.; Fairman, R.; Carey, J. The DNA-binding domain of the hexameric arginine repressor. *J. Mol. Biol.,* **1995**, *254*(2), 150-162.
[http://dx.doi.org/10.1006/jmbi.1995.0607] [PMID: 7490739]

[136] Makarova, K. S.; Mironov, A. A.; Gelfand, M. S. Conservation of the binding site for the arginine repressor in all bacterial lineages *Genome Biol,* **2001**, *2*(4) RESEARCH0013

[137] Schneider, J.; Niermann, K.; Wendisch, V.F. Production of the amino acids l-glutamate, l-lysine, l-ornithine and l-arginine from arabinose by recombinant *Corynebacterium glutamicum. J. Biotechnol.,* **2011**, *154*(2-3), 191-198.
[http://dx.doi.org/10.1016/j.jbiotec.2010.07.009] [PMID: 20638422]

[138] Vyas, S.; Maas, W.K. Feedback inhibition of acetylglutamate synthetase by arginine in *Escherichia coli. Arch. Biochem. Biophys.,* **1963**, *100*, 542-546.
[http://dx.doi.org/10.1016/0003-9861(63)90124-3] [PMID: 13998082]

[139] Eckhardt, T.; Leisinger, T. Isolation and characterization of mutants with a feedback resistant N-acetylglutamate synthase in *Escherichia coli* K 12. *Mol. Gen. Genet.,* **1975**, *138*(3), 225-232.
[http://dx.doi.org/10.1007/BF00269349] [PMID: 1102931]

[140] Rajagopal, B.S.; DePonte, J., III; Tuchman, M.; Malamy, M.H. Use of inducible feedback-resistant N-acetylglutamate synthetase (*argA*) genes for enhanced arginine biosynthesis by genetically engineered *Escherichia coli* K-12 strains. *Appl. Environ. Microbiol.,* **1998**, *64*(5), 1805-1811.
[PMID: 9572954]

[141] Dou, W.; Xu, M.; Cai, D.; Zhang, X.; Rao, Z.; Xu, Z. Improvement of L-arginine production by overexpression of a bifunctional ornithine acetyltransferase in *Corynebacterium crenatum. Appl. Biochem. Biotechnol.,* **2011**, *165*(3-4), 845-855.
[http://dx.doi.org/10.1007/s12010-011-9302-3] [PMID: 21785983]

[142] Xu, M.; Rao, Z.; Yang, J.; Xia, H.; Dou, W.; Jin, J.; Xu, Z. Heterologous and homologous expression of the arginine biosynthetic *argC~H* cluster from *Corynebacterium crenatum* for improvement of (L) - arginine production. *J. Ind. Microbiol. Biotechnol.,* **2012**, *39*(3), 495-502.
[http://dx.doi.org/10.1007/s10295-011-1042-4] [PMID: 22009057]

[143] Hwang, J.H.; Hwang, G.H.; Cho, J.Y. Effect of increased glutamate availability on L-ornithine production in *Corynebacterium glutamicum. J. Microbiol. Biotechnol.,* **2008**, *18*(4), 704-710.
[PMID: 18467864]

[144] Lu, D.L. J.; Mao, Z. Engineering of *Corynebacterium glutamicum* to Enhance L-ornithine Production by Gene Knockout and Comparative Proteomic Analysis. *Chin. J. Chem. Eng.,* **2012**, *20*(4), 731-739.
[http://dx.doi.org/10.1016/S1004-9541(11)60242-5]

[145] Tuchman, M.; Rajagopal, B.S.; McCann, M.T.; Malamy, M.H. Enhanced production of arginine and urea by genetically engineered *Escherichia coli* K-12 strains. *Appl. Environ. Microbiol.,* **1997**, *63*(1), 33-38.
[PMID: 8979336]

[146] Ikeda, M.; Mitsuhashi, S.; Tanaka, K.; Hayashi, M. Reengineering of a *Corynebacterium glutamicum* L-arginine and L-citrulline producer. *Appl. Environ. Microbiol.,* **2009**, *75*(6), 1635-1641. [http://dx.doi.org/10.1128/AEM.02027-08] [PMID: 19139237]

[147] Cho, J-M. Mutant strain for producing L-ornithine or L-arginine, and method for producing the same. *E,* **2010**, *A2*, 2522720.

[148] Rahman, M.M. Z.Q.; Dou, W.; Zhiming, R.; Xu, Z. Over-expression of NAD kinase in *Corynebacterium crenatum* and its Impact on L-arginine Biosynthesis. *Trop. J. Pharm. Res.,* **2012**, *11*(6)

[149] Lindner, S.N.; Niederholtmeyer, H.; Schmitz, K.; Schoberth, S.M.; Wendisch, V.F. Polyphosphate/ATP-dependent NAD kinase of *Corynebacterium glutamicum*: biochemical properties and impact of *ppnK* overexpression on lysine production. *Appl. Microbiol. Biotechnol.,* **2010**, *87*(2), 583-593. [http://dx.doi.org/10.1007/s00253-010-2481-y] [PMID: 20180116]

[150] Vrljic, M.; Garg, J.; Bellmann, A.; Wachi, S.; Freudl, R.; Malecki, M.J.; Sahm, H.; Kozina, V.J.; Eggeling, L.; Saier, M.H., Jr; Eggeling, L.; Saier, M.H., Jr The LysE superfamily: topology of the lysine exporter LysE of *Corynebacterium glutamicum*, a paradyme for a novel superfamily of transmembrane solute translocators. *J. Mol. Microbiol. Biotechnol.,* **1999**, *1*(2), 327-336. [PMID: 10943564]

[151] Bellmann, A.; Vrljić, M.; Pátek, M.; Sahm, H.; Krämer, R.; Eggeling, L. Expression control and specificity of the basic amino acid exporter *LysE* of *Corynebacterium glutamicum. Microbiology,* **2001**, *147*(Pt 7), 1765-1774. [http://dx.doi.org/10.1099/00221287-147-7-1765] [PMID: 11429454]

[152] Eggeling, L.S. H. The Cell Wall Barrier of *Corynebacterium glutamicum* and Amino Acid Efflux. *J. Biosci. Bioeng.,* **2001**, *92*(3), 201-213. [http://dx.doi.org/10.1016/S1389-1723(01)80251-6]

[153] Peeters, E.; Nguyen Le Minh, P.; Foulquié-Moreno, M.; Charlier, D. Competitive activation of the *Escherichia coli argO* gene coding for an arginine exporter by the transcriptional regulators Lrp and ArgP. *Mol. Microbiol.,* **2009**, *74*(6), 1513-1526. [http://dx.doi.org/10.1111/j.1365-2958.2009.06950.x] [PMID: 19906180]

[154] Xu, M.; Rao, Z.; Yang, J.; Dou, W.; Xu, Z. The effect of a LYSE exporter overexpression on L-arginine production in *Corynebacterium crenatum. Curr. Microbiol.,* **2013**, *67*(3), 271-278. [http://dx.doi.org/10.1007/s00284-013-0358-x] [PMID: 23559017]

[155] Wissenbach, U.; Keck, B.; Unden, G. Physical map location of the new artPIQMJ genes of Escherichia coli, encoding a periplasmic arginine transport system. *J. Bacteriol.,* **1993**, *175*(11), 3687-3688. [PMID: 8501075]

[156] Wissenbach, U.; Six, S.; Bongaerts, J.; Ternes, D.; Steinwachs, S.; Unden, G. A third periplasmic transport system for L-arginine in Escherichia coli: molecular characterization of the artPIQMJ genes, arginine binding and transport. *Mol. Microbiol.,* **1995**, *17*(4), 675-686. [http://dx.doi.org/10.1111/j.1365-2958.1995.mmi_17040675.x] [PMID: 8801422]

[157] Nicoloff, H.H. J.-C.; Bringel, F. Carbamoyl-phosphate synthetases (CPS) in lactic acid bacteria and

other Gram-positive bacteria. *Lait,* **2001**, *81*(1-2), 151-159.
[http://dx.doi.org/10.1051/lait:2001119]

[158] Xu, M.; Rao, Z.; Xu, H.; Lan, C.; Dou, W.; Zhang, X.; Xu, H.; Jin, J.; Xu, Z. Enhanced production of L-arginine by expression of Vitreoscilla hemoglobin using a novel expression system in Corynebacterium crenatum. *Appl. Biochem. Biotechnol.,* **2011**, *163*(6), 707-719.
[http://dx.doi.org/10.1007/s12010-010-9076-z] [PMID: 20835781]

[159] Adrio, J.L.; Demain, A.L. Recombinant organisms for production of industrial products. *Bioeng. Bugs,* **2010**, *1*(2), 116-131.
[http://dx.doi.org/10.4161/bbug.1.2.10484] [PMID: 21326937]

[160] List, B. Proline-catalyzed asymmetric reactions. *Tetrahedron,* **2002**, *58*(28), 5573-5590.
[http://dx.doi.org/10.1016/S0040-4020(02)00516-1]

[161] Bach, T.M.; Hara, R.; Kino, K.; Ohtsu, I.; Yoshida, N.; Takagi, H. Microbial production of N-acetyl cis-4-hydroxy-L-proline by coexpression of the Rhizobium L-proline cis-4-hydroxylase and the yeast N-acetyltransferase Mpr1. *Appl. Microbiol. Biotechnol.,* **2013**, *97*(1), 247-257.
[http://dx.doi.org/10.1007/s00253-012-4204-z] [PMID: 22707053]

[162] Nakanishi, T.; Yonekura, H.; Hattori, K.; Hirao, T.; Azuma, T. *Processes for producing L-proline by fermentation,* **1989**. EP0098122

[163] Yoshinaga, F.; Konishi, S.; Okumura, S.; Katsuya, N. Studies on the fermentative production of L-proline. I. Production of L-proline by an isoleucine auxotrophic mutant of *Brevibacterium flavum* 2247. *J. Gen. Appl. Microbiol.,* **1966**, *12*, 219-228.
[http://dx.doi.org/10.2323/jgam.12.219]

[164] Kato, J.; Horie, S.; Komatsubara, S.; Kisumi, M.; Chibata, I. Production of L-proline by *Kurthia catenaforma. Appl. Microbiol.,* **1968**, *16*(8), 1200-1206.
[PMID: 5675509]

[165] Bayer, A.S.; Coulter, S.N.; Stover, C.K.; Schwan, W.R. Impact of the high-affinity proline permease gene (*putP*) on the virulence of *Staphylococcus aureus* in experimental endocarditis. *Infect. Immun.,* **1999**, *67*(2), 740-744.
[PMID: 9916085]

[166] Culham, D.E.; Dalgado, C.; Gyles, C.L.; Mamelak, D.; MacLellan, S.; Wood, J.M. Osmoregulatory transporter ProP influences colonization of the urinary tract by Escherichia coli. *Microbiology,* **1998**, *144*(Pt 1), 91-102.
[http://dx.doi.org/10.1099/00221287-144-1-91] [PMID: 9467901]

[167] Schwan, W.R.; Coulter, S.N.; Ng, E.Y.; Langhorne, M.H.; Ritchie, H.D.; Brody, L.L.; Westbrock-Wadman, S.; Bayer, A.S.; Folger, K.R.; Stover, C.K. Identification and characterization of the PutP proline permease that contributes to *in vivo* survival of *Staphylococcus aureus* in animal models. *Infect. Immun.,* **1998**, *66*(2), 567-572.
[PMID: 9453610]

[168] Csonka, L.N. Physiological and genetic responses of bacteria to osmotic stress. *Microbiol. Rev.,* **1989**, *53*(1), 121-147.
[PMID: 2651863]

[169] Sleator, R.D.; Gahan, C.G.; Hill, C. Mutations in the listerial *proB* gene leading to proline overproduction: effects on salt tolerance and murine infection. *Appl. Environ. Microbiol.,* **2001**, *67*(10), 4560-4565.
[http://dx.doi.org/10.1128/AEM.67.10.4560-4565.2001] [PMID: 11571156]

[170] Adams, E.; Frank, L. Metabolism of proline and the hydroxyprolines. *Annu. Rev. Biochem.,* **1980**, *49*, 1005-1061.
[http://dx.doi.org/10.1146/annurev.bi.49.070180.005041] [PMID: 6250440]

[171] Umbarger, H.E. Amino acid biosynthesis and its regulation. *Annu. Rev. Biochem.,* **1978**, *47*, 532-606.
[http://dx.doi.org/10.1146/annurev.bi.47.070178.002533] [PMID: 354503]

[172] Hu, C.A.; Delauney, A.J.; Verma, D.P. A bifunctional enzyme (delta 1-pyrroline-5-carboxylate synthetase) catalyzes the first two steps in proline biosynthesis in plants. *Proc. Natl. Acad. Sci. USA,* **1992**, *89*(19), 9354-9358.
[http://dx.doi.org/10.1073/pnas.89.19.9354] [PMID: 1384052]

[173] Smith, C.J.; Deutch, A.H.; Rushlow, K.E. Purification and characteristics of a gamma-glutamyl kinase involved in *Escherichia coli* proline biosynthesis. *J. Bacteriol.,* **1984**, *157*(2), 545-551.
[PMID: 6319365]

[174] Serebrijski, I.; Wojcik, F.; Reyes, O.; Leblon, G. Multicopy suppression by *asd* gene and osmotic stress-dependent complementation by heterologous *proA* in *proA* mutants. *J. Bacteriol.,* **1995**, *177*(24), 7255-7260.
[PMID: 8522535]

[175] Tsuchida, T.; Kubota, K.; Yoshinaga, F. Improvement of L-proline production by sulfaguanidine resistant mutants derived from L-glutamic acid-producing bacteria. *Agric. Biol. Chem.,* **1986**, *50*, 2201-2207.
[http://dx.doi.org/10.1271/bbb1961.50.2201]

[176] Sekine, T.; Kawaguchi, A.; Hamano, Y.; Takagi, H. Desensitization of feedback inhibition of the *Saccharomyces cerevisiae* gamma-glutamyl kinase enhances proline accumulation and freezing tolerance. *Appl. Environ. Microbiol.,* **2007**, *73*(12), 4011-4019.
[http://dx.doi.org/10.1128/AEM.00730-07] [PMID: 17449694]

[177] Sugiura, M.; Kisumi, M. Proline-hyperproducing strains of *Serratia marcescens*: enhancement of proline analog-mediated growth inhibition by increasing osmotic stress. *Appl. Environ. Microbiol.,* **1985**, *49*(4), 782-786.
[PMID: 2860849]

[178] Dandekar, A.M.; Uratsu, S.L. A single base pair change in proline biosynthesis genes causes osmotic stress tolerance. *J. Bacteriol.,* **1988**, *170*(12), 5943-5945.
[PMID: 3056927]

[179] Hans, S.; Bathe, B.; Thierbach, G. *Mutant of the proB gene from coryneform bacteria. ,* **2006**. WO 2006/066758 A1

[180] Kosuge, T.; Hoshino, T. Construction of a proline-producing mutant of the extremely thermophilic eubacterium *Thermus thermophilus* HB27. *Appl. Environ. Microbiol.,* **1998**, *64*(11), 4328-4332.
[PMID: 9797285]

[181] Jensen, J.V.; Wendisch, V.F. Ornithine cyclodeaminase-based proline production by *Corynebacterium glutamicum*. *Microb. Cell Fact.,* **2013**, *12*, 63.
[http://dx.doi.org/10.1186/1475-2859-12-63] [PMID: 23806148]

[182] Díaz, E. Bacterial degradation of aromatic pollutants: a paradigm of metabolic versatility. *Int. Microbiol.,* **2004**, *7*(3), 173-180.
[PMID: 15492931]

[183] Shen, X.H.; Liu, Z.P.; Liu, S.J. Functional identification of the gene locus (ncg12319 and characterization of catechol 1,2-dioxygenase in *Corynebacterium glutamicum*. *Biotechnol. Lett.,* **2004**, *26*(7), 575-580.
[http://dx.doi.org/10.1023/B:BILE.0000021958.86258.08] [PMID: 15168857]

[184] Brinkrolf, K.; Brune, I.; Tauch, A. Transcriptional regulation of catabolic pathways for aromatic compounds in *Corynebacterium glutamicum*. *Genet. Mol. Res.,* **2006**, *5*(4), 773-789.
[PMID: 17183485]

[185] Lee, S.Y.; Kim, Y.H.; Min, J. Conversion of phenol to glutamate and proline in *Corynebacterium glutamicum* is regulated by transcriptional regulator ArgR. *Appl. Microbiol. Biotechnol.,* **2010**, *85*(3), 713-720.
[http://dx.doi.org/10.1007/s00253-009-2206-2] [PMID: 19707750]

[186] Kelle, R.; Hermann, T.; Bathe, B. L-Lysine production. In: *Handbook of Corynebacterium glutamicum*; Eggeling, L.; Bott, M., Eds.; CRC Press: Boca Raton, USA, **2005**; pp. 465-488.

[187] Wittmann, C.; Becker, J. The L-Lysine Story: From Metabolic Pathways to Industrial Production. In: *Amino Acid Biosynthesis – Pathways, Regulation and Metabolic Engineering*; Wendisch, V.F., Ed.; Springer: Heidelberg, Germany, **2007**.
[http://dx.doi.org/10.1007/7171_2006_089]

[188] Graaf, A.A. Metabolic Flux Analysis of *Corynebacterium glutamicum*. In: *Bioreaction Engineering - Modeling and Control*; Springer: Berlin, Germany, **2000**; pp. 506-555.

[189] Eggeling, L.; Sahm, H. L-glutamate and L-lysine: traditional products with impetuous developments. *Appl. Microbiol. Biotechnol.,* **1999**, *52*, 146-153.
[http://dx.doi.org/10.1007/s002530051501]

[190] Velasco, A.M.; Leguina, J.I.; Lazcano, A. Molecular evolution of the lysine biosynthetic pathways. *J. Mol. Evol.,* **2002**, *55*(4), 445-459.
[http://dx.doi.org/10.1007/s00239-002-2340-2] [PMID: 12355264]

[191] Bartlett., A. T. M., White, P. J. Species of *Bacillus* That Make a Vegetative Peptidoglycan Containing Lysine Lack Diaminopimelate Epimerase but Have Diaminopimelate Dehydrogenase. *J. Gen. Microbiol.,* **1985**, *131*(Sep), 2145-2152.

[192] Malumbres, M.; Martín, J.F. Molecular control mechanisms of lysine and threonine biosynthesis in amino acid-producing corynebacteria: redirecting carbon flow. *FEMS Microbiol. Lett.,* **1996**, *143*(2-3), 103-114.
[http://dx.doi.org/10.1111/j.1574-6968.1996.tb08468.x] [PMID: 8837462]

[193] Schrumpf, B.; Schwarzer, A.; Kalinowski, J.; Pühler, A.; Eggeling, L.; Sahm, H. A functionally split pathway for lysine synthesis in *Corynebacterium glutamicium*. *J. Bacteriol.,* **1991**, *173*(14), 4510-

4516.
[PMID: 1906065]

[194] Sahm, H.; Eggeling, L.; de Graaf, A.A. Pathway analysis and metabolic engineering in *Corynebacterium glutamicum*. *Biol. Chem.,* **2000**, *381*(9-10), 899-910.
[http://dx.doi.org/10.1515/BC.2000.111] [PMID: 11076021]

[195] Ohnishi, J.; Mitsuhashi, S.; Hayashi, M.; Ando, S.; Yokoi, H.; Ochiai, K.; Ikeda, M. A novel methodology employing Corynebacterium glutamicum genome information to generate a new L-lysine-producing mutant. *Appl. Microbiol. Biotechnol.,* **2002**, *58*(2), 217-223.
[http://dx.doi.org/10.1007/s00253-001-0883-6] [PMID: 11876415]

[196] Kalinowski, J.; Bathe, B.; Bartels, D.; Bischoff, N.; Bott, M.; Burkovski, A.; Dusch, N.; Eggeling, L.; Eikmanns, B.J.; Gaigalat, L.; Goesmann, A.; Hartmann, M.; Huthmacher, K.; Krämer, R.; Linke, B.; McHardy, A.C.; Meyer, F.; Möckel, B.; Pfefferle, W.; Pühler, A.; Rey, D.A.; Rückert, C.; Rupp, O.; Sahm, H.; Wendisch, V.F.; Wiegräbe, I.; Tauch, A. The complete *Corynebacterium glutamicum* ATCC 13032 genome sequence and its impact on the production of L-aspartate-derived amino acids and vitamins. *J. Biotechnol.,* **2003**, *104*(1-3), 5-25.
[http://dx.doi.org/10.1016/S0168-1656(03)00154-8] [PMID: 12948626]

[197] Cremer, J.; Eggeling, L.; Sahm, H. Control of the lysine biosynthetic sequence in *Corynebacterium glutamicum* as analyzed by overexpression of the individual corresponding genes. *Appl. Environ. Microbiol.,* **1991**, *57*(6), 1746-1752.
[PMID: 16348510]

[198] Follettie, M.T.; Peoples, O.P.; Agoropoulou, C.; Sinskey, A.J. Gene structure and expression of the *Corynebacterium flavum* N13 ask-asd operon. *J. Bacteriol.,* **1993**, *175*(13), 4096-4103.
[PMID: 8100567]

[199] Kalinowski, J.; Cremer, J.; Bachmann, B.; Eggeling, L.; Sahm, H.; Pühler, A. Genetic and biochemical analysis of the aspartokinase from *Corynebacterium glutamicum*. *Mol. Microbiol.,* **1991**, *5*(5), 1197-1204.
[http://dx.doi.org/10.1111/j.1365-2958.1991.tb01893.x] [PMID: 1956296]

[200] Kalinowski, J.; Bachmann, B.; Thierbach, G.; Pühler, A. Aspartokinase genes lysC alpha and lysC beta overlap and are adjacent to the aspartate beta-semialdehyde dehydrogenase gene asd in *Corynebacterium glutamicum*. *Mol. Gen. Genet.,* **1990**, *224*(3), 317-324.
[http://dx.doi.org/10.1007/BF00262424] [PMID: 1980002]

[201] Sugimoto, M.; Tanaka, A.; Suzuki, T.; Matsui, H.; Nakamori, S.; Takagi, H. Sequence analysis of functional regions of homoserine dehydrogenase genes from L-lysine and L-threonine-producing mutants of *Brevibacterium lactofermentum*. *Biosci. Biotechnol. Biochem.,* **1997**, *61*(10), 1760-1762.
[http://dx.doi.org/10.1271/bbb.61.1760] [PMID: 9362124]

[202] Vrljic, M.; Sahm, H.; Eggeling, L. A new type of transporter with a new type of cellular function: L-lysine export from *Corynebacterium glutamicum*. *Mol. Microbiol.,* **1996**, *22*(5), 815-826.
[http://dx.doi.org/10.1046/j.1365-2958.1996.01527.x] [PMID: 8971704]

[203] Becker, J.; Wittmann, C. Bio-based production of chemicals, materials and fuels -*Corynebacterium glutamicum* as versatile cell factory. *Curr. Opin. Biotechnol.,* **2011**.
[PMID: 22138494]

[204] Blombach, B.; Hans, S.; Bathe, B.; Eikmanns, B.J. Acetohydroxyacid synthase, a novel target for improvement of L-lysine production by Corynebacterium glutamicum. *Appl. Environ. Microbiol.*, **2009**, *75*(2), 419-427.
[http://dx.doi.org/10.1128/AEM.01844-08] [PMID: 19047397]

[205] Becker, J.; Klopprogge, C.; Herold, A.; Zelder, O.; Bolten, C.J.; Wittmann, C. Metabolic flux engineering of L-lysine production in *Corynebacterium glutamicum*--over expression and modification of G6P dehydrogenase. *J. Biotechnol.*, **2007**, *132*(2), 99-109.
[http://dx.doi.org/10.1016/j.jbiotec.2007.05.026] [PMID: 17624457]

[206] Kiefer, P.; Heinzle, E.; Wittmann, C. Influence of glucose, fructose and sucrose as carbon sources on kinetics and stoichiometry of lysine production by *Corynebacterium glutamicum. J. Ind. Microbiol. Biotechnol.*, **2002**, *28*(6), 338-343.
[http://dx.doi.org/10.1038/sj.jim.7000252] [PMID: 12032807]

[207] Peters-Wendisch, P.G.; Kreutzer, C.; Kalinowski, J.; Pátek, M.; Sahm, H.; Eikmanns, B.J. Pyruvate carboxylase from *Corynebacterium glutamicum*: characterization, expression and inactivation of the *pyc* gene. *Microbiology,* **1998**, *144*(Pt 4), 915-927.
[http://dx.doi.org/10.1099/00221287-144-4-915] [PMID: 9579065]

[208] Sauer, U.; Eikmanns, B.J. The PEP-pyruvate-oxaloacetate node as the switch point for carbon flux distribution in bacteria. *FEMS Microbiol. Rev.*, **2005**, *29*(4), 765-794.
[http://dx.doi.org/10.1016/j.femsre.2004.11.002] [PMID: 16102602]

[209] Riedel, C.; Rittmann, D.; Dangel, P.; Möckel, B.; Petersen, S.; Sahm, H.; Eikmanns, B.J. Characterization of the phosphoenolpyruvate carboxykinase gene from *Corynebacterium glutamicum* and significance of the enzyme for growth and amino acid production. *J. Mol. Microbiol. Biotechnol.*, **2001**, *3*(4), 573-583.
[PMID: 11565516]

[210] Kind, S.; Becker, J.; Wittmann, C. Increased lysine production by flux coupling of the tricarboxylic acid cycle and the lysine biosynthetic pathway--metabolic engineering of the availability of succinyl-CoA in *Corynebacterium glutamicum. Metab. Eng.,* **2013**, *15*, 184-195.
[http://dx.doi.org/10.1016/j.ymben.2012.07.005] [PMID: 22871505]

[211] Becker, J.; Zelder, O.; Hafner, S.; Schroder, H.; Wittmann, C. From zero to hero--design-based systems metabolic engineering of for L-lysine production. *Metab. Eng.,* **2011**, *13*(2), 159-168.
[http://dx.doi.org/10.1016/j.ymben.2011.01.003] [PMID: 21241816]

[212] Dominguez, H.; Rollin, C.; Guyonvarch, A.; Guerquin-Kern, J.L.; Cocaign-Bousquet, M.; Lindley, N.D. Carbon-flux distribution in the central metabolic pathways of *Corynebacterium glutamicum* during growth on fructose. *Eur. J. Biochem.,* **1998**, *254*(1), 96-102.
[http://dx.doi.org/10.1046/j.1432-1327.1998.2540096.x] [PMID: 9652400]

[213] Wendisch, V.F.; de Graaf, A.A.; Sahm, H.; Eikmanns, B.J. Quantitative determination of metabolic fluxes during coutilization of two carbon sources: comparative analyses with *Corynebacterium glutamicum* during growth on acetate and/or glucose. *J. Bacteriol.,* **2000**, *182*(11), 3088-3096.
[http://dx.doi.org/10.1128/JB.182.11.3088-3096.2000] [PMID: 10809686]

[214] Wittmann, C.; Kiefer, P.; Zelder, O. Metabolic fluxes in *Corynebacterium glutamicum* during lysine production with sucrose as carbon source. *Appl. Environ. Microbiol.,* **2004**, *70*(12), 7277-7287.

[http://dx.doi.org/10.1128/AEM.70.12.7277-7287.2004] [PMID: 15574927]

[215] Becker, J.; Klopprogge, C.; Zelder, O.; Heinzle, E.; Wittmann, C. Amplified expression of fructose 1,6-bisphosphatase in *Corynebacterium glutamicum* increases *in vivo* flux through the pentose phosphate pathway and lysine production on different carbon sources. *Appl. Environ. Microbiol.,* **2005**, *71*(12), 8587-8596. [http://dx.doi.org/10.1128/AEM.71.12.8587-8596.2005] [PMID: 16332851]

[216] Georgi, T.; Rittmann, D.; Wendisch, V.F. Lysine and glutamate production by *Corynebacterium glutamicum* on glucose, fructose and sucrose: roles of malic enzyme and fructose-1,6-bisphosphatase. *Metab. Eng.,* **2005**, *7*(4), 291-301. [http://dx.doi.org/10.1016/j.ymben.2005.05.001] [PMID: 15979917]

[217] Wittmann, C.; de Graaf, A.A. Metabolic flux analysis in *Corynebacterium glutamicum*. In: *Handbook of Corynebacterium glutamicum*; Eggeling, L.; Bott, M., Eds.; CRC Press: Boca Raton, USA, **2005**; pp. 277-304.

[218] Kiefer, P.; Heinzle, E.; Zelder, O.; Wittmann, C. Comparative metabolic flux analysis of lysine-producing Corynebacterium glutamicum cultured on glucose or fructose. *Appl. Environ. Microbiol.,* **2004**, *70*(1), 229-239. [http://dx.doi.org/10.1128/AEM.70.1.229-239.2004] [PMID: 14711646]

[219] Bartek, T.; Blombach, B.; Zönnchen, E.; Makus, P.; Lang, S.; Eikmanns, B.J.; Oldiges, M. Importance of NADPH supply for improved L-valine formation in *Corynebacterium glutamicum. Biotechnol. Prog.,* **2010**, *26*(2), 361-371. [PMID: 20014412]

[220] Pompejus, M.; Kröger, B.; Schröder, H.; Zelder, O.; Haberhauer, G.; Haefner, S.; Klopprogge, C. *C. Corynebacterium* glutamicum genes encoding regulatory proteins involved in the production of methionine, **2011**. US20110229937

[221] Ohnishi, J.; Katahira, R.; Mitsuhashi, S.; Kakita, S.; Ikeda, M. A novel *gnd* mutation leading to increased L-lysine production in *Corynebacterium glutamicum. FEMS Microbiol. Lett.,* **2005**, *242*(2), 265-274. [http://dx.doi.org/10.1016/j.femsle.2004.11.014] [PMID: 15621447]

[222] Marx, A.; Hans, S.; Möckel, B.; Bathe, B.; de Graaf, A.A.; McCormack, A.C.; Stapleton, C.; Burke, K.; O'Donohue, M.; Dunican, L.K. Metabolic phenotype of phosphoglucose isomerase mutants of *Corynebacterium glutamicum. J. Biotechnol.,* **2003**, *104*(1-3), 185-197. [http://dx.doi.org/10.1016/S0168-1656(03)00153-6] [PMID: 12948638]

[223] Takeno, S.; Murata, R.; Kobayashi, R.; Mitsuhashi, S.; Ikeda, M. Engineering of *Corynebacterium glutamicum* with an NADPH-generating glycolytic pathway for L-lysine production. *Appl. Environ. Microbiol.,* **2010**, *76*(21), 7154-7160. [http://dx.doi.org/10.1128/AEM.01464-10] [PMID: 20851994]

[224] Kabus, A.; Georgi, T.; Wendisch, V.F.; Bott, M. Expression of the *Escherichia coli pntAB* genes encoding a membrane-bound transhydrogenase in *Corynebacterium glutamicum* improves L-lysine formation. *Appl. Microbiol. Biotechnol.,* **2007**, *75*(1), 47-53. [http://dx.doi.org/10.1007/s00253-006-0804-9] [PMID: 17216441]

[225] Hirao, T.; Nakano, T.; Azuma, T.; Sugimoto, M.; Nakanishi, T. L-lysine production in continuous

culture of an l-lysine hyperproducing mutant of *Corynebacterium glutamicum. Appl. Microbiol. Biotechnol.,* **1989**, *32*(3), 269.
[http://dx.doi.org/10.1007/BF00184972]

[226] Hermann, T. Industrial production of amino acids by coryneform bacteria. *J. Biotechnol.,* **2003**, *104*(1-3), 155-172.
[http://dx.doi.org/10.1016/S0168-1656(03)00149-4] [PMID: 12948636]

[227] Miwa, H.; Tamura, K.; Koyama, H.; Tsuruta, M.; Tosaka, O.; Shimazaki, K.; Nakamura, T.; Nakayama, T. Process and apparatus for regulating the concentration of the carbon source in the aerobic culture of a microorganism. 1992. France Patent No. 2669935

[228] Pfefferle, W.; Lotter, H.; Friedrich, H.; Degener, W. Process for the fermentative production of amino acids. *E,* **1993**, *B1*, 0532867.

[229] Weuster-Botz, D.; Karutz, M.; Joksch, B.; Schärtges, D.; Wandrey, C. Integrated development of fermentation and downstream processing for L-isoleucine production with *Corynebacterium glutamicum. Appl. Microbiol. Biotechnol.,* **1996**, *46*, 209-219.
[http://dx.doi.org/10.1007/s002530050807]

[230] Sassi, A.H.; Fauvart, L.; Deschamps, A.M.; Lebeault, J.M. Fed-batch production of L-lysine by *Corynebacterium glutamicum. Biochem. Eng. J.,* **1998**, *1*(1), 85-90.
[http://dx.doi.org/10.1016/S1369-703X(97)00014-4]

[231] Nakanishi, K. Process for producing L-lysine. E.U. 2002.Patent Application No. EP1253195,

[232] Takors, R.; Bathe, B.; Rieping, M.; Hans, S.; Kelle, R.; Huthmacher, K. Systems biology for industrial strains and fermentation processes--example: amino acids. *J. Biotechnol.,* **2007**, *129*(2), 181-190.
[http://dx.doi.org/10.1016/j.jbiotec.2007.01.031] [PMID: 17367886]

[233] Dong, X.; Quinn, P.J.; Wang, X. Metabolic engineering of *Escherichia coli* and *Corynebacterium glutamicum* for the production of L-threonine. *Biotechnol. Adv.,* **2011**, *29*(1), 11-23.
[http://dx.doi.org/10.1016/j.biotechadv.2010.07.009] [PMID: 20688145]

[234] Lee, M.H.; Lee, H.W.; Park, J.H.; Ahn, J.O.; Jung, J.K.; Hwang, Y.I. Improved L-threonine production of *Escherichia coli* mutant by optimization of culture conditions. *J. Biosci. Bioeng.,* **2006**, *101*(2), 127-130.
[http://dx.doi.org/10.1263/jbb.101.127] [PMID: 16569607]

[235] Yang, H.; Liao, Y.; Wang, B.; Lin, Y.; Pan, L. Draft genome sequence of *Escherichia coli* XH001, a producer of L-threonine in industry. *J. Bacteriol.,* **2011**, *193*(22), 6406-6407.
[http://dx.doi.org/10.1128/JB.06099-11] [PMID: 22038967]

[236] Kim, Y.H.; Park, J.S.; Cho, J.Y.; Cho, K.M.; Park, Y.H.; Lee, J. Proteomic response analysis of a threonine-overproducing mutant of *Escherichia coli. Biochem. J.,* **2004**, *381*(Pt 3), 823-829.
[http://dx.doi.org/10.1042/BJ20031763] [PMID: 15104539]

[237] Miyajima, R.; Shiio, I. Regulation of aspartate family amino acid biosynthesis in *Brevibacterium flavum*. 3. Properties of homoserine dehydrogenase. *J. Biochem.,* **1970**, *68*(3), 311-319.
[PMID: 4394351]

[238] Archer, J.A.; Solow-Cordero, D.E.; Sinskey, A.J. A C-terminal deletion in *Corynebacterium glutamicum* homoserine dehydrogenase abolishes allosteric inhibition by L-threonine. *Gene,* **1991**,

107(1), 53-59.
[http://dx.doi.org/10.1016/0378-1119(91)90296-N] [PMID: 1743520]

[239] Reinscheid, D.J.; Eikmanns, B.J.; Sahm, H. Analysis of a *Corynebacterium glutamicum hom* gene coding for a feedback-resistant homoserine dehydrogenase. *J. Bacteriol.,* **1991**, *173*(10), 3228-3230. [PMID: 1902466]

[240] Shiio, I.; Miyajima, R.; Nakamori, S. Homoserine dehydrogenase genetically desensitized to the feedback inhibition in *Brevibacterium flavum. J. Biochem.,* **1970**, *68*(6), 859-866. [PMID: 5497450]

[241] Lee, K.H.; Park, J.H.; Kim, T.Y.; Kim, H.U.; Lee, S.Y. Systems metabolic engineering of *Escherichia coli* for L-threonine production. *Mol. Syst. Biol.,* **2007**, *3*, 149.
[http://dx.doi.org/10.1038/msb4100196] [PMID: 18059444]

[242] Ogawa-Miyata, Y.; Kojima, H.; Sano, K. Mutation analysis of the feedback inhibition site of aspartokinase III of *Escherichia coli* K-12 and its use in L-threonine production. *Biosci. Biotechnol. Biochem.,* **2001**, *65*(5), 1149-1154.
[http://dx.doi.org/10.1271/bbb.65.1149] [PMID: 11440130]

[243] Lee, J.H.; Lee, D.E.; Lee, B.U.; Kim, H.S. Global analyses of transcriptomes and proteomes of a parent strain and an L-threonine-overproducing mutant strain. *J. Bacteriol.,* **2003**, *185*(18), 5442-5451.
[http://dx.doi.org/10.1128/JB.185.18.5442-5451.2003] [PMID: 12949096]

[244] Debabov, V.G. The threonine story. *Adv. Biochem. Eng. Biotechnol.,* **2003**, *79*, 113-136.
[http://dx.doi.org/10.1007/3-540-45989-8_4] [PMID: 12523390]

[245] Möckel, B.; Eggeling, L.; Sahm, H. Threonine dehydratases of Corynebacterium glutamicum with altered allosteric control: their generation and biochemical and structural analysis. *Mol. Microbiol.,* **1994**, *13*(5), 833-842.
[http://dx.doi.org/10.1111/j.1365-2958.1994.tb00475.x] [PMID: 7815942]

[246] Reinscheid, D.J.; Kronemeyer, W.; Eggeling, L.; Eikmanns, B.J.; Sahm, H. Stable Expression of *hom-1-thrB* in *Corynebacterium glutamicum* and Its Effect on the Carbon Flux to Threonine and Related Amino Acids. *Appl. Environ. Microbiol.,* **1994**, *60*(1), 126-132.
[PMID: 16349146]

[247] Yamamoto, K.; Ishihama, A. Two different modes of transcription repression of the *Escherichia coli* acetate operon by IclR. *Mol. Microbiol.,* **2003**, *47*(1), 183-194.
[http://dx.doi.org/10.1046/j.1365-2958.2003.03287.x] [PMID: 12492863]

[248] Ishida, M.; Sato, K.; Hashiguchi, K.; Ito, H.; Enei, H.; Nakamori, S. High fermentative production of L-threonine from acetate by a Brevibacterium flavum stabilized strain transformed with a recombinant plasmid carrying the Escherichia coli thr operon. *Biosci. Biotechnol. Biochem.,* **1993**, *57*(10), 1755-1756.
[http://dx.doi.org/10.1271/bbb.57.1755] [PMID: 7764273]

[249] Zakataeva, N.P.; Aleshin, V.V.; Tokmakova, I.L.; Troshin, P.V.; Livshits, V.A. The novel transmembrane *Escherichia coli* proteins involved in the amino acid efflux. *FEBS Lett.,* **1999**, *452*(3), 228-232.
[http://dx.doi.org/10.1016/S0014-5793(99)00625-0] [PMID: 10386596]

[250] Kruse, D.; Krämer, R.; Eggeling, L.; Rieping, M.; Pfefferle, W.; Tchieu, J.H.; Chung, Y.J.; Jr Saier, M.H.; Burkovski, A. Influence of threonine exporters on threonine production in *Escherichia coli. Appl. Microbiol. Biotechnol.,* **2002**, *59*(2-3), 205-210. [http://dx.doi.org/10.1007/s00253-002-0987-7] [PMID: 12111147]

[251] Palmieri, L. Threonine diffusion and threonine transport in *Corynebacterium glutamicum* and their role in threonine production. *Arch. Microbiol.,* **1996**, *165*, 48-54. [http://dx.doi.org/10.1007/s002030050295]

[252] Simic, P.; Sahm, H.; Eggeling, L. L-threonine export: use of peptides to identify a new translocator from *Corynebacterium glutamicum. J. Bacteriol.,* **2001**, *183*(18), 5317-5324. [http://dx.doi.org/10.1128/JB.183.18.5317-5324.2001] [PMID: 11514515]

[253] Diesveld, R.; Tietze, N.; Furst, O.; Reth, A.; Bathe, B.; Sahm, H.; Eggeling, L. Activity of Exporters of *Escherichia coli* in *Corynebacterium glutamicum*, and Their Use to Increase L-Threonine Production. *J. Mol. Microbiol. Biotechnol.,* **2008**. [PMID: 18594129]

[254] Eggeling, L.; Sahm, H. New ubiquitous translocators: amino acid export by *Corynebacterium glutamicum* and *Escherichia coli. Arch. Microbiol.,* **2003**, *180*(3), 155-160. [http://dx.doi.org/10.1007/s00203-003-0581-0] [PMID: 12879215]

[255] Sumantran, V.N.; Schweizer, H.P.; Datta, P. A novel membrane-associated threonine permease encoded by the *tdcC* gene of *Escherichia coli. J. Bacteriol.,* **1990**, *172*(8), 4288-4294. [PMID: 2115866]

[256] Lee, J.W.; Choi, S.; Park, J.H.; Vickers, C.E.; Nielsen, L.K.; Lee, S.Y. Development of sucrose-utilizing Escherichia coli K-12 strain by cloning β-fructofuranosidases and its application for L-threonine production. *Appl. Microbiol. Biotechnol.,* **2010**, *88*(4), 905-913. [http://dx.doi.org/10.1007/s00253-010-2825-7] [PMID: 20711572]

[257] Conrado, R.J.; Varner, J.D.; DeLisa, M.P. Engineering the spatial organization of metabolic enzymes: mimicking nature's synergy. *Curr. Opin. Biotechnol.,* **2008**, *19*(5), 492-499. [http://dx.doi.org/10.1016/j.copbio.2008.07.006] [PMID: 18725290]

[258] Bashor, C.J.; Horwitz, A.A.; Peisajovich, S.G.; Lim, W.A. Rewiring cells: synthetic biology as a tool to interrogate the organizational principles of living systems. *Annu. Rev. Biophys.,* **2010**, *39*, 515-537. [http://dx.doi.org/10.1146/annurev.biophys.050708.133652] [PMID: 20192780]

[259] Dueber, J.E.; Wu, G.C.; Malmirchegini, G.R.; Moon, T.S.; Petzold, C.J.; Ullal, A.V.; Prather, K.L.; Keasling, J.D. Synthetic protein scaffolds provide modular control over metabolic flux. *Nat. Biotechnol.,* **2009**, *27*(8), 753-759. [http://dx.doi.org/10.1038/nbt.1557] [PMID: 19648908]

[260] Lee, J.H.; Jung, S.C.; Bui, M.; Kang, K.H.; Song, J.J.; Kim, S.C. Improved production of L-threonine in Escherichia coli by use of a DNA scaffold system. *Appl. Environ. Microbiol.,* **2013**, *79*(3), 774-782. [http://dx.doi.org/10.1128/AEM.02578-12] [PMID: 23160128]

[261] Conrado, R.J.; Wu, G.C.; Boock, J.T.; Xu, H.; Chen, S.Y.; Lebar, T.; Turnšek, J.; Tomšič, N.; Avbelj, M.; Gaber, R.; Koprivnjak, T.; Mori, J.; Glavnik, V.; Vovk, I.; Benčina, M.; Hodnik, V.; Anderluh, G.;

Dueber, J.E.; Jerala, R.; DeLisa, M.P. DNA-guided assembly of biosynthetic pathways promotes improved catalytic efficiency. *Nucleic Acids Res.,* **2012**, *40*(4), 1879-1889.
[http://dx.doi.org/10.1093/nar/gkr888] [PMID: 22021385]

[262] Szczesiul, M.; Wampler, D.E. Regulation of a metabolic system *in vitro*: synthesis of threonine from aspartic acid. *Biochemistry,* **1976**, *15*(10), 2236-2244.
[http://dx.doi.org/10.1021/bi00655a033] [PMID: 179564]

[263] Chassagnole, C.; Raïs, B.; Quentin, E.; Fell, D.A.; Mazat, J.P. An integrated study of threonine-pathway enzyme kinetics in *Escherichia coli. Biochem. J.,* **2001**, *356*(Pt 2), 415-423.
[http://dx.doi.org/10.1042/bj3560415] [PMID: 11368768]

[264] Chen, N.; Huang, J.; Feng, Z.B.; Yu, L.; Xu, Q.Y.; Wen, T.Y. Optimization of fermentation conditions for the biosynthesis of L-threonine by Escherichia coli. *Appl. Biochem. Biotechnol.,* **2009**, *158*(3), 595-604.
[http://dx.doi.org/10.1007/s12010-008-8385-y] [PMID: 18931947]

[265] Liu, S.; Liang, Y.; Liu, Q.; Tao, T.; Lai, S.; Chen, N.; Wen, T. Development of a two-stage feeding strategy based on the kind and level of feeding nutrients for improving fed-batch production of L-threonine by *Escherichia coli. Appl. Microbiol. Biotechnol.,* **2013**, *97*(2), 573-583.
[http://dx.doi.org/10.1007/s00253-012-4317-4] [PMID: 22965189]

[266] Radmacher, E.; Vaitsikova, A.; Burger, U.; Krumbach, K.; Sahm, H.; Eggeling, L. Linking central metabolism with increased pathway flux: L-valine accumulation by *Corynebacterium glutamicum. Appl. Environ. Microbiol.,* **2002**, *68*(5), 2246-2250.
[http://dx.doi.org/10.1128/AEM.68.5.2246-2250.2002] [PMID: 11976094]

[267] Elisáková, V.; Pátek, M.; Holátko, J.; Nesvera, J.; Leyval, D.; Goergen, J.L.; Delaunay, S. Feedback-resistant acetohydroxy acid synthase increases valine production in *Corynebacterium glutamicum. Appl. Environ. Microbiol.,* **2005**, *71*(1), 207-213.
[http://dx.doi.org/10.1128/AEM.71.1.207-213.2005] [PMID: 15640189]

[268] Blombach, B.; Schreiner, M.E.; Bartek, T.; Oldiges, M.; Eikmanns, B.J. *Corynebacterium glutamicum* tailored for high-yield L-valine production. *Appl. Microbiol. Biotechnol.,* **2008**, *79*(3), 471-479.
[http://dx.doi.org/10.1007/s00253-008-1444-z] [PMID: 18379776]

[269] Hashiguchi, K.; Takesada, H.; Suzuki, E.; Matsui, H. Construction of an L-isoleucine overproducing strain of Escherichia coli K-12. *Biosci. Biotechnol. Biochem.,* **1999**, *63*(4), 672-679.
[http://dx.doi.org/10.1271/bbb.63.672] [PMID: 10361680]

[270] Katashkina, J. Y.; Lunts, M. G.; Doroshenko, V. G.; Fomina, S. A.; Skorokhodova, A. Y.; Ivanovskaya, L. V.; Mashko, S. V. *Method for producing an L-amino acid using a bacterium with an optimized level of gene expression,* **2006**. US 7,604,979 B2

[271] Kartsumata, R.; Hashimoto, S. Process for producing L-valine 1996. US5521074

[272] Blombach, B.; Schreiner, M.E.; Holátko, J.; Bartek, T.; Oldiges, M.; Eikmanns, B.J. L-valine production with pyruvate dehydrogenase complex-deficient *Corynebacterium glutamicum. Appl. Environ. Microbiol.,* **2007**, *73*(7), 2079-2084.
[http://dx.doi.org/10.1128/AEM.02826-06] [PMID: 17293513]

[273] Marienhagen, J.; Kennerknecht, N.; Sahm, H.; Eggeling, L. Functional analysis of all aminotransferase

proteins inferred from the genome sequence of *Corynebacterium glutamicum. J. Bacteriol.,* **2005**, *187*(22), 7639-7646.
[http://dx.doi.org/10.1128/JB.187.22.7639-7646.2005] [PMID: 16267288]

[274] Fotheringham, I.G.; Dacey, S.A.; Taylor, P.P.; Smith, T.J.; Hunter, M.G.; Finlay, M.E.; Primrose, S.B.; Parker, D.M.; Edwards, R.M. The cloning and sequence analysis of the aspC and tyrB genes from Escherichia coli K12. Comparison of the primary structures of the aspartate aminotransferase and aromatic aminotransferase of E. coli with those of the pig aspartate aminotransferase isoenzymes. *Biochem. J.,* **1986**, *234*(3), 593-604.
[http://dx.doi.org/10.1042/bj2340593] [PMID: 3521591]

[275] Changeux, J.P. Allosteric Interactions on Biosynthetic L-Threonine Deaminase from *E. coli* K12. *Cold Spring Harb. Symp. Quant. Biol.,* **1963**, *28*, 497.
[http://dx.doi.org/10.1101/SQB.1963.028.01.066]

[276] Eisenstein, E. Cloning, expression, purification, and characterization of biosynthetic threonine deaminase from *Escherichia coli. J. Biol. Chem.,* **1991**, *266*(9), 5801-5807.
[PMID: 2005118]

[277] Fisher, K.E.; Eisenstein, E. An efficient approach to identify ilvA mutations reveals an amino-terminal catalytic domain in biosynthetic threonine deaminase from *Escherichia coli. J. Bacteriol.,* **1993**, *175*(20), 6605-6613.
[PMID: 8407838]

[278] Möckel, B.; Eggeling, L.; Sahm, H. Functional and structural analyses of threonine dehydratase from *Corynebacterium glutamicum. J. Bacteriol.,* **1992**, *174*(24), 8065-8072.
[PMID: 1459955]

[279] I, E., Cordes, C., Eggeling, L., Sahm, H. Regulation of Acetohydroxy Acid Synthase in Corynebacterium-Glutamicum during Fermentation of Alpha-Ketobutyrate to L-Isoleucine. *Appl. Microbiol. Biotechnol.,* **1987**, *25*(4), 346-351.

[280] Vinogradov, V.; Vyazmensky, M.; Engel, S.; Belenky, I.; Kaplun, A.; Kryukov, O.; Barak, Z.; Chipman, D.M. Acetohydroxyacid synthase isozyme I from *Escherichia coli* has unique catalytic and regulatory properties. *Biochim. Biophys. Acta,* **2006**, *1760*(3), 356-363.
[http://dx.doi.org/10.1016/j.bbagen.2005.10.008] [PMID: 16326011]

[281] Lawther, R.P.; Wek, R.C.; Lopes, J.M.; Pereira, R.; Taillon, B.E.; Hatfield, G.W. The complete nucleotide sequence of the *ilvGMEDA* operon of *Escherichia coli* K-12. *Nucleic Acids Res.,* **1987**, *15*(5), 2137-2155.
[http://dx.doi.org/10.1093/nar/15.5.2137] [PMID: 3550695]

[282] Vyazmensky, M.; Sella, C.; Barak, Z.; Chipman, D.M. Isolation and characterization of subunits of acetohydroxy acid synthase isozyme III and reconstitution of the holoenzyme. *Biochemistry,* **1996**, *35*(32), 10339-10346.
[http://dx.doi.org/10.1021/bi9605604] [PMID: 8756689]

[283] Mendel, S.; Elkayam, T.; Sella, C.; Vinogradov, V.; Vyazmensky, M.; Chipman, D.M.; Barak, Z. Acetohydroxyacid synthase: a proposed structure for regulatory subunits supported by evidence from mutagenesis. *J. Mol. Biol.,* **2001**, *307*(1), 465-477.
[http://dx.doi.org/10.1006/jmbi.2000.4413] [PMID: 11243831]

[284] Mendel, S.; Vinogradov, M.; Vyazmensky, M.; Chipman, D.M.; Barak, Z. The N-terminal domain of the regulatory subunit is sufficient for complete activation of acetohydroxyacid synthase III from *Escherichia coli. J. Mol. Biol.,* **2003**, *325*(2), 275-284. [http://dx.doi.org/10.1016/S0022-2836(02)01142-7] [PMID: 12488095]

[285] Leyval, D.; Uy, D.; Delaunay, S.; Goergen, J.L.; Engasser, J.M. Characterisation of the enzyme activities involved in the valine biosynthetic pathway in a valine-producing strain of *Corynebacterium glutamicum. J. Biotechnol.,* **2003**, *104*(1-3), 241-252. [http://dx.doi.org/10.1016/S0168-1656(03)00162-7] [PMID: 12948642]

[286] Pátek, M.; Krumbach, K.; Eggeling, L.; Sahm, H. Leucine synthesis in *Corynebacterium glutamicum*: enzyme activities, structure of *leuA*, and effect of *leuA* inactivation on lysine synthesis. *Appl. Environ. Microbiol.,* **1994**, *60*(1), 133-140. [PMID: 8117072]

[287] Wessler, S.R.; Calvo, J.M. Control of *leu* operon expression in *Escherichia coli* by a transcription attenuation mechanism. *J. Mol. Biol.,* **1981**, *149*(4), 579-597. [http://dx.doi.org/10.1016/0022-2836(81)90348-X] [PMID: 6171647]

[288] Pátek, M.; Hochmannová, J.; Jelínková, M.; Nesvera, J.; Eggeling, L. Analysis of the *leuB* gene from *Corynebacterium glutamicum. Appl. Microbiol. Biotechnol.,* **1998**, *50*(1), 42-47. [http://dx.doi.org/10.1007/s002530051254] [PMID: 9720199]

[289] Lange, C.; Rittmann, D.; Wendisch, V.F.; Bott, M.; Sahm, H. Global expression profiling and physiological characterization of Corynebacterium glutamicum grown in the presence of L-valine. *Appl. Environ. Microbiol.,* **2003**, *69*(5), 2521-2532. [http://dx.doi.org/10.1128/AEM.69.5.2521-2532.2003] [PMID: 12732517]

[290] Holátko, J.; Elisáková, V.; Prouza, M.; Sobotka, M.; Nesvera, J.; Pátek, M. Metabolic engineering of the L-valine biosynthesis pathway in *Corynebacterium glutamicum* using promoter activity modulation. *J. Biotechnol.,* **2009**, *139*(3), 203-210. [http://dx.doi.org/10.1016/j.jbiotec.2008.12.005] [PMID: 19121344]

[291] Cusyatiner, M. M.; Voroshilova, E. B.; Rostova, Y. G.; Ivanovskaya, L. V.; Lunts, M. G.; Khourges, E. M. *Method for producing L-leucine,* **2004**. US20040091980 A1

[292] Hasegawa, S.; Uematsu, K.; Natsuma, Y.; Suda, M.; Hiraga, K.; Jojima, T.; Inui, M.; Yukawa, H. Improvement of the redox balance increases L-valine production by Corynebacterium glutamicum under oxygen deprivation conditions. *Appl. Environ. Microbiol.,* **2012**, *78*(3), 865-875. [http://dx.doi.org/10.1128/AEM.07056-11] [PMID: 22138982]

[293] Park, J.H.; Lee, K.H.; Kim, T.Y.; Lee, S.Y. Metabolic engineering of *Escherichia coli* for the production of L-valine based on transcriptome analysis and in silico gene knockout simulation. *Proc. Natl. Acad. Sci. USA,* **2007**, *104*(19), 7797-7802. [http://dx.doi.org/10.1073/pnas.0702609104] [PMID: 17463081]

[294] Livshits, V. A.; Doroshenko, V. G.; Gorshkova, N. V.; Belaryeva, A. V.; Ivanovskaya, L. V.; Khourges, E. M.; Akhverdian, V. Z.; Gusyatiner, M. M.; Kozlov, Y. I. Y. I. Mutant ilvH gene and method for producing L-valine 2004. US 6,737,255 B2

[295] Morbach, S.; Sahm, H.; Eggeling, L. Use of Feedback-Resistant Threonine Dehydratases of

Corynebacterium glutamicum To Increase Carbon Flux towards l-Isoleucine. *Appl. Environ. Microbiol.,* **1995**, *61*(12), 4315-4320.
[PMID: 16535185]

[296] Morbach, S.; Sahm, H.; Eggeling, L. l-Isoleucine Production with *Corynebacterium glutamicum*: Further Flux Increase and Limitation of Export. *Appl. Environ. Microbiol.,* **1996**, *62*(12), 4345-4351.
[PMID: 16535457]

[297] Kennerknecht, N.; Sahm, H.; Yen, M.R.; Pátek, M.; Saier Jr, M.H., Jr; Eggeling, L. Export of L-isoleucine from *Corynebacterium glutamicum*: a two-gene-encoded member of a new translocator family. *J. Bacteriol.,* **2002**, *184*(14), 3947-3956.
[http://dx.doi.org/10.1128/JB.184.14.3947-3956.2002] [PMID: 12081967]

[298] Kutukova, E.A.; Livshits, V.A.; Altman, I.P.; Ptitsyn, L.R.; Zyiatdinov, M.H.; Tokmakova, I.L.; Zakataeva, N.P. The *yeaS* (*leuE*) gene of *Escherichia coli* encodes an exporter of leucine, and the Lrp protein regulates its expression. *FEBS Lett.,* **2005**, *579*(21), 4629-4634.
[http://dx.doi.org/10.1016/j.febslet.2005.07.031] [PMID: 16098526]

[299] Guardiola, J.; De Felice, M.; Klopotowski, T.; Iaccarino, M. Mutations affecting the different transport systems for isoleucine, leucine, and valine in Escherichia coli K-12. *J. Bacteriol.,* **1974**, *117*(2), 393-405.
[PMID: 4590465]

[300] Tauch, A.; Hermann, T.; Burkovski, A.; Krämer, R.; Pühler, A.; Kalinowski, J. Isoleucine uptake in *Corynebacterium glutamicum* ATCC 13032 is directed by the brnQ gene product. *Arch. Microbiol.,* **1998**, *169*(4), 303-312.
[http://dx.doi.org/10.1007/s002030050576] [PMID: 9531631]

[301] Ebbighausen, H.; Weil, B.; Krämer, R. Transport of branched-chain amino acids in *Corynebacterium glutamicum. Arch. Microbiol.,* **1989**, *151*(3), 238-244.
[http://dx.doi.org/10.1007/BF00413136] [PMID: 2705860]

[302] Brik Ternbach, M.; Bollman, C.; Wandrey, C.; Takors, R. Application of model discriminating experimental design for modeling and development of a fermentative fed-batch L-valine production process. *Biotechnol. Bioeng.,* **2005**, *91*(3), 356-368.
[http://dx.doi.org/10.1002/bit.20504] [PMID: 15984033]

[303] Colón, G.E.; Nguyen, T.T.; Jetten, M.S.; Sinskey, A.J.; Stephanopoulos, G. Production of isoleucine by overexpression of *ilvA* in a *Corynebacterium lactofermentum* threonine producer. *Appl. Microbiol. Biotechnol.,* **1995**, *43*(3), 482-488.
[http://dx.doi.org/10.1007/BF00218453] [PMID: 7632398]

[304] Hashiguchi, K.; Kojima, H.; Sato, K.; Sano, K. Effects of an Escherichia coli ilvA mutant gene encoding feedback-resistant threonine deaminase on L-isoleucine production by Brevibacterium flavum. *Biosci. Biotechnol. Biochem.,* **1997**, *61*(1), 105-108.
[http://dx.doi.org/10.1271/bbb.61.105] [PMID: 9028039]

[305] Kelle, R.; Hermann, T.; Weuster-Botz, D.; Eggeling, L.; Krämer, R.; Wandrey, C. Glucose-controlled Lisoleucine fed-batch production with recombinant strains of *Corynebacterium glutamicum. J. Biotechnol.,* **1996**, *50*, 123-136.
[http://dx.doi.org/10.1016/0168-1656(96)01554-4]

[306] Gusyatiner, M. M.; Lunts, M. G.; Kozlov, Y. I.; Ivanovskaya, L. V.; Voroshilova, E. B. DNA coding for mutant isopropylmalate synthase, l-leucine-producing microorganism and method for producing l-leucine 2002. EP 1 568 776 A2

[307] Bongaerts, J.; Krämer, M.; Müller, U.; Raeven, L.; Wubbolts, M. Metabolic engineering for microbial production of aromatic amino acids and derived compounds. *Metab. Eng.,* **2001**, *3*(4), 289-300. [http://dx.doi.org/10.1006/mben.2001.0196] [PMID: 11676565]

[308] Ogino, T.; Garner, C.; Markley, J.L.; Herrmann, K.M. Biosynthesis of aromatic compounds: 13C NMR spectroscopy of whole *Escherichia coli* cells. *Proc. Natl. Acad. Sci. USA,* **1982**, *79*(19), 5828-5832. [http://dx.doi.org/10.1073/pnas.79.19.5828] [PMID: 6136965]

[309] Pittard, J. Biosynthesis of the aromatic amino acids. In: *Escherichia coli and Salmonella. Cellular and molecular biology, al, N. e*; Press, A.S., Ed.; Washington, DC, **1996**; Vol. 1, pp. 458-484.

[310] Li, P.P.; Liu, Y.J.; Liu, S.J. Genetic and biochemical identification of the chorismate mutase from *Corynebacterium glutamicum. Microbiology,* **2009**, *155*(Pt 10), 3382-3391. [http://dx.doi.org/10.1099/mic.0.029819-0] [PMID: 19589834]

[311] Tribe, D.E.; Camakaris, H.; Pittard, J. Constitutive and repressivle enzymes of the common pathway of aromatic biosynthesis in *Escherichia coli* K-12: regulation of enzyme synthesis at different growth rates. *J. Bacteriol.,* **1976**, *127*(3), 1085-1097. [PMID: 8426]

[312] Ikeda, M. Towards bacterial strains overproducing L-tryptophan and other aromatics by metabolic engineering. *Appl. Microbiol. Biotechnol.,* **2006**, *69*(6), 615-626. [http://dx.doi.org/10.1007/s00253-005-0252-y] [PMID: 16374633]

[313] Michel, G.; Roszak, A.W.; Sauvé, V.; Maclean, J.; Matte, A.; Coggins, J.R.; Cygler, M.; Lapthorn, A.J. Structures of shikimate dehydrogenase AroE and its Paralog YdiB. A common structural framework for different activities. *J. Biol. Chem.,* **2003**, *278*(21), 19463-19472. [http://dx.doi.org/10.1074/jbc.M300794200] [PMID: 12637497]

[314] Benach, J.; Lee, I.; Edstrom, W.; Kuzin, A.P.; Chiang, Y.; Acton, T.B.; Montelione, G.T.; Hunt, J.F. The 2.3-A crystal structure of the shikimate 5-dehydrogenase orthologue *YdiB* from *Escherichia coli* suggests a novel catalytic environment for an NAD-dependent dehydrogenase. *J. Biol. Chem.,* **2003**, *278*(21), 19176-19182. [http://dx.doi.org/10.1074/jbc.M301348200] [PMID: 12624088]

[315] Whipp, M.J.; Pittard, A.J. A reassessment of the relationship between *aroK*- and *aroL*-encoded shikimate kinase enzymes of *Escherichia coli. J. Bacteriol.,* **1995**, *177*(6), 1627-1629. [PMID: 7883721]

[316] Gibson, M.I.; Gibson, F. Preliminary studies on the isolation and metabolism of an intermediate in aromatic biosynthesis: chorismic acid. *Biochem. J.,* **1964**, *90*(2), 248-256. [http://dx.doi.org/10.1042/bj0900248] [PMID: 5834234]

[317] Ikeda, M. Tryptophan production. In: *Handbook of Corynebacterium glutamicum*; Eggeling, L.; Bott, M., Eds.; CRC Press: Boca Raton, USA, **2005**; pp. 491-511. [http://dx.doi.org/10.1201/9781420039696.ch21]

[318] Hagino, H.; Nakayama, K. Regulatory properties of anthranilate synthase from *Corynebacterium glutamicum. Agric. Biol. Chem.,* **1975**, *39*, 323-330.
[http://dx.doi.org/10.1271/bbb1961.39.323]

[319] Hagino, H.; Nakayama, K. The biosynthetic control in aromatic amino acid producing mutants of *Corynebacterium glutamicum. Agric. Biol. Chem.,* **1975**, *39*, 351-361.
[http://dx.doi.org/10.1271/bbb1961.39.351]

[320] Shiio, I.; Miyajima, R.; Nakagawa, M. Regulation of aromatic amino acid biosynthesis in *Brevibacterium flavum.* I. Regulation of anthranilate synthetase. *J. Biochem.,* **1972**, *72*(6), 1447-1455.
[PMID: 4664742]

[321] Sugimoto, S.; Shiio, I. Regulation of tryptophan biosynthesis by feedback inhibition of the second-step enzyme, anthranilate phosphoribosyltransferase, in *Brevibacterium flavum. Agric. Biol. Chem.,* **1983**, *47*, 2295-2305.
[http://dx.doi.org/10.1271/bbb1961.47.2295]

[322] Sugimoto, M.; Shiio, I. Tryptophan synthase and production of L-tryptophan in regulatory mutants. *Agric. Biol. Chem.,* **1982**, *46*, 2711-2718.
[http://dx.doi.org/10.1271/bbb1961.46.2711]

[323] Sugimoto, S.; Shiio, I. Enzymes of the tryptophan synthetic pathway in *Brevibacterium flavum. J. Biochem.,* **1977**, *81*(4), 823-833.
[PMID: 881418]

[324] Sprenger, G.A. From scratch to value: engineering *Escherichia coli* wild type cells to the production of L-phenylalanine and other fine chemicals derived from chorismate. *Appl. Microbiol. Biotechnol.,* **2007**, *75*(4), 739-749.
[http://dx.doi.org/10.1007/s00253-007-0931-y] [PMID: 17435995]

[325] Ikeda, M.; Katsumata, R. Metabolic Engineering To Produce Tyrosine or Phenylalanine in a Tryptophan-Producing *Corynebacterium glutamicum* Strain. *Appl. Environ. Microbiol.,* **1992**, *58*(3), 781-785.
[PMID: 16348670]

[326] Hsu, S.K.; Lin, L.L.; Lo, H.H.; Hsu, W.H. Mutational analysis of feedback inhibition and catalytic sites of prephenate dehydratase from *Corynebacterium glutamicum. Arch. Microbiol.,* **2004**, *181*(3), 237-244.
[http://dx.doi.org/10.1007/s00203-004-0649-5] [PMID: 14749915]

[327] Backman, K.; O'Connor, M.J.; Maruya, A.; Rudd, E.; McKay, D.; Balakrishnan, R.; Radjai, M.; DiPasquantonio, V.; Shoda, D.; Hatch, R. Genetic engineering of metabolic pathways applied to the production of phenylalanine. *Ann. N. Y. Acad. Sci.,* **1990**, *589*, 16-24.
[http://dx.doi.org/10.1111/j.1749-6632.1990.tb24231.x] [PMID: 2192656]

[328] Takagi, M.; Nishio, Y.; Oh, G.; Yoshida, T. Control of L-phenylalanine production by dual feeding of glucose and L-tyrosine. *Biotechnol. Bioeng.,* **1996**, *52*(6), 653-660.
[http://dx.doi.org/10.1002/(SICI)1097-0290(19961220)52:6<653::AID-BIT3>3.3.CO;2-3] [PMID: 18629944]

[329] Gerigk, M.; Bujnicki, R.; Ganpo-Nkwenkwa, E.; Bongaerts, J.; Sprenger, G.; Takors, R. Process

control for enhanced L-phenylalanine production using different recombinant *Escherichia coli* strains. *Biotechnol. Bioeng.,* **2002**, *80*(7), 746-754.
[http://dx.doi.org/10.1002/bit.10428] [PMID: 12402320]

[330] Maiti, T.K.; Roy, A.; Mukherjee, S.K.; Chatterjee, S.P. Microbial production of L-tyrosine: a review. *Hindustan Antibiot. Bull.,* **1995**, *37*(1-4), 51-65.
[PMID: 8972142]

[331] Flores, N.; Xiao, J.; Berry, A.; Bolivar, F.; Valle, F. Pathway engineering for the production of aromatic compounds in *Escherichia coli. Nat. Biotechnol.,* **1996**, *14*(5), 620-623.
[http://dx.doi.org/10.1038/nbt0596-620] [PMID: 9630954]

[332] Frost, J.W.; Snell, K.D.; Frost, K.M. *Deblocking the common pathway of aromatic amino acid synthesis*; US, **1998**, p. 5776736.

[333] Krämer, M.; Bongaerts, J.; Bovenberg, R.; Kremer, S.; Müller, U.; Orf, S.; Wubbolts, M.; Raeven, L. Metabolic engineering for microbial production of shikimic acid. *Metab. Eng.,* **2003**, *5*(4), 277-283.
[http://dx.doi.org/10.1016/j.ymben.2003.09.001] [PMID: 14642355]

[334] Liao, J.C.; Hou, S.Y.; Chao, Y.P. Pathway analysis, engineering, and physiological considerations for redirecting central metabolism. *Biotechnol. Bioeng.,* **1996**, *52*(1), 129-140.
[http://dx.doi.org/10.1002/(SICI)1097-0290(19961005)52:1<129::AID-BIT13>3.0.CO;2-J] [PMID: 18629859]

[335] Wu, Y.Q.; Jiang, P.H.; Fan, C.S.; Cai, Y.R.; Song, D.X.; Huang, W.D. Cloning and co-expression of *ppsA* and *pckA* genes in *Escherichia coli. Journal of Fudan University,* **2002**, *41*, 31-35.

[336] Báez-Viveros, J.L.; Flores, N.; Juárez, K.; Castillo-España, P.; Bolivar, F.; Gosset, G. *Metabolic transcription analysis of engineered Escherichia coli strains that overproduce L-phenylalanine*; Microb. Cell Factories, **2007**.

[337] Miller, J.E.; Backman, K.C.; O'Connor, M.J.; Hatch, R.T. Production of phenylalanine and organic acids by phosphoenolpyruvate carboxylase-deficient mutants of *Escherichia coli. J. Ind. Microbiol. Biotechnol.,* **1987**, *2*, 143-149.

[338] Patnaik, R.; Liao, J.C. Engineering of *Escherichia coli* central metabolism for aromatic metabolite production with near theoretical yield. *Appl. Environ. Microbiol.,* **1994**, *60*(11), 3903-3908.
[PMID: 7993080]

[339] Gosset, G.; Yong-Xiao, J.; Berry, A. A direct comparison of approaches for increasing carbon flow to aromatic biosynthesis in *Escherichia coli. J. Ind. Microbiol.,* **1996**, *17*(1), 47-52.
[http://dx.doi.org/10.1007/BF01570148] [PMID: 8987689]

[340] Patnaik, R.; Spitzer, R.G.; Liao, J.C. Pathway engineering for production of aromatics in *Escherichia coli*: Confirmation of stoichiometric analysis by independent modulation of AroG, TktA, and Pps activities. *Biotechnol. Bioeng.,* **1995**, *46*(4), 361-370.
[http://dx.doi.org/10.1002/bit.260460409] [PMID: 18623323]

[341] Backman, K. C. Method of biosynthesis of phenylalanine 1992. US5169768 A

[342] Liao, J. C. Microorganisms and methods for overproduction of DAHP by cloned pps gene 1996.WO 1996008567 B1,

[343] Sprenger, G.; Siewe, R.; Sahm, H.; Karutz, M.; Sonke, T. Microbial preparation of substances from aromatic metabolism 1998.WO9,818,937,

[344] Koehn, S. J.; Evans, T. M.; Nelson, R. A.; Taylor, P. P. Methods for increasing carbon conversion efficiency in microorganisms. 1994.WO9,428,154,

[345] Babitzke, P.; Romeo, T. CsrB sRNA family: sequestration of RNA-binding regulatory proteins. *Curr. Opin. Microbiol.,* **2007**, *10*(2), 156-163. [http://dx.doi.org/10.1016/j.mib.2007.03.007] [PMID: 17383221]

[346] Yakandawala, N.; Romeo, T.; Friesen, A.D.; Madhyastha, S. Metabolic engineering of *Escherichia coli* to enhance phenylalanine production. *Appl. Microbiol. Biotechnol.,* **2008**, *78*(2), 283-291. [http://dx.doi.org/10.1007/s00253-007-1307-z] [PMID: 18080813]

[347] Lu, J.L.; Liao, J.C. Metabolic engineering and control analysis for production of aromatics: Role of transaldolase. *Biotechnol. Bioeng.,* **1997**, *53*(2), 132-138. [http://dx.doi.org/10.1002/(SICI)1097-0290(19970120)53:2<132::AID-BIT2>3.0.CO;2-P] [PMID: 18633957]

[348] Frost, J. W. Enhanced production of common aromatic pathway compounds. 1992.US 5168056 A,

[349] Mascarenhas, D.; Ashworth, D.J.; Chen, C.S. Deletion of *pgi* alters tryptophan biosynthesis in a genetically engineered strain of *Escherichia coli. Appl. Environ. Microbiol.,* **1991**, *57*(10), 2995-2999. [PMID: 1746959]

[350] Shen, T.; Liu, Q.; Xie, X.; Xu, Q.; Chen, N. Improved production of tryptophan in genetically engineered Escherichia coli with TktA and PpsA overexpression *J. Biomed. Biotechnol,* **2012**, *2012*, 605219..

[351] Ikeda, M.; Okamoto, K.; Katsumata, R. Cloning of the transketolase gene and the effect of its dosage on aromatic amino acid production in *Corynebacterium glutamicum. Appl. Microbiol. Biotechnol.,* **1999**, *51*(2), 201-206. [http://dx.doi.org/10.1007/s002530051382] [PMID: 10091326]

[352] Katsumata, R.; Kino, K. Process for producing amino acids by fermentation. 1989.WO 2008044409 A1,

[353] Olson, M.M.; Templeton, L.J.; Suh, W.; Youderian, P.; Sariaslani, F.S.; Gatenby, A.A.; Van Dyk, T.K. Production of tyrosine from sucrose or glucose achieved by rapid genetic changes to phenylalanine-producing *Escherichia coli strains. Appl. Microbiol. Biotechnol.,* **2007**, *74*(5), 1031-1040. [http://dx.doi.org/10.1007/s00253-006-0746-2] [PMID: 17216463]

[354] Gavini, N.; Pulakat, L. Role of translation of the *pheA* leader peptide coding region in attenuation regulation of the *Escherichia coli pheA* gene. *J. Bacteriol.,* **1991**, *173*(15), 4904-4907. [PMID: 1856183]

[355] Lütke-Eversloh, T.; Stephanopoulos, G. Combinatorial pathway analysis for improved L-tyrosine production in *Escherichia coli*: identification of enzymatic bottlenecks by systematic gene overexpression. *Metab. Eng.,* **2008**, *10*(2), 69-77. [http://dx.doi.org/10.1016/j.ymben.2007.12.001] [PMID: 18243023]

[356] Rodriguez, A.; Martínez, J.A.; Báez-Viveros, J.L.; Flores, N.; Hernández-Chávez, G.; Ramírez, O.T.; Gosset, G.; Bolivar, F. Constitutive expression of selected genes from the pentose phosphate and

aromatic pathways increases the shikimic acid yield in high-glucose batch cultures of an Escherichia coli strain lacking PTS and pykF. *Microb. Cell Fact.,* **2013**, *12*(1), 86. [http://dx.doi.org/10.1186/1475-2859-12-86] [PMID: 24079972]

[357] Doroshenko, V.; Airich, L.; Vitushkina, M.; Kolokolova, A.; Livshits, V.; Mashko, S. YddG from *Escherichia coli* promotes export of aromatic amino acids. *FEMS Microbiol. Lett.,* **2007**, *275*(2), 312-318. [http://dx.doi.org/10.1111/j.1574-6968.2007.00894.x] [PMID: 17784858]

[358] Wang, J.; Cheng, L.K.; Wang, J.; Liu, Q.; Shen, T.; Chen, N. Genetic engineering of Escherichia coli to enhance production of L-tryptophan. *Appl. Microbiol. Biotechnol.,* **2013**, *97*(17), 7587-7596. [http://dx.doi.org/10.1007/s00253-013-5026-3] [PMID: 23775271]

[359] Ikeda, M.; Ozaki, A.; Katsumata, R. Phenylalanine production by metabolically engineered *Corynebacterium glutamicum* with the *pheA* gene of *Escherichia coli. Appl. Microbiol. Biotechnol.,* **1993**, *39*(3), 318-323. [http://dx.doi.org/10.1007/BF00192085] [PMID: 7763713]

[360] McEvoy, J.J.; Joyce, A. Production of L-phenylalanine by DL-phenylalanine hydroxamate-resistant Tyr- mutants of *Bacillus subtilis. Mol. Cell. Biochem.,* **1974**, *4*(3), 191-195. [http://dx.doi.org/10.1007/BF01731480] [PMID: 4214999]

[361] Hagino, H.; Nakayama, K. Production of Aromatic Amino-Acids by Microorganisms. 4. L-Phenylalanine Production by Analog-Resistant Mutants of *Corynebacterium-Glutamicum. Agric. Biol. Chem.,* **1974**, *38*(1), 157-161. [http://dx.doi.org/10.1271/bbb1961.38.157]

[362] Tsuchida, T.; Kubota, K.; Morinaga, Y.; Matsui, H.; Enei, H.; Yoshinaga, F. Production of L-Phenylalanine by a Mutant of *Brevibacterium lactofermentum* 2256. *Agric. Biol. Chem.,* **1987**, *51*(8), 2095-2101. [http://dx.doi.org/10.1271/bbb1961.51.2095]

[363] Shiio, I.; Sugimoto, S.; Kawamura, K. Breeding of phenylalanine-producing *Brevibacterium flavum* strains by removing feedback-regulation of both the 2 key enzymes in its biosynthesis. *Agric. Biol. Chem.,* **1988**, *52*(9), 2247-2253. [http://dx.doi.org/10.1271/bbb1961.52.2247]

[364] Choi, Y.J.; Tribe, D.E. Continuous production of phenylalanine using an *Escherichia coli* regulatory mutant. *Biotechnol. Lett.,* **1982**, *4*(4), 223-228. [http://dx.doi.org/10.1007/BF00132390]

[365] Hwang, S.O.; Gil, G.H.; Cho, Y.J.; Kang, K.R.; Lee, J.H.; Bae, J.C. The fermentation process for L-phenylalanine production using an auxotrophic regulatory mutant of *Escherichia coli. Appl. Microbiol. Biotechnol.,* **1985**, *22*(2), 108-113. [http://dx.doi.org/10.1007/BF00250029]

[366] Schmid, A.; Dordick, J.S.; Hauer, B.; Kiener, A.; Wubbolts, M.; Witholt, B. Industrial biocatalysis today and tomorrow. *Nature,* **2001**, *409*(6817), 258-268. [http://dx.doi.org/10.1038/35051736] [PMID: 11196655]

[367] Liu, D.X.; Fan, C.S.; Tao, J.H.; Liang, G.X.; Gao, S.E.; Wang, H.J.; Li, X.; Song, D.X. Integration of *E. coli aroG-pheA* tandem genes into *Corynebacterium glutamicum tyrA* locus and its effect on L-

phenylalanine biosynthesis. *World J. Gastroenterol.,* **2004**, *10*(24), 3683-3687. [http://dx.doi.org/10.3748/wjg.v10.i24.3683] [PMID: 15534933]

[368] Chan, M.S.; Hsu, W.H. Cloning of m-fluorophenylalanine-resistant gene and mutational analysis of feedback-resistant prephenate dehydratase from *Corynebacterium glutamicum. Biochem. Biophys. Res. Commun.,* **1996**, *219*(2), 537-542. [http://dx.doi.org/10.1006/bbrc.1996.0269] [PMID: 8605023]

[369] Brunhuber, N.M.; Blanchard, J.S. The biochemistry and enzymology of amino acid dehydrogenases. *Crit. Rev. Biochem. Mol. Biol.,* **1994**, *29*(6), 415-467. [http://dx.doi.org/10.3109/10409239409083486] [PMID: 7705101]

[370] Khamduang, M.; Packdibamrung, K.; Chutmanop, J.; Chisti, Y.; Srinophakun, P. Production of L-phenylalanine from glycerol by a recombinant *Escherichia coli. J. Ind. Microbiol. Biotechnol.,* **2009**, *36*(10), 1267-1274. [http://dx.doi.org/10.1007/s10295-009-0606-z] [PMID: 19565282]

[371] Thongchuang, M.; Pongsawasdi, P.; Chisti, Y.; Packdibamrung, K. Design of a recombinant *Escherichia coli* for producing L-phenylalanine from glycerol. *World J. Microbiol. Biotechnol.,* **2012**, *28*(10), 2937-2943. [http://dx.doi.org/10.1007/s11274-012-1104-4] [PMID: 22806734]

[372] Zhang, C.; Zhang, J.; Kang, Z.; Du, G.; Yu, X.; Wang, T.; Chen, J. Enhanced production of L-phenylalanine in *Corynebacterium glutamicum* due to the introduction of *Escherichia coli* wild-type gene *aroH. J. Ind. Microbiol. Biotechnol.,* **2013**, *40*(6), 643-651. [http://dx.doi.org/10.1007/s10295-013-1262-x] [PMID: 23526182]

[373] Jossek, R.; Bongaerts, J.; Sprenger, G.A. Characterization of a new feedback-resistant 3-deoxy-D-arabino-heptulosonate 7-phosphate synthase AroF of *Escherichia coli. FEMS Microbiol. Lett.,* **2001**, *202*(1), 145-148. [PMID: 11506923]

[374] Rüffer, N.; Heidersdorf, U.; Kretzers, I.; Sprenger, G.A.; Raeven, L.; Takors, R. Fully integrated L-phenylalanine separation and concentration using reactive-extraction with liquid-liquid centrifuges in a fed-batch process with *E. coli. Bioprocess Biosyst. Eng.,* **2004**, *26*(4), 239-248. [http://dx.doi.org/10.1007/s00449-004-0354-4] [PMID: 15045576]

[375] Zhou, H.; Liao, X.; Liu, L.; Wang, T.; Du, G.; Chen, J. Enhanced L-phenylalanine production by recombinant *Escherichia coli* BR-42 (pAP-B03) resistant to bacteriophage BP-1 *via* a two-stage feeding approach. *J. Ind. Microbiol. Biotechnol.,* **2011**, *38*(9), 1219-1227. [http://dx.doi.org/10.1007/s10295-010-0900-9] [PMID: 21104105]

[376] Chen, R.; Hatzimanikatis, V.; Yap, W.M.; Postma, P.W.; Bailey, J.E. Metabolic consequences of phosphotransferase (PTS) mutation in a phenylalanine-producing recombinant *Escherichia coli. Biotechnol. Prog.,* **1997**, *13*(6), 768-775. [http://dx.doi.org/10.1021/bp970060h] [PMID: 9413135]

[377] Berry, A. Improving production of aromatic compounds in *Escherichia coli* by metabolic engineering. *Trends Biotechnol.,* **1996**, *14*(7), 250-256. [http://dx.doi.org/10.1016/0167-7799(96)10033-0] [PMID: 8771798]

[378] Gosset, G. Production of aromatic compounds in bacteria. *Curr. Opin. Biotechnol.,* **2009**, *20*(6), 651-

658.
[http://dx.doi.org/10.1016/j.copbio.2009.09.012] [PMID: 19875279]

[379] Gu, P.; Yang, F.; Kang, J.; Wang, Q.; Qi, Q. One-step of tryptophan attenuator inactivation and promoter swapping to improve the production of L-tryptophan in *Escherichia coli. Microb. Cell Fact.,* **2012**, *11*, 30.
[http://dx.doi.org/10.1186/1475-2859-11-30] [PMID: 22380540]

[380] Gunsalus, R.P.; Yanofsky, C. Nucleotide sequence and expression of *Escherichia coli trpR*, the structural gene for the *trp* aporepressor. *Proc. Natl. Acad. Sci. USA,* **1980**, *77*(12), 7117-7121.
[http://dx.doi.org/10.1073/pnas.77.12.7117] [PMID: 7012834]

[381] Hu, C.; Jiang, P.; Xu, J.; Wu, Y.; Huang, W. Mutation analysis of the feedback inhibition site of phenylalanine-sensitive 3-deoxy-D-arabino-heptulosonate 7-phosphate synthase of *Escherichia coli. J. Basic Microbiol.,* **2003**, *43*(5), 399-406.
[http://dx.doi.org/10.1002/jobm.200310244] [PMID: 12964183]

[382] Spraggon, G.; Kim, C.; Nguyen-Huu, X.; Yee, M.C.; Yanofsky, C.; Mills, S.E. The structures of anthranilate synthase of Serratia marcescens crystallized in the presence of (i) its substrates, chorismate and glutamine, and a product, glutamate, and (ii) its end-product inhibitor, L-tryptophan. *Proc. Natl. Acad. Sci. USA,* **2001**, *98*(11), 6021-6026.
[http://dx.doi.org/10.1073/pnas.111150298] [PMID: 11371633]

[383] Chan, E.C.; Tsai, H.L.; Chen, S.L.; Mou, D.G. Amplification of the tryptophan operon gene in *Escherichia coli* chromosome to increase L-tryptophan biosynthesis. *Appl. Microbiol. Biotechnol.,* **1993**, *40*, 301-305.
[http://dx.doi.org/10.1007/BF00170384]

[384] Yajima, Y.; Sakimoto, K.; Takahashi, K.; Miyao, K.; Kudome, Y.; Aichi, K. L-Tryptophan-producing microorganism and production of L-tryptophan 1990.JP02190182,

[385] Camakaris, H.; Cowan, P.; James, P. Production of tryptophan by the bacterium *Escherichia coli. E,* **1997**, *B1*, 0789073.

[386] Ikeda, M.; Nakanishi, K.; Kino, K.; Katsumata, R. Fermentative production of tryptophan by a stable recombinant strain of *Corynebacterium glutamicum* with a modified serine-biosynthetic pathway. *Biosci. Biotechnol. Biochem.,* **1994**, *58*(4), 674-678.
[http://dx.doi.org/10.1271/bbb.58.674] [PMID: 7764859]

[387] Shi, S.; Chen, T.; Zhang, Z.; Chen, X.; Zhao, X. Transcriptome analysis guided metabolic engineering of *Bacillus subtilis* for riboflavin production. *Metab. Eng.,* **2009**, *11*(4-5), 243-252.
[http://dx.doi.org/10.1016/j.ymben.2009.05.002] [PMID: 19446032]

[388] Nahku, R.; Valgepea, K.; Lahtvee, P.J.; Erm, S.; Abner, K.; Adamberg, K.; Vilu, R. Specific growth rate dependent transcriptome profiling of *Escherichia coli* K12 MG1655 in accelerostat cultures. *J. Biotechnol.,* **2010**, *145*(1), 60-65.
[http://dx.doi.org/10.1016/j.jbiotec.2009.10.007] [PMID: 19861135]

[389] Gu, P.; Kang, J.; Yang, F.; Wang, Q.; Liang, Q.; Qi, Q. The improved L-tryptophan production in recombinant *Escherichia coli* by expressing the polyhydroxybutyrate synthesis pathway. *Appl. Microbiol. Biotechnol.,* **2013**, *97*(9), 4121-4127.
[http://dx.doi.org/10.1007/s00253-012-4665-0] [PMID: 23321909]

[390] Kumari, S.; Tishel, R.; Eisenbach, M.; Wolfe, A.J. Cloning, characterization, and functional expression of *acs*, the gene which encodes acetyl coenzyme A synthetase in *Escherichia coli. J. Bacteriol.,* **1995**, *177*(10), 2878-2886.
[PMID: 7751300]

[391] Huang, J.; Shi, J.; Liu, Q.; Xu, Q.; Xie, X.; Wen, T.; Chen, N. Effects of gene *pta* disruption on L-tryptophan fermentation. *Wei Sheng. Wu Xue Bao,* **2011**, *51*(4), 480-487.

[392] Lin, H.; Castro, N.M.; Bennett, G.N.; San, K.Y. Acetyl-CoA synthetase overexpression in *Escherichia coli* demonstrates more efficient acetate assimilation and lower acetate accumulation: a potential tool in metabolic engineering. *Appl. Microbiol. Biotechnol.,* **2006**, *71*(6), 870-874.
[http://dx.doi.org/10.1007/s00253-005-0230-4] [PMID: 16496143]

[393] Castaño-Cerezo, S.; Pastor, J.M.; Renilla, S.; Bernal, V.; Iborra, J.L.; Cánovas, M. An insight into the role of phosphotransacetylase (*pta*) and the acetate/acetyl-CoA node in *Escherichia coli. Microb. Cell Fact.,* **2009**, *8*, 54.
[http://dx.doi.org/10.1186/1475-2859-8-54] [PMID: 19852855]

[394] Gu, P.; Yang, F.; Li, F.; Liang, Q.; Qi, Q. Knocking out analysis of tryptophan permeases in *Escherichia coli* for improving L-tryptophan production. *Appl. Microbiol. Biotechnol.,* **2013**, *97*(15), 6677-6683.
[http://dx.doi.org/10.1007/s00253-013-4988-5] [PMID: 23695779]

[395] Zhao, Z.; Chen, S.; Wu, D.; Wu, J.; Chen, J. [Effect of mufti-gene knockout of L-tryptophan transport system on L-tryptophan production in Escherichia coli]. *Sheng Wu Gong Cheng Xue Bao,* **2011**, *27*(12), 1765-1772.
[PMID: 22506417]

[396] Liu, Q.; Cheng, Y.; Xie, X.; Xu, Q.; Chen, N. Modification of tryptophan transport system and its impact on production of L-tryptophan in *Escherichia coli. Bioresour. Technol.,* **2012**, *114*, 549-554.
[http://dx.doi.org/10.1016/j.biortech.2012.02.088] [PMID: 22456235]

[397] Heatwole, V.M.; Somerville, R.L. Cloning, nucleotide sequence, and characterization of *mtr*, the structural gene for a tryptophan-specific permease of *Escherichia coli* K-12. *J. Bacteriol.,* **1991**, *173*(1), 108-115.
[PMID: 1987112]

[398] Whipp, M.J.; Pittard, A.J. Regulation of aromatic amino acid transport systems in *Escherichia coli* K-12. *J. Bacteriol.,* **1977**, *132*(2), 453-461.
[PMID: 334742]

[399] Cheng, L.K.; Wang, J.; Xu, Q.Y.; Zhao, C.G.; Shen, Z.Q.; Xie, X.X.; Chen, N. Strategy for pH control and pH feedback-controlled substrate feeding for high-level production of L-tryptophan by Escherichia coli. *World J. Microbiol. Biotechnol.,* **2013**, *29*(5), 883-890.
[http://dx.doi.org/10.1007/s11274-012-1243-7] [PMID: 23283691]

[400] Bell, A.A.; Wheeler, M.H. Biosynthesis and Functions of Fungal Melanins. *Annu. Rev. Phytopathol.,* **1986**, *24*, 411-451.
[http://dx.doi.org/10.1146/annurev.py.24.090186.002211]

[401] Qi, W. W.; Sariaslani, S.; Tang, X. S. Methods for the production of tyrosine, cinnamic acid and para-

hydroxycinnamic acid. 2002.WO 02/090523,

[402] Ito, H.; Sakurai, S.; Tanaka, T.; Sato, K.; Enei, H. Genetic breeding of L-tyrosine producer from *Brevibacterium lactofermentum*. *Agric. Biol. Chem.,* **1990**, *54*(3), 699-705. [http://dx.doi.org/10.1271/bbb1961.54.699] [PMID: 1369436]

[403] Roy, A.; Mukhopadhyay, S.K.; Chatterjee, S.P. Production of tyrosine by auxotrophic and analogue resistant mutants of *Arthrobacter globiformis*. *J. Sci. Ind. Res.,* **1997**, *56*, 727-733.

[404] Katsumata, R.; Ikeda, M. Process for producing L-tryptophan, L-tyrosine or L-phenylalanine 1993.EP0600463B1,

[405] Lütke-Eversloh, T.; Stephanopoulos, G. L-tyrosine production by deregulated strains of *Escherichia coli*. *Appl. Microbiol. Biotechnol.,* **2007**, *75*(1), 103-110. [http://dx.doi.org/10.1007/s00253-006-0792-9] [PMID: 17221195]

[406] Santos, C.N.; Xiao, W.; Stephanopoulos, G. Rational, combinatorial, and genomic approaches for engineering L-tyrosine production in *Escherichia coli*. *Proc. Natl. Acad. Sci. USA,* **2012**, *109*(34), 13538-13543. [http://dx.doi.org/10.1073/pnas.1206346109] [PMID: 22869698]

[407] Leonard, E.; Lim, K.H.; Saw, P.N.; Koffas, M.A. Engineering central metabolic pathways for high-level flavonoid production in *Escherichia coli*. *Appl. Environ. Microbiol.,* **2007**, *73*(12), 3877-3886. [http://dx.doi.org/10.1128/AEM.00200-07] [PMID: 17468269]

[408] Juminaga, D.; Baidoo, E.E.; Redding-Johanson, A.M.; Batth, T.S.; Burd, H.; Mukhopadhyay, A.; Petzold, C.J.; Keasling, J.D. Modular engineering of L-tyrosine production in *Escherichia coli*. *Appl. Environ. Microbiol.,* **2012**, *78*(1), 89-98. [http://dx.doi.org/10.1128/AEM.06017-11] [PMID: 22020510]

[409] Chávez-Béjar, M.I.; Lara, A.R.; López, H.; Hernández-Chávez, G.; Martinez, A.; Ramírez, O.T.; Bolívar, F.; Gosset, G. Metabolic engineering of *Escherichia coli* for L-tyrosine production by expression of genes coding for the chorismate mutase domain of the native chorismate mutase-prephenate dehydratase and a cyclohexadienyl dehydrogenase from *Zymomonas mobilis*. *Appl. Environ. Microbiol.,* **2008**, *74*(10), 3284-3290. [http://dx.doi.org/10.1128/AEM.02456-07] [PMID: 18344329]

[410] Patnaik, R.; Zolandz, R.R.; Green, D.A.; Kraynie, D.F. L-tyrosine production by recombinant Escherichia coli: fermentation optimization and recovery. *Biotechnol. Bioeng.,* **2008**, *99*(4), 741-752. [http://dx.doi.org/10.1002/bit.21765] [PMID: 18069696]

[411] Wada, M.; Takagi, H. Metabolic pathways and biotechnological production of L-cysteine. *Appl. Microbiol. Biotechnol.,* **2006**, *73*(1), 48-54. [http://dx.doi.org/10.1007/s00253-006-0587-z] [PMID: 17021879]

[412] Martens, J.; Offermanns, H.; Scherberich, P. Facile Synthesis of Racemic Cysteine. *Angew. Chem. Int. Ed. Engl.,* **1981**, *20*(8), 668-668. [http://dx.doi.org/10.1002/anie.198106681]

[413] Sharp, S. The Human Hair Additive in Your Food http://www.motherjones.com/ blue-marble/2010/03/human-hair-additive-your-food [15.09.2013];

[414] Dassler, T.; Maier, T.; Winterhalter, C.; Böck, A. Identification of a major facilitator protein from

Escherichia coli involved in efflux of metabolites of the cysteine pathway. *Mol. Microbiol.,* **2000**, *36*(5), 1101-1112.
[http://dx.doi.org/10.1046/j.1365-2958.2000.01924.x] [PMID: 10844694]

[415] Pizer, L.I. The Pathway and Control of Serine Biosynthesis in *Escherichia Coli. J. Biol. Chem.,* **1963**, *238*, 3934-3944.
[PMID: 14086727]

[416] Peters-Wendisch, P.; Stolz, M.; Etterich, H.; Kennerknecht, N.; Sahm, H.; Eggeling, L. Metabolic engineering of *Corynebacterium glutamicum* for L-serine production. *Appl. Environ. Microbiol.,* **2005**, *71*(11), 7139-7144.
[http://dx.doi.org/10.1128/AEM.71.11.7139-7144.2005] [PMID: 16269752]

[417] Kredich, N. M. Biosynthesis of cysteine *Escherichia coli and Salmonella,* **1996**, 514-527.

[418] Haitani, Y.; Awano, N.; Yamazaki, M.; Wada, M.; Nakamori, S.; Takagi, H. Functional analysis of L-serine O-acetyltransferase from *Corynebacterium glutamicum. FEMS Microbiol. Lett.,* **2006**, *255*(1), 156-163.
[http://dx.doi.org/10.1111/j.1574-6968.2005.00068.x] [PMID: 16436075]

[419] Takagi, H.; Kobayashi, C.; Kobayashi, S.; Nakamori, S. PCR random mutagenesis into *Escherichia coli* serine acetyltransferase: isolation of the mutant enzymes that cause overproduction of L-cysteine and L-cystine due to the desensitization to feedback inhibition. *FEBS Lett.,* **1999**, *452*(3), 323-327.
[http://dx.doi.org/10.1016/S0014-5793(99)00679-1] [PMID: 10386615]

[420] Wirtz, M.; Hell, R. Production of cysteine for bacterial and plant biotechnology: application of cysteine feedback-insensitive isoforms of serine acetyltransferase. *Amino Acids,* **2003**, *24*(1-2), 195-203.
[PMID: 12624753]

[421] Nakamori, S.; Kobayashi, S.I.; Kobayashi, C.; Takagi, H. Overproduction of L-cysteine and L-cystine by *Escherichia coli* strains with a genetically altered serine acetyltransferase. *Appl. Environ. Microbiol.,* **1998**, *64*(5), 1607-1611.
[PMID: 9572924]

[422] Takagi, H.; Awano, N.; Kobayashi, S.; Noji, M.; Saito, K.; Nakamori, S. Overproduction of L-cysteine and L-cystine by expression of genes for feedback inhibition-insensitive serine acetyltransferase from *Arabidopsis thaliana* in *Escherichia coli. FEMS Microbiol. Lett.,* **1999**, *179*(2), 453-459.
[PMID: 10518750]

[423] Awano, N.; Wada, M.; Kohdoh, A.; Oikawa, T.; Takagi, H.; Nakamori, S. Effect of cysteine desulfhydrase gene disruption on L-cysteine overproduction in *Escherichia coli. Appl. Microbiol. Biotechnol.,* **2003**, *62*(2-3), 239-243.
[http://dx.doi.org/10.1007/s00253-003-1262-2] [PMID: 12883870]

[424] Awano, N.; Wada, M.; Mori, H.; Nakamori, S.; Takagi, H. Identification and functional analysis of *Escherichia coli* cysteine desulfhydrases. *Appl. Environ. Microbiol.,* **2005**, *71*(7), 4149-4152.
[http://dx.doi.org/10.1128/AEM.71.7.4149-4152.2005] [PMID: 16000837]

[425] Wada, M.; Awano, N.; Haisa, K.; Takagi, H.; Nakamori, S. Purification, characterization and identification of cysteine desulfhydrase of *Corynebacterium glutamicum*, and its relationship to cysteine production. *FEMS Microbiol. Lett.,* **2002**, *217*(1), 103-107.

[http://dx.doi.org/10.1111/j.1574-6968.2002.tb11462.x] [PMID: 12445652]

[426] Franke, I.; Resch, A.; Dassler, T.; Maier, T.; Böck, A. YfiK from *Escherichia coli* promotes export of O-acetylserine and cysteine. *J. Bacteriol.,* **2003**, *185*(4), 1161-1166. [http://dx.doi.org/10.1128/JB.185.4.1161-1166.2003] [PMID: 12562784]

[427] Pittman, M.S.; Corker, H.; Wu, G.; Binet, M.B.; Moir, A.J.; Poole, R.K. Cysteine is exported from the *Escherichia coli* cytoplasm by CydDC, an ATP-binding cassette-type transporter required for cytochrome assembly. *J. Biol. Chem.,* **2002**, *277*(51), 49841-49849. [http://dx.doi.org/10.1074/jbc.M205615200] [PMID: 12393891]

[428] Maier, T.H. Semisynthetic production of unnatural L-alpha-amino acids by metabolic engineering of the cysteine-biosynthetic pathway. *Nat. Biotechnol.,* **2003**, *21*(4), 422-427. [http://dx.doi.org/10.1038/nbt807] [PMID: 12640465]

[429] Moran, E.T., Jr Response of broiler strains differing in body fat to inadequate methionine: live performance and processing yields. *Poult. Sci.,* **1994**, *73*(7), 1116-1126. [http://dx.doi.org/10.3382/ps.0731116] [PMID: 7937473]

[430] Wagner, T. Production of L-methionine from D,L-5-(2-methylthioethyl)hydantoin by resting cells of a new mutant strain of Arthrobacter species DSM 7330. *J. Biotechnol.,* **1996**, *46*, 63-68. [http://dx.doi.org/10.1016/0168-1656(95)00193-X]

[431] Katz, R.S.; Baker, D.H. Efficacy of D-, L- and DL-methionine for growth of chicks fed crystalline amino acid diets. *Poult. Sci.,* **1975**, *54*(5), 1667-1674. [http://dx.doi.org/10.3382/ps.0541667] [PMID: 1187522]

[432] Mondal, S.; Das, Y.B.; Chatterjee, S.P. Methionine production by microorganisms. *Folia Microbiol. (Praha),* **1996**, *41*(6), 465-472. [http://dx.doi.org/10.1007/BF02814659] [PMID: 9131801]

[433] Krömer, J.O.; Wittmann, C.; Schröder, H.; Heinzle, E. Metabolic pathway analysis for rational design of L-methionine production by Escherichia coli and Corynebacterium glutamicum. *Metab. Eng.,* **2006**, *8*(4), 353-369. [http://dx.doi.org/10.1016/j.ymben.2006.02.001] [PMID: 16621639]

[434] Dischert, W.; Vasseur, P.; Boisart, C.; Figge, R.M. Increasing NADPH availability for methionine production. *E,* **2013**, *A1*, 2633037.

[435] Zelder, O.; Schröder, H.; Klopprogge, C.; Herold, A. Method of Producing Methionine in Corynebacteria by Over-Expressing Enzymes of the Pentose Phosphate Pathway 2012.20120288901,

[436] Zelder, O.; Haefner, S.; Herold, A.; Klopprogge, C.; Schroder, H.; Yocum, R.; Williams, M. use of dimethyl disulfide for methionine production in microorganisms 2010.EP1907558,

[437] Figge, R.; Lux, F.; Raynaud, C.; Chateau, M.; Soucaille, P. Methods for producing methionine by culturing a microorganism modified to enhance production of cysteine 2013.US8389250 B2,

[438] Trötschel, C.; Krämer, R.; Burkovski, A.; Bathe, B. method for fermentative preparation of methionine by use of recombinant coryneform bacteria. 2010.EP1745138,

[439] Bathe, B.; Schischka, N.; Pfefferle, W. Process for the production of L-methionine using coryneform bacteria. *E,* **2012**, *B1*, 2218789.

[440] Figge, R.M. Methionine Biosynthesis in *Escherichia coli* and *Corynebacterium glutamicum*. In: *Amino Acid Biosynthesis – Pathways, Regulation and Metabolic Engineering*; Wendisch, V.F., Ed.; Springer: Heidelberg, Germany, **2007**; pp. 163-193.
[http://dx.doi.org/10.1007/7171_2006_059]

[441] Lee, H.S.; Hwang, B.J. Methionine biosynthesis and its regulation in *Corynebacterium glutamicum*: parallel pathways of transsulfuration and direct sulfhydrylation. *Appl. Microbiol. Biotechnol.,* **2003**, *62*(5-6), 459-467.
[http://dx.doi.org/10.1007/s00253-003-1306-7] [PMID: 12845493]

[442] Kumar, D.; Gomes, J. Methionine production by fermentation. *Biotechnol. Adv.,* **2005**, *23*(1), 41-61.
[http://dx.doi.org/10.1016/j.biotechadv.2004.08.005] [PMID: 15610965]

[443] Nakamori, S.; Kobayashi, S.; Nishimura, T.; Takagi, H. Mechanism of L-methionine overproduction by Escherichia coli: the replacement of Ser-54 by Asn in the MetJ protein causes the derepression of L-methionine biosynthetic enzymes. *Appl. Microbiol. Biotechnol.,* **1999**, *52*(2), 179-185.
[http://dx.doi.org/10.1007/s002530051506] [PMID: 10499257]

[444] Bestel-Corre, G.; Chateau, M.; Figge, R.; Raynaud, C.; Soucaille, P. Recombinant enzyme with altered feedback sensitivity 2009.US20090029424,

[445] Maier, D.T. Process for the fermentative production of L-methionine 2004. EP 1445310 A1,

[446] Park, S.D.; Lee, J.Y.; Sim, S.Y.; Kim, Y.; Lee, H.S. Characteristics of methionine production by an engineered *Corynebacterium glutamicum* strain. *Metab. Eng.,* **2007**, *9*(4), 327-336.
[http://dx.doi.org/10.1016/j.ymben.2007.05.001] [PMID: 17604670]

[447] Hädrich, D. B. A. Nucleotide sequences which code for the metY gene. 2009.US 20020110878 A1,

[448] Tosa, T.; Mori, T.; Fuse, N.; Chibata, I. Studies on continuous enzyme reactions. 3. Enzymatic properties of the DEAE-cellulose-aminoacylase complex. *Enzymologia,* **1967**, *32*(3), 153-168.
[PMID: 5622071]

[449] Cheng, Y.; Zhou, Y.; Yang, L.; Zhang, C.; Xu, Q.; Xie, X.; Chen, N. Modification of histidine biosynthesis pathway genes and the impact on production of L-histidine in Corynebacterium glutamicum. *Biotechnol. Lett.,* **2013**, *35*(5), 735-741.
[http://dx.doi.org/10.1007/s10529-013-1138-1] [PMID: 23355034]

[450] Eggeling, L.; Sahm, H. Amino acid production: principles of metabolic engineering. In: *Metabolic Engineering*; Lee, S.Y.; Papoutsakis, E.T., Eds.; Marcel Dekker Inc: New York, USA, **1999**; pp. 153-176.

[451] Shimizu, H.; Shimizu, N.; Shioya, S. Roles of Glucose and Acetate as Carbon Sources in L-Histidine Production with *Brevibacterium flavum* FERM1564 Revealed by Metabolic Flux Analysis. *Biotechnol. Bioprocess Eng.,* **2002**, *7*, 171-177.
[http://dx.doi.org/10.1007/BF02932915]

[452] Mizukami, T.; Hamu, A.; Ikeda, M.; Oka, T.; Katsumata, R. Cloning of the ATP phosphoribosyl transferase gene of *Corynebacterium glutamicum* and application of the gene to L-histidine production. *Biosci. Biotechnol. Biochem.,* **1994**, *58*(4), 635-638.
[http://dx.doi.org/10.1271/bbb.58.635] [PMID: 7764856]

[453] Araki, K.; Nakayama, K. Studies on histidine fermentation: Part I. L-histidine production by histidine analog-resistant mutants from several bacteria. *Agric. Biol. Chem.,* **1971**, *35*, 2081-2088. [http://dx.doi.org/10.1271/bbb1961.35.2081]

[454] Kisumi, M.; Nakanishi, N.; Takagi, T.; Chibata, I. L-Histidine production by histidase-less regulatory mutants of *Serratia marcescens* constructed by transduction. *Appl. Environ. Microbiol.,* **1977**, *34*(5), 465-472. [PMID: 337893]

[455] Araki, K.; Nakayama, K. Histidine production by *Corynebacterium glutamicum* mutants, multiresistant to analogs of histidine, tryptophan, purine and pyrimidine. *Agric. Biol. Chem.,* **1974**, *38*, 2209-2218. [http://dx.doi.org/10.1271/bbb1961.38.2209]

[456] Alifano, P.; Fani, R.; Liò, P.; Lazcano, A.; Bazzicalupo, M.; Carlomagno, M.S.; Bruni, C.B. Histidine biosynthetic pathway and genes: structure, regulation, and evolution. *Microbiol. Rev.,* **1996**, *60*(1), 44-69. [PMID: 8852895]

[457] Kulis-Horn, R.K.; Persicke, M.; Kalinowski, J. Histidine biosynthesis, its regulation and biotechnological application in *Corynebacterium glutamicum. Microb. Biotechnol.,* **2013**. [PMID: 23617600]

[458] Zhang, Y.; Shang, X.; Deng, A.; Chai, X.; Lai, S.; Zhang, G.; Wen, T. Genetic and biochemical characterization of *Corynebacterium glutamicum* ATP phosphoribosyltransferase and its three mutants resistant to feedback inhibition by histidine. *Biochimie,* **2012**, *94*(3), 829-838. [http://dx.doi.org/10.1016/j.biochi.2011.11.015] [PMID: 22172596]

[459] Klyachko, E. V.; Shakulov, R. S.; Kozlov, Y. I. Mutant phosphoribosylpyrophosphate synthetase and method for producing L-histidine. 2008. US 20050176033 A1 ,

[460] Klyachko, E. V.; Shakulov, R. S.; Kozlov, Y. I. Method for producing L-histidine using bacteria of enterobacteriaceae family. 2008.US 20050048631 A1,

[461] Klyachko, E. V.; Shakulov, R. S.; Kozlov, Y. I. Method for producing L-histidine using Enterobacteriaceae bacteria which has an enhanced purH gene produced 2010.US 7785860 B2,

[462] Chim-anage, P.; Shioya, S.; Suga, K. Optimum conditions for histidine production by fed-batch culture of *Brevibacterium flavum. J. Ferment. Bioeng.,* **1990**, *70*, 386-391. [http://dx.doi.org/10.1016/0922-338X(90)90119-H]

[463] Araki, K.; Nakayama, K. Studies on Histidine Fermentation. 1. L-Histidine Production by Histidine Analog-Resistant Mutants from Several Bacteria. *Agric. Biol. Chem.,* **1971**, *35*(13), 2081. [http://dx.doi.org/10.1271/bbb1961.35.2081]

[464] Eggeling, L. L-Serine and Glycine. In: *Amino Acid Biosynthesis~ Pathways, Regulation and Metabolic Engineering*; Springer, **2007**; pp. 259-272. [http://dx.doi.org/10.1007/7171_2006_068]

[465] Peters-Wendisch, P.; Netzer, R.; Eggeling, L.; Sahm, H. 3-Phosphoglycerate dehydrogenase from *Corynebacterium glutamicum*: the C-terminal domain is not essential for activity but is required for inhibition by L-serine. *Appl. Microbiol. Biotechnol.,* **2002**, *60*(4), 437-441.

[http://dx.doi.org/10.1007/s00253-002-1161-y] [PMID: 12466884]

[466] Li, Y.; Chen, G.K.; Tong, X.W.; Zhang, H.T.; Liu, X.G.; Liu, Y.H.; Lu, F.P. Construction of Escherichia coli strains producing L-serine from glucose. *Biotechnol. Lett.,* **2012**, *34*(8), 1525-1530. [http://dx.doi.org/10.1007/s10529-012-0937-0] [PMID: 22547037]

[467] Stolz, M.; Peters-Wendisch, P.; Etterich, H.; Gerharz, T.; Faurie, R.; Sahm, H.; Fersterra, H.; Eggeling, L. Reduced folate supply as a key to enhanced L-serine production by Corynebacterium glutamicum. *Appl. Environ. Microbiol.,* **2007**, *73*(3), 750-755. [http://dx.doi.org/10.1128/AEM.02208-06] [PMID: 17142381]

[468] Kisumi, M.; Ashikaga, Y.; Chibata, I. Studies on the Fermentative Preparation of L-Aspartic Acid from Fumaric Acid. *Bulletin of the Agricultural Chemical Society of Japan,* **1960**, *24*(3), 296-305. [http://dx.doi.org/10.1271/bbb1924.24.296]

[469] Fusee, M.C.; Swann, W.E.; Calton, G.J. Immobilization of *Escherichia coli* Cells Containing Aspartase Activity with Polyurethane and Its Application for l-Aspartic Acid Production. *Appl. Environ. Microbiol.,* **1981**, *42*(4), 672-676. [PMID: 16345865]

[470] Umemura, I.; Takamatsu, S.; Sato, T.; Tosa, T.; Chibata, I. Improvement of Production of L-Aspartic Acid Using Immobilized Microbial-Cells. *Appl. Microbiol. Biotechnol.,* **1984**, *20*(5), 291-295. [http://dx.doi.org/10.1007/BF00270588]

[471] Suzuki, Y.; Yasui, T.; Mino, Y.; Abe, S. Production of L-Aspartic Acid from Fumaric-Acid by a Fumaric-Acid Assimilating Obligate Thermophile, Bacillus Stearothermophilus-Kp1041 *European Journal of Appl.ied Microbiology and Biotechnology,* **1980**, *11*(1), 23-27.

[472] Chibata, I.; Kakimoto, T.; Kato, J. Enzymatic production of L-alanine by *Pseudomonas dacunhae. Appl. Microbiol.,* **1965**, *13*(5), 638-645. [PMID: 5867644]

[473] Shibatani, T.; Kakimoto, T.; Chibata, I. Stimulation of L-asparate beta-decarboxylase formation by L-glutamate in *Pseudomonas dacunhae* and Improved production of L-alanine. *Appl. Environ. Microbiol.,* **1979**, *38*(3), 359-364. [PMID: 533269]

[474] Takamatsu, S.; Umemura, I.; Yamamoto, K.; Sato, T.; Tosa, T.; Chibata, I. Production of L-Alanine from Ammonium Fumarate Using 2 Immobilized Microorganisms - Elimination of Side Reactions. *European Journal of Appl.ied Microbiology and Biotechnology,* **1982**, *15*(3), 147-152.

[475] Chibata, I.; Tosa, T.; Sato, T. Continuous Production of L-Aspartic Acid - Improvement of Productivity by Both Development of Immobilization Method and Construction of New *Escherichia coli* Strain. *Appl. Biochem. Biotechnol.,* **1986**, *13*(3), 231-240. [http://dx.doi.org/10.1007/BF02798461]

[476] Jojima, T.; Fujii, M.; Mori, E.; Inui, M.; Yukawa, H. Engineering of sugar metabolism of Corynebacterium glutamicum for production of amino acid L-alanine under oxygen deprivation. *Appl. Microbiol. Biotechnol.,* **2010**, *87*(1), 159-165. [http://dx.doi.org/10.1007/s00253-010-2493-7] [PMID: 20217078]

[477] Chambers, I.; Frampton, J.; Goldfarb, P.; Affara, N.; McBain, W.; Harrison, P.R. The structure of the

mouse glutathione peroxidase gene: the selenocysteine in the active site is encoded by the 'termination' codon, TGA. *EMBO J.,* **1986**, *5*(6), 1221-1227. [PMID: 3015592]

[478] Zinoni, F.; Birkmann, A.; Stadtman, T.C.; Böck, A. Nucleotide sequence and expression of the selenocysteine-containing polypeptide of formate dehydrogenase (formate-hydrogen-lyase-linked) from *Escherichia coli. Proc. Natl. Acad. Sci. USA,* **1986**, *83*(13), 4650-4654. [http://dx.doi.org/10.1073/pnas.83.13.4650] [PMID: 2941757]

[479] Hao, B.; Gong, W.; Ferguson, T.K.; James, C.M.; Krzycki, J.A.; Chan, M.K. A new UAG-encoded residue in the structure of a methanogen methyltransferase. *Science,* **2002**, *296*(5572), 1462-1466. [http://dx.doi.org/10.1126/science.1069556] [PMID: 12029132]

[480] Srinivasan, G.; James, C.M.; Krzycki, J.A. Pyrrolysine encoded by UAG in Archaea: charging of a UAG-decoding specialized tRNA. *Science,* **2002**, *296*(5572), 1459-1462. [http://dx.doi.org/10.1126/science.1069588] [PMID: 12029131]

[481] Lobanov, A.V.; Kryukov, G.V.; Hatfield, D.L.; Gladyshev, V.N. Is there a twenty third amino acid in the genetic code? *Trends Genet.,* **2006**, *22*(7), 357-360. [http://dx.doi.org/10.1016/j.tig.2006.05.002] [PMID: 16713651]

[482] Böck, A.; Thanbichler, M.; Rother, M.; Resch, A. *"Selenocysteine," in Aminoacyl-tRNA Synthetases*; Ibba, M.; Francklyn, C.S.; Cusack, S., Eds.; Landes Bioscience: Georgetown, Wash, USA, **2005**, pp. 320-327.

[483] Fischer, N.; Paleskava, A.; Gromadski, K.B.; Konevega, A.L.; Wahl, M.C.; Stark, H.; Rodnina, M.V. Towards understanding selenocysteine incorporation into bacterial proteins. *Biol. Chem.,* **2007**, *388*(10), 1061-1067. [http://dx.doi.org/10.1515/BC.2007.108] [PMID: 17937620]

[484] Rother, M.; Krzycki, J. A. Selenocysteine, pyrrolysine, and the unique energy metabolism of methanogenic archaea *Archaea,* **2010**, *2010* [http://dx.doi.org/10.1155/2010/453642]

[485] Longstaff, D.G.; Larue, R.C.; Faust, J.E.; Mahapatra, A.; Zhang, L.; Green-Church, K.B.; Krzycki, J.A. A natural genetic code expansion cassette enables transmissible biosynthesis and genetic encoding of pyrrolysine. *Proc. Natl. Acad. Sci. USA,* **2007**, *104*(3), 1021-1026. [http://dx.doi.org/10.1073/pnas.0610294104] [PMID: 17204561]

[486] Gaston, M.A.; Jiang, R.; Krzycki, J.A. Functional context, biosynthesis, and genetic encoding of pyrrolysine. *Curr. Opin. Microbiol.,* **2011**, *14*(3), 342-349. [http://dx.doi.org/10.1016/j.mib.2011.04.001] [PMID: 21550296]

[487] Gaston, M.A.; Zhang, L.; Green-Church, K.B.; Krzycki, J.A. The complete biosynthesis of the genetically encoded amino acid pyrrolysine from lysine. *Nature,* **2011**, *471*(7340), 647-650. [http://dx.doi.org/10.1038/nature09918] [PMID: 21455182]

[488] Mukai, T.; Kobayashi, T.; Hino, N.; Yanagisawa, T.; Sakamoto, K.; Yokoyama, S. Adding l-lysine derivatives to the genetic code of mammalian cells with engineered pyrrolysyl-tRNA synthetases. *Biochem. Biophys. Res. Commun.,* **2008**, *371*(4), 818-822. [http://dx.doi.org/10.1016/j.bbrc.2008.04.164] [PMID: 18471995]

[489] Shi, L.; Ying, G.; Tang, Z.; Chen, J.; Xiong, W.; Wang, H. Continuous enzymatic production of 5′-nucleotides using free nuclease P1 in ultrafiltration membrane reactor. *J. Membr. Sci.,* **2009**, *345*, 217-222.
[http://dx.doi.org/10.1016/j.memsci.2009.09.001]

[490] Yoshikawa, M.; Kato, T.; Takenishi, T. Studies of phosphorylation. III. Selective phosphorylation of unprotected nucleosides. *Bull. Chem. Soc. Jpn.,* **1969**, *42*, 3505-3508.
[http://dx.doi.org/10.1246/bcsj.42.3505]

[491] Mori, H.; Iida, A.; Fujio, T.; Teshiba, S. A novel process of inosine 5′-monophosphate production using overexpressed guanosine/inosine kinase. *Appl. Microbiol. Biotechnol.,* **1997**, *48*(6), 693-698.
[http://dx.doi.org/10.1007/s002530051117] [PMID: 9457797]

[492] Mihara, Y.; Utagawa, T.; Yamada, H.; Asano, Y. Phosphorylation of nucleosides by the mutated acid phosphatase from *Morganella morganii. Appl. Environ. Microbiol.,* **2000**, *66*(7), 2811-2816.
[http://dx.doi.org/10.1128/AEM.66.7.2811-2816.2000] [PMID: 10877772]

[493] Rittmann, D.; Sorger-Herrmann, U.; Wendisch, V.F. Phosphate starvation-inducible gene *ushA* encodes a 5′ nucleotidase required for growth of *Corynebacterium glutamicum* on media with nucleotides as the phosphorus source. *Appl. Environ. Microbiol.,* **2005**, *71*(8), 4339-4344.
[http://dx.doi.org/10.1128/AEM.71.8.4339-4344.2005] [PMID: 16085822]

[494] Chambert, R.; Pereira, Y.; Petit-Glatron, M.F. Purification and characterization of YfkN, a trifunctional nucleotide phosphoesterase secreted by *Bacillus subtilis. J. Biochem.,* **2003**, *134*(5), 655-660.
[http://dx.doi.org/10.1093/jb/mvg189] [PMID: 14688230]

[495] Ledesma-Amaro, R.; Jiménez, A.; Santos, M.A.; Revuelta, J.L. Biotechnological production of feed nucleotides by microbial strain improvement. *Process Biochem.,* **2013**, *48*(9), 1263-1270.
[http://dx.doi.org/10.1016/j.procbio.2013.06.025]

[496] Kotani, Y.; Yamaguchi, K.; Kato, F.; Furuya, A. Inosine accumulation by mutants of *Brevibacterium (Corynebacterium) ammoniagenes*: strain improvement and culture conditions. *Agric. Biol. Chem.,* **1978**, *42*, 399-405.
[http://dx.doi.org/10.1271/bbb1961.42.399]

[497] Fujio, T.; Nishi, T.; Ito, S.; Maruyama, A. High level expression of XMP aminase in *Escherichia coli* and its application for the industrial production of 5′-guanylic acid. *Biosci. Biotechnol. Biochem.,* **1997**, *61*(5), 840-845.
[http://dx.doi.org/10.1271/bbb.61.840] [PMID: 9178561]

[498] Peifer, S.; Barduhn, T.; Zimmet, S.; Volmer, D.A.; Heinzle, E.; Schneider, K. Metabolic engineering of the purine biosynthetic pathway in *Corynebacterium glutamicum* results in increased intracellular pool sizes of IMP and hypoxanthine. *Microb. Cell Fact.,* **2012**, *11*, 138.
[http://dx.doi.org/10.1186/1475-2859-11-138] [PMID: 23092390]

[499] Asahara, T.; Mori, Y.; Zakataeva, N.P.; Livshits, V.A.; Yoshida, K.; Matsuno, K. Accumulation of gene-targeted *Bacillus subtilis* mutations that enhance fermentative inosine production. *Appl. Microbiol. Biotechnol.,* **2010**, *87*(6), 2195-2207.
[http://dx.doi.org/10.1007/s00253-010-2646-8] [PMID: 20524113]

[500] Matsui, H.; Kawasaki, H.; Shimaoka, M.; Kurahashi, O. Investigation of various genotype characteristics for inosine accumulation in *Escherichia coli* W3110. *Biosci. Biotechnol. Biochem.,* **2001**, *65*(3), 570-578.
[http://dx.doi.org/10.1271/bbb.65.570] [PMID: 11330670]

[501] Shimaoka, M.; Kawasaki, H.; Takenaka, Y.; Kurahashi, O.; Matsui, H. Effects of *edd* and *pgi* disruptions on inosine accumulation in *Escherichia coli. Biosci. Biotechnol. Biochem.,* **2005**, *69*(7), 1248-1255.
[http://dx.doi.org/10.1271/bbb.69.1248] [PMID: 16041126]

[502] Kamada, N.; Yasuhara, A.; Takano, Y.; Nakano, T.; Ikeda, M. Effect of transketolase modifications on carbon flow to the purine-nucleotide pathway in *Corynebacterium ammoniagenes. Appl. Microbiol. Biotechnol.,* **2001**, *56*(5-6), 710-717.
[http://dx.doi.org/10.1007/s002530100738] [PMID: 11601619]

[503] Kutukova, E. A.; Livshits, V. A.; Zakataeva, N. P. A method for producing purine nucleosides and nucleotides by fermentation using a bacterium belonging to the genus *escherichia or bacillus* 2007.WO 2008084629 A1,

[504] Johansen, L.E.; Nygaard, P.; Lassen, C.; Agersø, Y.; Saxild, H.H. Definition of a second *Bacillus subtilis pur* regulon comprising the *pur* and *xpt-pbuX* operons plus *pbuG, nupG (yxjA),* and *pbuE (ydhL). J. Bacteriol.,* **2003**, *185*(17), 5200-5209.
[http://dx.doi.org/10.1128/JB.185.17.5200-5209.2003] [PMID: 12923093]

[505] Zakataeva, N.P.; Gronskiy, S.V.; Sheremet, A.S.; Kutukova, E.A.; Novikova, A.E.; Livshits, V.A. A new function for the *Bacillus* PbuE purine base efflux pump: efflux of purine nucleosides. *Res. Microbiol.,* **2007**, *158*(8-9), 659-665.
[http://dx.doi.org/10.1016/j.resmic.2007.08.003] [PMID: 17935948]

CHAPTER 4

Biotechnological Production of Organic Acids

Hélia Harumi Sato* and **Haroldo Yukio Kawaguti**

Department of Food Science, School of Food Engineering, University of Campinas, Campinas, Brazil

Abstract: Organic acids are the intermediates or products of many metabolic pathways, such as the Krebs cycle and lactic and acetic fermentations, and are important contributors to the taste and flavor of many fruits, vegetables and also fermented foods. Organic acids are widely applied in the food, pharmaceutical and chemical industries and can be produced by microbial fermentation, chemical synthesis or obtained using enzymes. The demand for organic acids is growing continuously, and numerous efforts and research investments have been made to increase the yield and productivity through the selection of new strains, by obtaining genetically modified microorganisms, optimizing fermentation processes and improving the recovery and purification processes. This chapter focuses on the production and application of citric, acetic, lactic, fumaric, malic, gluconic and ascorbic acids.

Keywords: Acetic acid, *Acetobacter sp*, Acidulants, Ascorbic acid, *Aspergillus niger*, *Aureobasidium pullulans*, *Brevibacterium* sp , Citric acid, Fumaric acid, Gluconic acid, *Gluconobacter* sp, *Ketogulonigenium vulgare*, Lactic acid, *Lactobacillus* sp, Malic acid, Organic acids, *Penicillium* sp, *Rhizopus* sp, *Saccharomyces cerevisiae*, *Yarrowia lipolytica*.

INTRODUCTION

The market for organic acids increased from US $2.5 billion in 2008 to US $4.0 billion in 2013 [1]. The primary organic acids used in food applications are citric, acetic, lactic, malic, fumaric and gluconic acids. Citric, malic and fumaric acids

* **Address correspondence to Hélia Harumi Sato:** School of Food Engineering, State University of Campinas, UNICAMP, Campinas, Estado de São Paulo, Brazil; CEP 13086 862; E-mail: heliah@fea.unicamp.br.

Juliano Lemos Bicas, Mário Roberto Maróstica, Glaucia Maria Pastore (Eds.)

are naturally found in fruits and vegetables and have been used as acidulants in beverages, canned fruit, juices, powdered drinks, gelatin powder, and others. Some fruits are known for the presence of high concentrations of certain organic acids. With some exceptions, citrus fruit juices contain high concentrations of citric acid and smaller amounts of malic acid. Malic acid is the main organic acid of apples, while grapes have a high concentration of tartaric acid, and the star fruit (*Averroha carambola*) has a high oxalic acid content.

Organic acids can be applied in foods, pharmaceuticals, detergents, cosmetics and other products, such as acidulants, chelating agents, polymers and chemical compound precursors (Table **1)** [2]. Organic acids are added to foods to lower the pH and adjust the acid taste of products such as fruit juices, jellies, candies, gelatin powders, carbonated drinks, powdered drinks, yogurts, desserts, fruits, canned vegetables, pickles, olives, cheeses and other foods, as well as to control microbial growth. Acidulants and their salts aid in food preservation, enabling a reduction in the time used in the sterilization processes and preventing the germination of microbial spores at low pH values. Organic acids can act in synergy with antioxidants, preventing rancidity and other reactions by forming complexes with copper and iron ions. They are also used as buffer systems in foods to modify the consistency and melting point of candies and cheeses. Trisodium citrate is used as a stabilizer in cheese, ice cream and milk, and citric acid, in combination with ascorbic acid and other antioxidants, for the inhibition of enzymatic browning in juices and fruit pulps [3].

Citric acid is the most important acidulant for the food and beverage industries and is widely used due to its easy assimilation, palatability, high solubility, low toxicity and low cost. Malic acid is also found in fruits and vegetables and is mainly used in beverages, candies and foods. Fumaric acid is used in the food industry and is an intermediate in the production of malic acid and L-aspartic acid, the latter being used to obtain aspartame. The use of organic acids depends on their ease of handling and storage, degree of hygroscopicity and solubility, on their properties, such as flavor modification, compatibility with other acids and food components, and their ability to act as Ca^{2+} chelators.

Cost is the most important factor in the production of organic acids due to the competition between industries. Currently citric acid is primarily produced by the submerged fermentation of media containing sucrose, molasses, glucose or hydrolyzed starch using *Aspergillus niger* strains [7, 11]. The fermentation process should be optimized by selecting a microbial strain showing a high production of the metabolites of interest by using optimized medium cultivation and optimal conditions of temperature, pH, agitation and aeration to obtain high yields (g acid/g glucose), productivity (g acid/L/h) and final acid concen-trations (g/L).

Table 1. Production and applications of organic acids.

Organic Acid	Structural formula/ Estimated price	Annual production metric tons/year	Production method	Applications
Citric Acid $C_6H_8O_7$	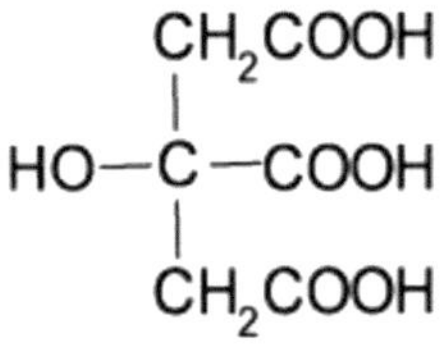 US $1.0 - 10.0/Kg	1, 600, 000 [2]	Fermentation	Food; beverages; pharmaceutical industry; pH stabilizer in cosmetics; chemical and metallurgical industries; metal cleaning [4, 5, 6]
Acetic Acid $C_2H_4O_2$	**H_3C — COOH** **US $0.5 - 0.6/Kg**	190, 000 [6]	Fermentation	Food industry (vinegar).
		7, 000, 000 [6]	Chemical synthesis	Vinylacetate for polymers, ethyl acetate [6]
Lactic Acid $C_3H_6O_3$	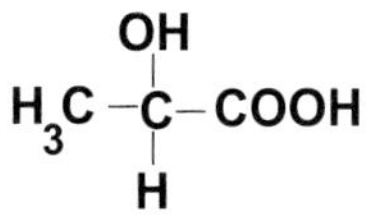 US $1.0 - 1.3/Kg	250, 000 [7]	Fermentation	Food; beverages; biodegradable plastic PLA [6, 8]
Fumaric acid $C_4H_4O_4$	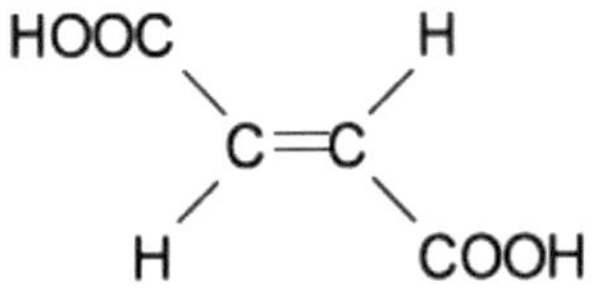 US $1.1 - 1.6/Kg	90, 000 [2]	Chemical synthesis	Food acidulant; medicine; raw material for polyester resins [2, 6]

(Table 1) contd.....

Organic Acid	Structural formula/ Estimated price	Annual production metric tons/year	Production method	Applications
Gluconic Acid $C_6H_{12}O_7$	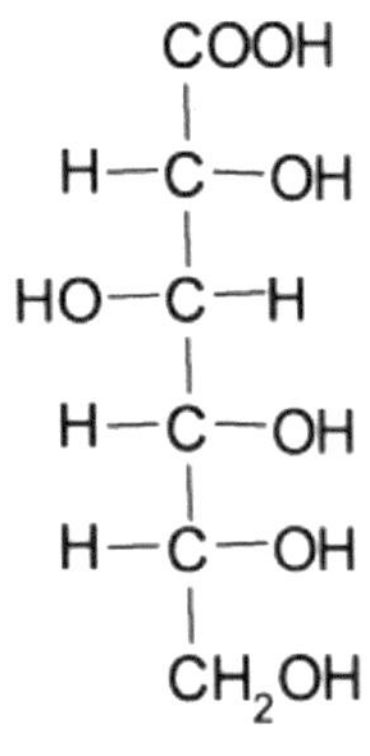 US $1.0 – 3.0/Kg	87, 000 [6]	Fermentation	Food; feed; pharmaceutical; chelating agent [6]
Malic Acid $C_4H_6O_5$	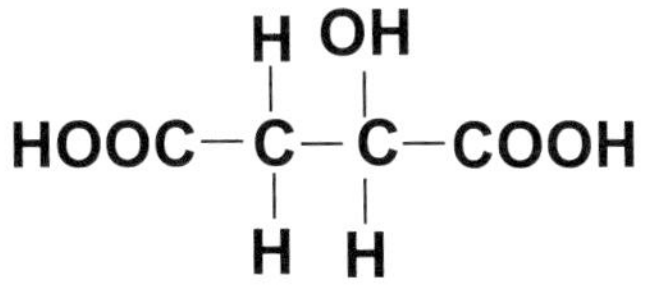 US $1.5 – 2.3/Kg	200, 000 [2]	Chemical synthesis	Food; beverages; polymer intermediate [2, 6]
Ascorbic Acid $C_6H_8O_6$	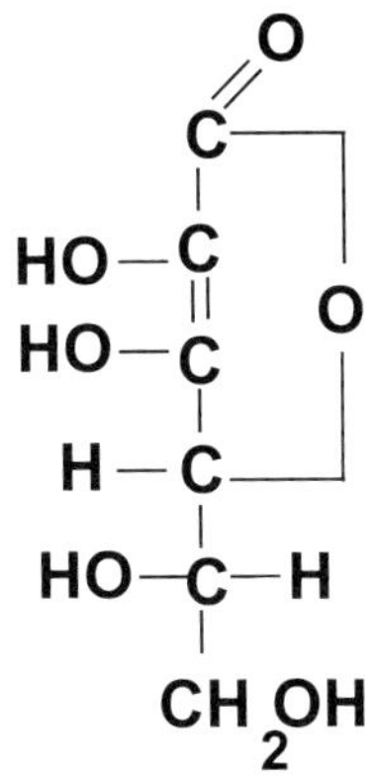 US $2.5 – 3.5/Kg	110, 000 [9]	Chemical synthesis and bioconversion step	Pharmaceutical industry; food industry; animal feed [9, 10]

The use of carbon sources, nitrogen sources and low cost, abundant substrates, such as molasses, hydrolyzed starch, corn steep liquor and waste matters from the food industries, is essential to optimize the fermentation processes.

CITRIC ACID

Citric acid is a Krebs cycle intermediate and is found in plants, animals and microorganisms (Fig. **1**). Karadeniz [12] observed that citric acid was the main organic acid in citrus fruit juices, except for Clementine tangerines (*Citrus reticulata*), which showed a higher malic acid content. Malic acid is usually the second most abundant acid in citrus fruit juices. Sweet orange (*Citrus sinensis*) varieties have 13.28 g/L-citric acid, 7.79 g/L of L-malic acid and 0.373 g/L of L-fumaric acid, while lemon varieties have 55.11 g/L of L-citric acid and 6 g/L of L-malic acid. Grapefruit, lemon and sour oranges do not have fumaric acid. Cámara and co-workers [13] analyzed the organic acids of fresh pineapple juice and found 0.529 g citric acid/100 mL and 0.243 g of malic acid/100 mL.

Citric acid is a white crystal with an acid taste and little odor. It can be produced in anhydrous ($C_6H_8O_7$; 192.13 MM) or hydrated ($C_6H_8O_7.H_2O$; 210.14 MM) forms, with solubility values at 20°C of 1,330 g/L and 1,630 g/L, respectively.

Citric acid is primarily used in the food (70%), pharmaceutical (12%) and other industries (18%). Citric acid and its salts are used in many industrial applications as acidulants, chelating agents and derivatization agents.

Citric acid shows high solubility, is biodegradable and, due to its ease of assimilation and its low toxicity, is widely used in the food and drink, pharmaceutical, chemical, cleaning product, cosmetics, detergent and fertilizer industries. In foods, citric acid is used in juices, fruit jellies, jams, candies, canned fruit and vegetables, tomato sauces, and dry soups and gravies, amongst others [2, 4, 7, 11].

Citric acid is used in combination with ascorbic acid and other antioxidants to inhibit the enzymatic browning of fruits and vegetables and prevent the oxidation reactions catalyzed by trace metals, aiding in color preservation. Sodium citrate is used in dairy products as a stabilizer of the UHT processes in milk and cream cheese. Sodium citrate can be used in the pharmaceutical industry as an anticoagulant and stabilizing agent and to adjust the pH of detergents and cosmetics, such as shampoos, to chelate metal ions and prevent discoloration [2, 5, 14].

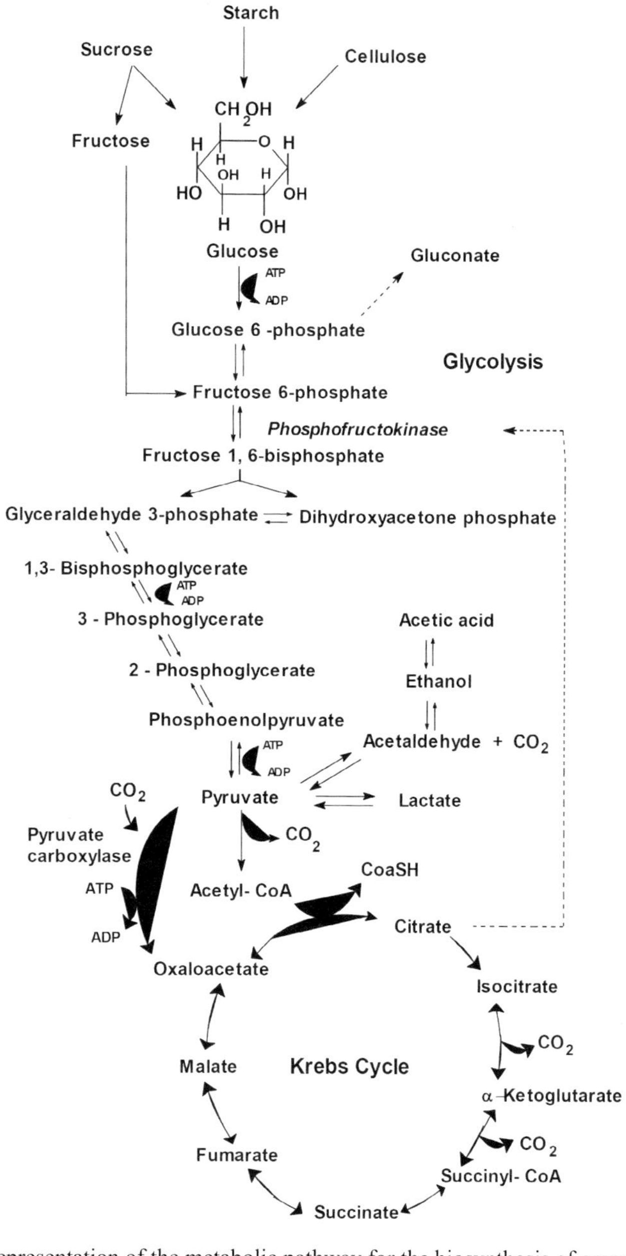

Fig. (1). Schematic representation of the metabolic pathway for the biosynthesis of organic acids.

Citric acid is one of the most important commercially available products obtained by microbial fermentation. Currently, citric acid is commercially produced *via* submerged fermentation using *Aspergillus niger* grown in media containing molasses, glucose or sucrose [2, 15]. The production of citric acid was estimated at 1,600,000 metric tons per year, and China is the largest citric acid producing country, accounting for ~40% of the worldwide output [2]. Other fungal strains, such as *Penicillium*, *Mucor*, *Trichoderma* and other species of *Aspergillus*, bacteria such as *Actinomyces* and *Arthrobacter*, and yeasts, such as *Candida* spp. and *Yarrowia lipolytica*, can also be used to produce citric acid [7, 14].

The filamentous fungus *Aspergillus niger* is the one of the most used microorganisms in the production of citric acid due to its properties such as high availability and growth, adjustment of the microorganism to various substrates, and the accumulation of large amounts of citric acid. It also has the ability to adapt to different environments and control the secretion of citric acid from the mitochondria and cytosol, and then facilitates the secretion and decreases the degradation of citric acid [14]. The *Aspergillus niger* strains accumulate high concentrations of citric acid when grown in media containing high concentrations (15 - 25%) of carbohydrates; optimized concentrations of ammonium nitrate or ammonium sulfate as the nitrogen source; the restriction of trace metal ions, mainly iron, copper, zinc and manganese; limited phosphate concentrations; and high concentrations of dissolved oxygen by stirring constantly at low pH (1.5 - 2.8) in the temperature range from 28-33°C. Molasses, sucrose, glucose and fructose can be used as carbon sources after removal of the cations [3, 4]. In addition to a carbon source such as sucrose, other factors affect the yield of citric acid from *Aspergillus niger* strains, such as the pH of the culture medium, fermentation time and temperature, oxygen flow rate, moisture content, particle size, and the density and age of the inoculum [16].

A typical medium composition for citric acid production contains a carbon source (120-250 g/L of molasses), nitrogen source (0.1-2.0 g/L of $(NH_4)_2SO_4$ or NH_4NO_3), phosphorus source (0.2-1.0 g/L of KH_2PO_4), salts (0.2-0.5 g/L of $MgSO_4$; 0.5-1.5 g/L of $CaCl_2$; $< 10^{-6}$ mol/L of $ZnSO_4$; $< 10^{-4}$ mol/L of $FeSO_4$; $< 10^{-8}$ mol/L of $MnSO_4$), and other supplements (1-4% of methanol or ethanol (v/m), and 0.5-3 g/L of oils) and pH 2.0 [2]. The parameters required for citric acid production by *A. niger* (approximately 200 g/L of citric acid from 240 g/L of glucose) include a high carbohydrate concentration, high dissolved oxygen, constant agitation and low pH (1.7 - 2.0) [17]. *A. niger* requires certain essential trace metal ions for growth [18]; however, a limitation in the amount of some trace metal ions is necessary for the production of citric acid. The metals ions that must be limited include Mn^{2+}, Zn^{2+}, Fe^{2+}, Cu^{2+}, heavy metals and alkaline metals [19], with Mn^{2+} as the major required limitation [20]. The addition of manganese at concentrations as low as 2 ppb to ferrocyanide-treated beet molasses during acid fermentation by *Aspergillus niger* NRC A-1-233 resulted in a 10% citric acid yield. Concentrations of manganese up to 100 ppb have shown to adversely affect the rate of citric acid production and the yield at 100ppb was less than 25% of that obtained at 1ppb or less [21]. *A. niger* is known to produce oxalic acid, but the low pH suppresses the formation of oxalic acid and gluconic acid [11]. The optimum pH for citric acid production by *A. niger* is 1.7 - 2.0. At pH values higher than 3.0, oxalic and gluconic acid are produced instead of citric acid. Low pH inactivates the glucose oxidase enzyme and prevents gluconic acid production [22].

The main advantage of using *Aspergillus niger* strains in comparison with other microorganisms is their ease of handling, their ability to grow in media containing organic residues and by-products, and the ease of recovery of the product from the fermentation broth [14]. Table **2** illustrates citric acid production using different substrates and fermentation processes.

An increased yield of citric acid can be obtained by screening new strains, modifying microorganisms by genetic engineering, using enzyme regulation control mechanisms, and optimizing the fermentation conditions, recovery and purification of the product [23, 28 - 31].

The phosphofructokinase, pyruvate carboxylase, citrate synthase and alpha-ketoglutarate dehydrogenase enzymes play a regulatory role in citrate biosynthesis, and the main control step in the production of citric acid involves the inhibition of phosphofructokinase. Inhibition of this enzyme by citric acid can be overcome with high concentrations of intracellular NH_4^+, enabling the efficient operation of the glycolytic pathway and the overproduction of citric acid. Screening for phosphofructokinase producing strains that are not allosterically inhibited by citric acid can increase the production of the latter. A deficiency of manganese ions causes a change in the composition of the fungal cell wall, forming small pellets of mycelium. This is the best morphology for the production of citric acid, and also results in higher concentrations of intracellular NH_4^+ ions that reverse the inhibition of citric acid by phosphofructokinase. Other factors contributing to an increase in citric acid yield are the inhibition of isocitrate dehydrogenase by citric acid and low pH values (1.7 to 2.0). After 4 to 5 days of fermentation, most of the carbohydrate (80%) has been converted to citric acid, reaching 100 g/L [7]. One of the reasons why several fungi accumulate citric, L-malic and fumaric acids is the presence of the pyruvic acid carboxylase enzyme in the cytosol [4]. Pyruvic acid carboxylase, which catalyzes the carboxylation of pyruvic acid to oxaloacetic acid, is important in the replacement of this Krebs cycle intermediate by promoting the biosynthesis of citric acid.

Table 2. Citric acid production using different substrates and fermentation processes.

Microorganism and Process	Substrate	Citric Acid Production	Fermentation parameters	References
Aspergillus niger CCT 7716 *Aspergillus niger* CCT 7717 and CCT 7718 (obtained by mutagenesis) Solid state fermentation in Erlenmeyer flasks and glass columns	Citric pulp and addition of a nutritive solution containing sugarcane molasses (120 g/L of total sugars) and methanol (4.0%)	278.4, 294.9 and 261.1 g citric acid/kg of dry citric pulp from *Aspergillus niger* CCT 7716, CCT 7718 and CCT 7717, using glass columns (forced aeration), respectively, after 4 days 410.7, 446.8 and 492.7 g /kg of dry citric pulp from *Aspergillus niger* CCT 7716, CCT 7718 and CCT 7717, respectively, using Erlenmeyer flasks (aeration by diffusion) citric acid reached after fourth days	Adjust the initial moisture to 65%, temperature of 30 °C, pH 5.5	[23]
Aspergillus niger Superficial liquid fermentation in 250-mL Duran flasks	Sweet potato starch hydrolysate (153.77 g/L), (NH4)2HPO4 (3.55 g/L), KH2PO4 (2.58 g/L)	83.01 g/L of citric acid after eight days.	pH 6.0	[24]
Aspergillus niger Submerged fermentation in 1.2-L fermentor	Olive oil (28.7 g/L), silicone oil (30.0 g/L), almond oil 28.7 g/L	68.5 g/L of citric acid after 7 days.	30 °C	[16]

(Table 2) contd.....

Microorganism and Process	Substrate	Citric Acid Production	Fermentation parameters	References
Aspergillus niger van Tiegh 1867 Submerged fermentation in 250-mL Erlenmeyer flasks	Orange peel medium fortified with sugar cane molasses at a final sugar concentration of 14% in the presence of 3.5% methanol.	640 g citric acid/kg orange peel after 72 hours.	Moisturized of 65%, with bed loading of 20%, initial pH 5.0, temperature 30 °C, 250 rpm	[25]
Aspergillus niger NRRL-567 Submerged fermentation in 500 mL Erlenmeyer flasks	Industrial wastes such as brewery spent liquid, lactoserum and starch industry water sludge.	The optimal concentration of 11.34 g/L of citric acid was obtained using brewery spent liquid and 3% methanol, after 120 hours incubation. Addition of methanol resulted in an increase of 56% citric acid production. A higher concentration of 18.34 g/L of citric acid was obtained with the addition of brewery spent liquid with 40% apple pomace ultrafiltration sludge having suspended solids at a concentration of 30 g/L	The process was performed by varying the temperature (25-35 °C), pH (3-5), 200 rpm, incubation time, addition of inducers, and addition of different proportions of apple pomace ultrafiltration sludge.	[26]
Yarrowia lipolytica IMUFRJ 50682 Submerged fermentation in 1-L flasks	Crude glycerol stemming from biodiesel industries. Media also contained minerals, such as KH2PO4 (12.0 g/L), Na2HPO4.7H2O (22.66 g/L), MgSO4.7H2O (1.5 g/L), CaCl2.2H2O (0.2 g/L), FeCl3.6H2O (0.15 g/L), ZnSO4.7H2O (0.02 g/L), MnSO4.H2O (0.06 g/L)	12.94 g/L of citric and 6.66 g/L isocitric acids, respectively, in 160 h of fermentation from medium containing 45 g/L of glycerol. Addition of ammonium sulfate to the culture medium resulted in 1.46 g/L citric acid and 16.79 g/L isocitric acid in 93 hours fermentation.	Temperature of 28 °C, 250 rpm	[27]

(Table 2) contd.....

Microorganism and Process	Substrate	Citric Acid Production	Fermentation parameters	References
Yarrowia lipolytica NG40/UV7 Submerged Feed Batch Fermentation in 10 L fermentor	Glycerol (waste or pure glycerol) (2.0 g/L), (NH4)2SO4 (4.5 g/L), MgSO4.7H2O (1.4 g/L), NaCl (0.5 g/L), Ca(NO3)2 (0.8 g/L), KH2PO4 (2.0 g/L), K2HPO4 (0.2 g/L).	115 g/L of citric acid and mass yield coefficient of 0.64 g/g and 4.6 g/L of isocitric acid in the medium with pure glycerol 112 g/L and mass yield coefficient of 0.90 g/g of citric acid and 5.3 g/L of isocitric acid in the medium with raw glycerol	Temperature of 28 °C, pH 5.0, *p* O2 of 50% (air saturation), agitation rate of 800 rpm, addition of glycerol (20-80 g/L) was performed as the *p* O2 value changed by 10%.	[28]

Aspergillus strains can also use lipids as their carbon source. An *Aspergillus niger* strain was used to ferment various oils containing unsaturated fatty acids, and it was found that the addition of 2.87% olive oil, 3.00% silicone oil and 2.87% almond oil resulted in the production of citric acid with a 68.5 g/L yield [16].

Fermentation by *Aspergillus niger* in solid culture media using low cost industrial residues, such as wheat bran, soybean meal, apple processing waste and cassava, has been studied as an alternative means of producing citric acid, and it also reduces the amount of waste discarded in the environment. Dhillon and co-workers [32] studied the production of citric acid by *Aspergillus niger* NRRL – 567 in a 7.5 L bench reactor using apple pomace ultrafiltration sludge as the substrate. The maximum yield obtained was 23.67 g/L after 120 hours in a control culture medium with no inducer, and 40.34 g/L after 132 hours of fermentation in medium containing 3% (v/v) methanol as an inducing agent. The fermented control culture medium showed pseudoplastic characteristics, which were likely due to the development of filamentous fungi. On the other hand, the culture broth supplemented with methanol as the inducer also showed non-Newtonian characteristics, but had lower viscosity due to the development of fungi in the pellet form.

Betiku and Adesina [24] studied the production of citric acid by *Aspergillus niger* using response surface methodology. *Aspergillus niger* was inoculated into a medium containing hydrolyzed sweet potato starch as an alternative carbon source, producing a maximum concentration of 83.01 g/L citric acid.

The production of citric acid using by-products or organic residues as the substrates has the advantages of adding value to these substrates and providing a destination for these wastes, which would otherwise be treated or discarded, thus reducing the costs for disposal of the waste and for the production of the citric acid. Hamdy [25] proposed the use of agro-industrial by-products and waste for the production of citric acid to reduce the fermentation costs and also as an alternative for the use of such residues, reducing their accumulation in the environment. A strain of *Aspergillus niger* was fermented in a media composed of orange peel supplemented with sugar cane molasses. The influence of the different parameters on the fermentation process was verified. The highest citric acid production, 640 g/kg of orange peel, was obtained after 72 hours of fermentation in shaken Erlenmeyer flasks, using a culture medium containing 20% orange peel with molasses at a final sugar concentration of 14%, an initial pH of 5.0, 30 °C and 250 rpm. The tests also demonstrated that when sugar cane molasses was added to the culture medium to increase the sugar content, the medium did not require pretreatment with EDTA or iron-cyanide.

The solid state fermentation of *Aspergillus niger* LPB 21 using cassava bagasse as the substrate resulted in 220, 309, 263 and 269 g citric acid/Kg cassava bagasse in Erlenmeyer flasks, column fermenters, a tray bioreactor and a horizontal drum bioreactor, respectively. The fermentation of *Aspergillus niger* LPB 21 in a column bioreactor using optimum conditions for the aeration rate, 60 mL/min (3 mL/[g.min]), and an initial humidity of 60%, as determined by factorial experimental design, resulted in 265.7 g citric acid/kg of dry cassava bagasse [15].

Silva and co-workers [27] studied the production of citric acid by *Yarrowia lipolytica* IMUFRJ 50682 using crude glycerol, a byproduct from biodiesel production, as the carbon source. The composition of the culture medium influenced the production of citric and isocitric acids. The concentration of citric acid was favored by a lower initial glycerol concentration and the absence of ammonium sulfate. The production of citric and isocitric acids was 12.94 g/L and 6.66 g/L, respectively, using a medium containing 45 g/L of glycerol, 0.1 g/L yeast extract and salts, after 160 h of fermentation at 28 ºC and 250 rpm. In tests performed with the addition of ammonium sulfate, isocitric acid was produced in larger amounts than citric acid, 16.79 g/L and 1.46 g/L, respectively, after 93 hours of fermentation.

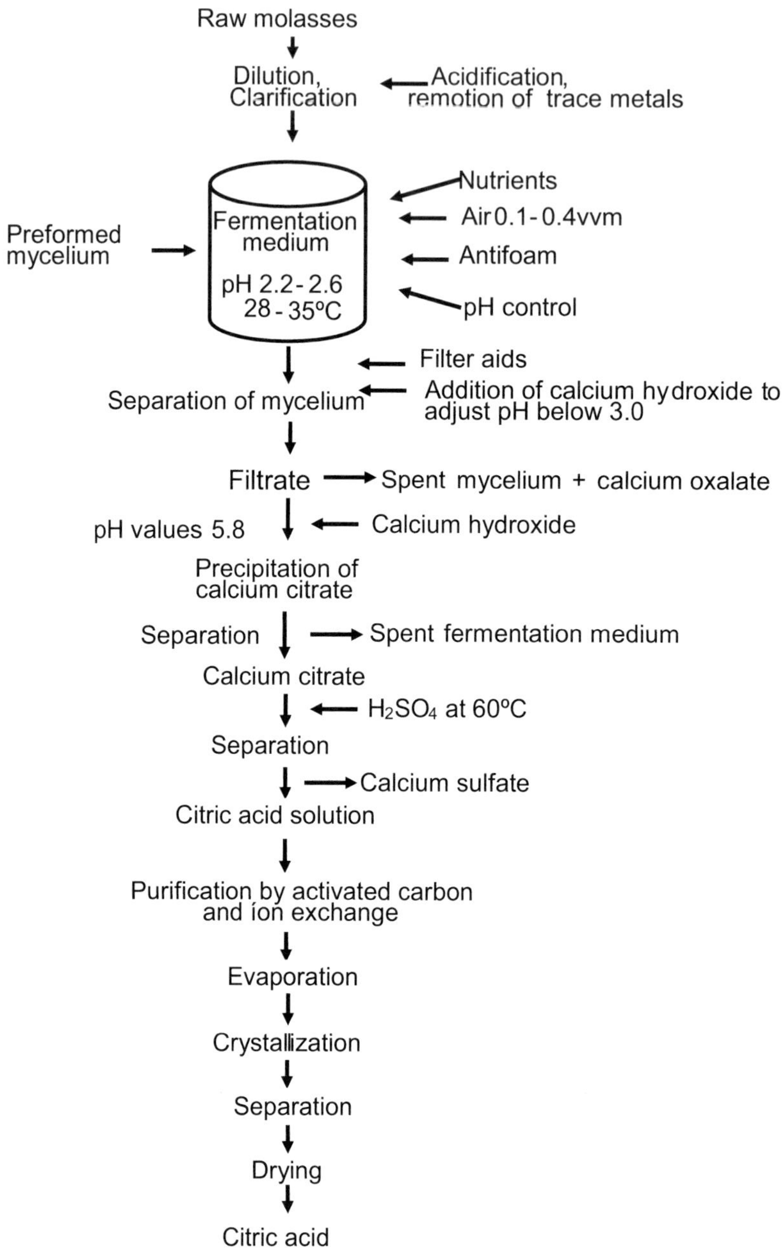

Fig. (2). Citric acid production and recovery.

Rodrigues and co-workers [23] studied the production of citric acid by three strains of *Aspergillus niger*, one wild type and two mutant strains, in solid state fermentation using citrus pulp as the substrate, and media supplemented with sugarcane molasses (120 g/L of total sugars) and methanol (4% w/v). The highest yields were obtained when the strains were fermented in Erlenmeyer flasks and aerated by diffusion compared to fermentation in glass column reactors. The citric acid yield reached 410.7, 446.8 and 492.7 g citric acid/Kg of dry pulp in Erlenmeyer flasks (aeration by diffusion) using *Aspergillus niger* 7716, CCT 7718 and CCT 7717, respectively.

Selected or genetically modified strains of the yeast *Yarrovia lipolytica* have been reported to produce high yields of citric acid from sucrose, molasses or glycerol. Modified yeast strains with low aconitase activity should be used to prevent the undesirable formation of isocitric acid [28, 29, 33]. Morgunov and co-workers [28] used a mutant strain of *Yarrowia lipolytica*, NG40/UV7, to obtain citric acid in a liquid culture medium containing pure glycerol or an industrial waste containing glycerol. The *Yarrowia lipolytica* strain NG40/UV7 was fermented in the optimized culture medium, in a batch process, and produced 115 g/L of citric acid. It was obtained a mass yield coefficient of 0.64 g/L, and 4.6 g/L of isocitric acid. In the medium with raw glycerol, the production of citric acid was 112 g/L with a mass yield coefficient of 0.90 g/L, and that of isocitric acid was 5.3 g/L at 28°C, pH 5.0, and 50% oxygen saturation.

Rywinska and co-workers studied the biosynthesis of citric acid from glycerol by *Yarrowia lipolytica* mutants in batch-fed fermentation [34]. The fermentation of the *Yarrowia lipolytica* Wratislavia AW G7 strain in a medium containing 200 g/L of glycerol reached a maximum citric acid production of 139 g/L after 120 h, with a yield of 0.69 gram of citric acid per gram of glycerol consumed [34].

Citric acid can be recovered by precipitation, extraction and adsorption methods [2, 14, 35 - 38]. After fermentation, the mycelium can be removed by filtration and the citric acid is recovered from the filtrate by the addition of calcium hydroxide or calcium carbonate (Fig. **2**). The precipitate can be recovered by vacuum filtration and the impurities removed by washing the obtained precipitate. The precipitate can be treated with a solution of 60-70% sulfuric acid at

50 – 60 °C to form calcium sulfate and free citric acid. The citric acid solution can be concentrated under vacuum and the calcium sulfate removed. The citric acid can be purified using activated carbon and ion exchange resins, concentrated and then centrifuged to obtain the crystals. The recrystallization process may be repeated several times to remove the impurities [35].

ACETIC ACID

Acetic acid is produced either by chemical synthesis or by fermentation with microorganisms. The annual production of acetic acid has been estimated at 7 million tons [6]. Most glacial acetic acid is produced by chemical synthesis from petroleum [39]. Acetic acid is used in the production of polymers such as vinyl acetate, solvents such as ethyl acetate, and for defrosting salts for use on roads and runways [6].

Approximately 10% of the world's acetic acid production is from fermentation to produce vinegar for use in food [39]. Vinegar has been a part of the human diet as a condiment and food preservative, and has been the basis for simple remedies for people and animals since ancient times [40]. It is primarily produced by the fermentation of wines, cider, cereal, malt distillate, sugarcane ethanol, grape juice, apple juice, rice, *etc.*, and used in mayonnaise, mustard, pickles, pickled sausages, salad dressings, ketchup, marinades and canned vegetables, and other foods.

Vinegar is produced by acetic acid bacteria by the conversion of ethanol and oxygen to acetic acid and water. It is produced by either submerged fermentation or solid state fermentation methods. The first step in acetic acid production by microorganisms is the fermentation of the carbohydrate to ethanol by yeast, and the second step is the oxidation of ethanol by bacteria of the genus *Acetobacter* and *Gluconoacetobacter*. Acetic fermentation is an exothermic aerobic oxidation process, where the substrate, containing 50-100 g/L of ethanol, is partially oxidized by acetic bacteria to produce acetic acid, water, glycerol and other minor products. Oxygen is essential for the oxidation of ethanol to acetaldehyde and then to acetic acid. The pyrroquinoline quinone-dependent alcohol dehydrogenase (PQQ-ADH) is a cell membrane-bound enzyme that oxidizes ethanol to acetaldehyde, which is coupled to the reduction of ubiquinone Q. This, in turn, deoxidizes the ubiquinol oxidase and reduces oxygen to form water (Fig. **3**). The cell membrane-bound aldehyde dehydrogenase then oxidizes the acetaldehyde to acetic acid [41].

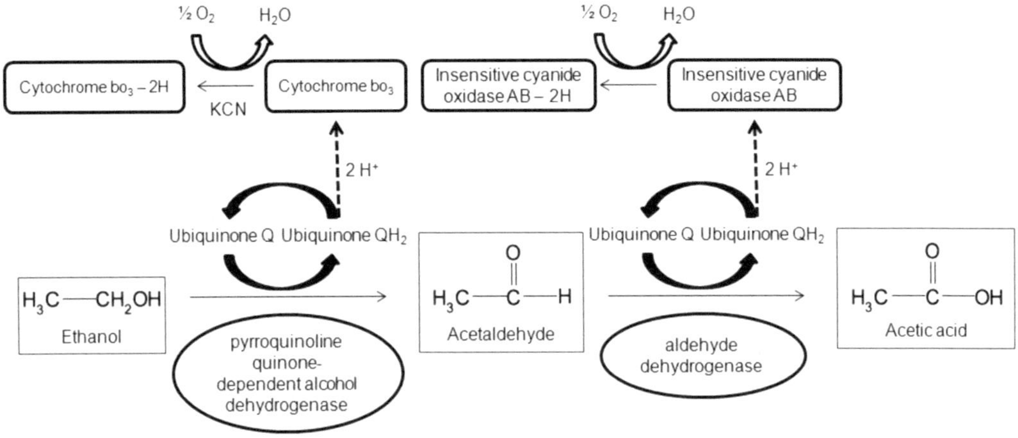

Fig. (3). Oxidation of ethanol to acetic acid by acetic bacteria (adapted from Yakushi and Matsushita [41]).

The quality of the product depends on the raw material (wine or some other substrate), the metabolism of the acetic acid bacteria, which produces some additional transformations (mostly oxidation reactions but also ester formations), the interaction between the vinegar and the wood from the barrels, and the aging process. The acidity of wine vinegar (acetification obtained exclusively from wine) must be at least 6% (w/v), and the maximum residual ethanol allowed is 1.5% (v/v) [40]. Vinegar may contain small amounts of glycerol, esters, acids, sugars, anthocyanins, phenolic compounds, salts and other compounds, depending on the feedstock and the aging process in barrels [42].

There are three main vinegar production processes: 1) the Slow, Orleans or French Process, 2) the Fast (or German) Process and 3) the Submerged Process. The Slow Process, Orleans or French Process is the earliest process, in which the wine only partially fills the barrels to improve the area-to-volume ratio. The acetic bacteria grow on the surface of the liquid and produce a gelatinous film known as the mother vinegar. The Fast Process, also known as the Schüzenback system, arose in Germany. This system consists of a generator in the form of a cylindrical tank that has three parts: the upper section, the larger section (middle part), which is filled with wood shavings, bamboo, corn cobs or other material to provide a large surface area, and the lower section. The wine or liquid is placed on the top of the generator through a feeder or spray device. It then passes to the drip section, which contains the largest chips, where ethanol is oxidized to acetic acid by the acetic bacteria growing on the support film. The liquid, or vinegar, passes through the drip to the lower section, where it is collected. A rapid improvement of this process occurred when a system of forced aeration and temperature control was added. Approximately 98% of a 12% (v/v) ethanol starting solution is converted to acetic acid in 5 days. The main advantages of this process are its low cost, ease of control, high acetic acid yield, reduced space requirement for the equipment and a lower evaporation loss. The third vinegar production process is called the Submerged Process. In submerged fermentation, an adequate and continuous supply of oxygen is required in the tank, and brief interruptions of the supply of oxygen, even for a few minutes, affect the acetic acid yield. Submerged fermentation is distinguished by high acetic acid productivity, with the ethanol oxidation being approximately 30 times faster than other methods. Moreover, the process can be automated, although this requires a large investment [40, 43 - 45].

At the industrial scale, vinegar is primarily produced by submerged fermentation, an aerobic process in which the ethanol in wine or cider is oxidized to acetic acid by strictly aerobic acid bacteria [46]. Fig. (**4**) illustrates the submerged fermentation in a semi-continuous process for vinegar production.

Wine vinegar is a product of low commercial value, but balsamic vinegar has a higher value. The balsamic vinegar from Italy is produced by the traditional method, a slow fermentation surface process using cooked grapes containing soluble solids in the range from 20-60 °Brix and pH value in the range from 2.3 to 3.2. Like other vinegars, balsamic vinegar is obtained by fermentation in a two-step process. In the first step, the fermentable sugars are converted to ethanol by yeast, obtaining an ethanol content of 4-10%, and, in the second step, the acetic bacteria oxidize the ethanol to acetic acid [47, 48].

Wang and co-workers [49] described the use of a mixed culture of *Saccharomyces cerevisiae* and *Acetobacter pasteurianus* CICC7015 to obtain a high yield of acetic acid. The process was divided into three steps, where the first step was fermentation by *Saccharomyces cerevisiae* to produce alcohol. The second step consisted of the simultaneous fermentation by *Saccharomyces cerevisiae* and *Acetobacter pasteurianus*, followed by *Acetobacter pasteurianus* alone in the third step for the production of acetic acid. The average acetic acid concentration obtained was 52.5 g/L. A glucose feeding strategy was employed to enhance the production, and the acetic acid concentration reached 66.00 g/L.

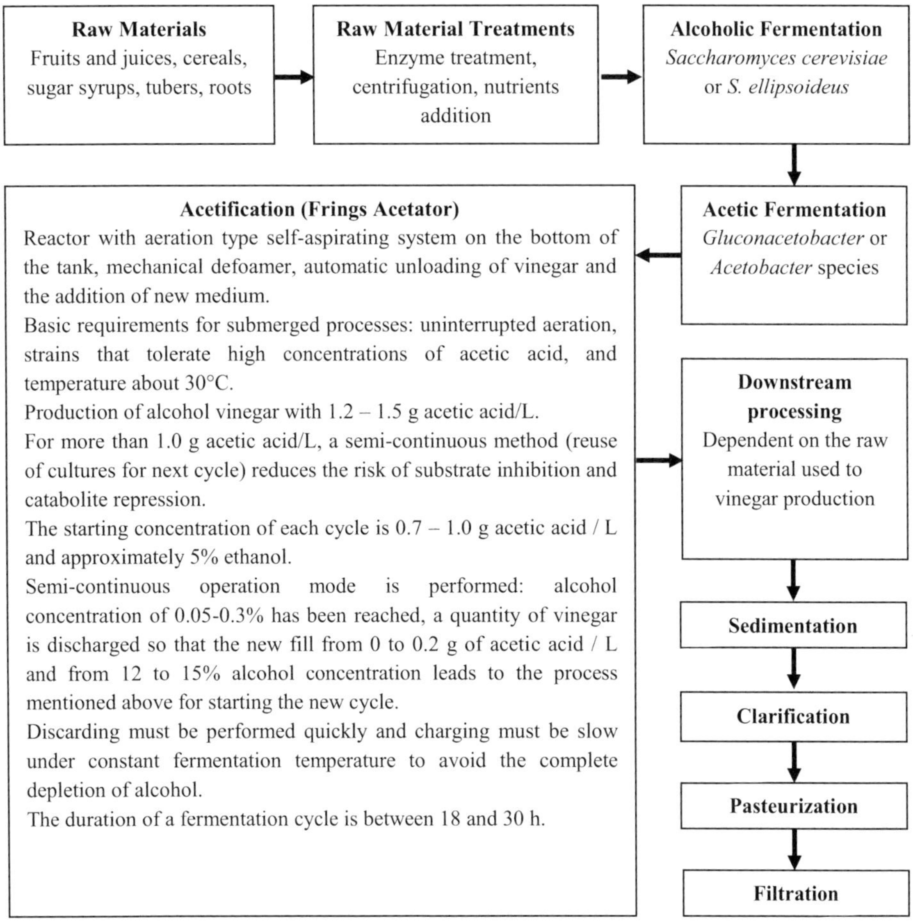

Fig. (4). Schematic representation of submerged fermentation in semi-continuous process for vinegar production [adapted from Greenshields [43]; Ebner and Follmann [44]; Gullo and co-workers [46]].

The carbon, nitrogen, inorganic ions and vitamin sources influence the performance of the acetic acid bacteria, and glucose, yeast extract and salts, such as KH_2PO_4, $MgSO_4$ and NaH_2PO_4, have been used to test the fermentation to vinegar in Erlenmeyer flasks [50]. *Acetobacter pasteurianus* HN101 was used in semi-continuous batch and fed-batch fermentation processes in 10 L fermentation tanks to producc vinegar. The average rate of acetylation in the semi-continuous fermentation was 2 g of acetic acid/L/h, where each cycle lasted 12 h and the average stoichiometric yield was 94%. The tests served as a basis for the success of fed-batch fermentation. A high vinegar concentration of approximately 116 g/L was obtained at the end of the fermentation (85 hours), without the appearance of a lag phase. Highly concentrated vinegar containing 170 g acetic acid/L was obtained by fed-batch fermentation using *Acetobacter* sp [51].

Cloning of the membrane-bound aldehyde dehydrogenase gene of *Acetobacter polyoxogenes* NBI 1028 improved the acetic acid production. The pAL25 plasmid containing the full-length structural gene was obtained and expressed in both *Escherichia coli* and *Acetobacter aceti* to produce a fused protein (78 kDa) with a short N-terminal beta-galactosidase peptide. Transformation of *Acetobacter aceti* subsp. *xylinum* with pAL25 doubled the acetic acid production rate (1.8 to 4 g/L/h) and also increased the concentration of acetic acid (68 to 97 g/L) in submerged fermentation [52].

It is especially important to optimize the fermentation conditions and follow the kinetics of microbial growth and the parameters for acetification of the medium in the batch acetic fermentation process for vinegar production [53]. The fermentation of a liquid medium to produce an aerobic acetic acid bacteria culture may result in reduced productivity and yield if the oxygen supply is interrupted during the process; it is important to maintain a constant oxygen supply in both the initial and final stages of fermentation [54]. There are several parameters that can limit the production of acetic acid production that must be controlled, such as oxygen availability and a suitable initial acidity of the fermentation medium. Qi and co-workers [53] used *Acetobacter pasteurianus* FS1 to produce vinegar in a semi-continuous process in a 10 L Fring reactor, and found that acetification rate and oxygen consumption of *Acetobacter pasteurianus* FS1 was influenced by the initial acidity during alcohol vinegar fermentation. The initial acidity of 40 g/L was used to adjust the bacterial metabolism. An average acetification of 1.81 g of acid/L h was obtained in the fermentation process and, under these conditions, the stoichiometric yield of the vinegar was also improved to 94.3%. The airflow rate was also reduced from 0.22 vvm to 0.13 vvm, economizing the use of oxygen.

Schlepütz and co-workers [54] reported that the production of acetic acid using submerged fermentation is sensitive to the pre-culturing or pre-inoculum culturing step, causing a lack of reproducibility in the fermentation process at the industrial level. The traditional method, which uses a pipette in the pre-cultivation step, is not suitable for industrial fermentations with high concentrations of alcohol and acetic acid. A continuous oxygen supply is important. For example, the use of a gas bubbles column and direct transfer of the pre-inoculum to the main culture medium prevents attenuation of the activity or mutations. The authors were searching for a methodology to produce acetic acid by submerged fermentation on a small scale, which was reproducible in high concentrations of ethanol and acetic acid, and had the potential for applications on an industrial level, with the guarantee of a constant oxygen supply during the inoculation process. Two methods for acetic acid production were tested. In the traditional method, the bacteria were pre-cultured in a 9 L bioreactor, drained and placed in a container, and then inoculated into stirred Erlenmeyer flasks containing the culture medium with the aid of a pipette. In the other method, the bacteria were transferred to bubble columns using a syringe and then injected into the Erlenmeyer flasks without any interruption of the oxygen supply. The latter method showed higher and more reproducible bacterial respiration curves, unlike the traditional method in which the curves were low and not reproducible.

The fermentation parameters for vinegar production from sugarcane were optimized on a 50 L scale and validated for 3 years (2010-2012) using indigenous *S. cerevisiae* 35 and *A. aceti* ACI strains for the ethanolic and acetic acid fermentations, respectively. The ethanolic fermentation was complete in 4 days, producing between 8 and 11% ethanol (v/v). The acetic acid fermentation was performed by first shaking for 3 days, and then incubating for 23 ± 2 days to produce vinegar containing 4.18-6.43% (w/v) acetic acid. Recycling (2/3 volume) of the previous fermentation batches decreased the fermentation time to 5-9 days, with an increase in the acetification rate [55].

LACTIC ACID

Lactic acid is industrially produced by the fermentation of carbohydrates using lactic acid bacteria. Pyruvic acid, an intermediate of the glycolytic pathway, is reduced to lactic acid by lactate dehydrogenase. The production of lactic acid has been estimated at 250,000 tons per year [7].

Lactic acid is a bacterial fermentation product used in the food industry to manufacture cheeses, cultured sausages, yogurt, pickles, and buttermilk. Lactic acid can also be used as a food additive, such as an acidulant, preservative, and flavor-enhancer in beverages, jellies and cheese [56]. Acidification with the direct addition of lactic acid can be used to achieve the sensory characteristics of the final product and to avoid the risk of proliferation of undesirable microorganisms [57]. It is also used in the production of emulsifiers, such as sodium or calcium stearoyl lactylate, and in the production of biodegradable plastics, such as polylactic biopolymers (PLA) [6, 7, 39, 58 - 61]. Calcium-sodium stearoyl lactate and sodium stearoyl lactylate are used as a dough conditioner in bread [60, 61] and are also used in the pharmaceutical, cosmetic, leather and textile industries.

Several microorganisms, such as fungi, yeasts, algae and bacteria, can produce lactic acid. Strains with improved characteristics, such as improving the yield and productivity, improvement of L-lactic acid production with optical purity, reduced nutritional requirements and the ability to assimilate various substrates, are desirable for the industrial production of lactic acid.

Lactic acid-producing bacteria can be divided into four main groups: lactic acid bacteria (LAB), *Bacillus* strains, *Escherichia coli* and *Corynebacterium glutamicum* [59]. In general, fermentation by lactic acid bacteria has several limitations, including the production of L and D-lactic acids, low yields due to the formation of byproducts, the requirement for nutritionally rich media and a high risk of infection by bacteriophages, resulting in cell lysis and a reduction in lactic acid production [62]. The lactic acid bacterium *Lactobacillus delbrueckii*, including the subspecies *L. delbrueckii* subsp. *delbrueckii*, *L. delbrueckii* subsp. *bulgaricus* and *L. delbrueckii* subsp. *lactis*, are used to produce lactic acid.

The *Lactobacillus delbrueckii* ATCC 9649 mutant DP3 strain produced 77 g/L lactic acid in a pH-controlled (6.0) stirred-tank-batch fermentation using 12% glucose, 1% yeast extract, and mineral salt/oleic medium (Demirci and Pometto III, 1992). A fed-batch fermentation of *Lactobacillus lactis* BME5-18M in low glucose medium was used as a control strategy to produce large quantities of L-lactic acid. Using low glucose and continually supplying the sugar during fermentation, the inhibitory effect of glucose was avoided and up to 210 g/L of L-lactic acid (97% yield) was obtained. The L-lactic acid productivity was 2.2 g/L/h [63].

Nguyen and co-workers [64] produced L and D- lactic acids using *Lactobacillus paracasei* and *Lactobacillus coryniformis* by simultaneous saccharification and co-fermentation. Biomass waste from the industrial processing of *Curcuma longa* (WCL), which contains high starch and protein contents, was also used. L. *paracasei* LA104 and L. *coryniformis* ATCC 25600 are homofermentative L-lactic acid and D-lactic acid producing strains, respectively. Assays aimed at optimizing the fermentation conditions were performed in Erlenmeyer flasks with stirring. The culture medium contained 140 g of residue/L, 56 g $CaCO_3$/L, 179.2 FPU cellulase/L, 870.8 U cellobiase/L, 1120 FAU alpha-amylase/L and 420 U amyloglucosidase/L, supplemented with 10 g soy bean meal/L (batch fermentation by strain *L.paracasei* LA104) or 15 g soy bean meal/L (*L. coryniformis* ATCC 256004). The culture medium containing 353.54 g of WCL biomass waste was fermented in 7 L bioreactor in two different assays. One test was performed with *L.paracasei* LA104 with the addition of 20 g of soy bean meal to the culture medium at 37 °C, and the other test used the *L. coryniformis* ATCC 25600 strain with the addition of 30 g of soy bean meal at 34 °C. The fermentations were performed at 200 rpm, a dissolved oxygen level of less than 0.5 ppm and pH 6.0. The enzyme mixture (16.8 mL) was added to the culture medium and inoculated with a 10% seed culture. For the production of L-lactic acid, the maximum lactic acid concentrations, average productivities, conversion to reducing sugars and lactic acid yields were 97.13 g/L, 2.7 g/L/h, 95.99% and 69.38 g/100 g dry material, respectively. For the production of D-lactic acid, the corresponding values were 91.61 g/L, 2.08 g/L/h 90.53% and 65.43 g/100 g dry material, respectively.

Zhao and co-workers [65] studied the process of simultaneous saccharification and lactic acid fermentation of pretreated corn stover using the *Pediococcus acidilactic* DQ2 strain. Corn stover was obtained from corn straw, pretreated with dilute sulfuric acid and biodetoxified, producing a high solids content. This material was used as the lignocellulosic substrate for the lactic acid bacteria, which are tolerant of high temperatures. The simultaneous saccharification and fermentation was performed in a 5 L bioreactor, and the scale-up experiment was performed in a 50 L bioreactor. The fermentation processes were performed at 48 °C, pH 5.5, and approximately 30% (w/w) solid waste. In this study, 101.9 g/L lactic acid, 77.2% yield, and 1.06 g/L/hr productivity were obtained.

Although many lactic acid bacteria produce L-lactic acid with high yields and productivity [58], they usually require complex nutrients. One genetically engineered wine yeast, *Saccharomyces cerevisiae* OC-2T T165R, produced 122 g/L lactic acid and the optical purity was 99.9% or higher using the fermentation of a cane juice-based medium [8].

L-lactic acid was produced from glucose and hydrolyzed starch using the thermophilic *Bacillus coagulans* WCP10-4, which is highly tolerant of glucose concentrations. In the batch fermentation of *Bacillus coagulans* WCP10-4 in culture medium containing 240 g/L of glucose, adjusted to pH 6.0, 210 g/L of lactic acid with a 95% yield and 3.5 g/L/h productivity were obtained. Using a temperature of 50 ° C for the simultaneous saccharification and fermentation of culture medium containing 200 g/L of unsterilized corn starch 202.0 g/L-lactic acid was obtained [66].

FUMARIC ACID

Fumaric acid is a dicarboxylic acid that is used as an acidulant in foods and beverages and in the feed industry. It is also used in chemical industries for the production of polyester resins, lubricating oils, lacquers, inks and also carboxylating agents for the production of styrene butadiene rubber [6, 7, 67, 68]. Fumaric acid is an intermediate chemical reagent used in the preparation of edible products such as candy coverings, emulsifiers for water in oil mixtures and additives for pasta. Fumaric acid is used in the production of the acidulant malic acid and the production of aspartic acid, which is used in the production of the sweetener aspartame. Fumaric acid has the advantage of being only slightly hygroscopic and is used in the production of soft drinks and gelatin powders because it stabilizes the other ingredients and extends the shelf life. Fumaric acid is a solid acid that confers increased acidity in aqueous solution, and can be used as a replacement for citric acid because a smaller amount is needed to achieve the same result as citric acid. The low solubility of fumaric acid in cold water (0.56 g/100 mL) limits its application, and it is more suitable for dry mixes, such as gelatin powder.

Fumaric acid is chemically produced by the isomerization of maleic acid or maleic anhydride from petrochemicals, including the catalytic oxidation of hydrocarbons (*e.g.,* benzene, *n*-butane or an *n*-butane/*n*-butene mixture) [69, 70]. However, with the increase in petroleum prices, the price of maleic anhydride has also increased because it is a petroleum derivative [71]. Fumaric acid production by fermentation has been studied as an alternative for obtaining the chemically synthesized product. The *Rhizopus* spp. strains have been studied the most and they are the best producers [72, 73]. The production of fumaric acid has been estimated at 90,000 metric tons per year [2]. The production of fumaric acid by biotechnological methods using *Rhizopus oryzae* strains offers several advantages such as the production of the acid using renewable materials, including glucose and molasses from sugar cane.

Fumaric acid can be produced from succinic acid in an intermediate step of the Krebs cycle. The accumulation of fumaric acid can occur under conditions of stress, such as nitrogen limitation. Under the conditions of nitrogen starvation, the microorganism produces fumaric acid instead of producing cell mass from glucose [3]. Apart from nitrogen limitation, $CaCO_3$ is also required in the culture medium. $CaCO_3$ converts fumaric acid into calcium fumarate, which precipitates, and also provides CO_2 for the carboxylation of pyruvic acid by the pyruvic acid carboxylase and the formation of oxaloacetic acid, thus replenishing the Krebs cycle intermediates. Increased formation and excretion of fumaric acid is obtained by adding surfactants, such as polysorbate, to the medium or by screening for strains with high pyruvic acid carboxylase activity. The addition of surfactants increases cell permeability and the excretion of fumaric acid [4].

Cao and co-workers [67] described a process for the production and recovery of fumaric acid using immobilized *Rhizopus oryzae* and an adsorption column. The growing *Rhizopus* mycelia were immobilized on the plastic discs of a rotary contactor where the biofilm was alternately exposed to a liquid medium containing glucose and air upon rotation of the discs in the horizontal fermentation vessel. The fumaric acid was simultaneously and continuously removed using a coupled adsorption column (polyvinyl pyridine PVP425 and Amberlite IRA-900), thereby moderating feedback inhibition, enhancing the fermentation rate and sustaining the viability of *Rhizopus oryzae*. The fermentation reactor, coupled with the adsorption column, was capable of producing fumaric acid with an average yield of 85 g/L from 100 g/L of glucose in repeated fed-batch cycles. On a weight yield basis, a 91% conversion of glucose to fumaric acid was obtained with a productivity of 4.25 g/L/h.

Zhou and co-workers [73] studied fumaric acid production by *Rhizopus delemar* NRRL 1526. The fermentation process was influenced by temperature, the agitation rate and the medium composition, and it affected the *Rhizopus delemar* pellet diameter, which had a significant influence on fumaric acid yield. The highest yields of fumaric acid (39.56 g/L in fermentation flasks and 35.42 g/L in stirred bioreactor tanks) were obtained from a *Rhizopus delemar* pellet with a diameter of 0.55 mm at 30 °C, 200 rpm and pH 5.0.

The production of ethanol and other by-products by the *Rhizopus oryzae* strain can affect the yield of fumaric acid from glucose in the fermentation process. This limitation was minimized by overexpressing the endogenous pyruvate carboxylase and the exogenous carboxylase phosphoenolpyruvate. The pyruvate carboxylase activity in the *pyc* transformants increased 56-83%. The *pepc* transformants exhibited significant phosphoenolpyruvate carboxylase activity (3 - 6 mU/mg) that was absent in the wild type strain. Overexpression of the *pepc* gene in *Rhyzopus oryzae* led to an increase in fumaric acid production (0.78 g/g glucose) compared to the wild type strain (0.62 g/g glucose). Due to the formation of large cell pellets that limited the oxygen supply, the *pyc* transformants grew poorly and had low fumaric yields (< 0.05 g/g glucose). This resulted in the accumulation of ethanol with a high yield of 0.13-0.36 g/g glucose [74].

A high fumaric acid producing strain of *Rhizopus oryzae* was obtained under femtosecond laser irradiation. The mutant *Rhizopus oryzae* FM 19 strain produced 49.4 g/L fumaric acid in a 7.5 L stirred fermentor, which was 1.56-fold higher than the parental strain. The obtained yield was 0.56 g fumaric acid per g glucose. Metabolites related to the Krebs cycle, the Embden-Meyerhof Pathway, and amino acid metabolism in the high-yield strain were increased, as was the unsaturated fatty acid/saturated fatty acid ratio. In particular, 4-aminobutyric acid (GABA) and 5-aminolevulinic acid (5-ALA) were increased 10.33- and 7.22-fold, respectively [75].

A typical medium for the production of fumaric acid contains a carbon source (80-130 g/L of glucose); a nitrogen source (0.25 g/L of $(NH_4)_2SO_4$ or none); a phosphorus source (0.3-0.6 g/L of KH_2PO_4); a salt mixture (0.25-0.4 g/L of $MgSO_4$; 0.044-0.088 g/L of $ZnSO_4$; and 0.0075 g/L of $FeSO_4$); a neutralizing agent (60-100 g/L of $CaCO_3$) and other supplements (0.5 mL/L of corn steep liquor, 15 mL/L of methanol) at pH 6.0 [2]. Research has shown that some stress factors contribute to the increase and accumulation of organic acids by microorganisms. Nitrogen limitation is an important stress factor for the accumulation of organic acids by several molds. Fungi fermentation in medium containing a limited source of nitrogen produces fumaric acid and/or L-malic acid from glucose and not biomass. High concentrations of fumaric or L-malic acids of above 100 g/L can be obtained using filamentous fungi [76, 77]. Addition of $CaCO_3$ to the medium is also required to obtain high concentrations of the acid salt because it maintains the pH of the medium at approximately 6.0 during the accumulation of fumaric acid and L-malic acid. $CaCO_3$ also supplies CO_2 to replenish tricarboxylic acid cycle intermediates by carboxylation of pyruvate to oxaloacetate. The added $CaCO_3$ promotes the precipitation of the calcium salts of the acids. The fumaric and L-malic acids accumulate as calcium salts [36].

The large-scale industrial fermentations of filamentous fungi are difficult due to their complex morphology, resulting in limitations during the submerged fermentation in relation to the mixing and aeration parameters. Another limitation is the difficulty in controlling the formation and size of the filamentous fungal pellets, which can cause problems regarding the limitation of oxygen and nutrient diffusion, reducing the acid production. To eliminate or reduce such limitations,

Gu and co-workers [70] developed a new method for producing fumaric acid using immobilized *Rhyzopus arrhizus* HR-07-13 cells. A device made of net and wire was developed to immobilize the filamentous fungi *Rhyzopus arrhizus* for fumaric acid production. The large surface area of the locking system was responsible for the rapid production of fumaric acid. A chain area of 150 cm^2 and 0.5 x 10^6/mL spores were the best parameters for fumaric acid production. The final concentration of fumaric acid using immobilized cells and the free cells were 32.03 g/L and 31.23 g/L, respectively. However, the fermentation time was reduced from 144 to 24 hours using immobilized cells and productivity was higher than the free-cells, 1.335 and 0.217 g/L h, respectively.

GLUCONIC ACID

Gluconic acid is found in fruits and honey and is used as an additive in foods, beverages, feed, pharmaceuticals, and also in cement in the construction industry. Sodium gluconate is used to sequester ions. D-glucono gamma-lactone acidifies slowly, providing a smooth taste to foods, and it can be used in dairy products, meats and salad dressings. The production of gluconic acid by fermentation was estimated at approximately 87,000 tons per year [6]. It was estimated that 34.5% (30,000 tons) was used in food industries, corresponding to a market of about US $150 million [1].

High yields of gluconic acid can be obtained from glucose by strains of *Aspergillus niger*, *Aureobasidium pululans* and *Penicillium* sp. Matsui and co-workers [78] studied simultaneous corn starch saccharification and the production of gluconic acid using immobilized *Aspergillus niger* cells. Starch liquefaction was performed by adding 0.005% of commercial α-amylase from *Bacillus subtilis* to 300 g corn starch, 0.15 g $MgSO_4 \cdot 7\ H_2O$, 0.2g KH_2PO_4 and 0.5 g $CaCl_2$ in 1 L deionized water at pH 6.0 and 86°C with a magnetic stirrer. Simultaneous saccharification was performed at 30 °C using liquefied starch with the simultaneous addition of commercial amyloglucosidase and *Aspergillus niger* JCM5549 for gluconic acid production at a DO of 150 mg/L by supplying pressurized oxygen gas. The production of gluconic acid in a repeated batch process was 272 g/L, with a yield of 6.1 g/L/h or above 90%, and a minor decrease in activity after 450 hours.

Kazemi and co-workers [79] performed tests to produce gluconic acid by an *Aspergillus niger* strain in an airlift loop bioreactor, using a semi-batch fermentation process with control of the aeration rate. The optimal substrate/initial microorganism concentration ratio was 55g of glucose per g of biomass leading to a 94.8% substrate conversion and the maximum production of gluconic acid. The dissolved oxygen concentration was maintained at a constant level of 2 mg/dm^3, which is considered ideal for the survival of the microorganism.

The cost of the gluconic acid production process can be reduced by replacing the conventional substrate and carbon source, such as glucose, by cheaper sources, such as corn starch hydrolysate, molasses, sugar beet, and the residual by-products from the juice processing industry. For instance, Purane and co-workers [80] used golden syrup as substrate, which is a by-product with a high glucose content that is obtained from the refining of sugarcane juice to obtain sugar or by acid treatment of a sugar solution. The *Aspergillus niger* NCIM 530 strain was used in a batch submerged fermentation process in a 50 L stirred tank reactor. The maximum conversion of glucose was 86.97%, with the production of 85.2 g gluconic acid/L in 44 hours of fermentation.

MALIC ACID

L-Malic acid is a dicarboxylic acid that is an intermediate in the Krebs cycle, and it is naturally found in many fruits and vegetables, such as apple, grape, orange, cherry, carrot and broccoli. Malic acid has no odor, a high solubility (558 g/L at 20 °C) and it is used as an acidulant and taste enhancer in the food and beverage industries (jams, jellies, candies, chewing gum, powdered drinks, pasta, pizza, beverages, *etc.*) [4, 81]. Malic acid is also used as a nutritional supplement and polymer intermediate.

The DL-malic acid racemic mixture is primarily obtained by chemical synthesis through the hydration of maleic acid or fumaric acid at high temperature and pressure. The annual production of malic acid has been estimated at 200,000 metric tons per year [2]. Malic acid can be produced by the fermentation of glucose using fungi, yeast or bacteria strains or by an enzymatic process, where fumaric acid is converted into malic acid by fumarase *via* an intermediate stage of the Krebs cycle. Pyruvic acid carboxylase, which catalyzes the carboxylation of pyruvic acid to oxaloacetic acid in the Krebs cycle, has been linked to an increased production of malic acid in filamentous fungi [82]. Strains of *Aspergillus*, *Brevibacterium*, *Saccharomyces cerevisiae, Torulopsis glabrata, Zygosaccharomyces rouxii* and *E. coli* can be used in the production of L-malic acid [83 - 86].

The conversion of high concentrations of fumaric acid (60 g/L) to L-malic acid by resting *Brevibacterium* sp B2D cells was described by Gong and co-workers [87]. The optimal temperature for the bioconversion of fumaric acid (60 g/L) to L-malic acid was 36 ºC at pH 6.4, with an 89.8% yield of L-malic acid from fumaric acid and an average productivity specificity of 0.41 g/g /h using resting *Brevibacterium* sp cells. The addition of 0.02% of surfactant increased the malic acid production rate two-fold [87]. Production of malic acid by permeabilized *Saccharomyces bayanus* cells was described by Presecki and Vasic-Racki [88]. The cells permeabilized with 0.2% (w/v) CTAB showed an increased fumarase activity and average conversion of fumaric acid of up to 82%. Yields of 21, 40, 83 and 175 mM of L-malic acid were obtained from 25, 50, 100 and 210 mM fumaric acid, respectively [88]. The engineered yeast strain *Saccharomyces cerevisiae* RWB 525 produced malic acid with titers of up to 59 g/L and yields of 0.42 mol (mol/ glucose) in batch grown cultures [85]. The yeast *Torulopsis glabrata* CCTCC M202019, a pyruvate producer, was transformed to establish a new pathway for the production of malate under aerobic conditions, with an increase in the accumulation of malate. The engineered strain *Torulopsis glabrata* CCTCC TG-PMS produced 8.5 g/L malate, 10 times higher than the control *Torulopsis glabrata* CCTCC TG-26 strain [89].

Ochsenreither and co-workers [90] studied the production of malic and fumaric acids by *Aspergillus oryzae* DSM 1863. The microorganism was used to ferment a medium composed of glucose and salts in a bioreactor, and produced 42 g/L of malic acid and fumaric acid as byproducts after 168 hours at 35 ºC.

The typical composition of a medium used to produce malic acid was described by Zhang and co-workers [2], containing a carbon source (100-150 g/L of glucose); a nitrogen source (0.2 g/L of $(NH_4)_2SO_4$ or urea); a phosphorus source

(0.15 g/L of KH_2PO_4, 0.15 g/L of K_2HPO4); salts (0.1 g/L of $MgSO_4$; 0.1 g/L of $CaCl_2$; 5-12 mg/L of $FeSO_4$ or 40 mg/L of Mn^{2+} /Al^{3+}/Cr^{3+}), a neutralizing agent (60-100 g/L of $CaCO_3$) and other supplements (pyruvic acid/ fumaric acid 10-20 g/L) at pH 6.0.

Several strains of the genus *Aspergillus* produce malic acid from pyruvate using a two-step pathway, also known as the reductive Krebs cycle. This simple and efficient pathway has a theoretical maximum yield of two moles of malate/mol glucose when the pyruvate is derived from the glycolytic pathway. The production of malic acid by *Aspergillus oryzae* NRRL 3488 was first improved by the over-expression of a native C4- dicarboxylate transporter, leading to a two-fold increase in the production rate of malate. Subsequently, the over-expression of the native alleles of cytosolic pyruvate carboxylase and malate dehydrogenase, which are present in the reductive Krebs cycle, in combination with the carrier, resulted in an additional 27% increase in the production rate of malate. Trials were performed in a batch process in a 2 L fermenter, with the temperature maintained at 34 °C, the inlet air flow at 1 vvm and the agitation rate at 500 rpm. The modified microorganism produced 154 g of malate/L in 164 hours, corresponding to a productivity rate of 0.94 g/L/h with a yield of 1.38 mol/mol (69% of the theoretical maximum) using glucose as the substrate [91].

Knuf and co-workers [92] investigated the production of malic acid by *Aspergillus oryzae* NRRL3488 under nitrogen starvation conditions by submerged fermentation in Erlenmeyer flasks and in a stirred 2.7 L reactor. The influence of two nitrogen sources on malic acid production was studied. The fermentation of *Aspergillus oryzae* NRRL3488 in a stirred fermentor containing medium composed of 50 g/ L glucose, 6 g/L peptone, salts and antifoam at 34 °C, 950 rpm, and an aeration rate of 1 vvm resulted in a higher final titer of 30.27 g/L malic acid. The cultivation in medium containing 50 g/ L glucose, 1.4 g/L ammonium sulfate, salts and antifoam resulted in a titer of 22.27 g/L malic acid.

The direct production of malic acid by microbial fermentation may be limited due to inhibition of the final product. Zou and co-workers [93] studied a new process for obtaining malic acid from polymalic acid by *Aureobasidium pullulans* ZX-10 fermentation. The fermentation of yeast in a stirred-tank bioreactor produced polymalic acid as the main product, which was equivalent to 87.6 g/L of malic acid and productivity of 0.61 g/L h. Fed-batch fermentations with cells immobilized in a fibrous-bed bioreactor achieved the highest concentration, 144.2 g/L, of polymalic acid and productivity of 0.74 g/L h. Malic acid was then produced from polymalic acid by hydrolysis with 2 M sulfuric acid at 85ºC.

ASCORBIC ACID

L-Ascorbic acid (vitamin C) is used as an antioxidant, nutritional supplement or additive in the food, pharmaceutical, feed and cosmetic industries [9, 94]. L-ascorbic acid is a vitamin for human and non-human primates and a few other mammals [95]. L-ascorbic acid is an antioxidant that acts as a cofactor for various metallo-enzymes involved in the hydroxylation of collagen and dopamine, carnitine biosynthesis, tyrosine metabolism, the alpha-amidation of peptide hormones. It is produced by the pharmaceutical industry as a vitamin supplement and is used as an additive by the cosmetic industry. L-ascorbic acid is used in foods to cure meat and inhibit the enzymatic browning of fruits and vegetables due to its reducing and antioxidant activities [96]. Ascorbic acid stabilizes the pigment of cured meat and prevents the formation of undesirable compounds, such as nitrosamines [3]. Fatty acid esters, such as ascorbyl palmitate and ascorbyl acetate, are soluble in lipids and can act as antioxidants in lipids [10, 97].

The production of ascorbic acid has been estimated at 110,000 tons per year and approximately 50% of the world's L-ascorbic acid production is applied in the pharmaceutical industry as a nutritional supplement. L-ascorbic acid is also used in the food industry (25%), in the production of beverages (15%) and in animal feed (10%) [9, 10].

L-ascorbic acid can be obtained by the Reichstein process [98] and by the two-step fermentation process [9, 10, 94, 99]. Both of these commercial processes for ascorbic acid production involve 2-keto-L-gulonic acid (2-KLGA) as a key intermediate and precursor of vitamin C.

In the Reichstein process, D-glucose is chemically hydrogenated to sorbitol in the initial step. The D-sorbitol is then oxidized to L-sorbose by bioconversion using *Gluconobacter oxydans* or *Gluconobacter melanogenes* strains. Because only L-ascorbic acid is biologically active, this biotransformation step performed by the stereo-specific bacterium *Gluconobacter* sp. is important. The L-sorbose product is crystallized, and, after condensation with acetone, diacetone sorbose is formed [10]. The diacetone sorbose is oxidized to 2-keto-L-gulonic acid (2-KLGA) using platinum as the acid catalyst, and the 2-KLGA then undergoes lactonization and enolization to produce L-ascorbic acid (Fig. **5**).

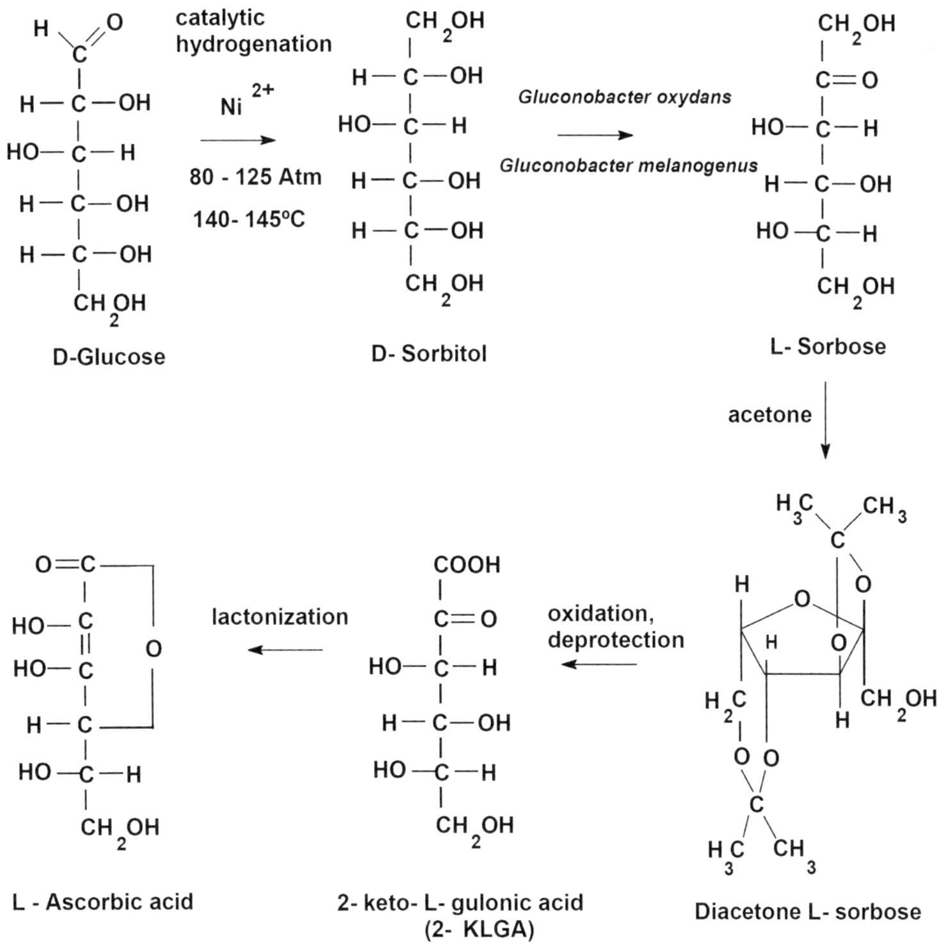

Fig. (5). Reactions involved in the Reichstein-Grussner process for the synthesis of ascorbic acid from glucose (Adapted from Rose [100]; Pappenberger and Hohmann [9]).

The biological reaction is catalyzed by *Gluconobacter oxydans*, which regioselectively oxidizes D-sorbitol to L-sorbose [101 - 103]. De Wulf and co-workers [103] were able to obtain a productivity of 200 g/L of L-sorbose from 200 g/L of D-sorbitol using a *Gluconobacter oxydans* mutant strain. Ascorbic acid cannot be produced directly from glucose by microorganisms. Thus, research on the use of microorganisms for the production of L-ascorbic acid involves the production of intermediates from the Reischtein Process. *Gluconobacter* sp and recombinant strains that produce large amounts of 2,5-diketo-L-gulonic acid from glucose and produce L-sorbose dehydrogenase and L-sorbosone dehydrogenase, which converts D-sorbitol to 2-keto-L-gulonic acid (2-KLGA), a precursor of ascorbic acid, have been reported [104 - 106].

2-Keto-L-gulonic acid (2-KLGA) has been produced from L-sorbose by a mixed culture of *Ketogulonigenium vulgare* and *Bacillus megaterium* [10]. *K. vulgare* converts L-sorbose into 2-keto-L-gulonic acid in this co-culture [107], but its poor growth rate when grown as a pure culture significantly decreases the yield of 2-KLGA. The growth of *K. vulgare* and its production of 2-KLGA, can be increased by co-cultivation with *Bacillus* spp, which can produce metabolites that are beneficial to the growth of *K. vulgare* [98, 108].

Folate derivatives are growth factors for *Ketogulonigenium vulgare.* Cai and co-workers [109] verified that when the folate biosynthesis gene cluster was introduced into *Ketogulonigenium vulgare,* it resulted in enhanced cell density and 2-keto-L-gulonic acid (2-KLGA) production. In fermenter studies, cell density was enhanced approximately 25% and 2-KLGA productivity was enhanced approximately 35% in comparison with the control, without over expressing the folate biosynthesis genes.

Du and co-workers [110] studied the expression of L-sorbose dehydrogenase and sorbosone dehydrogenase genes (*sdh/sndh*) and the synthesis (*pqq*ABCDEN) of the pyrroloquinoline (PQQ) genes in *Ketogulonigenium vulgare* to increase the production of 2-keto-L-gulonic acid (2-KLGA), the precursor of vitamin C, in *K. vulgare* and *Bacillus cereus* co-cultures. The overexpression of *sdh/sndh* and *pqq*ABCDEN enabled a 20% increase in the production of 2-KLGA (79.1 ± 0.6 g/L) in shaken flasks compared to the parental *K. vulgare* HB602 (6.59 ± 0.4 g/L).

CONCLUSIONS

Micro-organisms and enzymes can be used as biocatalysts for the production of several organic acids and other compounds that are naturally observed in fruits and vegetables from carbohydrate metabolism intermediates. There has been growing interest in the production of organic acids and other metabolites by fermentation with microorganisms due to several factors, such as the rising cost of oil, concern about environmental damage due to chemical residues, advances in genetic engineering and metabolic engineering to obtain mutant microorganisms, and increasing knowledge of the mechanisms regulating the biosynthesis of the metabolite flow and excretion mechanisms. The selection of new microbial strains producing organic acids, such as acetic acid, citric acid, lactic acid, malic acid, fumaric acid and succinic acid, that can be applied as acidulants in the food industry and as precursors of biodegradable polymers and other compounds contribute to the reduction of costs and the use of agro-industrial waste. The possibility of obtaining new organic acid producing strains by fermentation or enzymatic reactions using renewable carbon sources has aroused interest in fermentation and biotransformation processes. Optimization of the fermentation conditions, recovery and purification of organic acids can contribute to reducing the costs and increasing the applications of these products.

CONFLICT OF INTEREST

The authors confirm that they have no conflict of interest to declare for this publication.

ACKNOWLEDGEMENTS

Decleared none.

REFERENCES

[1] Business Communication CoBCC Research). *World markets for fermentation ingredients. Norwalk, CT, USA: GA-103R; Report code FOD020C,* **2004**. http:www.bcresearch.com.

[2] Zhang, K.; Zhang, B.; Yang, S.T. Production of citric, itaconic, fumaric and malic acids in filamentous fungal fermentations. , **2013**.
[http://dx.doi.org/10.1002/9781118642047.ch20]

[3] Goldberg, I.; Peleg, Y. – Citric, fumaric and malic acids. In: *Biotechnology and food ingredients*;

Goldberg, I.; Williams, R.; Rokem, J.S, Eds.; Van Nostrand Reinhold: New York, **1991**.

[4] Goldberg, I.; Rokem, J.S.; Pines, O. Review organic acids: old metabolites, new themes. *J. Chem. Technol. Biotechnol.,* **2006**, *81*, 1601-1661. [http://dx.doi.org/10.1002/jctb.1590]

[5] Soccol, C.R.; Vanderberghe, L.P.; Rodrigues, C.; Pandey, A. New perspectives for citric acid production and application. *Food Technol. Biotechnol.,* **2006**, *44*, 141-149.

[6] Sauer, M.; Porro, D.; Mattanovich, D.; Branduardi, P. Microbial production of organic acids: expanding the markets. *Trends Biotechnol.,* **2008**, *26*(2), 100-108. [http://dx.doi.org/10.1016/j.tibtech.2007.11.006] [PMID: 18191255]

[7] Demain, A.L. The business of biotechnology. *Ind. Biotechnol. (New Rochelle N.Y.),* **2007**, *3*, 269-283. [http://dx.doi.org/10.1089/ind.2007.3.269]

[8] Saitoh, S.; Ishida, N.; Onishi, T.; Tokuhiro, K.; Nagamori, E.; Kitamoto, K.; Takahashi, H. Genetically engineered wine yeast produces a high concentration of L-lactic acid of extremely high optical purity. *Appl. Environ. Microbiol.,* **2005**, *71*(5), 2789-2792. [http://dx.doi.org/10.1128/AEM.71.5.2789-2792.2005] [PMID: 15870375]

[9] Pappenberger, G.; Hohmann, H-P. Industrial production of L-ascorbic Acid (vitamin C) and D-isoascorbic acid. *Adv. Biochem. Eng. Biotechnol.,* **2014**, *143*, 143-188. [http://dx.doi.org/10.1007/10_2013_243] [PMID: 24258144]

[10] Bremus, C.; Herrmann, U.; Bringer-Meyer, S.; Sahm, H. The use of microorganisms in L-ascorbic acid production. *J. Biotechnol.,* **2006**, *124*(1), 196-205. [http://dx.doi.org/10.1016/j.jbiotec.2006.01.010] [PMID: 16516325]

[11] Jianlong, W.; Xianghua, W.; Ding, Z. Production of acid citric from molasses integrated with *in situ* product separation by ion-exchange resin adsorption. *Bioresour. Technol.,* **2000**, *75*, 231-234. [http://dx.doi.org/10.1016/S0960-8524(00)00067-5]

[12] Karadeniz, F. Main organic acid distribution of authentic citrus in Turkey. Tyrk J. *Agric. For.,* **2004**, *28*, 267-271.

[13] Cámara, M.M.; Diez, C. Toriza, M.E.; Cano, M.P. H`LC determination of organic acid in pineapple and nectars. *Z. Lebensm. Unters. Forsch.,* **1994**, *198*, 52-56. [http://dx.doi.org/10.1007/BF01195284]

[14] Dhillon, G.S.; Brar, S.K.; Verma, M.; Tyagi, R.D. Recent advances in citric acid bio-production and recovery. *Food Bioprocess Technol,* **2011**, *4*, 505-529. [http://dx.doi.org/10.1007/s11947-010-0399-0]

[15] Vandenberghe, L.P.; Soccol, C.R.; Prado, F.C.; Pandey, A. Comparison of citric acid production by solid-state fermentation in flask, column, tray, and drum bioreactors. *Appl. Biochem. Biotechnol.,* **2004**, *118*(1-3), 293-303. [http://dx.doi.org/10.1385/ABAB:118:1-3:293] [PMID: 15304757]

[16] Kishore, K.A.; Reddy, G.V. Modeling and optimization studies on the effect of stimulators on the fermentation of citric acid. *Asian-Am. J. Chem. .,* **2013**, *1*, 25-30.

[17] Demain, A.L.; Sanchez, S. Microbial synthesis of primary metabolites: current advances and future prospect. In: *Fermentation Microbiology and Biotechnology*; El-Mansi, E.M., Ed.; Taylor and Francis:

London, **2006**.

[18] Mattey, M. The production of organic acids. *Crit. Rev. Biotechnol.,* **1992**, *12*(1-2), 87-132. [http://dx.doi.org/10.3109/07388559209069189] [PMID: 1733523]

[19] Papagianni, M. Advances in citric acid fermentation by *Aspergillus niger*: biochemical aspects, membrane transport and modeling. *Biotechnol. Adv.,* **2007**, *25*(3), 244-263. [http://dx.doi.org/10.1016/j.biotechadv.2007.01.002] [PMID: 17337335]

[20] Sanchez, S.; Demain, A.L. Metabolic regulation and overproduction of primary metabolites. *Microb. Biotechnol.,* **2008**, *1*(4), 283-319. [http://dx.doi.org/10.1111/j.1751-7915.2007.00015.x] [PMID: 21261849]

[21] Clark, D.S.; Ito, K.; Horitsu, H. Effect of manganese and other heavy metals on submerged citric acid fermentation of molasses. *Biotechnol. Bioeng.,* **1966**, *III*, 465-471. [http://dx.doi.org/10.1002/bit.260080402]

[22] Kubicek, C.P.; Roehr, M. Citric acid fermentation. *Crit. Rev. Biotechnol.,* **1986**, *3*, 331-373. [http://dx.doi.org/10.3109/07388558509150788]

[23] Rodrigues, C.; Vandenberghe, L.P.; Sturm, W.; Dergint, D.E.; Spier, M.R.; de Carvalho, J.C.; Soccol, C.R. Effect of forced aeration on citric acid production by *Aspergillus* sp. mutants in SSF. *World J. Microbiol. Biotechnol.,* **2013**, *29*(12), 2317-2324. [http://dx.doi.org/10.1007/s11274-013-1397-y] [PMID: 23760557]

[24] Betiku, E.; Adesina, O.A. Statistical approach to the optimization of citric acid production using filamentous fungus *Aspergillus niger* grown on sweet potato starch hydrolyzate. *Biomass Bioenergy,* **2013**, *55*, 350-354. [http://dx.doi.org/10.1016/j.biombioe.2013.02.034]

[25] Hamdy, H.S. Citric acid production by *Aspergillus niger* grown on orange peel medium fortified with cane molasses. *Ann. Microbiol.,* **2013**, *63*, 267-278. [http://dx.doi.org/10.1007/s13213-012-0470-3]

[26] Dhillon, G.S.; Brar, S.K.; Verma, M. Biotechnological potential of industrial wastes for economical citric acid bioproduction by *Aspergillus niger* through submerged fermentation. *Int. J. Food Sci. Technol.,* **2012**, *47*, 542-548. [http://dx.doi.org/10.1111/j.1365-2621.2011.02875.x]

[27] Silva, L.V.; Tavares, C.B.; Amaral, P.F.; Coelho, M.A. Production of citric acid by *Yarrowia lipolytica* in different crude glycerol concentrations and in different nitrogen sources. *Chemical Engineering Transactions,* **2012**, *27*, 199-204.

[28] Morgunov, I.G.; Kamzolova, S.V.; Lunina, J.N. The citric acid production from raw glycerol by *Yarrowia lipolytica* yeast and its regulation. *Appl. Microbiol. Biotechnol.,* **2013**, *97*(16), 7387-7397. [http://dx.doi.org/10.1007/s00253-013-5054-z] [PMID: 23807667]

[29] Max, B.; Salgado, J.M.; Rodríguez, N.; Cortés, S.; Converti, A.; Domínguez, J.M. Biotechnological production of citric acid. *Braz. J. Microbiol.,* **2010**, *41*(4), 862-875. [http://dx.doi.org/10.1590/S1517-83822010000400005] [PMID: 24031566]

[30] Angumeenal, A.R.; Venkappayya, D. An overview of citric acid production. *LWR- . Food Sci. Technol. (Campinas.),* **2013**, *50*, 367-370.

[31] Wang, L.F.; Wang, Z.P.; Liu, X.Y.; Chi, Z.M. Citric acid production from extract of Jerusalem artichoke tubers by the genetically engineered yeast Yarrowia lipolytica strain 30 and purification of citric acid. *Bioprocess Biosyst. Eng.,* **2013**, *36*(11), 1759-1766. [http://dx.doi.org/10.1007/s00449-013-0951-1] [PMID: 23584740]

[32] Dhillon, G.S.; Brar, S.K.; Kaur, S.; Verma, M. Rheological studies during submerged citric acid fermentation by *Aspergillus niger* in stirred fermentor using apple pomace ultrafiltration sludge. *Food and Bioprocess Technology,* **2013**, *6*, 1240-1250. [http://dx.doi.org/10.1007/s11947-011-0771-8]

[33] Förster, A.; Jacobs, K.; Juretzek, T.; Mauersberger, S.; Barth, G. Overexpression of the ICL1 gene changes the product ratio of citric acid production by *Yarrowia lipolytica. Appl. Microbiol. Biotechnol.,* **2007**, *77*(4), 861-869. [http://dx.doi.org/10.1007/s00253-007-1205-4] [PMID: 17952432]

[34] Rywinska, A.; Rymowicz, W.; Zarowska, B.; Wojtatowicz, M. Biosynthesis of citric acid from glycerol by acetate mutants of *Yarrowia lipolytica* in fed-batch fermentation. *Food Technol. Biotechnol.,* **2009**, *47*, 1-6.

[35] Pazouki, M.; Panda, T. Recovery of citric acid - A review. *Bioprocess Eng.,* **1998**, *19*, 435-439. [http://dx.doi.org/10.1007/PL00009029]

[36] Wu, J.; Peng, Q.; Arlt, W.; Minceva, M. Model-based design of a pilot-scale simulated moving bed for purification of citric acid from fermentation broth. *J. Chromatogr. A,* **2009**, *1216*(50), 8793-8805. [http://dx.doi.org/10.1016/j.chroma.2009.03.028] [PMID: 19344909]

[37] Juang, R.S.; Chou, T.C. Sorption of citric acid from aqueous solutions by macroporous resins containing a tertiary amine equilibria. *Sep. Sci. Technol.,* **1996**, *31*, 1409-1425. [http://dx.doi.org/10.1080/01496399608001404]

[38] Chaudhary, N.; Raj, S. Optimization of process parameters of citric acid production from cane molasses using free and immobilized cells of manganese resistant mutant of *Aspergillus niger. Electronic J. of Environ. Agric. Food Chem,* **2012**, *11*, 180-187.

[39] Jarboe, LR; Zhan, X; Wang, X; Moore, JC; Shanmungam, KT; Ingram, LO Metabolic engineering for production of bioreneawble fuels and chemicals: contributions of synthetic biology. *Journal of Biomedicine and Biotechnology,* **2010**, *2010*, 18.

[40] Mas, A.; Torija, M.J.; García-Parrila, M.C.; Troncoso, A.M. Acetic acid bacteria and the production and quality of wine vinegar. *The Scientific Wourld Journal., article ID394671,* **2014**, 6. [http://dx.doi.org/10.1155/2014/394671]

[41] Yakushi, T.; Matsushita, K. Alcohol dehydrogenase of acetic acid bacteria: structure, mode of action, and applications in biotechnology. *Appl. Microbiol. Biotechnol.,* **2010**, *86*(5), 1257-1265. [http://dx.doi.org/10.1007/s00253-010-2529-z] [PMID: 20306188]

[42] Cerezo, A.B.; Tesfaye, W.; Soria, M.E.; Torija, M.J.; Mateo, E.; Garcia-Parrila, M.C.; Troncoso, A.M. effect of wood on the phenolic profile and sensory properties of wine vinegars during aging. *J. Food Compos. Anal.,* **2010**, *23*, 175-184. [http://dx.doi.org/10.1016/j.jfca.2009.08.008]

[43] Greenshields, R.N. Acetic Acid: Vinegar. In: *Economic Microbiology. Primary Products of*

Metabolism; Rose, A.H., Ed.; Academic Press, **1978**; Vol. 2, pp. 121-183.

[44] Ebner, H.; Follmann, H. Acetic Acid. In: *Biomass, Microorganisms for Especial Applications. Microbial Products I, Energy from Renewable Resources*; Academic Press: USA, **1983**; 3, pp. 387-407.

[45] Ghose, T.K.; Bhadra, A. Acetic acid. *Comprehensive biotechnology – The practice of biotechnology: Current commodity products,* **1985**, *3*, 701-719.

[46] Gullo, M.; Verzelloni, E.; Canonico, M. Aerobic submerged fermentation by acetic acid bacteria for vinegar production: process and biotechnological aspects. *Process Biochem.,* **2014**, *49*, 1571-1579. [http://dx.doi.org/10.1016/j.procbio.2014.07.003]

[47] Solieri, L.; Giudici, P. Yeasts associated to Traditional Balsamic Vinegar: ecological and technological features. *Int. J. Food Microbiol.,* **2008**, *125*(1), 36-45. [http://dx.doi.org/10.1016/j.ijfoodmicro.2007.06.022] [PMID: 17900732]

[48] Gullo, M.; Giudici, P. Acetic acid bacteria in traditional balsamic vinegar: phenotypic traits relevant for starter cultures selection. *Int. J. Food Microbiol.,* **2008**, *125*(1), 46-53. [http://dx.doi.org/10.1016/j.ijfoodmicro.2007.11.076] [PMID: 18177968]

[49] Wang, Z.; Yan, M.; Chen, X.; Li, D.; Quin, L.; Li, Z.; Yao, J.; Liang, X. Mixed culture of *Saccharomyces cerevisiae* and *Acetobacter pasteurianus* for acetic acid production. *Biochem. Eng. J.,* **2013**, *79*, 41-45. [http://dx.doi.org/10.1016/j.bej.2013.06.019]

[50] Yang, H-L.; Qi, Z-L.; Xi, X-L.; Xin, Y.; Zhang, L.; Leng, Y-W.; Quan, W.; Wang, W. An optimum medium designed and verified for alcohol vinegar fermentation. *Afr. J. Biotechnol.,* **2011**, *10*, 8421-8427.

[51] Berraud, C. Production of highly concentrated vinegar in fed-batch culture. *Biotechnol. Lett.,* **2000**, *22*, 451-454. [http://dx.doi.org/10.1023/A:1005639030971]

[52] Fukaya, M.; Tayama, K.; Tamaki, T.; Tagami, H.; Okumura, H.; Kawamura, Y.; Beppu, T. Cloning of the membrane-bound aldehyde dehydrogenase gene of *Acetobacter polyoxogenes* and improvement of acetic acid production by use of the cloned gene. *Appl. Environ. Microbiol.,* **1989**, *55*(1), 171-176. [PMID: 16347820]

[53] Qi, Z.; Yang, H.; Xia, X.; Xin, Y.; Zhang, L.; Wang, W.; Yu, X. A protocol for optimization vinegar fermentation according to the ratio of oxygen consumption *versus* acid yield. *J. Food Eng.,* **2013**, *116*, 304-309. [http://dx.doi.org/10.1016/j.jfoodeng.2012.12.029]

[54] Schlepütz, T.; Gerhards, J.P.; Büchs, J. Ensuring constant oxygen supply during inoculation is essential to obtain reproducible results with obligatory aerobic acetic acid bacteria in vinegar production. *Process Biochem.,* **2013**, *48*, 398-405. [http://dx.doi.org/10.1016/j.procbio.2013.01.009]

[55] Kocher, G.S.; Phutela, R.P.; Dhillon, H.K.; Uppal, S.K.; Arora, J.K.; Bakshi, D. Standardization of an economical bioprocess for production of natural vinegar from sugarcane. *Sugar Tech.,* **2014**, *16*, 15-21.

[http://dx.doi.org/10.1007/s12355-013-0245-6]

[56] Demirci, A. Pometto III, A.L. Enhanced production of D (-) –lactic acid by mutants of *Lactobacillus delbrueckii* ATTC 9649. *J. Ind. Microbiol.,* **1992**, *11*, 23-28. [http://dx.doi.org/10.1007/BF01583728]

[57] Martinez, F.A.; Balciunas, E.M.; Salgado, J.M.; González, J.M.; Oliveira, R.P. Lactic acid properties and production: A review. *Trends Food Sci. Technol.,* **2013**, *30*, 70-83. [http://dx.doi.org/10.1016/j.tifs.2012.11.007]

[58] Hofvendahl, K.; Hahn-Hägerdal, B. Factors affecting the fermentative lactic acid production from renewable resources(1). *Enzyme Microb. Technol.,* **2000**, *26*(2-4), 87-107. [http://dx.doi.org/10.1016/S0141-0229(99)00155-6] [PMID: 10689064]

[59] Abdel-Rahman, M.A.; Tashiro, Y.; Sonomoto, K. Recent advances in lactic acid production by microbial fermentation processes. *Biotechnol. Adv.,* **2013**, *31*(6), 877-902. [http://dx.doi.org/10.1016/j.biotechadv.2013.04.002] [PMID: 23624242]

[60] Zhang, Z.H.; Zeng, J.; Li, G.L.; Gao, H.Y. Effect of stearoyl lactate calcium-sodium as promising improvers for wheat bread. *Adv. Mat. Res.,* **2012**, *554-556*, 1406-1409. [http://dx.doi.org/10.4028/www.scientific.net/AMR.554-556.1406]

[61] Van Steertegem, B.; Pareyt, B.; Brijs, K.; Delcour, J.A. Impact of mixing time and sodium stearoyl lactylate on gluten polymerization during baking of wheat flour dough. *Food Chem.,* **2013**, *141*(4), 4179-4185. [http://dx.doi.org/10.1016/j.foodchem.2013.07.017] [PMID: 23993603]

[62] Budhavaram, N.K.; Fan, Z. Production of lactic acid from paper sludge using acid-tolerant, thermophilic Bacillus coagulan strains. *Bioresour. Technol.,* **2009**, *100*(23), 5966-5972. [http://dx.doi.org/10.1016/j.biortech.2009.01.080] [PMID: 19577925]

[63] Bai, D.M.; Wei, Q.; Yan, Z.H.; Zhao, X.M.; Li, X.G.; Xu, S.M. Fed-batch fermentation of *Lactobacillus lactis* for hyper-production of L-lactic acid. *Biotechnol. Lett.,* **2003**, *25*(21), 1833-1835. [http://dx.doi.org/10.1023/A:1026276925649] [PMID: 14677707]

[64] Nguyen, C.M.; Kim, J-S.; Nguyen, T.N.; Kim, S.K.; Choi, G.J.; Choi, Y.H.; Jang, K.S.; Kim, J.C. Production of L- and D-lactic acid from waste *Curcuma longa* biomass through simultaneous saccharification and cofermentation. *Bioresour. Technol.,* **2013**, *146*, 35-43. [http://dx.doi.org/10.1016/j.biortech.2013.07.035] [PMID: 23911815]

[65] Zhao, K.; Qiao, Q.; Chu, D.; Gu, H.; Dao, T.H.; Zhang, J.; Bao, J. Simultaneous saccharification and high titer lactic acid fermentation of corn stover using a newly isolated lactic acid bacterium *Pediococcus acidilactici* DQ2. *Bioresour. Technol.,* **2013**, *135*, 481-489. [http://dx.doi.org/10.1016/j.biortech.2012.09.063] [PMID: 23127836]

[66] Zhou, X.; Ye, L.; Wu, J.C. Efficient production of L-lactic acid by newly isolated thermophilic *Bacillus coagulans* WCP10-4 with high glucose tolerance. *Appl. Microbiol. Biotechnol.,* **2013**, *97*(10), 4309-4314. [http://dx.doi.org/10.1007/s00253-013-4710-7] [PMID: 23354450]

[67] Cao, N.; Du, J.; Gong, C.S.; Tsao, G.T. Simultaneous production and recovery of fumaric acid from immobilized *Rhizopus oryzae* with a rotary biofilm contactor and an adsorption column. *Appl.*

Environ. Microbiol., **1996**, *62*(8), 2926-2931.
[PMID: 16535381]

[68] Roa Engel, C.A.; Straathof, A.J.; Zijlmans, T.W.; van Gulik, W.M.; van der Wielen, L.A. Fumaric acid production by fermentation. *Appl. Microbiol. Biotechnol.,* **2008**, *78*(3), 379-389.
[http://dx.doi.org/10.1007/s00253-007-1341-x] [PMID: 18214471]

[69]

[70] Gu, C.; Zhou, Y.; Liu, L.; Tan, T.; Deng, L. Production of fumaric acid by immobilized *Rhizopus arrhizus* on net. *Bioresour. Technol.,* **2013**, *131*, 303-307.
[http://dx.doi.org/10.1016/j.biortech.2012.12.148] [PMID: 23360706]

[71] Roa Engel, C.A.; Straathof, A.J.; Zijlmans, T.W.; van Gulik, W.M.; van der Wielen, L.A. Fumaric acid production by fermentation. *Appl. Microbiol. Biotechnol.,* **2008**, *78*(3), 379-389.
[http://dx.doi.org/10.1007/s00253-007-1341-x] [PMID: 18214471]

[72] Oda, Y.; Yajima, Y.; Kinoshita, M.; Ohnishi, M. Differences of *Rhizopus oryzae* strains in organic acid synthesis and fatty acid composition. *Food Microbiol.,* **2003**, *20*, 371-375.
[http://dx.doi.org/10.1016/S0740-0020(02)00131-4]

[73] Zhou, Z.; Du, G.; Hua, Z.; Zhou, J.; Chen, J. Optimization of fumaric acid production by *Rhizopus delemar* based on the morphology formation. *Bioresour. Technol.,* **2011**, *102*(20), 9345-9349.
[http://dx.doi.org/10.1016/j.biortech.2011.07.120] [PMID: 21880482]

[74] Zhang, B.; Skory, C.D.; Yang, S-T. Metabolic engineering of *Rhizopus oryzae* : effects of overexpressing pyc and pepc genes on fumaric acid biosynthesis from glucose. *Metab. Eng.,* **2012**, *14*(5), 512-520.
[http://dx.doi.org/10.1016/j.ymben.2012.07.001] [PMID: 22814110]

[75] Yu, S.; Huang, D.; Wen, J.; Li, S.; Chen, Y.; Jia, X. Metabolic profiling of a *Rhizopus oryzae* fumaric acid production mutant generated by femtosecond laser irradiation. *Bioresour. Technol.,* **2012**, *114*, 610-615.
[http://dx.doi.org/10.1016/j.biortech.2012.03.087] [PMID: 22516248]

[76] Goldberg, I.; Steiglitz, B. Improved rate of fumaric acid production by Tweens and vegetable oils in *rhizopus arrhizus. Biotechnol. Bioeng.,* **1985**, *27*(7), 1067-1069.
[http://dx.doi.org/10.1002/bit.260270721] [PMID: 18553780]

[77] Battat, E.; Peleg, Y.; Bercovitz, A.; Rokem, J.S.; Goldberg, I. Optimization of L-malic acid production by Aspergillus flavus in a stirred fermentor. *Biotechnol. Bioeng.,* **1991**, *37*(11), 1108-1116.
[http://dx.doi.org/10.1002/bit.260371117] [PMID: 18597343]

[78] Matsui, T.; Tooyama, K.; Sato, S. Simultaneous saccharification of corn starch in gluconic acid production by *Aspergillus niger* immobilized on nonwoven fabric in a pressurized reactor. *J Microb Biochem Technol,* **2013**, *5*, 988-991.

[79] Kazemi, M.A.; Bamdad, H.; Papari, S.; Yaghmaei, S. Modeling and control of dissolved oxygen concentration in the fermentation of glucose to gluconic acid. *Periodica Polytechnic Chemical Engineering.,* **2013**, *57*, 63-70.
[http://dx.doi.org/10.3311/PPch.2172]

[80] Purane, N.K.; Sharma, S.K.; Salunkhe, P.D.; Labade, D.S.; Tondlikar, M.M. Gluconic acid production

from golden syrup by *Aspergillus niger* strain using semiautomatic stirred-tank fermenter. *Journal of Microbial and Biochemical Technology,* **2012**, *4*, 92-95. [http://dx.doi.org/10.4172/1948-5948.1000077]

[81] Zelle, R.M.; de Hulster, E.; Kloezen, W.; Pronk, J.T.; van Maris, A.J. Key process conditions for production of C(4) dicarboxylic acids in bioreactor batch cultures of an engineered Saccharomyces cerevisiae strain. *Appl. Environ. Microbiol.,* **2010**, *76*(3), 744-750. [http://dx.doi.org/10.1128/AEM.02396-09] [PMID: 20008165]

[82] Nelson, D.; Cox, M.M. *Lehninger Principles of Biochemistry,* 4th ed; Worth Publishers: New York, **2005**, p. 544.

[83] Bercovitz, A.; Peleg, Y.; Battat, E.; Rokem, J.S.; Goldberg, I. Localization of pyruvate carboxylase in organic acid-producing Aspergillus strains. *Appl. Environ. Microbiol.,* **1990**, *56*(6), 1594-1597. [PMID: 2383004]

[84] Taing, O.; Taing, K. Production of malic and succinic acid by sugar-tolerant yeast *Zygosachharomyces rouxii. Eur. Food Res. Technol.,* **2007**, *224*, 343-343. [http://dx.doi.org/10.1007/s00217-006-0323-z]

[85] Zelle, R.M.; de Hulster, E.; van Winden, W.A.; de Waard, P.; Dijkema, C.; Winkler, A.A.; Geertman, J.M.; van Dijken, J.P.; Pronk, J.T.; van Maris, A.J. Malic acid production by *Saccharomyces cerevisiae* : engineering of pyruvate carboxylation, oxaloacetate reduction, and malate export. *Appl. Environ. Microbiol.,* **2008**, *74*(9), 2766-2777. [http://dx.doi.org/10.1128/AEM.02591-07] [PMID: 18344340]

[86] Jantama, K.; Haupt, M.J.; Svoronos, S.A.; Zhang, X.; Moore, J.C.; Shanmugam, K.T.; Ingram, L.O. Combining metabolic engineering and metabolic evolution to develop nonrecombinant strains of Escherichia coli C that produce succinate and malate. *Biotechnol. Bioeng.,* **2008**, *99*(5), 1140-1153. [http://dx.doi.org/10.1002/bit.21694] [PMID: 17972330]

[87] Gong, C.S.; Cao, N.; Sun, Y.; Tsao, G.T. Production of L-malic acid from fumaric acid by resting cells of *Brevibacterium* sp. *Appl. Biochem. Biotechnol.,* **1996**, *57*, 481-487. [http://dx.doi.org/10.1007/BF02941728]

[88] Presecki, A.V.; Vasić-Racki, D. Production of L-malic acid by permeabilized cells of commercial *Saccharomyces* sp. strains. *Biotechnol. Lett.,* **2005**, *27*(23-24), 1835-1839. [http://dx.doi.org/10.1007/s10529-005-3890-3] [PMID: 16328976]

[89] Chen, X.; Xu, G.; Xu, N.; Zou, W.; Zhu, P.; Liu, L.; Chen, J. Metabolic engineering of *Torulopsis glabrata* for malate production. *Metab. Eng.,* **2013**, *19*, 10-16. [http://dx.doi.org/10.1016/j.ymben.2013.05.002] [PMID: 23707987]

[90] Ochsenreither, K.; Fischer, C.; Neumann, A.; Syldatk, C. Process characterization and influence of alternative carbon sources and carbon-to-nitrogen ratio on organic acid production by *Aspergillus oryzae* DSM1863. *Appl. Microbiol. Biotechnol.,* **2014**, *98*(12), 5449-5460. [http://dx.doi.org/10.1007/s00253-014-5614-x] [PMID: 24604500]

[91] Brown, S.H.; Bashkirova, L.; Berka, R.; Chandler, T.; Doty, T.; McCall, K.; McCulloch, M.; McFarland, S.; Thompson, S.; Yaver, D.; Berry, A. Metabolic engineering of *Aspergillus oryzae* NRRL 3488 for increased production of L-malic acid. *Appl. Microbiol. Biotechnol.,* **2013**, *97*(20), 8903-8912.

[http://dx.doi.org/10.1007/s00253-013-5132-2] [PMID: 23925533]

[92] Knuf, C.; Nookaew, I.; Brown, S.H.; McCulloch, M.; Berry, A.; Nielsen, J. Investigation of malic acid production in *Aspergillus oryzae* under nitrogen starvation conditions. *Appl. Environ. Microbiol.,* **2013**, *79*(19), 6050-6058.
[http://dx.doi.org/10.1128/AEM.01445-13] [PMID: 23892740]

[93] Zou, X.; Zhou, Y.; Yang, S.T. Production of polymalic acid and malic acid by *Aureobasidium pullulans* fermentation and acid hydrolysis. *Biotechnol. Bioeng.,* **2013**, *110*(8), 2105-2113.
[http://dx.doi.org/10.1002/bit.24876] [PMID: 23436475]

[94] Hancock, R.D. Recent patents on vitamin C: opportunities for crop improvement and single-step biological manufacture. *Recent Pat. Food Nutr. Agric.,* **2009**, *1*(1), 39-49.
[http://dx.doi.org/10.2174/2212798410901010039] [PMID: 20653525]

[95] Smirnoff, N. L-ascorbic acid biosynthesis. *Vitam. Horm.,* **2001**, *61*, 241-266.
[http://dx.doi.org/10.1016/S0083-6729(01)61008-2] [PMID: 11153268]

[96] Aguayo, E.; Requejo-Jackman, C.; Stanley, R.; Woolf, A. Effects of calcium ascorbate treatments and storage atmosphere on antioxidant activity and quality of fresh-cut apple slices. *Postharvest Biol. Technol.,* **2010**, *57*, 52-60.
[http://dx.doi.org/10.1016/j.postharvbio.2010.03.001]

[97] Hancock, R.D.; Viola, R. Biotechnological approaches for L-ascorbic acid production. *Trends Biotechnol.,* **2002**, *20*(7), 299-305.
[http://dx.doi.org/10.1016/S0167-7799(02)01991-1] [PMID: 12062975]

[98] Reichstein, T.; Grüssner, A. Eine ergiebige synthese der L-ascorbinsäusre (C Vitamin). *Helv. Chim. Acta,* **1934**, *17*, 311-328.
[http://dx.doi.org/10.1002/hlca.19340170136]

[99] Zhang, J.; Liu, J.; Shi, Z.; Liu, L.; Chen, J. Manipulation of *B. megaterium* growth for efficient 2-KGLG production by *K. vulgare. Process Biochem.,* **2010**, *45*, 602-606.
[http://dx.doi.org/10.1016/j.procbio.2009.11.016]

[100] Rose, A.H. History and scientific basis of commercial exploitation of microbial enzymes and bioconversions. In: *Microbial enzymes and bioconversions. Economic Microbiology*; Rose, A.H., Ed.; Academic Press, **1980**; Vol. 5, pp. 1-43.

[101] Giridhar, R.; Srivastava, A.K. Model based constant feed batch L-sorbose production process for improvement in L-sorbose productivity. *Chem. Biochem. Eng. Q.,* **2000**, *14*, 133-140.

[102] Deppenmeier, U.; Hoffmeister, M.; Prust, C. Biochemistry and biotechnological applications of *Gluconobacter* strains. *Appl. Microbiol. Biotechnol.,* **2002**, *60*(3), 233-242.
[http://dx.doi.org/10.1007/s00253-002-1114-5] [PMID: 12436304]

[103] De Wulf, P.; Soetaert, W.; Vandamme, E.J. Optimized synthesis of L-sorbose by C($_5$)-dehydrogenation of D-sorbitol with *Gluconobacter oxydans. Biotechnol. Bioeng.,* **2000**, *69*(3), 339-343.
[http://dx.doi.org/10.1002/1097-0290(20000805)69:3<339::AID-BIT12>3.0.CO;2-E] [PMID: 10861414]

[103] Sugisawa, T.; Hoshino, T.; Matsuda, S.; Nomura, S.; Setoguchi, Y.; Tazoe, M.; Shinjoh, M.; Someha, S.; Fujiwara, A. Microbial production of 2-ceto-L-gulonic acid from L-sorbose and D-sorbitol by

Gluconobacter melanogenus. *Agric. Biol. Chem.,* **1990**, *54*, 1201-1209.
[http://dx.doi.org/10.1271/bbb1961.54.1201]

[105] Saito, Y.; Ishii, Y.; Hayashi, H.; Imao, Y.; Akashi, T.; Yoshikawa, K.; Noguchi, Y.; Soeda, S.; Yoshida, M.; Niwa, M.; Hosoda, J.; Shimomura, K. Cloning of genes coding for L-sorbose and L-sorbosone dehydrogenases from Gluconobacter oxydans and microbial production of 2-keto-L-gulonate, a precursor of L-ascorbic acid, in a recombinant G. oxydans strain. *Appl. Environ. Microbiol.,* **1997**, *63*(2), 454-460.
[PMID: 9023923]

[106] Saito, Y.; Ishii, Y.; Hayashi, H.; Yoshikawa, K.; Noguchi, Y.; Yoshida, S.; Soeda, S.; Yoshida, M. Direct fermentation of 2-keto-L-gulonic acid in recombinant *Gluconobacter oxydans*. *Biotechnol. Bioeng.,* **1998**, *58*(2-3), 309-315.
[http://dx.doi.org/10.1002/(SICI)1097-0290(19980420)58:2/3<309::AID-BIT30>3.0.CO;2-4] [PMID: 10191408]

[107] Takagi, Y.; Sugisawa, T.; Hoshino, T. Continuous 2-keto-L-gulonic acid fermentation from L-sorbose by *Ketogulonigenium vulgare* DSM 4025. *Appl. Microbiol. Biotechnol.,* **2009**, *82*(6), 1049-1056.
[http://dx.doi.org/10.1007/s00253-008-1822-6] [PMID: 19137290]

[108] Sun, J.; Hao, A. The effect of different component produced by *Bacillus megaterium* to *Ketogulonigenium* sp. in the fermentation of vitamin C. *Hebei Chemical Engineering and Industry.,* **2006**, *12*, 27-28.

[109] Cai, L.; Yuan, M.Q. LI.Z.J.; Chen, J.C.; Chen, G.Q. Genetic engineering of *Ketogulonigenium vulgare* for enhanced production of 2-keto-L-gulonic acid. *J. Bacteriol.,* **2012**, *157*, 320-325.

[110] Du, J.; Bai, W.; Song, H.; Yuan, Y.J. Combinational expression of sorbose/sorbosone dehydrogenases and cofactor pyrroloquinoline quinone increases 2-keto-L-gulonic acid production in Ketogulonigenium vulgare-Bacillus cereus consortium. *Metab. Eng.,* **2013**, *19*, 50-56.
[http://dx.doi.org/10.1016/j.ymben.2013.05.006] [PMID: 23747604]

CHAPTER 5

Vitamins and Nutraceuticals

Júlio C. de Carvalho*, **Elia N. M. Ramírez**, **Ana M. O. Finco**, **Luis D. G. Mamani**, **Antônio I. Magalhães**, **Luis A. J. Letti** and **Carlos R. Soccol**

Bioprocess Engineering Laboratory, Federal University of Paraná, Curitiba, Brazil

Abstract: Vitamins are a class of molecules that play an essential role in metabolism. In order to enhance or to complete the nutritional value of a food, it is sometimes, necessary to add vitamins during processing. Most vitamins may be produced by chemical synthesis, but the natural sources are both more appealing to the consumer and, sometimes, more bioavailable. Natural and genetically modified microorganisms are a possible alternative to produce vitamins. Another broad class of molecules, nutraceuticals, includes substances which are not essential in diets, but which may have beneficial roles besides that of supplying building blocks, energy, coenzymes and minerals for the body. This chapter presents both the classes of compounds, vitamins and nutraceuticals, their use in processed food, and focuses on the classical and microbial sources of vitamins and then role in the human body, and some representative nutraceuticals. Aminoacids, organic acids, prebiotics, polyunsaturated oils, hydrocolloids, enzymes and other molecules which are bioactive or nutritionally important, but pertinent to other classes of additives, are not discussed in this chapter.

Keywords: Antioxidant, Coenzymes, Fermentation, Metabolism.

INTRODUCTION

During the last two decades, food industries have developed new products in order to meet customer's demand: practical, attractive and tasty products that also provide functional and nutraceutical components, are even seen by consumers as beneficial for health. This demand for healthier products, together with the inno-

* **Address correspondence to Julio Carvalho:** Department of Bioprocess Engineering and Biotechnology, Universidade Federal do Paraná, Cx Postal 19011, ACF Centro Politécnico, 81531-990 Curitiba, Brazil; Email: jccarvalho@ufpr.br.

Juliano Lemos Bicas, Mário Roberto Maróstica, Glaucia Maria Pastore (Eds.)

vation-driven marketing of processed foods, created a tendency of new product development where not only flavor, stability, or nutritional value, but also health issues drive the development of new products [1].

The development of a recommended "dietary reference intake" (DRI) for several classes of nutrients sheds light on the use of specific functional and nutraceutical components of food. The average levels of vitamin intake may be below the DRI in foods without supplementation [2, 3], because of segregation of parts of the raw food (*e.g.* separation of starch from cereals, or thermal degradation).

The importance of certain foods in maintaining healthy life is long known. Ancient cultures such as Greek and Egyptian reported that definite diseases appeared to be related to nutritional factors. However, it was only in the 20th century that vitamins were discovered and described as essential micronutrients for living organisms [4]. Casimir Funk has coined the term "vitamine" (from vital amine), later changed to "vitamin", in 1912. These substances may be defined as "organic compounds in food, needed in very small amounts for growth and maintaining good health". Vitamins are fundamental for animal survival because of their specific role in the physiology and metabolism [5, 6]. Several essential vitamins are not synthesized by the human body, and these have to be directly added to foods, as defined by regulating agencies [7]. Each vitamin has a specific set of roles on metabolism: as cofactors for the energy metabolism, or precursors to hormones, antioxidants, cofactors for several enzymes, and also play an important role in the oxidation-reduction reactions [5].

Given this importance for nutrition and health, which is distinctive from other functions of food components (such as sources of energy and biomolecule building blocks), vitamins may be seen as *nutraceutical components* of foods. Several foods or food components may be claimed to be "nutraceuticals", a term coined by DeFelice in 1989 to describe food or food components which provide health care improving the human immunological functions, and so on [8].

The term nutraceutical may encompass a wide range of dietetic supplements, functional foods, herbs and cereals – all that is needed to apply the term is that the food, substance or extract is not specifically a drug, and that it can be claimed to have activity beyond that of sustenance, containing beneficial vitamins, proteins, dietary fibers, *etc.* Moreover, a nutraceutical component could be present on beverages, medicines, supplements and foods [9, 10].

Nutraceutical supplements (*e.g.* concentrated carrot oil, rich in β-carotene) are still subjected to regulations as any other food, but not to specific pharma regulations. Because of this, claims regarding the bioactivity of a food or supplement are poorly regulated. Nutraceutical regulation is, however, being actively discussed [9 - 11] and a common ground will probably soon be reached.

Since there is a broad group of molecules that may be deemed "nutraceuticals", some of these are inevitably discussed in other chapters throughout this book: aminoacids (Chapter 2) may have an anabolic stimulating activity [12]; several colorants (Chapter 7) are also antioxidants or antimicrobials [13, 14]; microbial oils (Chapter 8), being rich in polyunsaturated fatty acids, are antioxidant and some are essential precursors of prostaglandins and cerebrosides [15, 16]; and hydrocolloids (Chapter 9) may be beneficial for the digestion and regulation of fat and glucose absorption [17, 18]; This chapter focuses on the activity and bioproduction of vitamins and selected nutraceuticals.

Vitamins and Nutraceuticals in Foods

Nutraceutical foods contain dietetic fibers, polyunsaturated fatty acids, proteins, peptides, aminoacids, minerals, vitamins and antioxidants such as glutathione or selenium [19]. Although nutraceuticals may be divided into *dietary supplements* and *functional foods,* when considered for their chemical composition or activity, these groups may be divided into more segments. Table **1** summarizes the most important nutraceuticals that are naturally present in foods – which makes their use easily defensible.

Regulation of Vitamins and Nutraceuticals in Foods

Regulatory or advisory agencies such as FDA (the USA Food and Drug Administration), AMA (American Medical Association), IFT (Institute of Food Technologists), FNB (Food and Nutrition Board) and FAO (Food and Agriculture Organization of the United Nations) have recommendations regarding the addition of nutraceutical components to food. The common effort of these agencies has generated the Codex Alimentarius, a standard international guidebook for additives (any substance added to processed food which is not an ingredient).

Table 1. Examples of foods with high content of specific nutraceutical substances*.

Group	Nutraceutical Substance/Family	Food sources
Aminoacids, peptides and proteins	Dairy peptides	Dairy products
Carotenoids	Lycopene	Tomatoes and tomato products, watermelon
	β-Carotene	Several (orange) fruits; vegetables; *Dunaliella*
	Lutein, zeaxanthin	Green leafy vegetables, corn, eggs
Minerals	**Mg, Fe, Zn, Ca, Se, Mn**	Case-dependent. Nuts are generally high in minerals.
Polyphenols and flavonoids	Isoflavones (*e.g.*, genistein, daidzein)	Soybeans and other legumes, apios
	Quercetin	Onion, red grapes, citrus, broccoli, yellow squash, clover
	Resveratrol	Grapes (skin), red wine
	Carnosol	Rosemary
	Catechins	Teas, berries, cocoa, apples, grapes
	Curcumin	Turmeric
	Ellagic acid	Grapes, strawberries, raspberries, walnuts
	Anthocyanates	Most berries, red wine
Probiotics	**Lactobacilli, Bifidobacteria**	Yogurt and other dairy
Carbohydrates and fibers	**β-Glucan**	Oat bran
	Lignans	Flax, rye
	Cellulose	All plants
	Inulin, Fructooligosaccharides (FOS)	Whole grains, onions, garlic, yacon, Jerusalem artichoke
	Psyllium mucilage	Psyllium husk
Fatty acids	**EPA and DHA**	Fish oils, *Schizochytrium,* several microalgae
	CLA	Grass-fed beef and dairy
	Monounsaturated fatty acids	Several vegetable oils, nuts.
Sterols and stanols	**Ergosterol**	Mushrooms, yeast

(Table 1) contd.....

Group	Nutraceutical Substance/Family	Food sources
Vitamins	**Listed in detail in Table 3**	
Other	**Allyl sulfur compounds**	Onions, garlic, shallots
	Capsaicinoids	Pepper (*Capsicum*)
	Isothiocyanates	Cruciferous vegetables
	Adenosine	Garlic, onion, yeast
	Indoles	Cabbage, broccoli, cauliflower, kale, brussels sprouts
	3-n-Butyl phthalide	Celery

Note: The substances listed in this table include those that are either accepted as, or claimed to be nutraceutical substances. [*Adapted mainly from [8, 20]].

However, despite the Codex Alimentarius being a general guide for control of the addition of functional and nutraceutical substances to food, each country or region has its own regulation [21]. In Brazil, these substances are regulated by the Brazilian Health Surveillance Agency (ANVISA), which only regulates functional components – apparently avoiding those claimed to be nutraceuticals. This is because despite ANVISA harmonizes its standards with the Codex Alimentarius, the agency is considering "functional components" as the same thing as "nutraceuticals" [22, 23]. Anyhow, many countries have adopted the regulations developed by agencies such as FDA. Table **2** shows the statutory or regulatory definitions of the products which FDA standardized [24].

Table 2. Classification of functional foods - between regular foods and regular drugs.

	Functional foods and nutritional supplements	Food for special dietary use (core nutraceuticals)	Medical nutrition
Target group	Healthy people seeking to preserve wellness	People with common health problems	Patients with special needs
Examples	Probiotic yogurts Weight-loss bars Isotonic sports drinks Vitamin and mineral supplements	Cholesterol-lowering products Products to slow progression of diabetes, dementia, or age-related muscle loss	Specializing infant feeding formulas Nutrition solutions for the frail or chronically ill Other clinical nutrition products
Channels	Supermarkets Internet	Supermarkets Pharmacies Internet	Pharmacies (often requiring some medical supervision)

(Table 2) contd.....

	Functional foods and nutritional supplements	Food for special dietary use (core nutraceuticals)	Medical nutrition
Regulation	Nutrition labeling required (according to NLEA, Nutrition Labeling and Education Act of 1990) Nutrition facts box required Supplement fact box required (according to DSHEA, Dietary Supplement and Health Educational Act of 1994)	Follows NLEA	Exempt form NLEA Market- regulated May be covered by medical insurance

[Adapted from [24, 25]].

Vitamins have an undisputed importance. Several processed foods have reduced vitamin content, which can be restored, depending on the product and claim. Table **3** presents the recommended average intake recommended for vitamins, as listed by the Codex Alimentarius.

Table 3. Overview of vitamins: recommended daily allowances, important food sources and possible microbial sources[d].

Vitamins	RNI[a], µg	EAR[b], µg	Sources[c]
Vitamin A	400	286	Food: milk and dairy products, glandular meats, liver and fish liver oils, egg yolk, [26]. Several carotenoids from green leafy vegetables and yellow fruits or legumes have provitamin–A activity. Microbial: *Blakeslea* [27], *Rhodotorula* [28] and *Dunaliella* [29].
B_1 Thiamine	500	400	Food: lean pork, wheat germ, cereals, legumes, potatoes
B_2 Riboflavin	500	400	Food: Milk and dairy products, egg yolk, legumes, green vegetables, whole cereals. Microbial: *Eremothecium ashbyii*, *Ashbya gossypii*, *Bacillus subtilis*, *Corynebacterium ammoniagenes* [30]
B_3 Niacin	6000	5000	Food: Meats, poultry, cereals, egg yolk
B_5[d] Pantothenic acid	2000	2000	Food: All animal sources, yeast extract, cauliflower, broccoli, tomato, cereals
B_6 Pyridoxine	500	400	Food: Yeast, wheat germ, meat, cereals, potato, banana Microbial: *Klebsiella sp*, *Flavobacterium sp.*, *Pichia guilliermondii*, *Bacillus subtilis*, *Rhizobium meliloti* [30]
B_7[d] Biotin	8	8	Food: vegetables, mushrooms, fruits, milk, egg
B_9 Folate	150	120	Food: Liver, cereals, green vegetables

(Table 3) contd.....

Vitamins	RNI[a], µg	EAR[b], µg	Sources[e]
Vitamin B_{12}	0.9	0.7	Food: liver, kidney, egg, milk, fish and poultry Microbial: *Propionibacterium sp* .
Vitamin D[c]	5	5	Food: fish, meat, milk and eggs [31]. Ergosterol (provitamin D): in mushrooms [31]. Microbial: yeasts [32, 33].
Vitamin E	500	400	Food: vegetable oils, nuts, seeds, green leafy vegetables Microbial: fermentations with *Euglena gracilis Z, Dunaliella tertiolecta, Tetraselmis suecica, Nannochloropsis oculata, Spirulina platensis*
Vitamin C	30000	25000	Food: Citrus fruits and green leafy vegetables [34]
Vitamin K	15	15	Food: mainly green leafy vegetables, certain plant oils. Microbial: fermentation with *Bacillus subtilis, Flavobacterium meningosepticum,* transgenic *Escherichia coli*

[a] RNI, Reference Nutrient Intake, or INL98, Individual Nutrient Levels - from FAO/WHO [35].

[b] EAR, Estimated Average Requirement (calculated values) based on FAO/WHO recommendations [36].

[c] Vitamin D should be added if there is inadequate exposure to sunlight.

[d] Adapted from Codex Alimentarius [37].

[e] Food sources from [38].

The effects of these substances is sometimes controversial (*e.g.*, polyphenols may have an irrelevant '*in vivo*' antioxidant effect [39]), and there is not a consensus on dietary levels for several of these substances. Table **4** shows WHO data for nutraceuticals other than vitamins.

Table 4. Recommend levels of common nutraceuticals (other than vitamins).

Nutraceuticals	Physiological Level	Dietary Level	Metabolite
Glucosamine [41]	-	-	Glucosaminic acid
Methylsulfonylmethane [42]	4mg/person	minimal	Methionine
Coenzyme Q9 [43]	0.5micro gm/ml	3-5mg/day	6-(5', 8', 6') carboxylated derivatives
Coenzyme Q10 [43]	0.5micro gm/ml	3-5mg/day	6-hydroxymelatonin
Carnitine [44]	9mg/L	3-97mg/100gm meat	Acetylcarnitine, Butyrobetaine
Flax lignans [45]	Diet dependent	Diet dependent	SDG-enterolactone
Resveratrol [46]	-	0.1-2.3 mg/L	Piceatannol, tetrahydroxystilbene, dihydroresveratrol

(Table 4) contd.....

Nutraceuticals	Physiological Level	Dietary Level	Metabolite
Grape seed [47]	-	-	Catechin/proanthocyanidin metabolites
Lycopene	-	2-5 mg/day	5, 6 dihydroxy 5, 6 dihydrolycopine
Lycopene	-		Lycopene epoxide
Soy isoflavones	-	15-20mg/day	18-demethylated

[Adapted from [48]].

The research in vitamins and nutraceuticals, and their applications to foods, is very active: several substances with bioactive properties have been tested as supplementary components for food manufacturing and food complementation [6 - 8, 48]. That is a good perspective for the biotechnology research development because food industries depend on in part on the technology created by this area to generate solutions which are directly "fit for use" (ISO:9000). In this context, a technology that is "fit for use" defines the quality through which industries will meet the customers' expectations as well as highlight how good are the regulations to the development of industries competencies, culture and marketplace [24].

VITAMINS: PROPERTIES, SOURCES AND BIOPRODUCTION

Most vitamins are relatively simple organic compounds, and of extreme importance in diets. This is why synthetic routes exist for most vitamins, although in few cases, the biotechnological route prevails. Table **5** lists the main sources for vitamins and coenzymes, and the prevailing production methods [30, 49]. There is an increasing interest in developing biotechnological methods that could substitute traditional procedures, which use food sources as raw material, are energy-intensive and generate a large amount of wastes that must be disposed off [49, 50].

Vitamin A

The role of foods rich in vitamin A in visual function has been known for thousands of years [51], although the vitamin itself was only isolated in 1913. This essential nutrient participates in several physiologic processes, such as growth, reproduction, immune response, and epithelial cell production and maintenance [52]. Its structure is depicted in Fig. (**1**).

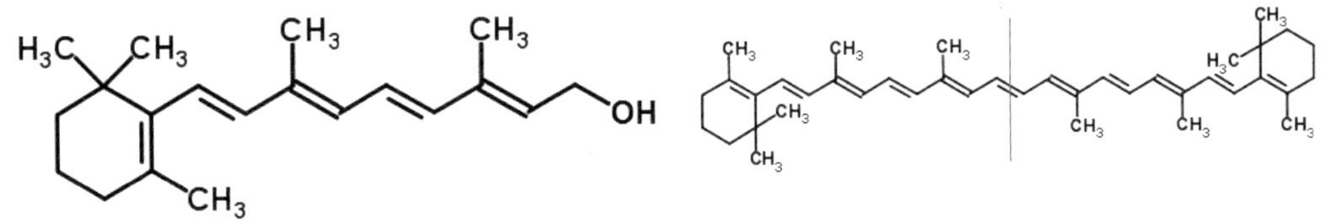

Fig. (1). Vitamin A (Retinol, left), and β-carotene (right). Cleavage of the carotenoid generates two molecules of the vitamin.

Table 5. Vitamins - production routes and estimated world production.

	Production method			**World production (tons)**			
	Biotechnological	Chemical	Extraction	1980[a]	1990[b]	2000[e]	2010[e]
Thiamine (B_1)		+		1700	4200	3700	n.a.
Riboflavin (B_2)	+			2000	2400	n.a	4400
Niacin	+	+		8500	22000	2800	40000 [53]
Panthothenic acid	+[c]	+		5000	7000	n.a.	n.a.
Pyridoxine (B_6)		+		1600	2550	3800	n.a.
Biotin	+[d]	+		2.7	25	88	112
Folic acid		+		100	400	534	
Vitamin B_{12}	+			12	10	25	175
Vitamin C	+[c]	+		40000	60000	10700	n.a.
Vitamin A		+	+	2500	2700	2700	n.a.
β-carotene	+		+	100	400	n.a.	n.a.
Ergosterol	+			25	38	n.a.	n.a.
Vitamin D_3		+	+	n.a.	5000	n.a.	n.a.
Tocopherol (E)	+[d]	+	+	6800	22000	30000	n.a.
Vitamin K_2		+		n.a.	500	n.a.	n.a.
Ubiquinone	+			n.a.	n.a.	n.a.	n.a.

[Adapted from [30]]. a, b: values compiled from Florent (1986) and Eggersdorfer (1996); c: biotransformations; d:pilot scale; e: compiled from [49, 50].

There are two groups of compounds with vitamin A activity: preformed retinoids (active form, from animal products) and the provitamin A betacarotenoids (precursor to vitamin A). These are listed in Table **6**. A stringent requisite is that the carotenoid contains at least one molecule of retinol, *i.e.* an unsubstituted β-ionone ring and a polyene chain containing 11 carbons [54].

Table 6. Groups of compounds with vitamin A activity.

Preformed Vitamin A: Retinoids1	Provitamin A carotenoid: predominately as beta-carotene2
Four categories of closely related compounds: an aldehyde (retinal), an alcohol (retinol), a carboxylic acid (retinoic acid), and esters of retinol (such as retinyl palmitate, hydrolyzed to retinol during digestion)	Precursor of vitamin A has to be converted into active vitamin A in the body.
Absorbed as retinol.	There are other provitamin A carotenoids: alpha-carotene and beta-cryptoxanthin.
Retinol is metabolized *in situ* to required forms: to retinal (for the eye) and retinoic acid (in the skin, bones and for gene expression).	Of more than 600 different carotenoids, only about 10% can be made into vitamin A [26]. Common carotenoids such as lycopene (tomato) and lutein (marigols) have NO vitamin activity.
Found in animal products: milk (including human), egg yolk, especially in liver and fish liver oils, and in dairy products. [26]	Found in green leafy vegetables (*e.g.*, spinach, and different young leafs), yellow vegetables (*e.g.*, pumpkins, and carrots), and yellow and orange noncitrus fruits (*e.g.*, mangoes, apricots, and papaya). Red palm oil is especially rich in provitamin A [60]

[Adapted from "Vitamin A and Carotenoids", in [61]].

A large number of geometrical isomers are theoretically possible for retinoids and carotenoids, but many mono-*cis* and poli-*cis* isomers exhibit stearic hindrance and are rarely encountered in nature (*e.g.* 7-*cis*- and 11-cis- β-carotene) [55, 56]. All-trans forms are predominant in natural foodstuffs and are often found along with *cis* isomers (*e.g.* 9-*cis*, 13-*cis*, 9,13-di-*cis* and only for carotenoids, 15-*cis*) [55, 57 - 59].

Foods containing provitamin A carotenoids tend to have less biologically available vitamin than animal products, but are more affordable. This explains why carotenoids are the main source of vitamin A in the diets of relatively poor populations. One of the international standards of measurement of vitamin A content is the retinol activity equivalent (RAE) (Table **7**).

Table 7. International standards of measurement for vitamin A.

Retinol activity equivalents (RAE) ratios for beta-carotene and other provitamin A carotenoids		
1 μg of dietary or supplemental vitamin A	1 μg of retinol*	1:1
2 μg of supplemental β-carotene	1 μg of retinol	2:1
12 μg of dietary β-carotene	1 μg of retinol	12:1
24 μg of dietary α-carotene	1 μg of retinol	24:1
24 μg of dietary β-cryptoxanthin	1 μg of retinol	24:1

* One IU is equivalent to 0.3 microgram (μg) of retinol, and one mcg of retinol is equivalent to 3.33 IU of retinol [52].

World and Regional Supply and Patterns

Pure vitamin A is mostly synthetically produced; it may also be extracted from natural sources. Nowadays, biotechnological routes have been developed, but these are still not competitive for industrial production.

In theory, vitamin A intake could be met worldwide by simply adopting a proper diet. What is observed, however, is that great differences exist between the effectively available sources and in the *per capita* consumption of vitamin across different countries, age categories, and socio-economic groups [62]. The world *per capita* availability of vitamin A for human consumption in the 80´s amounted to an estimated 790 μg RA / day. This is not, however, to be confused with intake [35].

Vitamin A Manufacturing Process

For the production of Vitamin A, four synthetic routes have been developed [63], two of which are used today: one based on Darzen's reaction to synthesize the key intermediate C14 aldehyde (used by Hoffmann-La Roche, Pfizer, Merck, Sharp & Dome, *etc.*) and another based on the use of 2,6,6-trimethylcyclohexanone, used by Takeda in Japan and Glaxo in U.K. and in India.

An alternative to the addition of Vitamin A to processed foods is to genetically transform staple foods in order to enhance the content in vitamins [64]. Using this approach, the total carotenoid content may be enhanced dramatically: Table **8** presents some species genetically manipulated, the wild compared with the transgenic types and their yield.

Several microorganisms may be used for provitamin A production, especially β-carotene, using *Blakeslea* [27], *Rhodotorula* [28], *Dunaliella* [29] and other microorganisms. Details about the production of β-carotene by *B. trispora* and *D. salina* are described in Chapter 7, since this carotenoid is also used as a natural color additive.

Table 8. Carotenoid and β-carotene levels in wild-type and transgenic plants.

		Carotenoid levels inμg/g of plant dry weight (and increase in transgenic plants)			
		Total carotenoids		**β-carotene**	
Species	**Genes (origin)**	**wild type**	**transgenic plants**	**wild type**	**transgenic plants***
Maize [68]	psy1 (*Zea mays* ; maize)	1.10	146.7 (133x)	0.14	57.35 (410x)
Maize [69]	psy1 (maize)	1.45	163.2 (112x)	0.35	59.32 (169x)
Potato [70]	crtB, crtI and crtY (*P. ananatis*)	5.8	114.4 (20x)	0.013	47.4 (3600x)
Cassava [71]	crtB (*P. ananatis*)	0.65	21.84 (33.6x)	0.41	6.67 (16x)
Carrot [72]	psy (*Arabidopsis*)	5.5	514.1 (93x)	1.27	214.63 (178x)
Canola [73]	idi, crtE, crtB, crtI and crtY (*P. ananatis*)	217	6567 (30x)	2	2142 (1070x)
Tomato [74]	Antisense DET-1 (tomato)	364	838 (2.3x)	163	130 (8x)

[Adapted from sources compiled by [54]].

Genetic engineering is also used for microorganisms: *E. coli* may reach high level production of carotenoids such as lycopene and β-carotene [65]. Indeed, *E. coli* natively harbors the basic pathways necessary for carotenoid synthesis; expression of two *Erwinia* carotenogenic genes crtB (encoding phytoene synthase) and crtI (encoding phytoene desaturase) in *E. coli* was sufficient for lycopene production at low levels [66].

Vitamins of the B-complex

Vitamins of the B-complex are a group of 8 water soluble molecules, usually found in the same food sources and play inter-related roles in cell metabolism [67]

Table **9** summarizes the most important data related to the B-complex vitamins: usual name, chemical formula, natural sources and function.

Table 9. B-complex vitamins, its functions and food sources.

Vitamin	Structure	Usual name	Formula	Function	Main sources
B_1	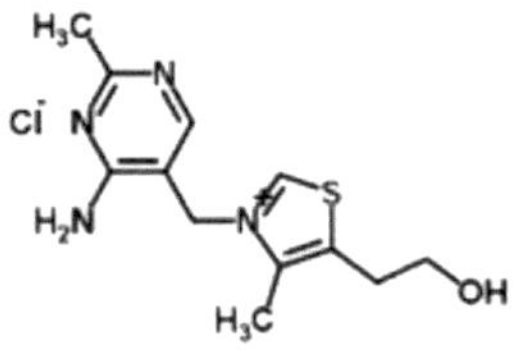	Thiamine	$C_{12}H_{17}N_4OS$	Component of the coenzyme thiamine pyrophosphate, essential for the oxidative decarboxilation of piruvate. Essential for catabolism and growth	Pork, wheat germ, breast milk, whole grain and cereals, legumes and potatoes
B_2	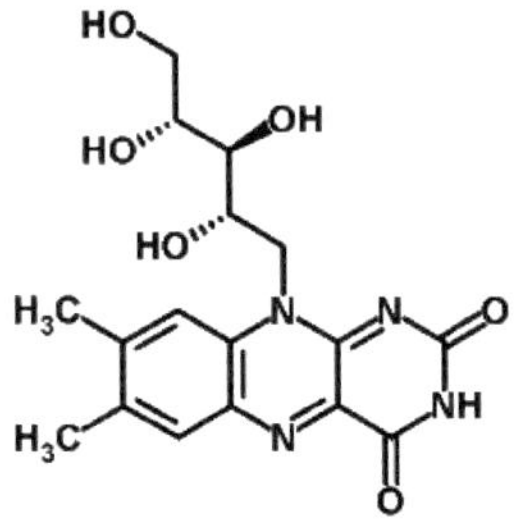	Riboflavin	$C_{17}H_{20}N_4O_6$	Component of coenzymes essential to oxidation-reduction reactions, such as in the electron transport in mitochondria (tissue respiration)	Milk, cheese Breast milk, Egg yolk, legumes, green vegetables, cereals
B_3	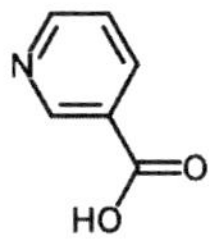	Niacin	$C_6NH_5O_2$	Component of coenzymes related to the cellular respiratory chain. Essential for glycogen synthesis and for fatty acid metabolism.	Lean meats, peanut Breast milk, poultry, fish, cereals, egg yolk
B_5	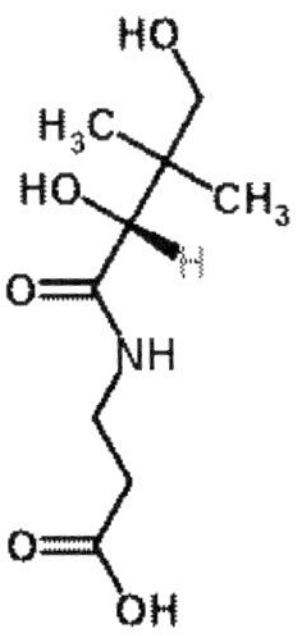	Pantothenic Acid	$C_9N_{17}HO_5$	Essential for the intermediary metabolism of carbohydrates, lipids and proteins	Egg, kidney, liver, salmon, yeast extract, cauliflower, broccoli, tomato, potato Breast milk, meat, fish, poultry, egg yolk, cereals

(Table 9) contd.....

Vitamin	Structure	Usual name	Formula	Function	Main sources
B_6	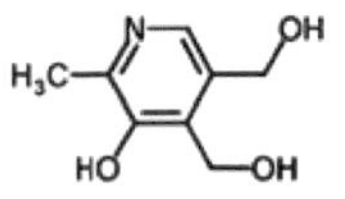	Pyridoxine	$C_8H_{11}NO_3$	Found in coenzymes essential for metabolism of carbohydrates, lipids and proteins. Aids in the synthesis of aminoacids and fatty acids	Yeast, wheat germ, meat, integral cereal grains, potato, banana, oat Breast milk
B_7	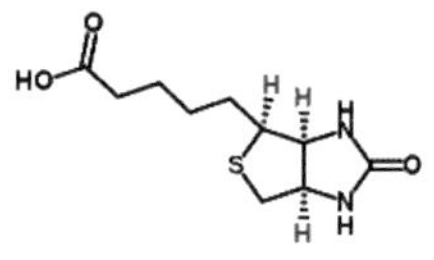	Biotin	$C_{10}H_{16}N_2O_3S$	Essential for several enzymatic processes of CO^2 fixation in active compounds of the metabolism . Important for metabolism of aminoacids and fatty acids	Vegetables in general, mushrooms, fruits in general, milk, egg, breast milk
B_9	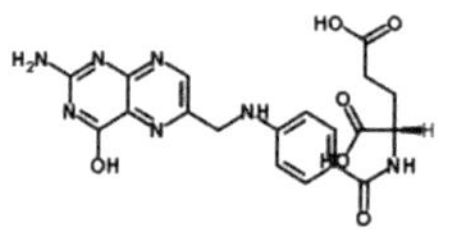	Folic Acid	$C_{19}H_{19}N_7O_6$	Transporter of groups with isolated carbon from one substance to another Essential for the synthesis of nucleic acids, important for maturation of red blood cells	Liver, purple beans and green vegetables Breast milk, cereals, cantaloupe
B_{12}	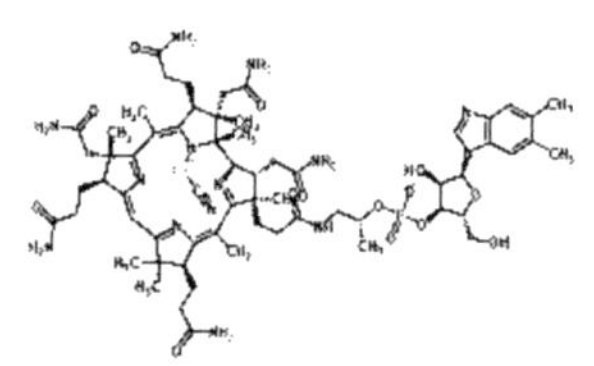	Cyanocobalamin	$C_{63}H_{88}CoN_{14}O_{14}P$	Important for the synthesis of nucleic acids, red blood cell maturation, and central nervous system metabolism	Liver, kidney, egg, milk, breast milk, fish, poultry

[Adapted from [38] (vitamin function and sources)]. Processed derivatives of these sources (*e.g.* whole grain bread) are also sources of B-complex vitamins.

The recommended dietary intake of B-complex vitamins is based on evidences such as: their content in the human milk; their content in human blood; clinical studies evaluating effects of their lack or excess. The dietary intakes vary according to several factors, and the most important ones are: age, sex, pregnancy and lactancy. Table **10** details the recommended diary intakes for adults (males and non-pregnant females) [35].

Table 10. Recommended daily intake for B-complex vitamins.

Vitamin	Recommended diary intake for adults (mg/day)	
	Male	Female
B_1	1.2	1.1
B_2	1.3	1.1
B_3	16*	14*
B_5	5.0	5.0
B_6	1.3**	1.3**
B_7	$30x10^{-3}$	$30x10^{-3}$
B_9	0.4	0.4
B_{12}	$2.4x10^{-3}$	$2.4x10^{-3}$

Source: World Health Organization [35] *mgNE/day (mg of Niacin equivalent per day)
** Above 51 years old the recommendations are 1.7mg/day and 1.5mg/day for males and females, respectively

Most vitamins of the B-complex are produced *via* chemical synthesis. Some may be obtained *via* fermentation processes (biological synthesis) and some can be obtained by both approaches.

Riboflavin (Vitamin B_2)

Riboflavin is produced either by chemical synthesis or by fermentative processes; through Ascomycetes *Eremothecium ashbyii* and *Ashbya gossypii* and also by genetically modified strains of *Bacillus subtilis* and *Corynebacterium ammoniagenes* [30].

Pantothenic acid (Vitamin B_5)

Pantothenic acid is produced on large scale *via* chemical synthesis. However, the racemic pantolactone derived from the conventional production requires optical resolution, with the use of expensive alkaloids. Some biotechnological approaches have been proposed as alternative to separate this racemic mixture [30].

Pyridoxine (Vitamin B_6)

This vitamin is produced exclusively by chemical synthesis, but search for microbial producers leads to potential producers: *Klebsiela sp*, *Flavobacterium sp.*, *Pichia guilliermondii*, *Bacillus subtilis*, and *Rhizobium meliloti* [30].

Cyanocobalamin (vitamin B_{12})

Cyanocobalamin is classically produced by fermentative processes. *Propionibacterium sp.* strains are usually employed to produce vitamin B_{12}, either in conventional CSTR reactors or in fixed bed reactors. *Pseudomonas denitrificans* and *Bacillus megaterium* are also known to have potential for cyanocobalamin production. The final concentrations of the extracted broths range from 2 mg/L to 200mg/L of vitamin B_{12} depending on the biological and physico-chemical characteristics of the system.

Yingnan *et al.* [75] used an anaerobic fixed bed reactor to produce methane from a mixture of CO_2 and H_2, with vitamin B_{12} as a coproduct. The production of Vitamin B_{12} reached 2.88mg/l for a hydraulic retention time (HRT) of 3 days, and 37.54mg/l for an HRT of 6 days. In another process of co-production, Wang *et al.* [76] evaluated the simultaneous production of propionic acid and Vitamin B_{12} using an expanded bed absorption reactor with an *in situ* product removal system. The strain was *Propionibacterium freudenreichii* CICC 10019 and the production of Vitamin B12 reached 43.04 mg/l in 160h, in a fed-batch fermentation.

Hugenschmidt *et al.* [77] evaluated the potential for the production of Vitamin B_{12} from 151 strains of lactic acid bacteria and 100 strains of propionic acid bacteria in supplemented whey permeate. The best producer was isolated from raw milk and identified as *Propionibacterium freudenreichii*, yielding 2.5mg/L of Vitamin B_{12}.

Kun-Tai *et al.* [78] showed that betaine has a negative effect over the cell growth of an industrial strain of *Pseudomonas denitrificans*, but a positive effect over the production of Vitamin. In a large scale fermentation (120m^3), the best production was reached when betaine concentration was maintained in a range from 5 to 7g/l, and in 162h of fermentation, a production of approximately 177mg/l of Vitamin

B_{12} was reached.

Kaiming *et al.* [79] used the strain *Propionibacterium freundreichii* IFO 12424 to compare its metabolism in an aerobic fermentation and in a DO shift fermentation. It was found that in the last case, the volumetric production of Vitamin B_{12} could be enhanced 2-fold when compared to the former case.

Moore *et al.* [80] used a strain of *Bacillus megaterium* (DSM 319) to produce cobalamin. They tested different concentrations of cobalt ion, trying to enhance biomass growth and vitamin production. In a medium with glycerol as carbon sources and in the presence of 10uM of cobalt, they observed production of cobalamin up to 200mg/L.

Vitamin C

This vitamin (or L-ascorbic acid) is an important redox cofactor that takes part in several biochemical reactions involved in metabolism of plants and animals [81]. Its main functions are the hydroxylation of proline and lysine in collagen and other connective tissue proteins, as a free radical scavenger, and in the synthesis of hormones and neurotransmitters [82]. Its structure is depicted in Fig. (**2**).

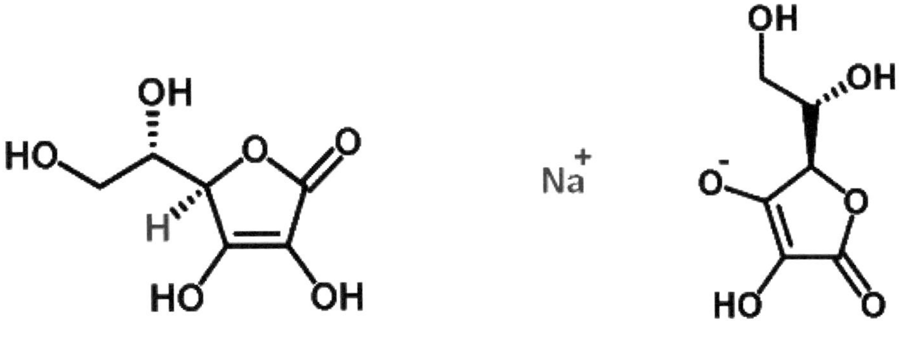

Fig. (2). Vitamin C (Ascorbic acid, left) and sodium ascorbate (right).

The severe deficiency of vitamin C can cause scurvy, a disease that causes bleeding and swollen gums, joint pain, non-healing wounds and destabilization of the teeth [82]. Recommended Daily Allowance (RDA) for vitamin C should be 75-90mg/day for adult humans [83]. Fruits and vegetables are the main sources of vitamin C, especially citrus fruits and green leafy vegetables [34].

L-ascorbic acid is commercially synthesized *via* the seven-step Reichstein process, involving six chemical steps and one fermentation step [84]. Advances in biotechnology are enabling the development of recombinant strains for the synthesis of Reichstein intermediates [85 - 87]. Studies of biosynthesis in plants, microalgae, and yeast are being conducted to develop new methods for the direct production of L-ascorbic acid [84, 88 - 90]).

Vitamin D

Vitamin D is formed by liposoluble compounds that can be transformed biologically into 1,25-dihydroxycholecalciferol (a more active form of the vitamin), through ultraviolet light irradiation and enzymes present in the liver and kidneys [91]. The activation of this hormone induces the synthesis of calcium absorption in the intestine by regulating the levels of calcium in the kidneys and bones. Rickets and poor bone formation problems are caused by vitamin D deficiency [82]. The structure of vitamin D and of its common precursor ergosterol is shown in Fig. (**3**).

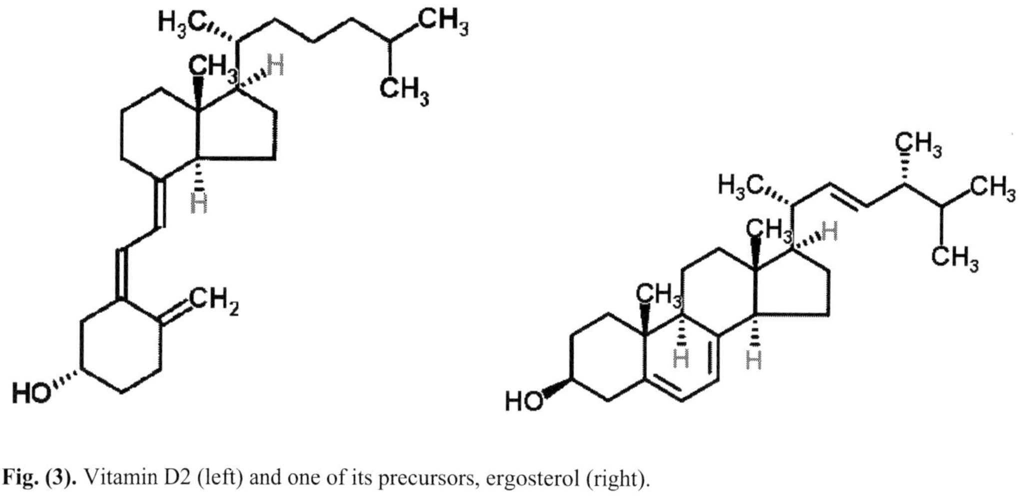

Fig. (3). Vitamin D2 (left) and one of its precursors, ergosterol (right).

The main forms of this group are vitamin D_3, cholecalciferol, and vitamin D_2, ergocalciferol [31, 91]. Ergocalciferol has a structure similar to that of cholecalciferol, differing only by one methyl group on C_{24} and a double bond between C_{22} and C_{23} [92]. The human body can get vitamin D in two ways: skin exposure to UV radiation converting 7-dehydrocholesterol to vitamin D_3 or by ingestion of food or supplements containing vitamin D [82].

The main sources of vitamin D_3 and its metabolites are animal foods such as fish, meat, milk and eggs [31]. This should be obtained by extracting 7-dehydrocholesterol, the precursor of cholecalciferol, of wool fat, followed by UV irradiation and purification [92]. Ergosterol, a precursor of vitamin D_2, is found in large amounts in mushrooms; animal foods may contain vitamin D_2 and its metabolites in lesser amounts [31].

Vitamin D_2 is produced industrially by irradiation of ergosterol extracted from yeast and then purifying it [32, 33]. Studies are being conducted to optimize the strain and fermentation conditions for the production of ergosterol [33, 93]. HE *et al.* [94] produced 1707 mg/L of ergosterol in bioreactors using cane molasses and recombinant strain of *Saccharomyces cerevisiae* YEH56 (pHXA42).

Vitamin E

In 1922, Evans and Bishop discovered vitamin E to be an essential dietary factor for reproduction in mammals. Since then, many reports have described health properties including antioxidant effects (by scavenging of reactive oxygen and nitrogen species) and modulation of signal transduction and gene expression [95].

Natural sources of vitamin E are found in lipid-rich vegetable such as nuts and some seeds (such as sunflower), and most vegetable oils, besides green leafy vegetables (such as spinach and broccoli). Microbial sources also exist (and will be discussed in the next section). Leaves of most common plants contain low levels of tocopherols (10-50 μg/g fresh weight), but they accumulate in higher concentrations in seeds (500-2000 μg/g) [96]. Several microorganisms can produce concentrations that surpass these reference values.

Natural vitamin E comprises eight different lipid-soluble compounds that could be classified into two related groups: tocopherols and tocotrienols, each one with species identified as α, β, γ and δ. These compounds are presented in Fig. (**4**), where substitutions in R_1 and R_2 correspond to methyl groups attached to the aromatic ring.

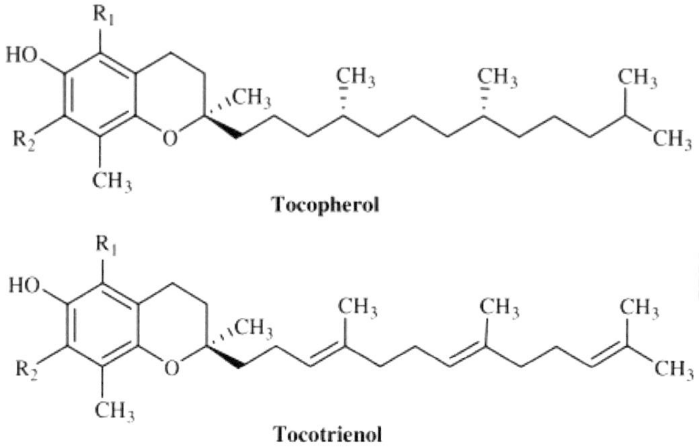

Type	R_1	R_2
α	CH_3	CH_3
β	CH_3	H
γ	H	CH_3
δ	H	H

Fig. (4). Structures of natural forms of Vitamin E.

α- tocopherol is the most common form of natural vitamin E in the human body, and it is exclusively produced by photosynthetic organisms. Because of its antioxidant properties, α- tocopherol is involved in the prevention of degenerative disorders (such as atherosclerosis, Alzheimer and Parkinson), cardiovascular diseases, light induced pathologies of skin and eyes, liver diseases and cancer [49, 97 - 99]. Vitamin E may have several stereoisomers, which make synthetic α-tocopherol different from the natural form. The eight synthetic stereoisomers are collectively known as *al-rac*-alpha-tocopherol, overall less bioavailable than natural forms [49]. This also drives the interest towards substituting synthetic vitamin E by the natural product.

In view of the fact that natural vitamin E production requires a photoautotrophic metabolism, new biotechnological processes are focused on the use of microalgae from freshwater and marine sources. It has been reported that several strains of *Euglena gracilis* Z. and *Dunaliella tertiolecta* produce α-tocopherol in concentrations higher than that of foods considered conventional sources of this vitamin [100].

Multiple studies have been conducted to optimize biological production of α-tocopherol, which involved screening of culture conditions, substrates, activation of heterotrophic metabolism, growing under high cell density and limited light availability [49, 101 - 104]. Furthermore, co-production of considerable quantities of β-carotene, vitamin C, tyrosine, paramylon and vitamin E has been reached with *E. gracilis* [100, 105, 106]. Table **11** summarizes main results obtained in biotechnological production of natural vitamin E.

Recovery of natural vitamin E from fermentation broth is of predominantly performed by extraction with organic solvents, because it lipid-soluble properties. However, supercritical fluid extraction (SFE) is also reported as one of the most important techniques for tocopherol isolation, mainly applied to recover vitamin E from different sources such as palm oil, rice bran and soybean, with results that reach 4-fold those obtained with solvent organic extraction [102].

Table 11. Average yields and final concentrations of vitamin E obtained by fermentation.

Microorganism, strain	Media	Conditions	P mg/L	X g/L	YP/X (mg/g)	Ref.
Dunaliella tertiolecta (wild)	ESW Nitrate as N source.	Air-lift 9-L	1.07	0.59	1.81	[105]
Dunaliella tertiolecta 19/6B	ASW	Bubble column 1.5 L	3	6	0.50	[101]
Tetraselmis suecica 66/4	EASW	Bubble column 1.5 L	2.8	7	0.40	[101]
Euglena gracilis IAM E-6(Z)	Ethanol as C-source. N2	STR 3-L Sequential heterotrophic autotrophic cultivation	8.05	4.8	1.7	[107]
Euglena gracilis (wild)	Glucose + ethanol as C-source	Air-lift 2 L (Fed-Batch)	23.43	19.7	1.19	[108]
Euglena gracilis Z	Glutamate + malate + ethanol	Flask	39.96	10.8	3.7	[100]
Nannochloropsis oculata CCAP 8491	ESW Nitrogen limitation	Polyethylene bags– 50 L	-	-	2.33	[109]
Trentepohlia aurea (wild)	5% CO2 enrichment air	Flask	1.17	3.9	0.3	[106]

SW: seawater; ASW artificial seawater, ESW enriched seawater; P:α-tocopherol; X: biomass, $Y_{P/X}$: yield of mg α-tocopherol/ g dry weight biomass.

Vitamin K

Another lipid-soluble vitamin and known as an anti-hemorrhagic factor is vitamin K. Discovered in 1935, it was described as a new organic compound that presented blood coagulation capacity. Nowadays, vitamin K is widely used in treatment for illness with bleeding symptoms like hemorrhagic disease of new-born infants and hemorrhagic symptoms caused by administration of antibiotics, salicylic acid or warfarin dosages. Recently, it was proposed also as key nutrient in the regulation of soft tissue calcification [50]. The structure of Vitamins K_1 and K_2 is presented in Fig. (**5**). A synthetic form of vitamin K, without the side chain, is vitamin K_3 (menadione) [110].

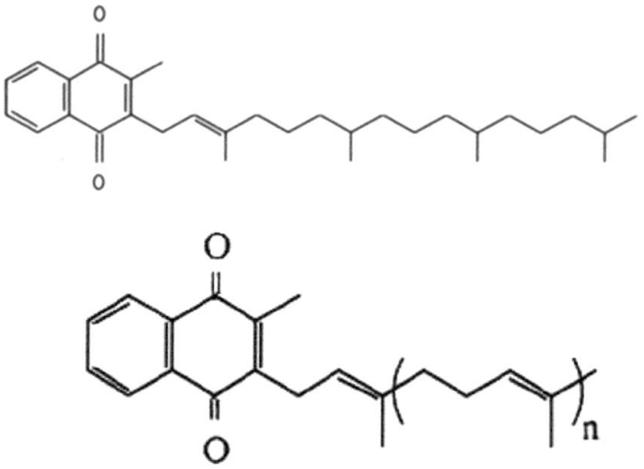

Vitamin K₁

Vitamin K₂

Fig. (5). Structure of natural vitamin K, n denotes the number of variable isoprene units in menaquinones [49].

Natural vitamin K presents two forms: vitamins K_1 (phylloquinone) and K_2 (menaquinone). The first is synthetized in plants, and is especially abundant in green leafy vegetables (kale, spinach, cabbage, broccoli, mustard green) reaching levels in the range 4-7 μg/g. Certain vegetable oils (such as soybean, rapeseed and olive oils) contain concentrations of 0.5-2 μg/g. On the other hand, menaquinones are mainly produced in animal tissues or through bacterial synthesis. Structural differences between these two forms of vitamin K govern many facets of metabolism, including transport pathways, being taken up by target tissues, excretion and final therapeutic activity, which is much higher for menaquinone.

In vivo, vitamin K is essential for the formation of specific proteins in blood and bone, acting as a cofactor for the post-translational carboxylation of glutamic acid residues. This property is strongly related to blood coagulation capacity described above. Furthermore, recent studies show that vitamin K consumption could reduce certain cardiovascular disorders and the risk of bone fractures [111].

Vitamin K_2, which shows higher therapeutic activity, refers to a series of naphthoquinones with isoprene chain of variable length (4-13 units). Fig.(**5**) shows general structure for phylloquinone (K_1) and menaquinone (K_2).

Table 12. Biotechnological producers of vitamin K_2.

Microorganism and strain	Medium	Conditions	MK mg/L	Cell growth (OD 600nm)	MK/ OD	MK mg/kg	Ref.
Bacillus subtilis natto	Glycerol as C-source	STR 3–L Fed-Batch	86, 48	34, 18	2, 53	-	[112]
B. subtilis natto	Yeast extract, soy peptone Glycerol	Flask Batch	62, 32	17	3, 67	-	[111]
B. subtilis natto	Yeast extract, soy peptone Glycerol	Flask Static Batch	53, 7	16	3, 35	-	[113]
B.s subtilis natto	Soybean	SSF 50% moisture	-	-	-	108, 1	[114]
B.subtilis natto	Mix of corn&; soybean	SSF, 60% moisture	-	-	-	57, 78	[115]
B. subtilis D200-41	Glycerol	Flask Batch	45, 1	12, 4	3, 64	-	[116]
B. subtilis K3-176	Soybean and glycerol	Jar Fermenter 7- L	35	15	2, 33	-	[117]
Flavobacterium meningosepticum HNA 350-22	Glycerol-polypeptone	Flask Batch	34	-	-	-	[118]
Flavobacterium sp. 238-7	Glycerol, detergent supplemented	Flask Batch	182	-	-	-	[119]

Nevertheless, considering that chemical synthesis of vitamin K_2 is rather difficult and the natural sources contain low concentrations, applications in medicine field have been focused on vitamin K_1 use.

Menaquinone has recently emerged as a dietary supplement, representing just a small part of the global market and worldwide production of vitamins nutraceuticals [50].

As it was previously discussed, vegetables and plant oils are the major dietary sources of vitamin K, specifically in phylloquinone form. Nonetheless, several bacteria isolated from human intestine have been identified as menaquinone producers, contributing to the vitamin K requirements. In some bacteria, plasma membranes contain menaquinones, which take an important part in electron and active transport, and in oxidative phosphorylation [111]. A majority of biotechnology researches have focused on *Bacillus subtilis* for vitamin K_2 production. *B. subtilis* has been used in Japanese cuisine for fermentation of soybean to obtain a traditional food known as *natto*, a classic source of vitamin K in Asiatic culture. Table **12** summarizes microbial producers of vitamin K.

Menaquinone produced by fermentation is currently recovered by extraction with organic solvents. Berenjian *et al.* [111] reported *n*-hexane:2-propanol (2:1, v/v) with 1:4 aqueous:organic phases (v/v), as an effective mixture for vitamin K_2 extraction from fermentation media.

NUTRACEUTICS: SOURCES AND BIOPRODUCTION

As previously stated, there is no consensus on recommended levels for nutraceutical intake. The term itself has no regulatory definition [120], but the perception of benefits for health, changes in eating habits, and concerns about quality of food supply [121, 122] have expanded the demand, and now functional options exist in all segments of the food market [123].

Any substance naturally present in foods and which is beneficial to health – beyond the obvious function as building block or energy source – may arguably be termed a nutraceutical. Therefore, there is a great variety of such substances; for example, carotenoids, which have inherent antioxidant properties, are more than 750 [124]; polyphenols, which may have multiple effects such as antioxidant and antitumoral activity, are over 5000 [125] and, considering the variety of the sugars to which they are usually bound, well over one million [126]. Most of these nutraceuticals are plant metabolites, and therefore are extracted from plants and vegetables, but several may be produced by cell cultures or microorganisms –

especially those present in mushrooms.

In order to produce a concentrated nutraceutical *supplement* or *additive*, it is necessary to process a food or vegetable rich in the desired active substance into a suitable formulation. This can be done by concentration of the liquid source (*e.g.* by spray drying), or extraction from the solid matrix, followed by concentration. [127 - 130].

Inositol

This polyol is essential for metabolism, and has been considered to be a vitamin (B_8) [131] until it was found to be produced in sufficient amounts in the body. Still, inositol derivatives are essential to regulate several aspects of membrane trafficking, usually as pyrophosphorilated molecules[132]. Actually, inositol is ubiquitous in the form of hexaphosphate phytic acid (Fig. **6**). However, this form is not well absorbed in the gut and phytates, being good cation chelators, are even considered antinutrients – although presenting other beneficial effects, such as hipocolesteremic and antitumoral [133]. Inositol is also present in the form of phospholipids in most foods.

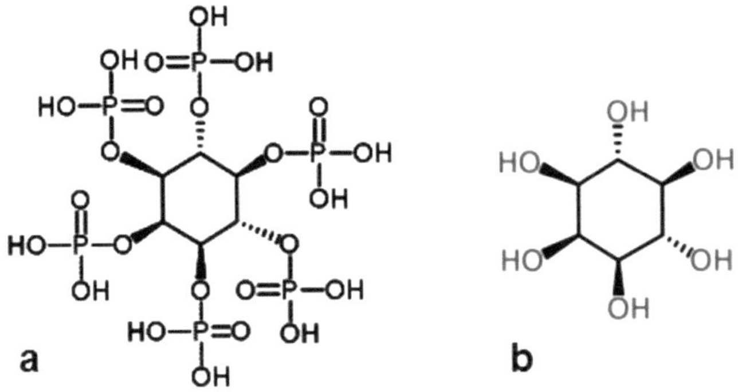

Fig. (6). Phytate (inositol hexaphosphate) (**a**) and inositol (**b**).

The production of inositol requires the recovery of phytates from vegetable sources. One of the most common is corn steep liquor, which contains 0.2% of inositol phosphates (dry basis) [134] and may be recovered by ion exchange [135]. Most of the inositol is present in the form of penta- and hexa phosphate, and further processing into more available requires hydrolysis using phosphatases. The phosphate may be precipitated using calcium, while the inositol solution may be concentrated. The following pages describe the sources and extraction of important nutraceuticals that do not fall into other nutrient or additive classes.

L(-)-carnitine

This is another substance which had a short life as "vitamin" (B_T), soon being found to be produced in microorganisms and higher animals. However, this molecule (Fig. **7**) is essential for the transport of fatty acids through the mitochondrial membrane of mammals [136], and is nowadays a popular supplement believed to enhance muscular response and recovery from exercise [137], and it is also antihyperlipoproteinemic [20].

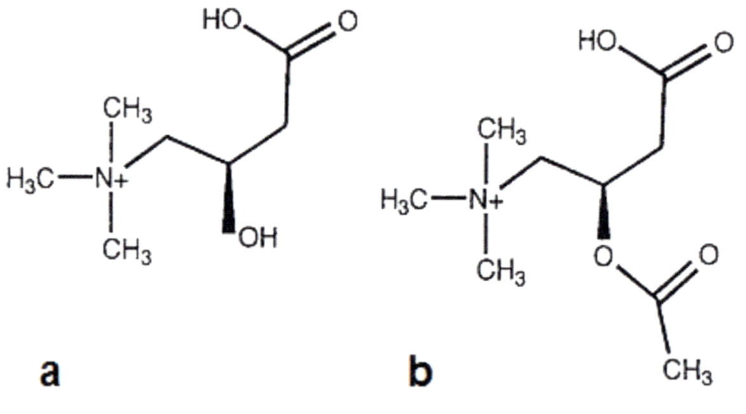

Fig. (7). L(-)-Carnitine (**a**) and its reversible form in muscle, acetyl-L(-)-Carnitine (**b**).

L(-)-Carnitine is synthesized in the body from the essential aminoacid L-lysine, and is also present in foods, especially animal sources – the name of the molecule is derived from the latin *carnis*, meat [138]. It may be easily produced synthetically, giving the DL- form – which is not desirable, and separating the D-isomer is expensive [139]. One alternative is the biotransformation of crotonobetaine or butyrobetaine, using *E. coli*. In that procedure, a productivity of $6g.L^{-1}.h^{-1}$ is achieved, with relatively high purity [140]. The quaternary ammonium compound may be recovered using ion exchange.

Glucosamine and Chondroitin Sulfate

These two substances are frequently used together as treatment for osteoarthritis, although the long-term benefit of the intake is disputed [141, 142]. The first Fig. (**8a**) is an amine-sugar which may be produced by hydrolysis of chitin, *e.g.* from shrimp shells, [143], by acid digestion with hydrochloric acid (6 N HCl at temperature of 100 °C for 4 h) [144], while the second may be extracted from shark fins or salmon, pig or cattle cartilage by protease treatment, through decolorization, filtration, precipitation and cation exchange chromatography [145].

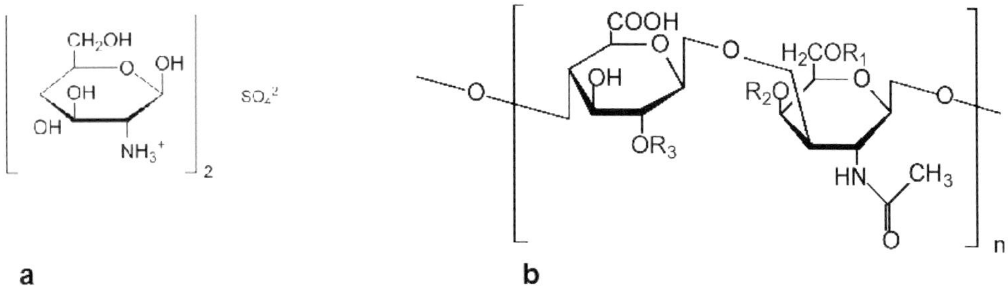

Fig. (8). Glucosamine sulfate (**a**) and chondroitin-A (**b**).

Glucosamine may also be produced by filamentous fungi such as *Aspergillus* sp.; fermentation reaches high concentrations (around 20% of the biomass, or 7-14g/L of glucosamine) [144, 146], and the amino sugar may be recovered from the biomass after filtration and hydrolysis, followed by ion exchange. This approach has been used by Cargill and has the advantages of being kosher [147] and permits the use of residual fungal biomass from other processes, *e.g.,* the production of enzymes. As for chondroitin, this sulfated polysaccharide is much more complex than the simple monomeric glucosamine; the positions and degree of sulfation vary greatly and change their properties, and a reliable process for production using microorganisms is possible [148, 149], but is still to be developed.

Coenzyme Q_{10}

Coenzymes Q_6 to Q_{10}, are lipophilic molecules with 6 to 10 isoprene residues attached to 2,3 dimethoxy, 5-methyl, 1,4 benzoquinone (Fig. **9**). These similar compounds are essential part of the mitochondrial electron transport system in several organisms [150]. In humans, the number of isoprenyl units in the final active coenzyme is 10. Most tissues are capable of producing this substance, hence the common name ubiquinone [151].

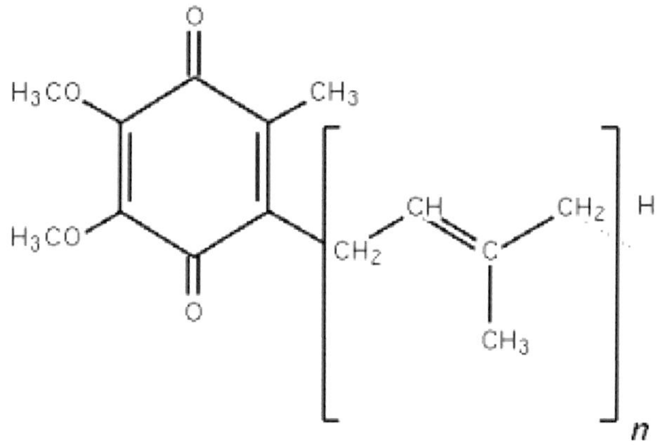

Fig. (9). Coenzymes Q. The number of isoprene residues (n) varies.

Coenzyme Q_{10}, or CoQ_{10}, is not an essential nutrient because it is synthesized in the body, but has essential *functions*. Its role in aerobic respiration and oxidative phosphorylation is so important [152] that this substance became popular not only as a drug, but also in dietary supplements and cosmetics [153, 154].

Several foods have a high content of CoQ_{10}, for example poultry, meat, fish, nuts and some oils [155, 156]. The production of Coenzyme Q_{10} may be done by chemical synthesis or by fermentation, using bacteria such as *Rhodobacter sphaeroides* [157] or recombinant *E coli* [158]; best results are reported using *Agrobacterium tumefaciens* [159], reaching 320 to 458mg/L in 96h. The product is intracellular, and the production consists of biomass separation, drying, alkaline solvent extraction and fractionation, and further adsorption or crystallization for eventual purification. Residual yeast or fish biomass may be similarly processed.

[160 - 163].

Resveratrol

This relatively simple polyphenol (Fig. **10**) has alleged anti-aging, anti-carcinogenic, anti-inflammatory, and anti-oxidant properties [164]. This compound exists in red wine in concentrations typically around 2-7mg/L, and has gained attention in the last two decades because of its possible role in prevention of cardiovascular disease and anticancer effects [165].

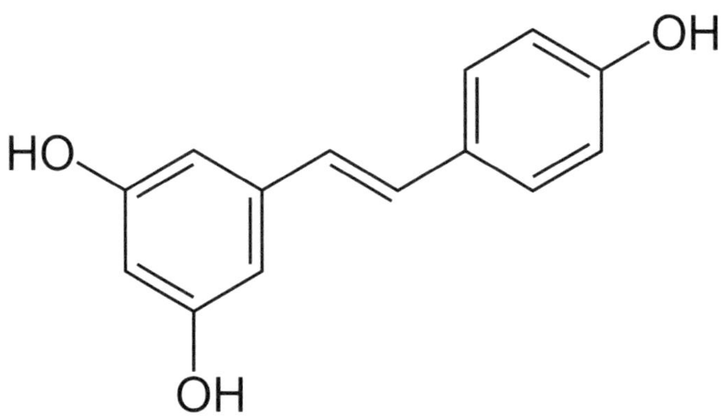

Fig. (10). Resveratrol (E)-3, 4', 5-trihydroxystilbene.

Natural resveratrol supplements usually contain powdered Japanese knotweed (*Fallopia japonica*) roots; it is suggested that Itadori tea, made from this plant, is a suitable substitute for wine as resveratrol source [166]. Other plants such as peanuts and soy also have minor amounts of resveratrol, but the only source where it is free (unglycosilated) is wine. Artificial resveratrol may be synthesized [167 - 169], while natural resveratrol may be produced using plant cell cultures or recombinant microorganisms. Plant cell cultures of *Vitis vinifera* (grape) with 50mM ciclodextrins and 100uM methyl jasmonate led to the production of 3g/L of resveratrol in 12 days [170]. The production using recombinant microorganisms still has lower final concentrations (up to 391 mg/L in 6 days were obtained with a recombinant *S. cerevisiae*, [171], and 2.4g/L in 25 hours by biotransformation with a recombinant *E. coli* [172]), and still small scales, but the processes are robust and have high productivity. Resveratrol may be recovered from the culture medium by solvent extraction or adsorption, followed by chromatography and / or crystallization [173, 174].

Choline

This compound, also known as vitamin B_4, is essential to metabolism and must be supplied through the diet, but is synthesized in small amounts in the body and thus, is not strictly regarded as a vitamin [175, 176]. Its most important dietary sources are soybeans, meat, eggs, and several other foods rich in phospholipids, where choline may take part (Fig. **11**). Besides being part of cell membranes in the form of phospholipids, choline derivatives such as acetylcholine are important neurotransmitters, whose supplementation may lead to improved memory [177].

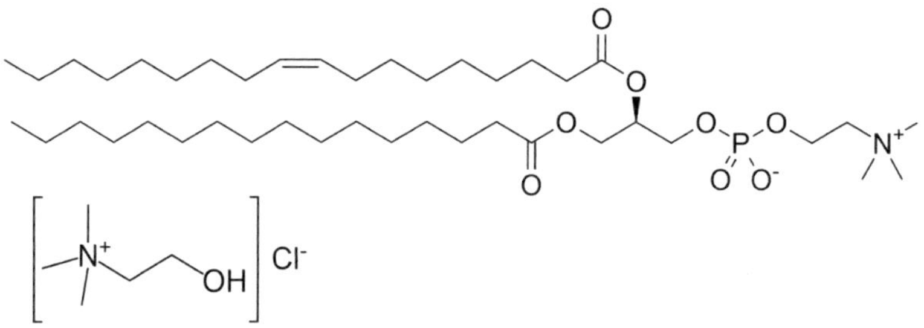

Fig. (11). A phospholipid (the large molecule), and choline chloride.

Although most organisms are capable of producing choline, its production by fermentation or cell cultures is uncommon; reports which describe its accumulation are usually concerned with its need as nutrient or its biosynthesis, as in [178 - 180]; the molecule is relatively simple and easily produced by synthesis, and natural phospholipids, by-products of oil extraction, will probably remain an important source of natural choline.

Carotenoids and Polyphenols

These large groups of molecules have, in general, an antioxidant activity which has a beneficial effect because of the reduction of free radicals, improving cell metabolism [181]. Common carotenoids such as beta-carotene, lycopene, lutein, zeaxanthin, and astaxanthin and polyphenols such as anthocyanidins, catechins, flavonoids, tannins, and procyanidins are employed in nutraceutical and cosmeceutical products because of a claimed antiaging effect [182]. Most carotenoids are also pigments, and the reader may consult Chapter 7 for further discussion.

As for polyphenols, the importance of these molecules cannot be stressed enough. However, its diversity is immense, and a thorough discussion of the class is beyond the scope of this text. Instead, we will cite three which are specially important (Fig. **12**), aside from the already discussed resveratrol: quercetin, a potent flavonoid antioxidant which is capable of protecting lipids form oxidation and has a free radical scavenging activity superior to that of ascorbic acid [183]; chlorogenic acid, an anti-inflammatory, analgesic and antipyretic phenolic compound, which is one of the most abundant antioxidants in foods and beverages [184]; and genistein, a estrogenic soy isoflavone recognized for its activity against breast carcinoma cell lines [185] (Fig. **12**).

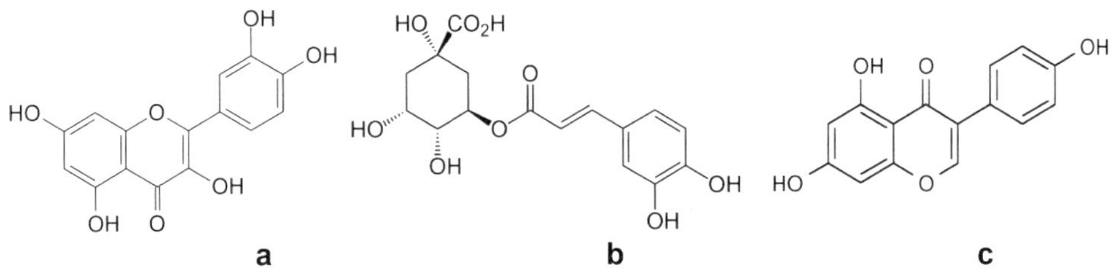

Fig. (12). Three examples of polyphenols, quercetin (**a**), chlorogenic acid (**b**), and genistein (**c**).

Most of the sources of these substances are either dietary (*e.g.* coffee, which has several antioxidant substances) [186] or from natural supplements based on vegetable extracts, such as dried berries or soy milk preparations. Two characteristics are common to the structures in Fig. (**12**), and to polyphenols in general: 1) these molecules are only moderately polar, and therefore the solubility in water may be limited. The extraction of the active principles from food matrixes may be done using a suitable solvent, *e.g.* an alcoholic or hydroalcoholic mixture [187, 188]. When the molecule is glycosylated – which is common – the solubility in water is higher. 2) These molecules are polyfunctional, and this leads to several derivatives: either by substitution of hydrogens in hydroxyl groups, or by more complex reactions and even polymerization. This leads to a large diversity of molecules, and although all will generally be antioxidant, each polyphenol may have a unique specific activity, which explains the active research on the field.

PERSPECTIVES

The vitamin market has been growing at a rate of 5 to 7% per year in the last decades, in a market worth US$ 23 billion in USA alone [189]. As the economy of less developed countries evolves, it is likely that the market will keep growing for the next decades. Part of the need for vitamins might be met by transformation of feedstocks, using biofortified GMO crops [190 - 193], although trust issues must be addressed for each cases. However, there will continue to be a demand for natural additives for food processing, and this may drive production efforts towards a) scaling up existing production platforms (*e.g.* vitamin E from *Euglena* sp. [107]), b) further screening of the microorganisms, especially those already involved in indigenous fermentation, such as dairy products and traditional Asian fermented products [194 - 197], or from the gut microbiome [198, 199] and c) the genetic modification of microorganisms by incorporation of microbial, plant or animal biosynthesis pathways to improve vitamin production by fermentation [200 - 204].

As for the application of vitamin and nutraceutical additives in foods and beverages, the development of encapsulation technologies might enhance the bioavailability of these substances [205], thus reducing the amount required in a given process or enhancing the shelf life of the product [206]. This is especially important for liposoluble vitamins such as E and A, or nutraceuticals such as carotenoids, CoQ_{10} and resveratrol, and may be achieved by the development of stabilized dispersions [207] and nanoencapsulation [205, 208].

CONFLICT OF INTEREST

The authors confirm that they have no conflict of interest to declare for this publication.

ACKNOWLEDGEMENTS

Decleared none.

REFERENCES

[1] Sun-Waterhouse, D. The development of fruit-based functional foods targeting the health and wellness market: a review. *Int. J. Food Sci. Technol.,* **2011**, *46*(5), 899-920. [http://dx.doi.org/10.1111/j.1365-2621.2010.02499.x]

[2] Batista, E.S.; Costa, A.G.; Pinheiro-Sant'ana, H.M. Adding vitamin E to foods: implications for the foods and for human health. *Revista de Nutrição,* **2007**, *20*(5), 525-535.

[3] Liberato, S.C. Fortification of industrialized foods with vitamins. *Revista de Nutrição,* **2006**, *19*(2), 215-231. [http://dx.doi.org/10.1590/S1415-52732006000200009]

[4] Zingg, J-M.; Vitamin, E. Vitamin E: an overview of major research directions. *Mol. Aspects Med.,* **2007**, *28*(5-6), 400-422. [http://dx.doi.org/10.1016/j.mam.2007.05.004] [PMID: 17624418]

[5] Combs, G.F., Jr *The vitamins*; Academic press, **2012**. [http://dx.doi.org/10.1016/B978-0-12-381980-2.00005-0]

[6] Mamede, A.C.; Tavares, S.D.; Abrantes, A.M.; Trindade, J.; Maia, J.M.; Botelho, M.F. The role of vitamins in cancer: a review. *Nutr. Cancer,* **2011**, *63*(4), 479-494. [http://dx.doi.org/10.1080/01635581.2011.539315] [PMID: 21541902]

[7] Fitzpatrick, T.B.; Basset, G.J.; Borel, P.; Carrari, F.; DellaPenna, D.; Fraser, P.D.; Hellmann, H.; Osorio, S.; Rothan, C.; Valpuesta, V.; Caris-Veyrat, C.; Fernie, A.R. Vitamin deficiencies in humans: can plant science help? *Plant Cell,* **2012**, *24*(2), 395-414. [http://dx.doi.org/10.1105/tpc.111.093120] [PMID: 22374394]

[8] Wildman, R.E. *Handbook of nutraceuticals and functional foods*; CRC press, **2006**. [http://dx.doi.org/10.1201/9781420006186]

[9] Kwak, N-s.; Jukes, D.J. Functional foods. Part 1: the development of a regulatory concept. *Food Contr.,* **2001**, *12*(2), 99-107. [http://dx.doi.org/10.1016/S0956-7135(00)00028-1]

[10] Gulati, O.P.; Berry Ottaway, P. Legislation relating to nutraceuticals in the European Union with a particular focus on botanical-sourced products. *Toxicology,* **2006**, *221*(1), 75-87. [http://dx.doi.org/10.1016/j.tox.2006.01.014] [PMID: 16487647]

[11] Coppens, P.; da Silva, M.F.; Pettman, S. European regulations on nutraceuticals, dietary supplements and functional foods: a framework based on safety. *Toxicology,* **2006**, *221*(1), 59-74. [http://dx.doi.org/10.1016/j.tox.2005.12.022] [PMID: 16469424]

[12] Shimomura, Y.; Yamamoto, Y.; Bajotto, G.; Sato, J.; Murakami, T.; Shimomura, N.; Kobayashi, H.; Mawatari, K. Nutraceutical effects of branched-chain amino acids on skeletal muscle. *J. Nutr.,* **2006**, *136*(2), 529S-532S. [PMID: 16424141]

[13] Delgado-Vargas, F.; Paredes-López, O. *Natural colorants for food and nutraceutical uses*; CRC Press, **2002**.
[http://dx.doi.org/10.1201/9781420031713]

[14] He, J.; Giusti, M.M. Anthocyanins: natural colorants with health-promoting properties. *Annu. Rev. Food Sci. Technol.,* **2010**, *1*, 163-187.
[http://dx.doi.org/10.1146/annurev.food.080708.100754] [PMID: 22129334]

[15] Ratledge, C. Single cell oils--have they a biotechnological future? *Trends Biotechnol.,* **1993**, *11*(7), 278-284.
[http://dx.doi.org/10.1016/0167-7799(93)90015-2] [PMID: 7763950]

[16] Gill, I.; Valivety, R. Polyunsaturated fatty acids, Part 1: Occurrence, biological activities and applications. *Trends Biotechnol.,* **1997**, *15*(10), 401-409.
[http://dx.doi.org/10.1016/S0167-7799(97)01076-7] [PMID: 9351284]

[17] Levrat-Verny, M-A.; Behr, S.; Mustad, V.; Rémésy, C.; Demigné, C. Low levels of viscous hydrocolloids lower plasma cholesterol in rats primarily by impairing cholesterol absorption. *J. Nutr.,* **2000**, *130*(2), 243-248.
[PMID: 10720177]

[18] Ellis, P.R.; Rayment, P.; Wang, Q. A physico-chemical perspective of plant polysaccharides in relation to glucose absorption, insulin secretion and the entero-insular axis. *Proc. Nutr. Soc.,* **1996**, *55*(3), 881-898.
[http://dx.doi.org/10.1079/PNS19960086] [PMID: 9004331]

[19] Andlauer, W.; Fürst, P. Nutraceuticals: a piece of history, present status and outlook. *Food Res. Int.,* **2002**, *35*(2), 171-176.
[http://dx.doi.org/10.1016/S0963-9969(01)00179-X]

[20] Budavari, S. *The Merck Index. An Encyclopedia of Chemicals, Drugs, and Biologicals.* , 12th; Merck & Co. Inc.: Whitehouse Station, NJ, **1996**.

[21] Flynn, A.; Moreiras, O.; Stehle, P.; Fletcher, R.J.; Müller, D.J.; Rolland, V. Vitamins and minerals: a model for safe addition to foods. *Eur. J. Nutr.,* **2003**, *42*(2), 118-130.
[http://dx.doi.org/10.1007/s00394-003-0391-9] [PMID: 12638033]

[22] ANVISA. Sanitária, V., Ed. *Diretrizes Básicas para Análise e Comprovação de Propriedades Funcionais e ou de Saúde Alegadas em Rotulagem de Alimentos in Resolução no 18, A.-A.N.d*; , **1999**.

[23] ANVISA. Sanitária, A-A., Ed. Regulamento de Procedimentos para Registro de Alimento com Alegação de Propriedades Funcionais e ou de Saúde em sua Rotulagem. In: *Resolução no 19*; , **1999**.

[24] Wrick, K.L. The impact of regulations on the business of nutraceuticals in the United States: Yesterday, today, and tomorrow *Reg. Func. Food. Nutrac. Global Pers.,* **2005**, 1-36.

[25] Jonathan Anscombe, M.T. Markus Stricker, Emmanuel Hembert, Antonella Leone-Kammler In: *Nutraceuticals: The Front Line of the Battle for Consumer Health, ATKearney, Editor*; , **2014**.

[26] Underwood, B.A.; Olson, J.A. *A brief guide to current methods of assessing vitamin A status. A report of the International Vitamin A Consultative Group*; IVACG, **1993**.

[27] Xu, F.; Yuan, Q-P.; Zhu, Y. Improved production of lycopene and β-carotene by *Blakeslea trispora*

with oxygen-vectors. *Process Biochem.,* **2007**, *42*(2), 289-293. [http://dx.doi.org/10.1016/j.procbio.2006.08.007]

[28] Bhosale, P.; Gadre, R.V. β-Carotene production in sugarcane molasses by a *Rhodotorula glutinis* mutant. *J. Ind. Microbiol. Biotechnol.,* **2001**, *26*(6), 327-332. [http://dx.doi.org/10.1038/sj.jim.7000138] [PMID: 11571614]

[29] Raja, R.; Hemaiswarya, S.; Rengasamy, R. Exploitation of *Dunaliella* for β-carotene production. *Appl. Microbiol. Biotechnol.,* **2007**, *74*(3), 517-523. [http://dx.doi.org/10.1007/s00253-006-0777-8] [PMID: 17225103]

[30] Shimizu, S. *Vitamins and related compounds: microbial production,* 2nd ed; Biotechnology Set, **2008**, pp. 318-340.

[31] Jakobsen, J.; Knuthsen, P. Stability of vitamin D in foodstuffs during cooking. *Food Chem.,* **2014**, *148*, 170-175. [http://dx.doi.org/10.1016/j.foodchem.2013.10.043] [PMID: 24262542]

[32] Green, J. Production of ergosterol from yeast. Patent US3006932., 1961.

[33] Tan, T.; Zhang, M.; Gao, H. Ergosterol production by fed-batch fermentation of *Saccharomyces cerevisiae. Enzyme Microb. Technol.,* **2003**, *33*(4), 366-370. [http://dx.doi.org/10.1016/S0141-0229(03)00132-7]

[34] Eitenmiller, R.R.; Landen, W., Jr; Ye, L. *Vitamin analysis for the health and food sciences*; CRC press, **2007**.

[35] Joint, F.; Organization, W.H. Vitamin and mineral requirements in human nutrition. *WHO,* **2005**, 341.

[36] Organization, W.H. *Guidelines on food fortification with micronutrients*; World Health Organization: Geneva, **2006**.

[37] Codex Standard, C. Guidelines for vitamin and mineral food supplements *The Codex Alimentarius Commission. CAC/GL,* **2005**, *2005*, 55-2005.

[38] L.K., MAHAN; Krause, S. Alimentos. *nutricao e dietoterapia,* **2005**, *9*, 111-118.

[39] Espín, J.C.; García-Conesa, M.T.; Tomás-Barberán, F.A. Nutraceuticals: facts and fiction. *Phytochemistry,* **2007**, *68*(22), 2986-3008. [http://dx.doi.org/10.1016/j.phytochem.2007.09.014] [PMID: 17976666]

[40] Imanaga, Y. Metabolism of D-Glucosamine. *J. Biochem.,* **1958**, *45*(8), 647-650.

[41] Anderson, J.W.; Nicolosi, R.J.; Borzelleca, J.F. Glucosamine effects in humans: a review of effects on glucose metabolism, side effects, safety considerations and efficacy. *Food Chem. Toxicol.,* **2005**, *43*(2), 187-201. [http://dx.doi.org/10.1016/j.fct.2004.11.006] [PMID: 15621331]

[42] ELY, A. What is the evidence for the safety and efficacy of dimethyl sulfoxide and methyl-sulfonylmethane in pain relief? *Pharm. J.,* **2002**, *269*(7223), 685-687.

[43] Turunen, M.; Olsson, J.; Dallner, G. Metabolism and function of coenzyme Q. *Biochimica et Biophysica Acta (BBA)- . Biomembranes,* **2004**, *1660*(1), 171-199. [http://dx.doi.org/10.1016/j.bbamem.2003.11.012]

[44] Evans, A.M.; Fornasini, G. Pharmacokinetics of L-carnitine. *Clin. Pharmacokinet.,* **2003**, *42*(11), 941-967.
[http://dx.doi.org/10.2165/00003088-200342110-00002] [PMID: 12908852]

[45] Touré, A.; Xueming, X. laxseed Lignans: Source, Biosynthesis, Metabolism, Antioxidant Activity, Bio-Active Components, and Health Benefits *Comprehensive reviews in food science and food safety,* **2010**, *9*(3), 261, 261.

[46] Zhu, Y. *In vitro* metabolism study of resveratrol and identification and determination of its main metabolite piceatannol by LC/EC and LC/MS/MS. *TIC,* **2004**, *4*, E7.

[47] Ward, N.C.; Croft, K.D.; Puddey, I.B.; Hodgson, J.M. Supplementation with grape seed polyphenols results in increased urinary excretion of 3-hydroxyphenylpropionic Acid, an important metabolite of proanthocyanidins in humans. *J. Agric. Food Chem.,* **2004**, *52*(17), 5545-5549.
[http://dx.doi.org/10.1021/jf049404r] [PMID: 15315398]

[48] Chaturvedi, S. Role of Nutraceuticals in health promotion. *Health,* **2011**, *4*, 5.

[49] Survase, S.A.; Bajaj, I.B.; Singhal, R.S. Biotechnological production of vitamins. *Food Technol. Biotechnol.,* **2006**, *44*(3), 381.

[50] Truong, J.T.; Booth, S.L. Emerging issues in vitamin K research. *J. Evid. Based Complementary Altern. Med.,* **2011**, *16*(1), 73-79.
[http://dx.doi.org/10.1177/1533210110392953]

[51] Wolf, G. A history of vitamin A and retinoids. *FASEB J.,* **1996**, *10*(9), 1102-1107.
[PMID: 8801174]

[52] Bennasir, H.; Sridhar, S.; Tech, M.; Abdel-Razek, T.T.; Vitamin, A. From physiology to disease prevention. *International Journal of Pharmaceutical Sciences Review and Research,* **2010**, *1*(1), 68-73.

[53] Ltd, F.C. *Niacin market and its production technology.,* **2011**. Available from: http://www.famachem.com/news/shownews.php?lang=en&id=31.

[54] Bai, C. A golden era—pro-vitamin A enhancement in diverse crops. *In Vitro Cell. Dev. Biol. Plant,* **2011**, *47*(2), 205-221.
[http://dx.doi.org/10.1007/s11627-011-9363-6]

[55] Ball, G.; Vitamin, A Vitamin A: retinoids and the provitamin A carotenoids. *Vitamins in foods: analysis, bioavailability and stability,* **2006**, 39-105.

[56] Britton, G. Structure and properties of carotenoids in relation to function. *FASEB J.,* **1995**, *9*(15), 1551-1558.
[PMID: 8529834]

[57] Chandler, L.; Schwartz, S. HPLC Separation of Cis Trans Carotene Isomers in Fresh and Processed Fruits and Vegetables. *J. Food Sci.,* **1987**, *52*(3), 669-672.
[http://dx.doi.org/10.1111/j.1365-2621.1987.tb06700.x]

[58] Godoy, H.T.; Rodriguez-Amaya, D.B. Occurrence of cis-isomers of provitamin A in Brazilian fruits. *J. Agric. Food Chem.,* **1994**, *42*(6), 1306-1313.
[http://dx.doi.org/10.1021/jf00042a011]

[59] Lessin, W.J.; Catigani, G.L.; Schwartz, S.J. Quantification of cis-trans isomers of provitamin A carotenoids in fresh and processed fruits and vegetables. *J. Agric. Food Chem.,* **1997**, *45*(10), 3728-3732.
[http://dx.doi.org/10.1021/jf960803z]

[60] Booth, S.; Johns, T.; Kuhnlein, H. Natural food sources of vitamin A and provitamin A. *Food Nutr. Bull.,* **1992**, *14*(1), 6-19.

[61] Shils, M.E.; Shike, M. *Modern nutrition in health and disease*; Lippincott Williams&; Wilkins, **2006**.

[62] Périssé, J.; Polacchi, W. Geographical distribution and recent changes in world supply of vitamin A. *Food Nutr. (Roma),* **1980**, *6*(1), 21-27.
[PMID: 7439468]

[63] DSIR. *Vitamin A Executive Summary.* , **2014**. http://www.dsir.gov.in/reports/techreps/tsr022.pdf.

[64] Unnevehr, L.; Pray, C.; Paarlberg, R. Addressing micronutrient deficiencies: alternative interventions and technologies **2008**, *10*(3), 124-134.

[65] Yoon, S-H.; Lee, S.H.; Das, A.; Ryu, H.K.; Jang, H.J.; Kim, J.Y.; Oh, D.K.; Keasling, J.D.; Kim, S.W. Combinatorial expression of bacterial whole mevalonate pathway for the production ofβ-carotene in *E. coli. J. Biotechnol.,* **2009**, *140*(3-4), 218-226.
[http://dx.doi.org/10.1016/j.jbiotec.2009.01.008] [PMID: 19428716]

[66] Lin, Y.; Jain, R.; Yan, Y. Microbial production of antioxidant food ingredients *via* metabolic engineering. *Curr. Opin. Biotechnol.,* **2014**, *26*, 71-78.
[http://dx.doi.org/10.1016/j.copbio.2013.10.004] [PMID: 24679261]

[67] Gropper, S.; Smith, J. *Advanced nutrition and human metabolism*; Cengage Learning, **2012**.

[68] Zhu, C.; Naqvi, S.; Breitenbach, J.; Sandmann, G.; Christou, P.; Capell, T. Combinatorial genetic transformation generates a library of metabolic phenotypes for the carotenoid pathway in maize. *Proc. Natl. Acad. Sci. USA,* **2008**, *105*(47), 18232-18237.
[http://dx.doi.org/10.1073/pnas.0809737105] [PMID: 19011084]

[69] Naqvi, S.; Farré, G.; Sanahuja, G.; Capell, T.; Zhu, C.; Christou, P. When more is better: multigene engineering in plants. *Trends Plant Sci.,* **2010**, *15*(1), 48-56.
[http://dx.doi.org/10.1016/j.tplants.2009.09.010] [PMID: 19853493]

[70] Diretto, G.; Al-Babili, S.; Tavazza, R.; Papacchioli, V.; Beyer, P.; Giuliano, G. Metabolic engineering of potato carotenoid content through tuber-specific overexpression of a bacterial mini-pathway. *PLoS One,* **2007**, *2*(4), e350.
[http://dx.doi.org/10.1371/journal.pone.0000350] [PMID: 17406674]

[71] Welsch, R.; Arango, J.; Bär, C.; Salazar, B.; Al-Babili, S.; Beltrán, J.; Chavarriaga, P.; Ceballos, H.; Tohme, J.; Beyer, P. Provitamin A accumulation in cassava (Manihot esculenta) roots driven by a single nucleotide polymorphism in a phytoene synthase gene. *Plant Cell,* **2010**, *22*(10), 3348-3356.
[http://dx.doi.org/10.1105/tpc.110.077560] [PMID: 20889914]

[72] Maass, D.; Arango, J.; Wüst, F.; Beyer, P.; Welsch, R. Carotenoid crystal formation in Arabidopsis and carrot roots caused by increased phytoene synthase protein levels. *PLoS One,* **2009**, *4*(7), e6373.
[http://dx.doi.org/10.1371/journal.pone.0006373] [PMID: 19636414]

[73] Fujisawa, M.; Takita, E.; Harada, H.; Sakurai, N.; Suzuki, H.; Ohyama, K.; Shibata, D.; Misawa, N. Pathway engineering of Brassica napus seeds using multiple key enzyme genes involved in ketocarotenoid formation. *J. Exp. Bot.,* **2009**, *60*(4), 1319-1332. [http://dx.doi.org/10.1093/jxb/erp006] [PMID: 19204032]

[74] Davuluri, G.R.; van Tuinen, A.; Fraser, P.D.; Manfredonia, A.; Newman, R.; Burgess, D.; Brummell, D.A.; King, S.R.; Palys, J.; Uhlig, J.; Bramley, P.M.; Pennings, H.M.; Bowler, C. Fruit-specific RNAi-mediated suppression of DET1 enhances carotenoid and flavonoid content in tomatoes. *Nat. Biotechnol.,* **2005**, *23*(7), 890-895. [http://dx.doi.org/10.1038/nbt1108] [PMID: 15951803]

[75] Yang, Y.; Zhang, Z.; Lu, J.; Maekawa, T. Continuous methane fermentation and the production of vitamin B_{12} in a fixed-bed reactor packed with loofah. *Bioresour. Technol.,* **2004**, *92*(3), 285-290. [http://dx.doi.org/10.1016/j.biortech.2003.09.011] [PMID: 14766162]

[76] Wang, P.; Wang, Y.; Liu, Y.; Shi, H.; Su, Z. Novel *in situ* product removal technique for simultaneous production of propionic acid and vitamin B_{12} by expanded bed adsorption bioreactor. *Bioresour. Technol.,* **2012**, *104*, 652-659. [http://dx.doi.org/10.1016/j.biortech.2011.10.047] [PMID: 22082511]

[77] Hugenschmidt, S. Screening of a natural biodiversity of lactic and propionic acid bacteria for folate and vitamin B_{12} production in supplemented whey permeate. *Int. Dairy J.,* **2010**, *20*(12), 852-857. [http://dx.doi.org/10.1016/j.idairyj.2010.05.005]

[78] Li, K-T.; Liu, D.H.; Li, Y.L.; Chu, J.; Wang, Y.H.; Zhuang, Y.P.; Zhang, S.L. Improved large-scale production of vitamin B_{12} by *Pseudomonas denitrificans* with betaine feeding. *Bioresour. Technol.,* **2008**, *99*(17), 8516-8520. [http://dx.doi.org/10.1016/j.biortech.2008.03.023] [PMID: 18440227]

[79] Ye, K. Efficient production of vitamin B_{12} from propionic acid bacteria under periodic variation of dissolved oxygen concentration. *J. Ferment. Bioeng.,* **1996**, *82*(5), 484-491. [http://dx.doi.org/10.1016/S0922-338X(97)86988-7]

[80] Moore, S.J.; Mayer, M.J.; Biedendieck, R.; Deery, E.; Warren, M.J. Towards a cell factory for vitamin B_{12} production in *Bacillus megaterium* : bypassing of the cobalamin riboswitch control elements. *N. Biotechnol.,* **2014**, *31*(6), 553-561. [http://dx.doi.org/10.1016/j.nbt.2014.03.003] [PMID: 24657453]

[81] Tarrago-Trani, M.T.; Phillips, K.M.; Cotty, M. Matrix-specific method validation for quantitative analysis of vitamin C in diverse foods. *J. Food Compos. Anal.,* **2012**, *26*(1), 12-25. [http://dx.doi.org/10.1016/j.jfca.2012.03.004]

[82] Lehninger, A. *Principles Of Biochemistry & EBook Author: Albert Lehninger, David L. Nelson, Michael M. Cox*; W. H., **2008**.

[83] Antioxidants, I.M.; Compounds, R. *Dietary Reference Intakes for Vitamin C, Vitamin E, Selenium, and Carotenoids: A Report of the Panel on Dietary Antioxidants and Related Compounds, Subcommittees on Upper Reference Levels of Nutrients and of Interpretation and Use of Dietary Reference Intakes, and the Standing Committee on the Scientific Evaluation of Dietary Reference Intakes, Food and Nutrition Board, Institute of Medicine*; National Academies Press, **2000**.

[84] Hancock, R.D.; Viola, R. Biotechnological approaches for L-ascorbic acid production. *Trends Biotechnol.,* **2002**, *20*(7), 299-305.
[http://dx.doi.org/10.1016/S0167-7799(02)01991-1] [PMID: 12062975]

[85] Grindley, J.F.; Payton, M.A.; van de Pol, H.; Hardy, K.G. Conversion of glucose to 2-keto-L-gulonate, an intermediate in L-ascorbate synthesis, by a recombinant strain of *Erwinia citreus. Appl. Environ. Microbiol.,* **1988**, *54*(7), 1770-1775.
[PMID: 16347687]

[86] De Wulf, P.; Soetaert, W.; Vandamme, E.J. Optimized synthesis of L-sorbose by C(5)-dehydrogenation of D-sorbitol with *Gluconobacter oxydans. Biotechnol. Bioeng.,* **2000**, *69*(3), 339-343.
[http://dx.doi.org/10.1002/1097-0290(20000805)69:3<339::AID-BIT12>3.0.CO;2-E] [PMID: 10861414]

[87] Shinjoh, M. Improved production of 2-keto-l-gulonic acid. Patent WO2010100236 A1., 2010.

[88] Doncheck, J.A. L-ascorbic acid containing biomass of Chlorella pyrenoidosa. Patent US5521090 A , 1996.

[89] Running, J. Process for the production of ascorbic acid with Prototheca. Patent US5900370 A., 1999.

[90] Kumar, M. Production of ascorbic acid using yeast. Patent US6630330 B1., 2011.

[91] Cook, L.S. A systematic literature review of vitamin D and ovarian cancer. *Am. J. Obstetrics Gynecology.,* **2010**, *203*(1), 70. e1-70. e8.
[http://dx.doi.org/10.1016/j.ajog.2010.01.062]

[92] Holick, M.F. The influence of vitamin D on bone health across the life cycle. *J. Nutr.,* **2005**, *135*(11), 2726S-2727S.
[PMID: 16251638]

[93] Shang, F.; Wang, Z.; Tan, T. High-cell-density cultivation for co-production of ergosterol and reduced glutathione by *Saccharomyces cerevisiae. Appl. Microbiol. Biotechnol.,* **2008**, *77*(6), 1233-1240.
[http://dx.doi.org/10.1007/s00253-007-1272-6] [PMID: 18071647]

[94] He, X.; Guo, X.; Liu, N.; Zhang, B. Ergosterol production from molasses by genetically modified *Saccharomyces cerevisiae. Appl. Microbiol. Biotechnol.,* **2007**, *75*(1), 55-60.
[http://dx.doi.org/10.1007/s00253-006-0807-6] [PMID: 17225097]

[95] Azzi, A. Molecular mechanism ofα-tocopherol action. *Free Radic. Biol. Med.,* **2007**, *43*(1), 16-21.
[http://dx.doi.org/10.1016/j.freeradbiomed.2007.03.013] [PMID: 17561089]

[96] Hirschberg, J. Production of high-value compounds: carotenoids and vitamin E. *Curr. Opin. Biotechnol.,* **1999**, *10*(2), 186-191.
[http://dx.doi.org/10.1016/S0958-1669(99)80033-0] [PMID: 10209146]

[97] Sen, C.K.; Khanna, S.; Rink, C.; Roy, S. Tocotrienols: the emerging face of natural vitamin E. *Vitam. Horm.,* **2007**, *76*, 203-261.
[http://dx.doi.org/10.1016/S0083-6729(07)76008-9] [PMID: 17628176]

[98] Theriault, A.; Chao, J.T.; Wang, Q.; Gapor, A.; Adeli, K. Tocotrienol: a review of its therapeutic potential. *Clin. Biochem.,* **1999**, *32*(5), 309-319.

[http://dx.doi.org/10.1016/S0009-9120(99)00027-2] [PMID: 10480444]

[99] Traber, M.G.; Atkinson, J.; Vitamin, E. Vitamin E, antioxidant and nothing more. *Free Radic. Biol. Med.,* **2007**, *43*(1), 4-15. [http://dx.doi.org/10.1016/j.freeradbiomed.2007.03.024] [PMID: 17561088]

[100] Rodríguez-Zavala, J.S.; Ortiz-Cruz, M.A.; Mendoza-Hernández, G.; Moreno-Sánchez, R. Increased synthesis ofα-tocopherol, paramylon and tyrosine by *Euglena gracilis* under conditions of high biomass production. *J. Appl. Microbiol.,* **2010**, *109*(6), 2160-2172. [http://dx.doi.org/10.1111/j.1365-2672.2010.04848.x] [PMID: 20854454]

[101] Carballo-Cárdenas, E.C.; Tuan, P.M.; Janssen, M.; Wijffels, R.H. Vitamin E (α-tocopherol) production by the marine microalgae *Dunaliella tertiolecta* and *Tetraselmis suecica* in batch cultivation. *Biomol. Eng.,* **2003**, *20*(4-6), 139-147. [http://dx.doi.org/10.1016/S1389-0344(03)00040-6] [PMID: 12919791]

[102] Mendiola, J.A. Enrichment of vitamin E from *Spirulina platensis* microalga by SFE. *J. Supercrit. Fluids,* **2008**, *43*(3), 484-489. [http://dx.doi.org/10.1016/j.supflu.2007.07.021]

[103] Kusmic, C. Euglena gracilis as source of the antioxidant vitamin E. Effects of culture conditions in the wild strain and in the natural mutant WZSL. *J. Appl. Phycol.,* **1998**, *10*(6), 555-559. [http://dx.doi.org/10.1023/A:1008022305865]

[104] Afiukwa, C.A.; Ogbonna, J.C. Effects of mixed substrates on growth and vitamin production by *Euglena gracilis. Afr. J. Biotechnol.,* **2007**, *6*(22)

[105] Abalde, J.; Fabregas, J.; Herrero, C. β-Carotene, vitamin C and vitamin E content of the marine microalga *Dunaliella tertiolecta* cultured with different nitrogen sources. *Bioresour. Technol.,* **1991**, *38*(2), 121-125. [http://dx.doi.org/10.1016/0960-8524(91)90142-7]

[106] Abe, K.; Nishimura, N.; Hirano, M. Simultaneous production ofβ-carotene, vitamin E and vitamin C by the aerial microalga *Trentepohlia aurea. J. Appl. Phycol.,* **1999**, *11*(4), 331-336. [http://dx.doi.org/10.1023/A:1008195710981]

[107] Ogbonna, J.C.; Tomiyama, S.; Tanaka, H. Production ofα-tocopherol by sequential heterotrophic-photoautotrophic cultivation of *Euglena gracilis. Prog. Ind. Microbiol.,* **1999**, *35*, 213-221. [http://dx.doi.org/10.1016/S0079-6352(99)80115-5]

[108] Fujita, T.; Aoyagi, H.; Ogbonna, J.C.; Tanaka, H. Effect of mixed organic substrate onα-tocopherol production by *Euglena gracilis* in photoheterotrophic culture. *Appl. Microbiol. Biotechnol.,* **2008**, *79*(3), 371-378. [http://dx.doi.org/10.1007/s00253-008-1443-0] [PMID: 18389233]

[109] Durmaz, Y.; Vitamin, E. (α-tocopherol) production by the marine microalgae *Nannochloropsis oculata* (Eustigmatophyceae) in nitrogen limitation. *Aquaculture,* **2007**, *272*(1), 717-722. [http://dx.doi.org/10.1016/j.aquaculture.2007.07.213]

[110] Scientific Opinion on the safety and efficacy of vitamin K3 (menadione sodium bisulphite and menadione nicotinamide bisulphite) as a feed additive for all animal species. *EFSA Journal,* **2014**, *1*(12), 29.

[111] Berenjian, A.; Mahanama, R.; Talbot, A.; Biffin, R.; Regtop, H.; Valtchev, P.; Kavanagh, J.; Dehghani, F. Efficient media for high menaquinone-7 production: response surface methodology approach. *N. Biotechnol.,* **2011**, *28*(6), 665-672. [http://dx.doi.org/10.1016/j.nbt.2011.07.007] [PMID: 21839866]

[112] Aydin Berenjian, R.M. Andrea Talbot, Hubert Regtop, John Kavanagh, Fariba Dehghani, Advances in menaquinone-7 production by *Bacillus subtilis* natto: fed-batch glycerol addition. *American Journal of Biochemistry and Biotechnology,* **2012**, *8*(2), 105-110. [http://dx.doi.org/10.3844/ajbbsp.2012.105.110]

[113] Berenjian, A.; Chan, N.L.; Mahanama, R.; Talbot, A.; Regtop, H.; Kavanagh, J.; Dehghani, F. Effect of biofilm formation by *Bacillus subtilis* natto on menaquinone-7 biosynthesis. *Mol. Biotechnol.,* **2013**, *54*(2), 371-378. [http://dx.doi.org/10.1007/s12033-012-9576-x] [PMID: 22740166]

[114] Mahanama, R. Modeling the effect of bed height and particle size for vitamin K2 production in a static bed fermenter. *Engineering Letters,* **2012**, *20*(1), 16.

[115] Mahanama, R. Enhanced production of menaquinone 7 *via* solid substrate fermentation from *Bacillus subtilis. International Journal of Food Engineering,* **2011**, *7*(5) [http://dx.doi.org/10.2202/1556-3758.2314]

[116] Sato, T.; Yamada, Y.; Ohtani, Y.; Mitsui, N.; Murasawa, H.; Araki, S. Production of menaquinone (vitamin K2)-7 by *Bacillus subtilis. J. Biosci. Bioeng.,* **2001**, *91*(1), 16-20. [http://dx.doi.org/10.1016/S1389-1723(01)80104-3] [PMID: 16232939]

[117] Sato, T.; Yamada, Y.; Ohtani, Y.; Mitsui, N.; Murasawa, H.; Araki, S. Efficient production of menaquinone (vitamin K2) by a menadione-resistant mutant of *Bacillus subtilis. J. Ind. Microbiol. Biotechnol.,* **2001**, *26*(3), 115-120. [http://dx.doi.org/10.1038/sj.jim.7000089] [PMID: 11420649]

[118] Tani, Y.; Asahi, S.; Yamada, H. Menaquinone (vitamin K2)-6 production by mutants of *Flavobacterium meningosepticum. J. Nutr. Sci. Vitaminol. (Tokyo),* **1986**, *32*(2), 137-145. [http://dx.doi.org/10.3177/jnsv.32.137] [PMID: 3761040]

[119] Tani, Y.; Taguchi, H. Extracellular production of menaquinone-4 by a mutant of *Flavobacterium* sp. 238-7 with a detergent-supplemented culture. *J. Ferment. Bioeng.,* **1989**, *67*(2), 102-106. [http://dx.doi.org/10.1016/0922-338X(89)90188-8]

[120] Kalra, E.K. Nutraceutical--definition and introduction. *AAPS PharmSci,* **2003**, *5*(3), E25. [http://dx.doi.org/10.1208/ps050325] [PMID: 14621960]

[121] Whitman, M. Understanding the perceived need for complementary and alternative nutraceuticals: lifestyle issues. *Clin. J. Oncol. Nurs.,* **2001**, *5*(5), 190-194. [PMID: 11899764]

[122] Henson, S.; Masakure, O.; Cranfield, J. The propensity for consumers to offset health risks through the use of functional foods and nutraceuticals: The case of lycopene. *Food Qual. Prefer.,* **2008**, *19*(4), 395-406. [http://dx.doi.org/10.1016/j.foodqual.2007.12.001]

[123] Siró, I.; Kápolna, E.; Kápolna, B.; Lugasi, A. Functional food. Product development, marketing and

consumer acceptance--a review. *Appetite,* **2008**, *51*(3), 456-467. [http://dx.doi.org/10.1016/j.appet.2008.05.060] [PMID: 18582508]

[124] Britton, G.; Liaaen-Jensen, S.; Pfander, H. *Carotenoids: handbook*; Springer, **2004**. [http://dx.doi.org/10.1007/978-3-0348-7836-4]

[125] Annegowda, H. Evaluation of phenolic content and antioxidant property of hydrolysed extracts of *Terminalia catappa* L. leaf. *Asian J. Plant Sci.,* **2010**, *9*(8), 479. [http://dx.doi.org/10.3923/ajps.2010.479.485]

[126] Sakakibara, H.; Honda, Y.; Nakagawa, S.; Ashida, H.; Kanazawa, K. Simultaneous determination of all polyphenols in vegetables, fruits, and teas. *J. Agric. Food Chem.,* **2003**, *51*(3), 571-581. [http://dx.doi.org/10.1021/jf020926l] [PMID: 12537425]

[127] Chiou, D.; Langrish, T. Development and characterisation of novel nutraceuticals with spray drying technology. *J. Food Eng.,* **2007**, *82*(1), 84-91. [http://dx.doi.org/10.1016/j.jfoodeng.2007.01.021]

[128] Moure, A. Advances in the manufacture, purification and applications of xylo-oligosaccharides as food additives and nutraceuticals. *Process Biochem.,* **2006**, *41*(9), 1913-1923. [http://dx.doi.org/10.1016/j.procbio.2006.05.011]

[129] Wang, L.; Weller, C.L. Recent advances in extraction of nutraceuticals from plants. *Trends Food Sci. Technol.,* **2006**, *17*(6), 300-312. [http://dx.doi.org/10.1016/j.tifs.2005.12.004]

[130] Rodríguez-Meizoso, I.; Marin, F.R.; Herrero, M.; Señorans, F.J.; Reglero, G.; Cifuentes, A.; Ibáñez, E. Subcritical water extraction of nutraceuticals with antioxidant activity from oregano. Chemical and functional characterization. *J. Pharm. Biomed. Anal.,* **2006**, *41*(5), 1560-1565. [http://dx.doi.org/10.1016/j.jpba.2006.01.018] [PMID: 16488570]

[131] Liu, Z.; Ji, T.; Yin, L.; Shen, J.; Shen, F.; Chen, G. Transcriptome sequencing analysis reveals the regulation of the hypopharyngeal glands in the honey bee, *Apis mellifera carnica* Pollmann. *PLoS One,* **2013**, *8*(12), e81001. [http://dx.doi.org/10.1371/journal.pone.0081001] [PMID: 24339892]

[132] Berridge, M.J.; Irvine, R.F. nositol phosphates and cell signalling **1989.**, *341*(6239), 197-205. [http://dx.doi.org/10.1038/341197a0]

[133] Urbano, G.; López-Jurado, M.; Aranda, P.; Vidal-Valverde, C.; Tenorio, E.; Porres, J. The role of phytic acid in legumes: antinutrient or beneficial function? *J. Physiol. Biochem.,* **2000**, *56*(3), 283-294. [http://dx.doi.org/10.1007/BF03179796] [PMID: 11198165]

[134] Liggett, R.W.; Koffler, H. Corn steep liquor in microbiology. *Bacteriol. Rev.,* **1948**, *12*(4), 297-311. [PMID: 16350125]

[135] Hull, S.R.; Montgomery, R. myo-Inositol phosphates in corn steep water. *J. Agric. Food Chem.,* **1995**, *43*(6), 1516-1523. [http://dx.doi.org/10.1021/jf00054a019]

[136] Kramer, K.; Hoppe, P-P.; Packer, L. *Nutraceuticals in health and disease prevention* ; , **2001**.

[137] Volek, J.S.; Kraemer, W.J.; Rubin, M.R.; Gómez, A.L.; Ratamess, N.A.; Gaynor, P. L-Carnitine L-tartrate supplementation favorably affects markers of recovery from exercise stress. *Am. J. Physiol. Endocrinol. Metab.,* **2002**, *282*(2), E474-E482. [http://dx.doi.org/10.1152/ajpendo.00277.2001] [PMID: 11788381]

[138] Karlic, H.; Lohninger, A. Supplementation of L-carnitine in athletes: does it make sense? *Nutrition,* **2004**, *20*(7-8), 709-715. [http://dx.doi.org/10.1016/j.nut.2004.04.003] [PMID: 15212755]

[139] Obón, J.M.; Maiquez, J.R.; Cánovas, M.; Kleber, H.P.; Iborra, J.L. High-density Escherichia coli cultures for continuous L(-)-carnitine production. *Appl. Microbiol. Biotechnol.,* **1999**, *51*(6), 760-764. [http://dx.doi.org/10.1007/s002530051459] [PMID: 10422223]

[140] Alvarez-Vasquez, F.; Cánovas, M.; Iborra, J.L.; Torres, N.V. Modeling, optimization and experimental assessment of continuous L-(-)-carnitine production by *Escherichia coli* cultures. *Biotechnol. Bioeng.,* **2002**, *80*(7), 794-805. [http://dx.doi.org/10.1002/bit.10436] [PMID: 12402325]

[141] Deal, C.L.; Moskowitz, R.W. Nutraceuticals as therapeutic agents in osteoarthritis. The role of glucosamine, chondroitin sulfate, and collagen hydrolysate. *Rheum. Dis. Clin. North Am.,* **1999**, *25*(2), 379-395. [http://dx.doi.org/10.1016/S0889-857X(05)70074-0] [PMID: 10356424]

[142] Simánek, V.; Kren, V.; Ulrichová, J.; Gallo, J. The efficacy of glucosamine and chondroitin sulfate in the treatment of osteoarthritis: are these saccharides drugs or nutraceuticals? *Biomed. Pap. Med. Fac. Univ. Palacky Olomouc Czech Repub.,* **2005**, *149*(1), 51-56. [http://dx.doi.org/10.5507/bp.2005.004] [PMID: 16170388]

[143] Quitain, A.T. Production of valuable materials by hydrothermal treatment of shrimp shells. *Ind. Eng. Chem. Res.,* **2001**, *40*(25), 5885-5888. [http://dx.doi.org/10.1021/ie010439f]

[144] Sitanggang, A.B.; Wu, H.S.; Wang, S.S.; Ho, Y.C. Effect of pellet size and stimulating factor on the glucosamine production using *Aspergillus sp* . BCRC 31742. *Bioresour. Technol.,* **2010**, *101*(10), 3595-3601. [http://dx.doi.org/10.1016/j.biortech.2009.12.084] [PMID: 20093019]

[145] Schiraldi, C.; Cimini, D.; De Rosa, M. Production of chondroitin sulfate and chondroitin. *Appl. Microbiol. Biotechnol.,* **2010**, *87*(4), 1209-1220. [http://dx.doi.org/10.1007/s00253-010-2677-1] [PMID: 20521042]

[146] Zhang, J.; Liu, L.; Li, J.; Du, G.; Chen, J. Enhanced glucosamine production by *Aspergillus* sp. BCRC 31742 based on the time-variant kinetics analysis of dissolved oxygen level. *Bioresour. Technol.,* **2012**, *111*, 507-511. [http://dx.doi.org/10.1016/j.biortech.2012.02.063] [PMID: 22401711]

[147] Barrow, C.; Shahidi, F. *Marine nutraceuticals and functional foods*; CRC Press, **2007**.

[148] Restaino, O.F.; Cimini, D.; De Rosa, M.; Catapano, A.; De Rosa, M.; Schiraldi, C. High cell density cultivation of *Escherichia coli* K4 in a microfiltration bioreactor: a step towards improvement of chondroitin precursor production. *Microb. Cell Fact.,* **2011**, *10*, 10.

[http://dx.doi.org/10.1186/1475-2859-10-10] [PMID: 21324163]

[149] Schiraldi, C.; Carcarino, I.L.; Alfano, A.; Restaino, O.F.; Panariello, A.; De Rosa, M. Purification of chondroitin precursor from *Escherichia coli* K4 fermentation broth using membrane processing. *Biotechnol. J.,* **2011**, *6*(4), 410-419.
[http://dx.doi.org/10.1002/biot.201000266] [PMID: 21381202]

[150] Hatefi, Y.; Coenzyme, Q. (ubiquinone). *Adv. Enzymol. Relat. Areas Mol. Biol.,* **1963**, *25*, 275-328.
[PMID: 14149679]

[151] Desbats, M.A.; Lunardi, G.; Doimo, M.; Trevisson, E.; Salviati, L. Genetic bases and clinical manifestations of coenzyme Q10 (CoQ10) deficiency. *J. Inherit. Metab. Dis.,* **2014**, **•••**, 1-12.
[PMID: 25091424]

[152] Hayashi, K.; Ogiyama, Y.; Yokomi, K.; Nakagawa, T.; Kaino, T.; Kawamukai, M. Functional conservation of coenzyme Q biosynthetic genes among yeasts, plants, and humans. *PLoS One,* **2014**, *9*(6), e99038.
[http://dx.doi.org/10.1371/journal.pone.0099038] [PMID: 24911838]

[153] Hathcock, J.N.; Shao, A. Risk assessment for coenzyme Q10 (Ubiquinone). *Regul. Toxicol. Pharmacol.,* **2006**, *45*(3), 282-288.
[http://dx.doi.org/10.1016/j.yrtph.2006.05.006] [PMID: 16814438]

[154] Hojerová, J. [Coenzyme Q10--its importance, properties and use in nutrition and cosmetics]. *Ceska Slov. Farm.,* **2000**, *49*(3), 119-123.
[PMID: 10953455]

[155] Weber, C.; Bysted, A.; Hłlmer, G. The coenzyme Q10 content of the average Danish diet. *Int. J. Vitam. Nutr. Res.,* **1997**, *67*(2), 123-129.
[PMID: 9129255]

[156] Pravst, I.; Žmitek, K.; Žmitek, J. Coenzyme Q10 contents in foods and fortification strategies. *Crit. Rev. Food Sci. Nutr.,* **2010**, *50*(4), 269-280.
[http://dx.doi.org/10.1080/10408390902773037] [PMID: 20301015]

[157] Kien, N.B.; Kong, I.S.; Lee, M.G.; Kim, J.K. Coenzyme Q10 production in a 150-l reactor by a mutant strain of *Rhodobacter sphaeroides. J. Ind. Microbiol. Biotechnol.,* **2010**, *37*(5), 521-529.
[http://dx.doi.org/10.1007/s10295-010-0699-4] [PMID: 20195885]

[158] Park, Y-C.; Kim, S.J.; Choi, J.H.; Lee, W.H.; Park, K.M.; Kawamukai, M.; Ryu, Y.W.; Seo, J.H. Batch and fed-batch production of coenzyme Q10 in recombinant Escherichia coli containing the decaprenyl diphosphate synthase gene from *Gluconobacter suboxydans. Appl. Microbiol. Biotechnol.,* **2005**, *67*(2), 192-196.
[http://dx.doi.org/10.1007/s00253-004-1743-y] [PMID: 15459799]

[159] Ha, S-J.; Kim, S.Y.; Seo, J.H.; Oh, D.K.; Lee, J.K. Optimization of culture conditions and scale-up to pilot and plant scales for coenzyme Q10 production by *Agrobacterium tumefaciens. Appl. Microbiol. Biotechnol.,* **2007**, *74*(5), 974-980.
[http://dx.doi.org/10.1007/s00253-006-0744-4] [PMID: 17124579]

[160] Cao, X-L.; Xu, Y.T.; Zhang, G.M.; Xie, S.M.; Dong, Y.M.; Ito, Y. Purification of coenzyme Q10 from fermentation extract: high-speed counter-current chromatography *versus* silica gel column chroma-

tography. *J. Chromatogr. A,* **2006**, *1127*(1-2), 92-96.
[http://dx.doi.org/10.1016/j.chroma.2006.05.083] [PMID: 16797569]

[161] Zhong, W.; Fang, J.; Liu, H.; Wang, X. Enhanced production of CoQ10 by newly isolated *Sphingomonas sp* . ZUTEO3 with a coupled fermentation-extraction process. *J. Ind. Microbiol. Biotechnol.,* **2009**, *36*(5), 687-693.
[http://dx.doi.org/10.1007/s10295-009-0538-7] [PMID: 19221819]

[162] Laplante, S.; Souchet, N.; Bryl, P. Comparison of low-temperature processes for oil and coenzyme Q10 extraction from mackerel and herring. *Eur. J. Lipid Sci. Technol.,* **2009**, *111*(2), 135-141.
[http://dx.doi.org/10.1002/ejlt.200800133]

[163] Zhao, S. Q.; Liu, Y. C.; Jiang, G. T. Extraction and Measurement Methods of Coenzyme Q10 from Biomass of Saccharomyces cerevisiae *Liquor-Making Science & Technology,* **2008**, *11*(2), 049.

[164] Smoliga, J.M.; Baur, J.A.; Hausenblas, H.A. Resveratrol and health - a comprehensive review of human clinical trials. *Mol. Nutr. Food Res.,* **2011**, *55*(8), 1129-1141.
[http://dx.doi.org/10.1002/mnfr.201100143] [PMID: 21688389]

[165] Higdon, J.; Drake, V.J.; Steward, W.P. Resveratrol. Micronutrient Information Center: Linus Pauling Institute. *Available from: http://lpi.oregonstate.edu/infocenter/phytochemicals/resveratrol/,* **2005**.

[166] Burns, J.; Yokota, T.; Ashihara, H.; Lean, M.E.; Crozier, A. Plant foods and herbal sources of resveratrol. *J. Agric. Food Chem.,* **2002**, *50*(11), 3337-3340.
[http://dx.doi.org/10.1021/jf0112973] [PMID: 12010007]

[167] Guiso, M.; Marra, C.; Farina, A. A new efficient resveratrol synthesis. *Tetrahedron Lett.,* **2002**, *43*(4), 597-598.
[http://dx.doi.org/10.1016/S0040-4039(01)02227-4]

[168] Farina, A.; Ferranti, C.; Marra, C. An improved synthesis of resveratrol. *Nat. Prod. Res.,* **2006**, *20*(3), 247-252.
[http://dx.doi.org/10.1080/14786410500059532] [PMID: 16401555]

[169] Sprouse, D. **2010**.

[170] Belchí-Navarro, S.; Almagro, L.; Lijavetzky, D.; Bru, R.; Pedreño, M.A. Enhanced extracellular production of trans-resveratrol in *Vitis vinifera* suspension cultured cells by using cyclodextrins and methyljasmonate. *Plant Cell Rep.,* **2012**, *31*(1), 81-89.
[http://dx.doi.org/10.1007/s00299-011-1141-8] [PMID: 21927985]

[171] Sydor, T.; Schaffer, S.; Boles, E. Considerable increase in resveratrol production by recombinant industrial yeast strains with use of rich medium. *Appl. Environ. Microbiol.,* **2010**, *76*(10), 3361-3363.
[http://dx.doi.org/10.1128/AEM.02796-09] [PMID: 20348297]

[172] Lim, C.G.; Fowler, Z.L.; Hueller, T.; Schaffer, S.; Koffas, M.A. High-yield resveratrol production in engineered *Escherichia coli. Appl. Environ. Microbiol.,* **2011**, *77*(10), 3451-3460.
[http://dx.doi.org/10.1128/AEM.02186-10] [PMID: 21441338]

[173] Mattivi, F. Solid phase extraction of trans-resveratrol from wines for HPLC analysis. *Z. Lebensm. Unters. Forsch.,* **1993**, *196*(6), 522-525.
[http://dx.doi.org/10.1007/BF01201331] [PMID: 8328217]

[174] Zhang, C. S.; Yang, Q. L.; Pan, L. J.; Yu, S. L. Research progress of extraction, isolation and purification of resveratrol *Science and Technology of Food Industry, ,* **2008**, *8*, 091.

[175] Blusztajn, J.K. Choline, a vital amine. *Science,* **1998**, *281*(5378), 794-795. [http://dx.doi.org/10.1126/science.281.5378.794] [PMID: 9714685]

[176] Zeisel, S.H.; da Costa, K.A. Choline: an essential nutrient for public health. *Nutr. Rev.,* **2009**, *67*(11), 615-623. [http://dx.doi.org/10.1111/j.1753-4887.2009.00246.x] [PMID: 19906248]

[177] Küllenberg, D.; Taylor, L.A.; Schneider, M.; Massing, U. Health effects of dietary phospholipids. *Lipids Health Dis.,* **2012**, *11*(3), 3. [http://dx.doi.org/10.1186/1476-511X-11-3] [PMID: 22221489]

[178] Ikawa, M.; Chakravarti, A.; Taylor, R.F. Occurrence of choline in *Lactobacillus plantarum. Can. J. Microbiol.,* **1972**, *18*(8), 1241-1245. [http://dx.doi.org/10.1139/m72-192] [PMID: 5052892]

[179] Kwong, E.; Fiala, G.; Barnes, R.H.; Kan, D.; Levenson, S.M. Choline biosynthesis in germfree rats. *J. Nutr.,* **1968**, *96*(1), 10-14.

[180] Sohlenkamp, C.; López-Lara, I.M.; Geiger, O. Biosynthesis of phosphatidylcholine in bacteria. *Prog. Lipid Res.,* **2003**, *42*(2), 115-162. [http://dx.doi.org/10.1016/S0163-7827(02)00050-4] [PMID: 12547654]

[181] Zoz, L. Torularhodin and Torulene: Bioproduction, Properties and Prospective Applications in Food and Cosmetics - a Review. *Braz. Arch. Biol. Technol.,* **2015**, *58*(2), 278-288. [http://dx.doi.org/10.1590/S1516-8913201400152]

[182] Anunciato, T.P.; da Rocha Filho, P.A. Carotenoids and polyphenols in nutricosmetics, nutraceuticals, and cosmeceuticals. *J. Cosmet. Dermatol.,* **2012**, *11*(1), 51-54. [http://dx.doi.org/10.1111/j.1473-2165.2011.00600.x] [PMID: 22360335]

[183] Pulido, R.; Bravo, L.; Saura-Calixto, F. Antioxidant activity of dietary polyphenols as determined by a modified ferric reducing/antioxidant power assay. *J. Agric. Food Chem.,* **2000**, *48*(8), 3396-3402. [http://dx.doi.org/10.1021/jf9913458] [PMID: 10956123]

[184] dos Santos, M.D.; Almeida, M.C.; Lopes, N.P.; de Souza, G.E. Evaluation of the anti-inflammatory, analgesic and antipyretic activities of the natural polyphenol chlorogenic acid. *Biol. Pharm. Bull.,* **2006**, *29*(11), 2236-2240. [http://dx.doi.org/10.1248/bpb.29.2236] [PMID: 17077520]

[185] Wang, H.J.; Murphy, P.A. Isoflavone content in commercial soybean foods. *J. Agric. Food Chem.,* **1994**, *42*(8), 1666-1673. [http://dx.doi.org/10.1021/jf00044a016]

[186] Yanagimoto, K.; Ochi, H.; Lee, K.G.; Shibamoto, T. Antioxidative activities of fractions obtained from brewed coffee. *J. Agric. Food Chem.,* **2004**, *52*(3), 592-596. [http://dx.doi.org/10.1021/jf030317t] [PMID: 14759154]

[187] Nawaz, H.; Shi, J.; Mittal, G.S.; Kakuda, Y. Extraction of polyphenols from grape seeds and concentration by ultrafiltration. *Separ. Purif. Tech.,* **2006**, *48*(2), 176-181. [http://dx.doi.org/10.1016/j.seppur.2005.07.006]

[188] Walter, E. Genistin (an isoflavone glucoside) and its aglucone, genistein, from soybeans. *J. Am. Chem. Soc.,* **1941**, *63*(12), 3273-3276.
[http://dx.doi.org/10.1021/ja01857a013]

[189] R.J.D, NG Serena With top lines dropping, firms reach for vitamins, in: The Wall Street Journal.: New York, **2013**.

[190] De Steur, H.; Mogendi, J. B.; Blancquaert, D.; Lambert, W.; Straeten, D.; Gellynck, X. Genetically modified rice with health benefits as a means to reduce micronutrient malnutrition: global status, consumer preferences and potential health impacts of rice biofortification *Wheat and rice in disease prevention and health: benefits, risks and mechanisms of whole grains in health promotion,* **2014**, 283-299.

[191] Zeigler, R.S. Biofortification: Vitamin A Deficiency and the Case for Golden Rice. In: *Plant Biotechnology*; Springer, **2014**; pp. 245-262.
[http://dx.doi.org/10.1007/978-3-319-06892-3_19]

[192] Winkler, J. *Biofortification: Improving the nutritional quality of staple crops*; Access Not Excess, **2011**, pp. 100-112.

[193] Blancquaert, D.; De Steur, H.; Gellynck, X.; Van Der Straeten, D. Present and future of folate biofortification of crop plants. *J. Exp. Bot.,* **2014**, *65*(4), 895-906.
[http://dx.doi.org/10.1093/jxb/ert483] [PMID: 24574483]

[194] Steinkraus, K.H. Bio-enrichment: production of vitamins in fermented foods In: *Microbiology of fermented foods*; Springer, **1997**; pp. 603-621.

[195] Keuth, S.; Bisping, B. Formation of vitamins by pure cultures of tempe moulds and bacteria during the tempe solid substrate fermentation. *J. Appl. Bacteriol.,* **1993**, *75*(5), 427-434.
[http://dx.doi.org/10.1111/j.1365-2672.1993.tb02798.x] [PMID: 8300444]

[196] Denter, J.; Bisping, B. Formation of B-vitamins by bacteria during the soaking process of soybeans for tempe fermentation. *Int. J. Food Microbiol.,* **1994**, *22*(1), 23-31.
[http://dx.doi.org/10.1016/0168-1605(94)90004-3] [PMID: 8060790]

[197] Rossi, M.; Amaretti, A.; Raimondi, S. Folate production by probiotic bacteria. *Nutrients,* **2011**, *3*(1), 118-134.
[http://dx.doi.org/10.3390/nu3010118] [PMID: 22254078]

[198] Lepage, P.; Leclerc, M.C.; Joossens, M.; Mondot, S.; Blottière, H.M.; Raes, J.; Ehrlich, D.; Doré, J. A metagenomic insight into our gut's microbiome. *Gut,* **2013**, *62*(1), 146-158.
[http://dx.doi.org/10.1136/gutjnl-2011-301805] [PMID: 22525886]

[199] LeBlanc, J.G.; Milani, C.; de Giori, G.S.; Sesma, F.; van Sinderen, D.; Ventura, M. Bacteria as vitamin suppliers to their host: a gut microbiota perspective. *Curr. Opin. Biotechnol.,* **2013**, *24*(2), 160-168.
[http://dx.doi.org/10.1016/j.copbio.2012.08.005] [PMID: 22940212]

[200] Gaspar, P.; Carvalho, A.L.; Vinga, S.; Santos, H.; Neves, A.R. From physiology to systems metabolic engineering for the production of biochemicals by lactic acid bacteria. *Biotechnol. Adv.,* **2013**, *31*(6), 764-788.
[http://dx.doi.org/10.1016/j.biotechadv.2013.03.011] [PMID: 23567148]

[201] Ma, Q.; Zhou, J.; Zhang, W.; Meng, X.; Sun, J.; Yuan, Y.J. Integrated proteomic and metabolomic analysis of an artificial microbial community for two-step production of vitamin C. *PLoS One,* **2011**, *6*(10), e26108. [http://dx.doi.org/10.1371/journal.pone.0026108] [PMID: 22016820]

[202] Ko, Y.; Ashok, S.; Ainala, S.K.; Sankaranarayanan, M.; Chun, A.Y.; Jung, G.Y.; Park, S. Coenzyme B12 can be produced by engineered *Escherichia coli* under both anaerobic and aerobic conditions. *Biotechnol. J.,* **2014**, *9*(12), 1526-1535. [http://dx.doi.org/10.1002/biot.201400221] [PMID: 25146562]

[203] Ye, V.M.; Bhatia, S.K. Pathway engineering strategies for production of beneficial carotenoids in microbial hosts. *Biotechnol. Lett.,* **2012**, *34*(8), 1405-1414. [http://dx.doi.org/10.1007/s10529-012-0921-8] [PMID: 22488437]

[204] Yan, G.L.; Wen, K.R.; Duan, C.Q. Enhancement of β-carotene production by over-expression of HMG-CoA reductase coupled with addition of ergosterol biosynthesis inhibitors in recombinant *Saccharomyces cerevisiae. Curr. Microbiol.,* **2012**, *64*(2), 159-163. [http://dx.doi.org/10.1007/s00284-011-0044-9] [PMID: 22086347]

[205] Gonnet, M.; Lethuaut, L.; Boury, F. New trends in encapsulation of liposoluble vitamins. *J. Control. Release,* **2010**, *146*(3), 276-290. [http://dx.doi.org/10.1016/j.jconrel.2010.01.037] [PMID: 20600399]

[206] Pezeshki, A.; Ghanbarzadeh, B.; Mohammadi, M.; Fathollahi, I.; Hamishehkar, H. Encapsulation of Vitamin A Palmitate in Nanostructured Lipid Carrier (NLC)-Effect of Surfactant Concentration on the Formulation Properties. *Adv. Pharm. Bull.,* **2014**, *4* Suppl. 2, 563-568. [PMID: 25671190]

[207] Khalid, N.; Kobayashi, I.; Neves, M.A.; Uemura, K.; Nakajima, M.; Nabetani, H. Formulation of monodisperse water-in-oil emulsions encapsulating calcium ascorbate and ascorbic acid 2-glucoside by microchannel emulsification. *Colloids Surf. A Physicochem. Eng. Asp.,* **2014**, *459*, 247-253. [http://dx.doi.org/10.1016/j.colsurfa.2014.07.014]

[208] Khayata, N.; Abdelwahed, W.; Chehna, M.F.; Charcosset, C.; Fessi, H. Stability study and lyophilization of vitamin E-loaded nanocapsules prepared by membrane contactor. *Int. J. Pharm.,* **2012**, *439*(1-2), 254-259. [http://dx.doi.org/10.1016/j.ijpharm.2012.09.032] [PMID: 23010283]

CHAPTER 6

Biotechnological Aroma Compounds

Gustavo B. Fanaro[a], **Pierre Fontanille**[b,c,d], **Mario R. Maróstica**[a] and **Juliano L. Bicas**[a,*]

[a] *School of Food Engineering, University of Campinas, Campinas, Brazil*

[b] *Clermont Université, Université Blaise Pascal, Institut Pascal, BP20206, F-63174 AUBIERE cedex, France*

[c] *CNRS, UMR 6602, Institut Pascal, F-63171 Aubière, France*

[d] *Clermont Université, Université Blaise Pascal, LABEX IMobS3, BP80026, F-63171 AUBIERE cedex, France*

Abstract: Aroma is one of the main attributes of foodstuff and thousands of volatile compounds are known, contributing to an innumerable number of aroma characters. Aroma compounds may be recovered from natural sources, chemically synthesized or produced by biotechnological means using microorganism and enzymes. This chapter is devoted to the use of biotechnological tools for aroma production, focusing on examples of commercially relevant aroma compounds which are currently produced biotechnologically and on emerging bioprocess approaches used for the microbial production of aroma compounds. Some case studies will be presented to illustrate some aspects of aroma production.

Keywords: Acetoin, Aroma, Bacteria, Bioaroma, Bioflavor, Bioreactor, Biotransformation, γ-decalactone, *De novo* synthesis, Diacetyl, Esters, Genetic engineering, Flavor, Fungi, Lactones, Natural, 2-phenylethanol, Vanillin, Volatile compounds, *Yarrowia lipolytica*.

INTRODUCTION

The sensory evaluation of foodstuff involves the combination of the main human

* **Address correspondence to Juliano Lemos Bicas:** Department of Food Science, University of Campinas. Rua Monteiro Lobato, 80. 13083-862, Brazil; E-mail: jlbicas@gmail.com.

senses. The first impression of a product is, in general, given by sight (color, appearance), which is usually followed by its odor perception (olfaction). During chewing and swallowing, gustation (taste), olfaction (aroma), touch (texture, mouth feel) and hearing (sound of crispness/crunchiness) are also involved. Complementarily, flavor is a perception given by a complex combination of olfactory, gustatory and trigeminal sensations perceived during tasting [1], also simplistically considered as the interaction of aroma and taste [2] (For the multisensory interactions underlying the flavor perception, it is recommended the reading of Auvray & Spence [3]). It is believed that aroma is more important than taste on determining the overall flavor [4]. The distinction of both aroma and taste is evident, at least in terms of complexity. Differently from the taste molecules, which result in a limited number of sensations (sweet, salty, sour, bitter and umami), volatile molecules perceived by olfactory system have numerous aroma descriptors. Therefore, considering that at least 8,000 volatile compounds have already been identified (Table **1**) there are countless possible combinations and intensities (*i.e.* aroma perceptions), resulting in a unique aroma description for each food product. In this sense, aroma may be considered the main element defining the characteristic flavor and the acceptance of food products. Thus, aroma compounds are frequently employed to reinforce or improve the sensory perception of food and beverages.

Table 1. List of all volatile compounds in food in the VCF database [12].

Compounds	Number of identified compounds
Hydrocarbons	985
Alcohols	834
Aldehydes	431
Ketones	742
Acids	438
Esters	1551
Lactones	134
Bases	675
Sulfur compounds	798
Acetals	118
Ethers	99

(Table 1) contd.....

Compounds	**Number of identified compounds**
Halogens	112
Nitriles and amides	112
Phenols	260
Furans	370
(Ep)oxides, pyrans, coumarins	302
Oxazol(in)es	90
Anhydrides and phtalides	42
Compounds from sources other than VCF Database	2814

Aroma compounds are volatile molecules (with a molecular weight rarely higher than 300 Da) which are able to sensitize specific cells located at human's olfactory cavity [5], where they arrive from inhaled air (nasal route) or from mouth (orthonasal route), resulting in perceptions called "odor" or "aroma", respectively [6]. These compounds are very potent, with detection thresholds generally in the range of parts per billion (ppb) [7]. As a result, the aroma molecules are usually present as minor components (about 50 ppm), although they may represent up to 50% of the cost of the final product depending on their quality and application (*e.g.* some beverages) [8]. Additionally, these substances do not contribute to the nutritional value of foods, although several biological activities (antimicrobial, antioxidant, somatic fat reducing, blood pressure regulating and anti-inflammatory properties) have been attributed to different aroma compounds [9].

In terms of chemical classification, aroma molecules may present different functional groups: hydrocarbons, alcohols, aldehydes, ketones, acids, esters/lactones and others (Fig. **1**). Among them, esters (and lactones) and carbonyl compounds (aldehydes plus ketones), followed by hydrocarbons and alcohols, are quantitatively the most important ones (Table **1**). The important contribution of volatile terpene hydrocarbons [$(C_5H_8)_n$], *i.e.* mono (n = 2) and sesquiterpenes (n = 3) and their oxygenated counterparts (see some examples in Fig. (**1**)) is worth mentioning, which are usually the main constituents of essential oils [10].

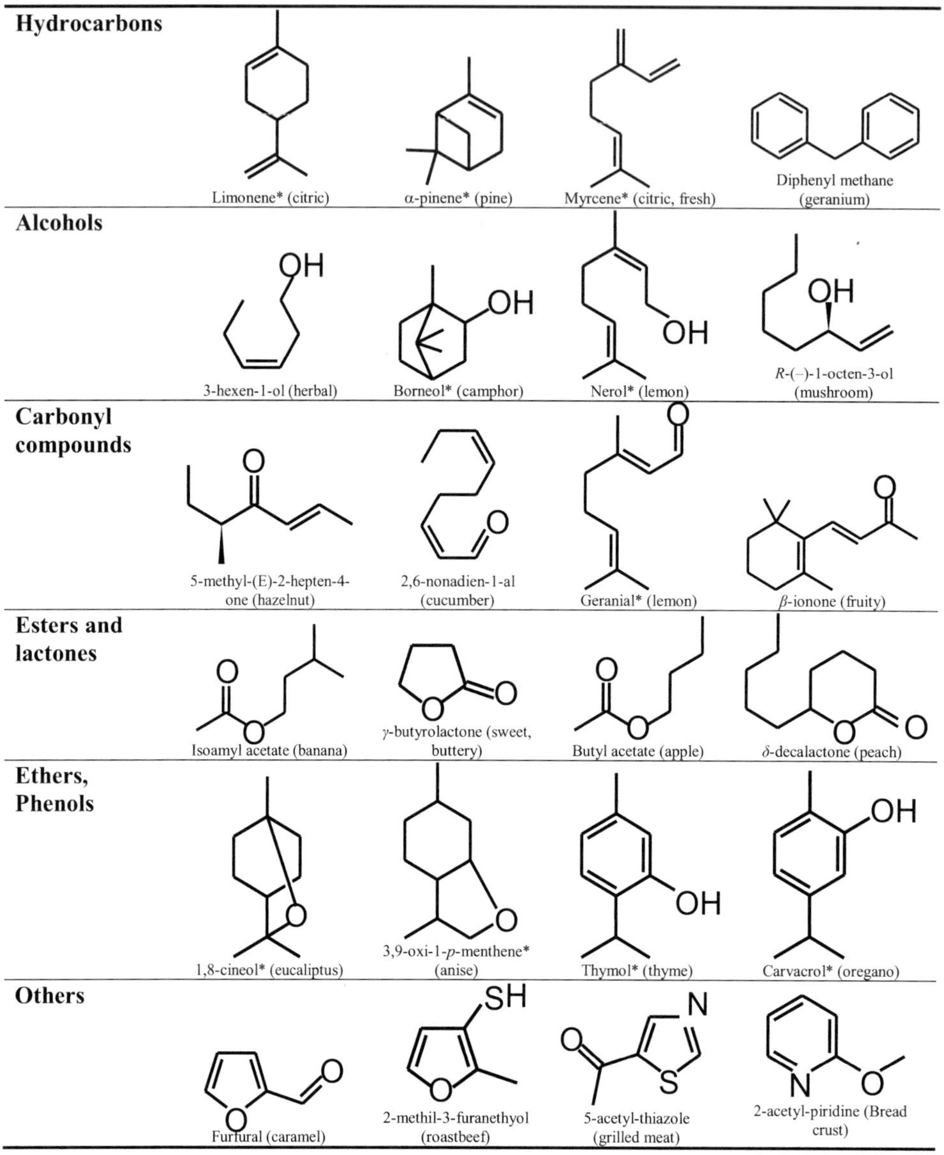

Fig. (1). Examples of some important aroma compounds and their respective sensorial notes [7, 11]. The compounds with an asterisk (*) are monoterpenes or their oxygenated derivatives.

The analysis, characterization and formulation of aroma mixtures may be a quite complex task. In some cases one single substance (character impact aroma), such as isoamyl acetate in banana, may be responsible to express the main sensory property of a food, but for other products, such as wine, coffee, cocoa, beer *etc.*, the combination of more than 800 volatile compounds, sometimes in trace amounts, is responsible for the definition of the characteristic aroma [13]. For these complex mixtures, objective descriptions of odors might be impossible to perform. Even the odor of a single chemical compound is extremely difficult to be described unequivocally. For this reason, the strategy employed to try to overcome this difficulty involves the use of adjectives (citrus, earthy, flowery, fruity, herbaceous, woody *etc.*), which relate the impression to other known products with similar odors [5] (see examples in Fig. (**1**)). Moreover, aroma threshold and sensory impression depends on: i) the molecule's chemical structure (the relation response x structure is so specific is so great that even enantiomers may have significant differences in their odor quality and threshold); ii) the concentration of the aroma substances; iii) the presence of other odorant molecules, which may provide additive or antagonistic effects, and, consequently, by food matrix [13]. In this sense, one of the most useful analytical techniques to evaluate the aroma profile of foodstuffs is gas chromatography-olfactometry [14].

As will be discussed in the sequence, aroma compounds may be recovered from natural sources (vegetables, animals), synthesized chemically or produced by biotechnological means. This chapter is focused on the methods, advantages and challenges related to the preparation of natural aroma compounds by biotechnological means.

Sources of Aroma Compounds

At the end of 19th century and beginning of 20th century, with the intense expansion of organic chemistry, chemical synthesis of aroma compounds faced a strong development and application, usually involving efficient processes and cheap petrochemical-derived raw materials. This began in 1870s with the synthesis of vanillin, as will be seen in the sequence, and about 50 years later chemically synthesized ingredients were first used in perfumery [15]. This is still the most relevant technique for producing aroma compounds: with some exceptions (*e.g.* in Germany), about 80% of the flavors and fragrances used worldwide are chemically synthesized and are thus declared "nature-identical"

[16]. The advantages of the chemical synthesis of aroma compounds are generally related to the satisfactory yields and to the good production rates, which result in a low price material. On the other hand, many drawbacks are associated to this procedure: harsh conditions (toxic catalysts, high pressure and temperature *etc.*) are usually required; a mixture of products resulting from the lack of adequate regio- and enantioselectivity to the substrate may be found; and the compounds generated are labeled as "artificial" or "nature identical", with lower market appeal [17]. In fact, the intense and growing demand for "natural"-labeled products is more and more influencing the market and food laws world-wide. The reflexes of such preference are mainly notable in Europe and USA, where 90% and 80%, respectively, of all beverage flavors are ''natural'' [9].

Natural aroma compounds may be prepared by two different techniques. The first refers to the most intuitive one: direct extraction from nature - essential oils, vegetable extracts and a few animal-derived products (amber, musk, zibet). The recovery of aroma compounds from plant materials is in fact the most traditional procedure, used since ancient times to produce perfumes and fragrances [18]. However, this procedure has few disadvantages: i) generally, the product of interest is usually present in low concentrations, what increases the costs of extraction and purification; ii) moreover, the amount and profile of volatile compounds is dependent on seasonal and climatic conditions, while the availability of the starting material may be subjected to political features; iii) finally, ecological problems may also be involved (extractive agriculture) [9, 19]. Thus, alternative techniques to generate natural aroma compounds are increasingly being pursued.

Besides this conventional method, natural aromas may be produced applying biotechnological approaches, *i.e.* using microorganisms or cell cultures as biocatalysts. These processes do not present the negative aspects related to the former methods, since they occur at mild conditions, present high chemo-, regio- and enantioselectivity, do not generate toxic wastes and the products may be generated uninterruptedly, under controlled and optimized conditions [19, 20]. Moreover, biotechnological aroma compounds have renewable origin, are virtually inextinguishable and may use low value feedstocks, such as agroindustrial residues and by-products, as starting material for their manufacture

[17, 21]. Finally, since they may be classified as natural compounds, the products obtained through this approach display favorable market appeal, contributing to an increase in their commercial prices. On the other hand, in most cases the yield and/or product concentration are typically too low to make the biotechnological process economically viable [22].

The different ways of obtaining vanillin, the world's most popular single flavor [8], are good examples to illustrate the main features related to those three methods of aroma production. This will be briefly described in the sequence.

Case Study: Methods for Vanillin Production

The use of vanilla (*Vanilla planifolia*, *V. tahitiensis* and *V. pompona*) pods as source of flavoring agents begun with the Aztecs (natives of Mexico), prior to 1500 AD. Mexico remained the only supplier of vanilla for the European market until mid-1800s, when the commercial production of vanilla became possible after the development of methods for manual pollination of their flowers [23]. Nowadays, the main producers of vanilla are Indonesia, Madagascar and China (Table **2**).

Table 2. Major producers of vanilla [24].

Country	Production in 2011 (metric tons)	Global production percentage (%)
Indonesia	3500	39.3
Madagascar	3000	33.7
China	1385	15.5
Mexico	362	4.1
Turkey	287	3.2

The production of vanilla pods is complex, laborious and time consuming. After fertilization (hand pollination), 10 to 12 months are required for the beans to become fully mature. These green vanilla pods are almost odorless and it is just during curing (takes from five weeks to almost 6 months, depending on the process) that the action of β-glycosidases releases the various vanilla flavor components [23]. Moreover, crop yields (thus, the price of vanilla) are subjected to fluctuations associated with the climate, political and economic decisions, intensive cultivation, pollination, harvesting and ripening of pods [25]. Therefore, this delicate, elaborate and desirable aroma of natural vanilla has a relatively high price, which limits its utilization to more value-added products. This has boosted the search for less expensive options to replace this extract.

In the 19th century, the structure of vanillin (Fig. **2**), the character impact compound of vanilla aroma, was identified and many efforts were made to synthesize it. This chemical process is relatively simple (and cheap), which is partially explained by the structure of vanillin, with no stereoisomers. In 1874 Haarman and Tiemann started the production of vanillin from coniferin. Just after that, in 1876, Tiemann developed a method for producing vanillin from eugenol, a constituent of clove oil. Currently, most vanillin is produced synthetically from petrochemicals, such as guaiacol, or lignin, present in sulphite liquor from the cellulose industry. This synthetic vanilla flavor dominates the market: only 20 tons, from the approximately 12,000 tons of vanillin consumed annually, is extracted from vanilla beans, while the rest comes from chemical synthesis. This is explained, at least in part, by the differences in prices, which are at least 100 times higher for the natural version [23, 25]. The difference on prices combined with the increasing demand for "natural", "healthy" and "bio" marketing claims, has resulted on a growing interest for the production of natural vanillin from other natural sources [16].

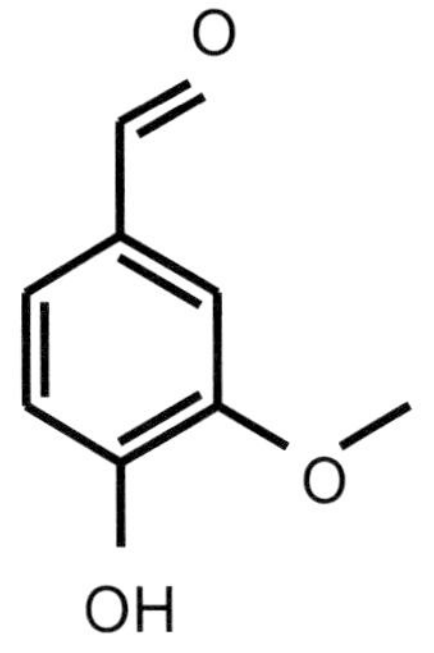

Fig. (2). Structure of vanillin.

The use of biotechnological tools to generate vanillin – in this case also known as "biovanillin" or "biotech" vanillin – is perhaps the most extensively exploited process for bioaroma production. Many different production strategies have already been considered, either by *de novo* synthesis or by biotransformation of a wide range of substrates (*e.g.* ferulic acid, eugenol, isoeugenol, vanillic acid and lignin), which can be mediated by plant tissue/cell cultures, microorganisms or even cell-free systems (enzymes) [23, 26]. According to Schrader *et al.* [27], the only commercial biocatalytic route is based on the microbial bioconversion of ferulic acid, a natural compound mainly isolated from rice bran. The bioconversion of ferulic acid into vanillin by *Amycolatopsis* sp. or *Streptomyces setonii* strains in a 10 L bioreactor produces, respectively, 11.5 g/L (with a conversion rate of 77.8%) after 32 h or 13.9 g/L (yield of 75%) after 17 hours of the substrate feed [19]. One example of a commercial product derived from such process is Rhovanil® Natural (Rhodia) [28], which is obtained by fermentation from ferulic acid using *Aspergillus niger* and *Pycnoporus cinnabarinus* [29]. Biovanillin has an intermediate price: it may cost US$500-1,000 per kg, whereas the vanillin originated from vanilla pods and the chemically synthesized have prices close to US$1,200-4,000 and US$16-20, respectively [22, 26, 30]. Therefore, it represents a less expensive option for natural vanillin supply. However, as mentioned by Havkin-Frenkel & Belanger [26], only a few companies are producing biovanillin in commercial scale – many of them abandoned this route due to the high production costs.

STRATEGIES FOR THE BIOTECHNOLOGICAL PRODUCTION OF AROMA COMPOUNDS

There are two main strategies for the biotechnological production of aroma compounds. The first, called *de novo* synthesis (fermentation), refers to volatiles production (generally a mixture) "from the new", *i.e.* they are biochemical intermediates (or end products) from the metabolism of a given microorganism grown in a culture medium. Whole cells are able to catabolize simple nutrients as carbohydrates, fats and proteins to complex molecules with high flavor impact. This property was first used empirically for the production of fermented foods [31]. However, an improvement in their production is often hampered by an inability to direct the metabolism of these microorganisms. Moreover, the products' concentration obtained, in this case, is usually below 100 mg/L. An example of such process is the production of fruity aroma by the cultivation of *Geotrichum fragrans* in cassava wastewater [32] (Fig. **3a**).

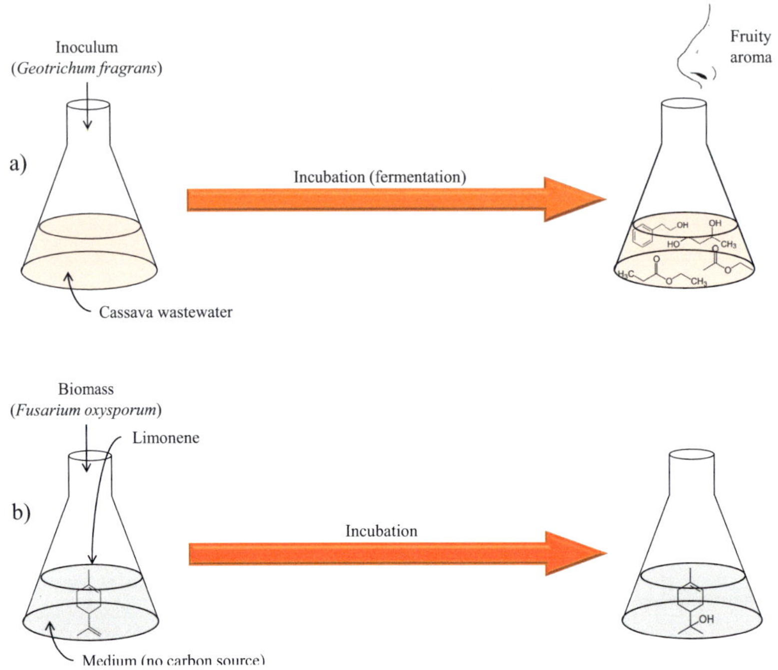

Fig. (3). Schematic representation of examples of "biotech" aroma production by: **a)** *de novo* synthesis and **b)** biotransformation.

The second, known as biotransformation, refers to the modification of a precursor molecule to form the desirable aroma compound through a single or few (in this case the process is called bioconversion) biocatalytic steps occurring within the micro-organism. In general, the products' concentration is much higher (in the order of g/L) than those observed for the *de novo* synthesis and appear as an attractive alternative for a larger production of aroma compounds. Highly valued flavors can be produced from inexpensive, readily available and renewable natural precursors by a biocatalyst which has a limited number of enzymatic steps. The enzymatic process is often used rather than chemical process because of the high specificity to attack specifics sites on the substrate. One example of biotransformation refers to the production of α-terpineol from limonene by *Fusarium oxysporum* (Fig. **3b**) [19].

In the sequence we will present some of the most commercially relevant processes for the biotechnological production of aroma.

COMMERCIALLY RELEVANT "BIOTECH" AROMA COMPOUNDS

There are several aroma compounds available commercially, some of them with probable biotechnological origin (Table **3**).

Table 3. Volatile compounds with probable biotechnological origin.

Acetaldehyde	4-Decanolide	4-Hexanolide	1-Octen-3-ol
Acetoin	5-Decanolide	*β*-Ionone	2-Pentanone
Benzaldehyde	2-Decenolactone	Methional	*i*-Pentyl alcohol
Benzyl butanoate	4-Dodecanolide	Methyl anthranilate	Phenylacetaldehyde
n-Butanol	5-Dodecanolide	2-Methylbutanoic acid	1-Phenylethyl acetate
i-Butyl alcohol	Ethyl acetate	3-Methylbutanal	*n*-Propanol
i-Butanal	Ethyl butanoate	*n*-Nonanal	Sclareolide
Cinnamic acid	Ethyl 2-methylbutanoate	*n*-Nonanol	4-Undecalactone
Cinnamic alcohol	Ethyl phenylacetate	2-Nonanone	2-Undecanone
Cinnamaldehyde	Furfuryl thiol	Nootkatone	Vanillin
n-Decanal	2-Heptanone	4-Octanolide	

Adapted from Berger [9].

In the sequence it will be presented the characteristic and proprieties of some of these compounds or their chemical classes.

Lactones

Lactones are a ubiquitous group of flavors and aromas that depending on the compound, exhibits oily and fruity properties like peach, apricot or coconut fragrance. These compounds are naturally found in several products fermented by yeasts, but most of lactones are difficult to produce. However, when hydroxy fatty acids are the feedstock, a group of lactones is easily obtained. Macrocyclic lactones possess musk notes are used in cosmetics and perfumes [33].

According to Romero-Guido [33] four main processes are used for the production of lactones flavors: 1) hydroxylation, β-oxidation, and lactonisation of fatty acids; 2) reduction of unsaturated lactones; 3) Baeyer–Villiger oxidation of cyclic ketone; 4) α,ω-oxidation of alkanes or fatty acids, depending on conditions, give rises to macrocyclic lactones.

γ-Decalactone, the lactone of 4-hydroxydecanoic acid, is the most widely flavor lactone used and exhibit an oily-peachy aroma. Okui *et al.* [34] were the first who observed the formation of γ-decalactone during the growth of a *Candida species* on ricinoleic acid. The use of microorganisms by industry led to a substantial price decrease of this compound from over US$ 10,000.00/kg in the early 1980s to about US$ 300.00/kg in 2004 [27].

Other industries applications are the antibacterial, antifungal or anti-inflammatory proprieties of the 2-deceno-δ-lactone, 6-pentyl-α-pyrone, and γ-decalactone and its precursor 4-hydroxydecanoic [35 - 37]. However, lactones have a toxic effect against yeasts, which can results in inhibition of yeast growth, that implies on the production rate limitation [38], and so, additional process in the lactones productions is required, as the removal the organic compounds by solvent extraction, trapping in vegetable oils, adsorption on active carbon or hydrophobic membranes [39].

The yeast *Sporobolomyces odorus* produces *de novo* up to 1.6 mg/L of 4-decalactone, resulting in an intense peach odor and the soil fungus, *Trichoderma viride*, generates a strong coconut aroma, with a yield of 170 mg/L of 6-penthyl-2-pyrone [40].

Esters

Esters are the most important group in fruity flavor and the industry uses in fruit-flavored products and dairy products. The production of volatile fruity aromas could occur during the growth of several microbial species. The molds *Saccharomyces* sp., *Hansenula* sp., and *Candida utilis* could be used in *de novo* synthesis to produce ethyl acetate in a medium containing ethanol [19].

Diacetyl

Diacetyl has a strong butter note and can be formed by *Streptococcus diacetylactis* from citric acid *via* pyruvic acid by condensation of acetyl-CoA molecules or by decarboxylation of acetolactic acid. The addition of pyruvate and citrate on the medium can increase the diacetyl formation. It can also be produced by *Lactobacillus* and *Streptococcus* species during fermentation from lactose and is largely used in dairy products [41].

2-Phenylethanol

This compound has a rose-like aroma and is naturally found in many kinds of flowers. In the final of the last decade, in all the world, the production of 2-phenylethanol was estimated to be approximately 10,000 tons, most of them obtained by chemical synthesis from benzene or styrene with a price of about US$ 5/kg. Natural 2-phenylethanol can be extracted from some flowers, however in very low concentration and in a very expensive process which results in a costly product (about US$ 1000/kg) [42].

2-phenylethanol can be obtained as an intermediate product by microorganisms' enzymatic transformation or microbial fermentation from phenylalanine and can be applied industrially with a low production cost. The process to obtain this compound, called Ehrlich pathway, consists of treating the substrate with transaminase and decarboxylase followed by three NAD^+-dependent dehydrogenases [43].

Similarly to the lactones, a large amount of 2-phenyilethanol is cytotoxic to the microorganisms which produce it. Concentrations between 2-3 g/L inhibit the development of several microorganism species, *e.g.* the growth rate of *Saccharomyces cerevisiae* decreases 75% in the presence of 2.5 g/L of 2-phenylethanol [44]. Thus, the yield in most biotechnological processes is usually below 4.0 g/L.

There are some possibilities for 2-phenylethanol removal from the bioreactor such as two-phase extraction, adsorption, pervaporation, and immobilized solvent extraction [45]. Also, studies are being conducted to improve production by means of strain mutagenesis or selection (including γ-radiation, base analogs, alkylating and intercalating agents, ion implantation, and random or rationalized selection of better strains), and medium composition or culture condition optimization (carbon sources, and phenylalanine concentrations), increasing the production of 2-phenylethanol considerably [42].

Mihal′ *et al.* [45] using *Saccharomyces cerevisiae* in a bioreactor fed with low concentration of glucose and 135 g of phenylalanine and a flow rate of air at 20 L/min. was able to convert the phenylalanine reaching yields of 93.2%. Using membrane extraction combined with continual extraction using pentane as solvent, the authors obtain a final concentration of 18.6 g/L of 2-phenylethanol.

Acetoin

Acetoin (3-hydroxy-2-butanone or acetyl methyl carbinol) has a pleasant yogurt odor and a fatty creamy butter taste. It is widely used in several industries and products such as fruit flavors; nuts; dairy, butters, and its derived; baked and confection goods, and beverages, among others. Besides that, it is very important on the beer maturation [46].

Acetoin can be produced by several living organisms, including mammals and humans (it is a key metabolite in the acetaldehyde detoxification mechanism), plants cells (carrot, maize, and rice), however, most of eukaryotic cells produce it in insignificant amounts. The yield is far than less 1 g/L for the most yeasts. On the other hand, bacteria have pathways that can produce higher amounts (up to 75.2 g/L) [46].

Due the fact the bacteria *Bacillus subtilis* being GRAS (considered generally regarded as safe) status, is often utilized for industrial production of several secondary metabolites, such as acetoin. The fermentation process pathways include the conversion of pyruvate to lactic acid, ethanol, acetic acid, acetoin and 2,3-butanediol [47].

Main strategies for improving microbial production of aroma compounds: A case study for γ-decalactone

There are many tools available for increasing the feasibility of industrial production of microbial aroma compounds. In this section some technologies and techniques that the industry can use in order to increase the yield of biotechnological γ-decalactone will be presented. It is expected that these and other optimizing tools could improve the bioprocesses and reduce the costs in the near future.

Modifying a microorganism

The yeast *Yarrowia lipolytica* is one of the most important microorganism used to produce γ-decalactone; However, the same strain that produces γ-decalactone may also degrade this compound by the oxidation pathway involving enzyme Aox3 (POX3 gene), which lowers the yields. Thus, by suppressing/disrupting the POX3 gene from *Y. lipolytica* and adding multiple copies of the POX2 gene, coding for the long-chain specific acyl-CoA oxidases, it is possible to increase the yield more than 3 times [48].

Other genetic technique involves the shuffling of desirable genes of two or more different DNA's from the same species coding for different qualities. This is a powerful approach for the rapid evolution of strains toward desirable phenotypes. Using five UV mutant strains with the characteristics of higher γ-decalactone production or shorter fermentation time, Zhao *et al.* [49] created four hereditarily stable recombinants with high γ-decalactone production through three rounds of genome shuffling. They were able to produce about 3.75 g/L of γ-decalactone in the fermentation medium after 64h. This value was 6.54-fold higher than the common industrial *Yarrowia* sp. yeast and the peak production was shortened by 8h.

Oxygen

The availability of oxygen is a limiting factor for the γ-decalactone production, since the main producing yeasts are aerobic, and so, the γ-decalactone production processes are run as batch and fed-batch cultivations. The fed-batch has the advantage of working at higher cell amount and lower residual substrate (which is inhibitory of all process); however the oxygen transfer is more complicated, once the solubility of oxygen decrease with the increase of cell densities due to a higher viscosity of the cell suspension [50]. Moradi *et al.* [51], using castor oil as raw material and *Y. lipolytica*, could increase the production of γ-decalactone by 60% in fed-batch reactors when pure oxygen was used compared to the use of atmospheric oxygen. The oxygen transfer is also important for increasing the concentration of volatiles in fermented *Agave tequilana* juice by *Saccharomyces cerevisiae* strains [52].

Oxygen may also be important for the profile of volatiles produced. Deetae *et al.* [53] showed that when the *Proteus vulgaris* is under anaerobic conditions, it produces a sulfur compound (rotten egg aroma) while the same bacterium produces pleasant aromas (fruity, fresh and cheese aromas) under aerobic conditions.

Cell permeability

Increasing the permeability of microorganism cells is another alternative that the industry can use to raise the production of aroma compounds, for instance γ-decalactone. An *et al.* [54] using strains of *Waltomyces lipofer* (yeast) and 10-hydroxystearic acid as substrate observed that the non-permeabilized cells produced about 12 g/L of γ-dodecalactone after 30h, with a yeld of 0.4 g/L/h, however when permeabilized cells were used, the production at the same time was about 46 g/L, with a productivity of 1.5 g/L/h (conversion yield of 76% wt/wt). The ideal permeabilization happened when the cells were treated with 50% of ethanol and 0.5% of Triton X-100.

Agroindustrial residues and by-products as substrate

Agricultural and forestry industries generate millions of tons of waste and by-products annually, *e.g.* aqueous residues, bagasse, pulps, peels, husks, and others, which could raise serious disposal issues and so, huge costs to various industries. Today, several companies have adopted an integrated management of their wastes according to the concept "reduce, reuse, and recycle" [19].

The use of agro-industrial residues and by-products as raw materials in the production of industrially relevant and value-added products is considered a viable and profitable alternative, since it can contribute to cost reduction and to lower the environmental impact resulting from their disposal. These residues and by-products are particularly suited to the field of enzyme and fermentation technology, especially applicable as substrates for biomass production [55].

Examples of volatiles that can be produced by microorganisms cultivated in waste substrate include: *Ceratocystis fimbriata* producing pineapple aroma from coffee husk supplemented with glucose and fruity aroma from wheat bran or sugarcane bagasse [56, 57], *Aspergillus niger* and *Pycnoporus cinnabarinus* producing vanillin from rice bran oil [58], *Bacillus subtilis* producing acetoin from molasses and soybean meal hydrolysate by [59] and *Rhizopus* strains producing fruity aroma in cassava bagasse combined with apple pomace or soybean in a medium supplemented with mineral salt [60].

One attractive compound that can be used as raw material for the production of lactones are the oil press cakes, once they have high residual oil content and are produced in great amounts. By using oil press cakes from olive, sunflower, linseed, and castor and the fungi *Ceratocystis moniliformis*, *Trichoderma harzianum*, *Moniliella suaveolens*, and *Pityrosporum ovale*, Laufenberg *et al.* [61] verified that neither olive, sunflower or linseed oil cakes resulted in γ-decalactone formation using GC/MS identification. However, the castor cake showed the presence of γ-decalactone formation due to its high residual of ricinoleic acid presence.

EMERGING BIOPROCESS DESIGNS FOR THE PRODUCTION OF AROMA COMPOUNDS

Engineering approach for a biotransformation process focuses on various ways to improve the economics of the overall system. This goal generally implies the optimization of the reaction rate and the yields of reactions but also the use of an efficient extraction process. Each time, environmental concerns have to be taken into account. Improvement of rates is mainly the result of biocatalyst engineering, while yields result from the biocatalyst selectivity and from mass transport between phases for the recovery of the product. This last phenomenon is of striking importance when compounds involved in the reaction are poorly soluble in water and/or they have a high vapor pressure. These chemical properties are often encountered when working with aroma compounds. Hence, the medium is often made of several phases which can be solid, liquid, supercritical fluid or gaseous. Many design for biotransformation processes can then be considered. In order to improve a biotransformation, it is essential to define the most appropriate process in function of the chemical properties of the molecule considered and its future use. Fig. (**4**) summarizes the different techniques which may be used in function of the priorities defined by the researcher.

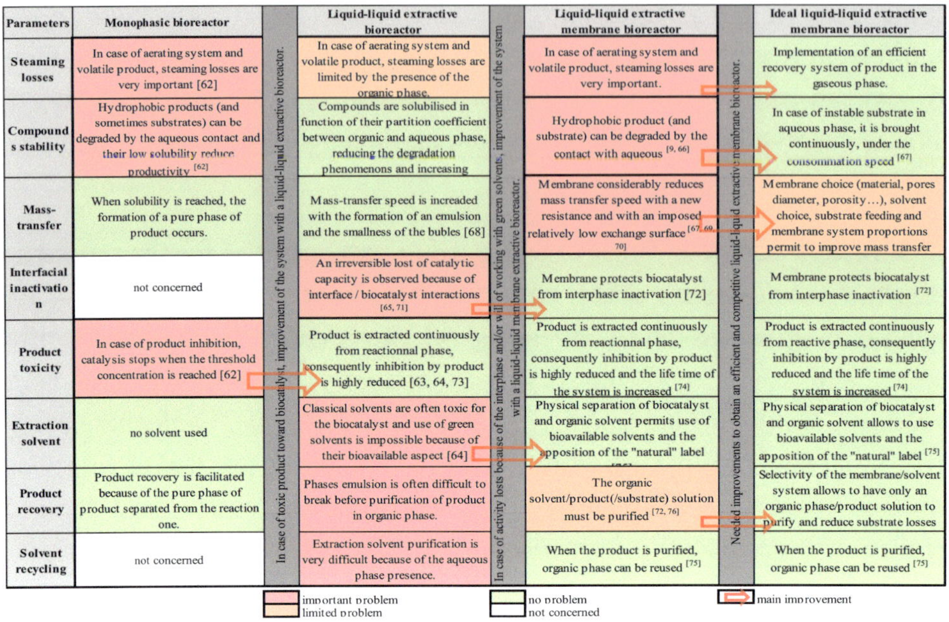

Parameters	Monophasic bioreactor	Liquid-liquid extractive bioreactor	Liquid-liquid extractive membrane bioreactor	Ideal liquid-liquid extractive membrane bioreactor
Steaming losses	In case of aerating system and volatile product, steaming losses are very important [62]	In case of aerating system and volatile product, steaming losses are limited by the presence of the organic phase.	In case of aerating system and volatile product, steaming losses are very important.	Implementation of an efficient recovery system of product in the gaseous phase.
Compounds stability	Hydrophobic products (and sometimes substrates) can be degraded by the aqueous contact and their low solubility reduce productivity [62]	Compounds are solubilised in function of their partition coefficient between organic and aqueous phase, reducing the degradation phenomenons and increasing	Hydrophobic product (and substrate) can be degraded by the contact with aqueous [9, 66]	In case of instable substrate in aqueous phase, it is brought continuously, under the consommation speed [67]
Mass-transfer	When solubility is reached, the formation of a pure phase of product occurs.	Mass-transfer speed is increaded with the formation of an emulsion and the smallness of the bubles [68]	Membrane considerably reduces mass transfer speed with a new resistance and with an imposed relatively low exchange surface [67, 69, 70]	Membrane choice (material, pores diameter, porosity…), solvent choice, substrate feeding and membrane system proportions permit to improve mass transfer
Interfacial inactivation	not concerned	An irreversible lost of catalytic capacity is observed because of interface / biocatalyst interactions [65, 71]	Membrane protects biocatalyst from interphase inactivation [72]	Membrane protects biocatalyst from interphase inactivation [72]
Product toxicity	In case of product inhibition, catalysis stops when the threshold concentration is reached [62]	Product is extracted continuously from reactionnal phase, consequently inhibition by product is highly reduced [63, 64, 73]	Product is extracted continuously from reactionnal phase, consequently inhibition by product is highly reduced and the life time of the system is increased [74]	Product is extracted continuously from reactive phase, consequently inhibition by product is highly reduced and the life time of the system is increased [74]
Extraction solvent	no solvent used	Classical solvents are often toxic for the biocatalyst and use of green solvents is impossible because of their bioavailable aspect [64]	Physical separation of biocatalyst and organic solvent permits use of bioavailable solvents and the apposition of the "natural" label	Physical separation of biocatalyst and organic solvent allows to use bioavailable solvents and the apposition of the "natural" label [75]
Product recovery	Product recovery is facilitated because of the pure phase of product separated from the reaction one.	Phases emulsion is often difficult to break before purification of product in organic phase.	The organic solvent/product(/substrate) solution must be purified [72, 76]	Selectivity of the membrane/solvent system allows to have only an organic phase/product solution to purify and reduce substrate losses
Solvent recycling	not concerned	Extraction solvent purification is very difficult because of the aqueous phase presence.	When the product is purified, organic phase can be reused [75]	When the product is purified, organic phase can be reused [75]

Fig. (4). Choice of hydrophobic molecules production process according to the significance of system parameters.

Single Liquid Phase Bioreactor

The simplest way to produce hydrophobic molecules by bioconversion is the use of a single liquid phase bioreactor. The unique aqueous phase contains biocatalyst and hydrophobic product forms a second immiscible pure phase on reaction medium surface when its solubility is reached in the reactive phase (Fig. **5**).

This type of production is easy to set up; there is no use of solvents and no problem of reprocessing plant. Purification of the product is facilitated by its purity in a phase easily separable from the medium [62].

Many compounds are produced with such a technique, namely aromas covered by patents. Gounaris [63] presents various applications of biotechnological methods namely production in single liquid phase bioreactors.

Nevertheless this process presents drawbacks. Low solubility in aqueous media of interest organic compounds leads to a poor volumetric productivity, which is not optimum for industrial production [64]. Pure product phase formation increase the loss of product by striping (notably when it is highly volatile) and cumbersome condensate systems can be necessary. And finally, substrate and/or product can be inhibitor or toxic for biocatalyst, reducing life-time of the system [65, 66].

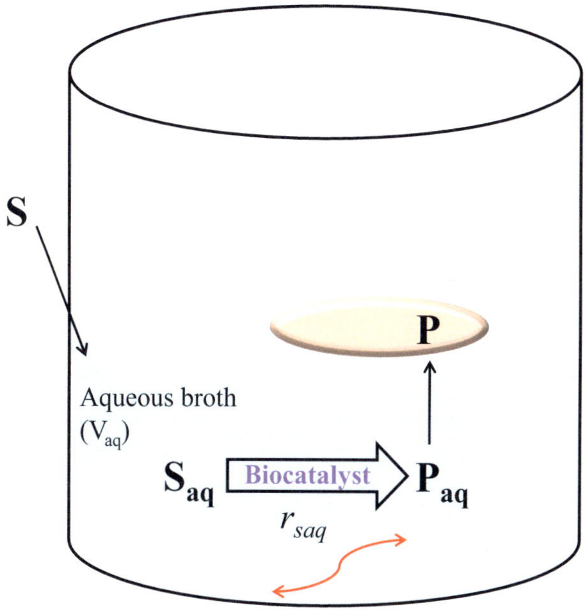

Fig. (5). Single liquid phase bioreactor.

Single liquid phase bioreactor is an interessant process for some molecules but in case of toxic effect of the subtrate and/or product for the biocatalyst and when compounds are hightly volatile, this process does not give satsfactory results. To overcome this, extractive bioconversions have been developed.

Liquid-liquid extractive bioreactor

Organic phase avoids the important problem of toxicity since the extraction of the product from the reaction phase occurs immediately after its formation. Consequently, toxicity of the product is highly limited and reaction is pushed in the production direction. Stability of the compounds is maintained with solutes partitioning between phases. And finally, the two phases mixing permits to develop an important exchange surface between the two phases. This large area allows quick mass transfers of substrate and/or product from one to the other phase increasing volumetric productivity [67 - 69]. These characteristics are very interesting when substrate or products are inhibitors for the bioconversion reaction or if they are toxic towards the biocatalyst. It is regularly the situation for the production of aromas and flavors.

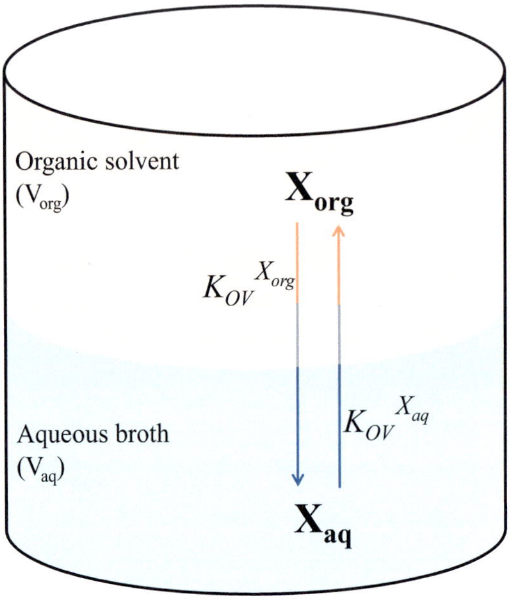

Fig. (6). Liquid-liquid extractive bioreactor.

However, the use of many organic solvents is not compatible for food use. The European Law on extraction solvents used in the production of foodstuffs and food ingredients [70] allows use of 18 extraction organic solvents which may be used during the processing of raw materials, food or food components and food ingredients. These Regulatory constraints may represent sometimes a hindrance to the development of new processes. The possible use of bio-available solvents such as vegetable oils as organic phase is increasingly studied for environmental purpose. It may lead to a production process of natural aroma. Bicas and coworkers [71] have developed an efficient bioprocess with a high conversion rate of limonene to α-terpineol. A biphasic medium system was used with an aqueous phase containing a concentrated resting cells of *Sphingobium* sp. and an organic phase with sunflower oil.

The efficiency of direct contact biphasic systems may present some limits. The first one is the harmful effect of organic phase on biocatalysts. Indeed, biocatalysts activity in aqueous medium could be reduced in contact of hydrophobic organic solvents. Enzymes are often damaged at the interface between water and hydrophobic solvent, which led to a degradation of their structure, a denaturing and finally, to a total loss of activity [72, 73]. The second main drawback is the difficulty in purifying the product. Organic phase in direct contact with fermentation broth often leads to the production of a stable emulsion. The extraction process requires then further technical systems to obtain purified compounds. Consequently, reusing the solvent after purification becomes a highly difficult task because of the important number of treatments to clean the organic phase. And finally stripping loses are reduced compared to single liquid phase bioreactors but are still present.

To overrule this drawbacks caused by the contact of the organic phase with biocatalyst, a physical separation must be considered between the two phases.

Membrane liquid-liquid extractive bioreactor

Membrane bioreactors have been investigated as an alternative to liquid-liquid extractive bioreactors [74]. Membranes are usually used for separation techniques: solid/liquid, liquid/liquid, gas/liquid or gas/gas. Theses separations are possible thanks to membrane selectivity (pores' diameter, surface characteristics...) coupled with the action of a force. Pressure, electrical potential, chemical potential (difference of concentration), gravity/centrifuge gradients are the commonly used forces. Others techniques are today still experimental: use of thermal gradients, light gradients and magnetic gradients. With these force action, the passage of components is made by convection (external mechanical constrains), migration (electrical constrain) or diffusion (concentration gradient and thermal agitation). The type of membrane used, the state of matter in the different compartments and the energy used define the characteristics of the separation considered. Table **4** sumarizes most of the current applications.

Research about membrane bioreactors began 35 years ago, and had been used mostly to treat industrial and municipal wastewater [75]. Recent advances show that such a technology can be attractive for a larger application spectrum.

Consequently today, more and more studies are running around the subject. For aroma production, the use of a membrane bioreactor with two liquid phases will be more particularly developed.

A semi-permeable membrane can be used between the two liquid phases to isolate biocatalyst in the aqueous medium. The extraction compartment allows the recovery of the product (Fig. 7).

Table 4. Membrane processes; driving force and selected compounds.

Driving force	Compounds crossing the membrane	Phase 1	Phase filling the membrane pores	Phase 2	Process
MECANICAL FORCE Hydrostatic, centrifuge or gravity pressure gradient	particles suspensions in a liquid	liquid 1	liquid 1	liquid 1	microfiltration
	macromolecules, colloids	liquid 1	liquid 1	liquid 1	ultrafiltration
	macromolecules	liquid 1	liquid 1	liquid 1	nanofiltration
	molecules	liquid 1	none (dense membrane)	liquid 1	reverse osmosis
	ions	liquid 1	none (dense membrane)	liquid 1	piezodialysis
	molecules	liquid 1	air (vacuum)	air (vacuum) and condensed compound	membrane distillation
	volatile compound	gas mix 1	gas mix 1	gas mix 1	gas permation
	volatile compound	gas mix 1	gas mix 1	gas mix 1	gas diffusion
	volatile compound	gas 1	phase 1 or phase 2 (affinity)	liquid 2	gas/liquid membrane contactor
ELECTRICAL FORCE Electrical gradient	ions	liquid 1	liquid 1	liquid 1	membrane electrolysis
	ions	liquid 1	2 membranes filled by liquid 1	liquid 1 (phase 3 also)	electrodialysis, electrodiarese

(Table 4) contd.....

Driving force	Compounds crossing the membrane	Phase 1	Phase filling the membrane pores	Phase 2	Process
CHEMICAL FORCE Solute or solvent concentration gradient	molecules, ions, colloids	liquid 1	phase 1 or phase 2 (affinity)	liquid 2	dialysis
	molecules, ions	gas or liquid 1	liquid 3	gas or liquid 2	liquid pertraction / supported liquid membrane
	molecules, ions	liquid 1	liquid 1	liquid 1	forward osmosis
	volatile compounds	gas 1	phase 1 or phase 2 (affinity)	liquid 2	gaz/liquid membrane contactor
	volatile compounds	liquid 1	phase 1 or phase 2 (affinity)	compound of interest (gas)	perdistillation/pervaporation
	volatile compound	gas 1	phase 1 or phase 2 (affinity)	compound of interest (gas)	vapor pervaporation

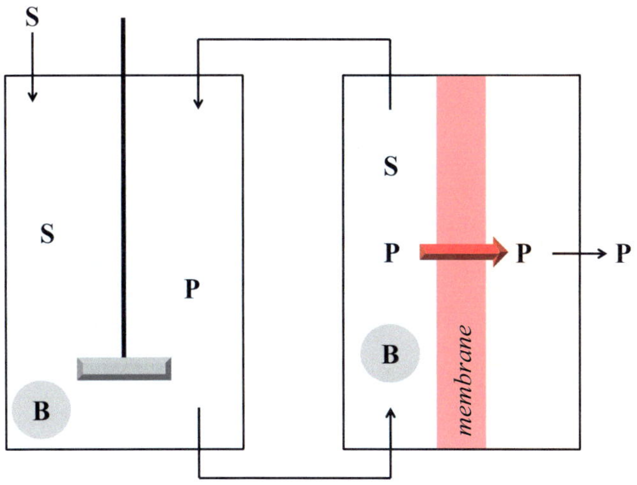

Fig. (7). Membrane liquid-liquid extractive bioreactor.

With membrane liquid-liquid extractive bioreactor, partition distribution of compounds and reservoir role of organic phase are still efficient. Product inhibition is reduced as soon as it is extracted in the organic phase like in liquid-

liquid extractive bioreactor [76]. The membrane protects biocatalyst from interface inactivation [74] which increases the life-time system. Continuous mode may be considered. Boontawan and Stuckey [77] have showed that the presence of the membrane increase the half-life of the biocatalyst *Pseudomonas fluorescens* NCIMB 11671 during the production of isonovalal from alpha-pinene oxide using fluorescent probes to determine the ratio alive / dead cells by microscopy.

Moreover, the separation of solvent and biocatalyst permits to have a larger choice of solvents [78]. And finally, purification steps are simplified. The product is diluted in a separated organic phase which can be reused after product recovery [74, 78, 79]. Leon and coworkers [80] used a three phases membrane contactor for the bioconversion of 2-Methyl-1,3-propanediol to (*R*)-β-Hydroxyisobutyric Acid. The aqueous phase contained Acetobacter as biocatalyst and an organic phase (trioctyl phosphine oxide in isooctane) was used to extract the hydrophobic compounds produced during bioconversion. Finally, a third liquid phase in a special hollow fiber membrane reaction system allowed extraction of the product from the organic phase. The use of three phases increased the extraction from 42% to 85% compared to classical biphasic system.

Although membrane contactor may bring technical solutions, this technology needs to be improved to consider its use on a larger scale. Striping loses of volatile compounds remains very important in aerated processes. The production rate remains also impacted by the mass transfer rate due to the reduced exchange surface and the resistance imposed by the membrane surface. However, advances in membrane technologies allow considering good prospects in the field of the production of aroma.

Supercritical Fluids as bioreaction media

Supercritical and near-supercritical fluids, such as carbon dioxide, freons (CHF_3), hydrocarbons (ethane, ethylene, propane) or inorganic compounds (SF_6, N_2O) can be used for enzymes as bioreaction media. Since high temperature could affect enzyme stability, the choice of supercritical fluids is often limited to compounds having a critical temperature between 0 and 60°C. The universally system used is supercritical carbon dioxide ($scCO_2$), explained by the fact that its critical point of

73.8 bars and 31.1°C makes equipment design and reaction set-up relatively simple. Moreover, it is non-toxic, non-flammable, safe for industrial use and can be removed easily after the reaction.

The major drawback of this reaction system is due to the use of high pressures leading to high energy needs and important equipment costs. In addition, the use of supercritical carbon dioxide can have negative effects on enzymes: i) a decrease of the pH of the microenvironment of the enzyme, ii) a covalent modification of free amino groups at the surface of the protein can form carbamates iii) a deactivation may occur during pressurization/depressurization cycles [81].

Ionic Liquids as solvent

Over the past ten years, an exciting development in biocatalysis in multiphasic media was the use of ionic liquids (ILs) to improve the activity, stability and selectivity of enzymes. These solvents have many useful properties, among them very low vapor pressure and excellent chemical and thermal stabilities, which make them environmentally friendly. Additionally their physical properties, such as density, viscosity, melting point, polarity, and miscibility with water or organic solvents, can be finely tuned by changing either the anion or the substituents in the cation or both. Zhang *et al.* [80] have described a comprehensive database on physical properties of ILs. For specific applications, literature reported other solvent properties such as polarity, hydrophobicity and solvent miscibility behavior of ILs [82].

A important variety of enzymes, particularly those that tolerate conventional organic solvents, can performed in ionic liquids. Their activities and stabilities are generally similar or higher than those observed in conventional organic solvents. Even if ionic liquids will clearly not provide advantages in all systems, improvements in reactivity or selectivity are notably observed for certain biotransformations of highly polar substrates, such as (poly)saccharides, amino acids and nucleotides, which cannot be performed in water owing to equilibrium limitations when the appropriate combination of cation and anion are selected. The subject attracts much interest and many interesting reviews reported progress on the topic [82, 83].

However, the recovery of non-volatile or low volatility products is one of the recurrent problems that occur when ionic liquid techniques are used for organic synthesis. One of the simplest methodologies to separate the products from the ILs phase consists to perform an extraction with solvents that are immiscible with the ILs, giving biphasic systems. Drawbacks are the extraction of small amounts of the IL and eventually of the catalyst, if a catalytic reaction is involved. Also, the partitioning of the solute between the phases limits the extent of solute extraction. The use of Volatile Organic Solvents (VOSs) remains a negative point for the integral green design of the process. Pervaporation, nanofiltration or extraction with supercritical carbon dioxide in ionic-liquid have also been proposed. These techniques represent interesting green industrial processes. Finally, ionic liquids remains today much more expensive than organic solvents, rendering them economically viable mainly for the production of high valuable compounds. However, the possibility of recycling these compounds should attenuate the cost impact [81].

CONCLUDING REMARKS

Many studies describing the biotechnological production of aroma compounds may be found in literature. Although, for most of them, the yields obtained are not yet high enough to assure economically feasible processes, it is expected that optimization tools associated to a rational use of resources may be used to maximize yields and reduce costs and, therefore, the flavors and fragrances supply derived from bioprocesses, which is currently lower than 10%, will certainly rise in future [9]. Emerging bioprocess technologies, such as those presented here, might be very useful to overcome some of the drawbacks still observed for the biotechnological processes.

CONFLICT OF INTEREST

The authors confirm that they have no conflict of interest to declare for this publication.

ACKNOWLEDGEMENTS

The authors acknowledge the Brazilian National Council of Technological and Scientific Development (CNPq) for the financial support (473981/2012-2).

REFERENCES

[1] Roudnitzky, N.; Bult, J.H.; de Wijk, R.A.; Reden, J.; Schuster, B.; Hummel, T. Investigation of interactions between texture and ortho- and retronasal olfactory stimuli using psychophysical and electrophysiological approaches. *Behav. Brain Res.,* **2011**, *216*(1), 109-115. [http://dx.doi.org/10.1016/j.bbr.2010.07.019] [PMID: 20655335]

[2] Hodgins, D. *Electronic nose: sensor array-based instruments that emulate human nose*; Marsili, R., Ed.; Marcel Dekker Inc: New York, **1997**, p. 400.

[3] Auvray, M.; Spence, C. The multisensory perception of flavor. *Conscious. Cogn.,* **2008**, *17*(3), 1016-1031. [http://dx.doi.org/10.1016/j.concog.2007.06.005] [PMID: 17689100]

[4] Taylor, A.J.; Linforth, R.S. Flavour release in the mouth. *Trends Food Sci. Technol.,* **1996**, *7*, 444-448. [http://dx.doi.org/10.1016/S0924-2244(96)10046-7]

[5] Surburg, H.; Panten, J. *Common Fragrance and Flavor Materials. Preparation, Properties and Uses*; , **2006**. [http://dx.doi.org/10.1002/3527608214]

[6] Linares, D. *Etude sur la voie de degradation de l'α-pinène chez Pseudomonas rhodesiae en milieu biphasique liquide-liquide*; Université Blaise Pascal, **2008**.

[7] Fontanille, P. *Biotransformation de α-pinene oxyde en cis-2-methyl-5-isopropylhexa-2,5-dienal (isonovalal) par Pseudomonas rhodesiae CIP 107491*; Université Blaise Pascal: France, **2002**.

[8] Berger, R.G. *Aroma Biotechnology*; Springer, **1995**. [http://dx.doi.org/10.1007/978-3-642-79373-8]

[9] Berger, R.G. Biotechnology of flavours--the next generation. *Biotechnol. Lett.,* **2009**, *31*(11), 1651-1659. [http://dx.doi.org/10.1007/s10529-009-0083-5] [PMID: 19609491]

[10] Başer, K.H.; Demirci, F. Chemistry of essential oils. In: *Flavours and fragrances: chemistry, bioprocessing and sustainability*; Springer, **2007**; pp. 43-86.

[11] Molina, G.; Bicas, J.L.; Maróstica, M.R., Jr; Pastore, G.M. Compostos de aroma. In: *Biotecnologia de Alimentos*; Pastore, G.M.; Bicas, J.L.; Maróstica, M.R., Jr, Eds.; Atheneu: São Paulo, **2013**.

[12] Volatile Compounds in Food. *http://vcf.tno.nl/VcfCompounds.cfm,*

[13] Belitz, H-D.; Grosch, W.; Schieberle, P. *Food Chemistry,* 4th ed; Springer, **2009**, p. 1070.

[14] Wagner, R. *Composição de voláteis e aroma de salames nacionais tipos Italiano e Milano*; University of Campinas, **2008**.

[15] Cheetham, P.S. Combining the technical push and the business pull for natural flavours. In: *Biotechnology of aroma compounds*; Berger, R.G., Ed.; Springer: Berlin, **1997**; p. 263. [http://dx.doi.org/10.1007/BFb0102061]

[16] Priefert, H.; Rabenhorst, J.; Steinbüchel, A. Biotechnological production of vanillin. *Appl. Microbiol. Biotechnol.,* **2001**, *56*(3-4), 296-314. [http://dx.doi.org/10.1007/s002530100687] [PMID: 11548997]

[17] Bicas, J.L.; Dionísio, A.P.; Silva, J.C.; Barros, F.F.; Pastore, G.M. Agro-industrial residues in biotechnological processes. In: *Industrial Fermentation: Food Processes, Nutrient Sources and Production Strategies*; Krause, J.; Fleischer, O., Eds.; Nova Science Publishers Inc: Hauppauge, **2010**; pp. 275-295.

[18] Pybus, D.H. The history of aroma chemistry and perfume. In: *The chemistry of fragrances: from perfumer to consumer*; Sell, C., Ed.; RSC Publishing, **2006**; pp. 3-23. [http://dx.doi.org/10.1039/9781847555342-00003]

[19] Bicas, J.L.; Silva, J.C.; Dionisio, A.P.; Pastore, G.M. Biotechnological production of bioflavors and functional sugars. *Ciência e Tecnol. Aliment.,* **2010**, *30*, 7-18.

[20] Bicas, J.L.; Dionísio, A.P.; Pastore, G.M. Bio-oxidation of terpenes: an approach for the flavor industry. *Chem. Rev.,* **2009**, *109*(9), 4518-4531. [http://dx.doi.org/10.1021/cr800190y] [PMID: 19645444]

[21] Zamzuri, N.A.; Abd-Aziz, S. Biovanillin from agro wastes as an alternative food flavour. *J. Sci. Food Agric.,* **2013**, *93*(3), 429-438. [http://dx.doi.org/10.1002/jsfa.5962] [PMID: 23208984]

[22] Xu, P.; Hua, D.; Ma, C. Microbial transformation of propenylbenzenes for natural flavour production. *Trends Biotechnol.,* **2007**, *25*(12), 571-576. [http://dx.doi.org/10.1016/j.tibtech.2007.08.011] [PMID: 17988755]

[23] Rao, S.R.; Ravishankar, G.A. Vanilla flavour: production by conventional and biotechnological routes. *J. Sci. Food Agric.,* **2000**, *80*, 289-304. [http://dx.doi.org/10.1002/1097-0010(200002)80:3<289::AID-JSFA543>3.0.CO;2-2]

[24] Food and Agriculture Organization. *No Title. http://faostat.fao.org/,* [accessed Jan 16, 2014]

[25] Converti, A.; Aliakbarian, B.; Domínguez, J.M.; Vázquez, G.B.; Perego, P. Microbial production of vanillin. Brazilian Journal of Microbiology. *Braz. J. Microbiol.,* **2010**, *41*(3), 519-530. [http://dx.doi.org/10.1590/S1517-83822010000300001] [PMID: 24031526]

[26] Havkin-Frenkel, D.; Belanger, F. Biotechnological production of vanillin. In: *Biotechnology in Flavor Production*; Havkin-Frenkel, D.; Belanger, F., Eds.; Wiley, **2009**; pp. 83-103.

[27] Schrader, J.; Etschmann, M.M.; Sell, D.; Hilmer, J.M.; Rabenhorst, J. Applied biocatalysis for the synthesis of natural flavour compounds--current industrial processes and future prospects. *Biotechnol. Lett.,* **2004**, *26*(6), 463-472. [http://dx.doi.org/10.1023/B:BILE.0000019576.80594.0e] [PMID: 15127786]

[28] Solvay. Rhovanil® Natural. *Available from: www.rhodia.com/en/markets_and_products/brands/rhovanil_natural.tcm,*

[29] Paterson, D. *Vanilla: natural or not?,* **2010**, 2-6.

[30] Hansen, E.H.; Møller, B.L.; Kock, G.R.; Bünner, C.M.; Kristensen, C.; Jensen, O.R.; Okkels, F.T.; Olsen, C.E.; Motawia, M.S.; Hansen, J. De novo biosynthesis of vanillin in fission yeast (Schizosaccharomyces pombe) and baker's yeast (Saccharomyces cerevisiae). *Appl. Environ. Microbiol.,* **2009**, *75*(9), 2765-2774. [http://dx.doi.org/10.1128/AEM.02681-08] [PMID: 19286778]

[31] Krings, U.; Berger, R.G. Biotechnological production of flavours and fragrances. *Appl. Microbiol. Biotechnol.,* **1998**, *49*(1), 1-8. [http://dx.doi.org/10.1007/s002530051129] [PMID: 9487704]

[32] Damasceno, S.; Cereda, M.P.; Pastore, G.M.; Oliveira, J.G. Production of volatile compounds by Geotrichum fragrans using cassava wastewater as substrate. *Process Biochem.,* **2003**, *39*(4), 411-414. [http://dx.doi.org/10.1016/S0032-9592(03)00097-9]

[33] Romero-Guido, C.; Belo, I.; Ta, T.M.; Cao-Hoang, L.; Alchihab, M.; Gomes, N.; Thonart, P.; Teixeira, J.A.; Destain, J.; Waché, Y. Biochemistry of lactone formation in yeast and fungi and its utilisation for the production of flavour and fragrance compounds. *Appl. Microbiol. Biotechnol.,* **2011**, *89*(3), 535-547. [http://dx.doi.org/10.1007/s00253-010-2945-0] [PMID: 20981417]

[34] Okui, S.; Uchiyama, M.; Mizugaki, M. Metabolism of hydroxyl fatty acids. II. Intermediates of the oxidative break down of ricinoleic acid by genus Candida. *J. Biochem.,* **1963**, *54*, 536-540. [PMID: 14099009]

[35] Bonnarme, P.; Djian, A.; Latrasse, A.; Féron, G.; Giniès, C.; Durand, A.; Le Quéré, J-L. Production of 6-pentyl-alpha-pyrone by Trichoderma sp. from vegetable oils. *J. Biotechnol.,* **1997**, *56*, 143-150. [http://dx.doi.org/10.1016/S0168-1656(97)00108-9]

[36] Feron, G.; Dufosse, L.; Pierard, E.; Bonnarme, P.; Quere, J.L.; Spinnler, H. Production, identification, and toxicity of gamma-decalactone and 4-hydroxydecanoic acid from Sporidiobolus spp. *Appl. Environ. Microbiol.,* **1996**, *62*(8), 2826-2831. [PMID: 16535376]

[37] Nago, H.; Matsumoto, M.; Nakai, S. 2-deceno-delta-lactoneproducing fungi, strains of Fusarium solani, isolated by using a medium containing decano-delta-lactone as the sole carbon source. *Biosci. Biotechnol. Biochem.,* **1993**, *57*, 2107-2110. [http://dx.doi.org/10.1271/bbb.57.2107]

[38] Alchihab, M.; Destain, J.; Aguedo, M.; Wathelet, J.P.; Thonart, P. The utilization of gum tragacanth to improve the growth of Rhodotorula aurantiaca and the production of gamma-decalactone in large scale. *Appl. Biochem. Biotechnol.,* **2010**, *162*(1), 233-241. [http://dx.doi.org/10.1007/s12010-009-8739-0] [PMID: 19680818]

[39] Bar, R. Cyclodextrin-aided bioconversions and fermentations. *Trends Biotechnol.,* **1989**, *7*, 2-4. [http://dx.doi.org/10.1016/0167-7799(89)90070-X]

[40] Welsh, F.W.; Murray, W.D.; Williams, R.E. Microbiological and enzymatic production of flavor and fragrance chemicals. *Crit. Rev. Biotechnol.,* **1989**, *9*(2), 105-169. [http://dx.doi.org/10.3109/07388558909040617]

[41] Labuda, I. Flavor compounds. In: *Encyclopedia of microbiology*; Schaechter, M., Ed.; Academic Press: Oxford, **2009**.
[http://dx.doi.org/10.1016/B978-012373944-5.00148-6]

[42] Hua, D.; Xu, P. Recent advances in biotechnological production of 2-phenylethanol. *Biotechnol. Adv.,* **2011**, *29*(6), 654-660.
[http://dx.doi.org/10.1016/j.biotechadv.2011.05.001] [PMID: 21601630]

[43] Zhang, H.; Cao, M.; Jiang, X.; Zou, H.; Wang, C.; Xu, X.; Xian, M. De-novo synthesis of 2-phenylethanol by *Enterobacter* sp. CGMCC 5087. *BMC Biotechnol.,* **2014**, *14*, 30.
[http://dx.doi.org/10.1186/1472-6750-14-30] [PMID: 24766677]

[44] Hua, D.L.; Lin, S.; Li, Y.F.; Chen, H.; Zhang, Z.B.; Du, Y.; Zhang, X.; Xu, P. Enhanced 2-phenylethanol production from L-phenylalanine *via in situ* product adsorption. *Biocatalysis Biotransform.,* **2010**, *28*, 259-266.
[http://dx.doi.org/10.3109/10242422.2010.500724]

[45] Mihal', M.; Krištofíková, L.; Markoŝ, J. Production of 2-phenylethanol in hybrid system using airlift reactorand immersed hollow fiber membrane module. *Chem. Eng. Process.,* **2013**, *72*, 144-152.
[http://dx.doi.org/10.1016/j.cep.2013.08.008]

[46] Xiao, Z.; Lu, J.R. Strategies for enhancing fermentative production of acetoin: a review. *Biotechnol. Adv.,* **2014**, *32*(2), 492-503.
[http://dx.doi.org/10.1016/j.biotechadv.2014.01.002] [PMID: 24412764]

[47] Nicholson, W.L. The Bacillus subtilis ydjL (bdhA) gene encodes acetoin reductase/2,3-butanediol dehydrogenase. *Appl. Environ. Microbiol.,* **2008**, *74*(22), 6832-6838.
[http://dx.doi.org/10.1128/AEM.00881-08] [PMID: 18820069]

[48] Guo, Y.; Song, H.; Wang, Z.; Ding, Y. Expression of POX2 gene and disruption of POX3 genes in the industrial Yarrowia lipolytica on the γ-decalactone production. *Microbiol. Res.,* **2012**, *167*(4), 246-252.
[http://dx.doi.org/10.1016/j.micres.2011.10.003] [PMID: 22115771]

[49] Zhao, Y.P.; Mu, X.Q.; Xu, Y. Improvement in γ-decalactone production by Yarrowia sp. after genome shuffling. *Chem. Pap.,* **2014**, *68*(8), 1030-1040.
[http://dx.doi.org/10.2478/s11696-014-0551-9]

[50] Glazyrina, J.; Materne, E.M.; Dreher, T.; Storm, D.; Junne, S.; Adams, T.; Greller, G.; Neubauer, P. High cell density cultivation and recombinant protein production with Escherichia coli in a rocking-motion-type bioreactor. *Microb. Cell Fact.,* **2010**, *9*(42), 42.
[http://dx.doi.org/10.1186/1475-2859-9-42] [PMID: 20509968]

[51] Moradi, H.; Asadollahi, M.A.; Nahvi, I. Improved γ-decalactone production from castor oil by fed-batch cultivation of Yarrowia lipolytica. *Biocatal. Agric. Biotechnol.,* **2013**, *2*, 64-68.

[52] Morán-Marroquín, G.A.; Córdova, J.; Valle-Rodríguez, J.O.; Estarrón-Espinosa, M.; Díaz-Montaño, D.M. Effect of dilution rate and nutrients addition on the fermentative capability and synthesis of aromatic compounds of two indigenous strains of Saccharomyces cerevisiae in continuous cultures fed with Agave tequilana juice. *Int. J. Food Microbiol.,* **2011**, *151*(1), 87-92.
[http://dx.doi.org/10.1016/j.ijfoodmicro.2011.08.008] [PMID: 21903290]

[53] Deetae, P.; Saint-Eve, A.; Spinnler, H.E.; Helinck, S. Critical effect of oxygen on aroma compound production by *Proteus vulgaris*. *Food Chem.,* **2011**, *126*, 134-139. [http://dx.doi.org/10.1016/j.foodchem.2010.10.089]

[54] An, J.U.; Joo, Y.C.; Oh, D.K. New biotransformation process for production of the fragrant compound γ-dodecalactone from 10-hydroxystearate by permeabilized Waltomyces lipofer cells. *Appl. Environ. Microbiol.,* **2013**, *79*(8), 2636-2641. [http://dx.doi.org/10.1128/AEM.02602-12] [PMID: 23396347]

[55] Orzua, M.C.; Mussatto, S.I.; Contreras-Esquivel, J.C.; Rodriguez, R.; de la Garza, H.; Teixeira, J.a.; Aguilar, C.N. Exploitation of agro industrial wastes as immobilization carrier for solid-state fermentation. *Ind. Crops Prod.,* **2009**, *30*(1), 24-27. [http://dx.doi.org/10.1016/j.indcrop.2009.02.001]

[56] Soares, M.; Christen, P.; Pandey, A.; Soccol, C.R. Fruity flavour production by Ceratocystis fimbriata grown on coffee husk in solid-state fermentation. *Process Biochem.,* **2000**, *35*(8), 857-861. [http://dx.doi.org/10.1016/S0032-9592(99)00144-2]

[57] Christen, P.; Meza, J.; Revah, S. Fruity aroma production in solid state fermentation by Ceratocystis fimbriata: influence of the substrate type and the presence of precursors. *Mycol. Res.,* **1997**, *101*(8), 911-919. [http://dx.doi.org/10.1017/S0953756297003535]

[58] Zheng, L.; Zheng, P.; Sun, Z.; Bai, Y.; Wang, J.; Guo, X. Production of vanillin from waste residue of rice bran oil by Aspergillus niger and Pycnoporus cinnabarinus. *Bioresour. Technol.,* **2007**, *98*(5), 1115-1119. [http://dx.doi.org/10.1016/j.biortech.2006.03.028] [PMID: 16782330]

[59] Xiao, Z.J.; Liu, P.H.; Qin, J.Y.; Xu, P. Statistical optimization of medium components for enhanced acetoin production from molasses and soybean meal hydrolysate. *Appl. Microbiol. Biotechnol.,* **2007**, *74*(1), 61-68. [http://dx.doi.org/10.1007/s00253-006-0646-5] [PMID: 17043817]

[60] Bramororski, A.; Soccol, C.R.; Christen, P.; Revah, S. Fruit aroma production by Ceratocystis fimbriata in static cultures from solid agro-industrial wastes. *Rev. Microbiol.,* **1998**, *28*(3), 208-211.

[61] Laufenberg, G.; Rosato, P.; Kunz, B. Adding value to vegetable waste: Oil press cakes as substrates for microbial decalactone production. *Eur. J. Lipid Sci. Technol.,* **2004**, *106*, 207-217. [http://dx.doi.org/10.1002/ejlt.200300898]

[62] ROSSIGNOL, C. *Combination of a membrane process based on liquid/liquid extraction with a bioreactor producing hydrophobic molecules*; Blaise Pascal University, **2013**.

[63] Gounaris, Y. Biotechnology for the production of essential oils, flavours and volatile isolates. A review. *Flavour Fragrance J.,* **2010**, *25*, 367-386. [http://dx.doi.org/10.1002/ffj.1996]

[64] Daugulis, A.J. Two-phase partitioning bioreactors: a new technology platform for destroying xenobiotics. *Trends Biotechnol.,* **2001**, *19*(11), 457-462. [http://dx.doi.org/10.1016/S0167-7799(01)01789-9] [PMID: 11602310]

[65] Tramper, J. Chemical *versus* biochemical conversion: when and how to use biocatalysts. *Biotechnol.*

Bioeng., **1996**, *52*(2), 290-295.
[http://dx.doi.org/10.1002/(SICI)1097-0290(19961020)52:2<290::AID-BIT8>3.0.CO;2-L] [PMID: 18629895]

[66] Marques, M.P.; de Carvalho, C.C.; Cabral, J.M.; Fernandes, P. Scaling-up of complex whole-cell bioconversions in conventional and non-conventional media. *Biotechnol. Bioeng.,* **2010**, *106*(4), 619-626.
[http://dx.doi.org/10.1002/bit.22711] [PMID: 20503299]

[67] León, R.; Fernandes, P.; Pinheiro, H.M.; Cabral, J.M. Whole-cell biocatalysis in organic media. *Enzyme Microb. Technol.,* **1998**, *23*, 483-500.
[http://dx.doi.org/10.1016/S0141-0229(98)00078-7]

[68] Heipieper, H.J.; Neumann, G.; Cornelissen, S.; Meinhardt, F. Solvent-tolerant bacteria for biotransformations in two-phase fermentation systems. *Appl. Microbiol. Biotechnol.,* **2007**, *74*(5), 961-973.
[http://dx.doi.org/10.1007/s00253-006-0833-4] [PMID: 17262209]

[69] Dafoe, J.; Daugulis, A. Bioproduction of cis-(1S,2R)-indandiol, a chiral pharmaceutical intermediate, using a solid-liquid two-phase partitioning bioreactor for enhanced removal of inhibitors. *J. Chem. Technol. Biotechnol.,* **2011**, *86*, 1379-1385.
[http://dx.doi.org/10.1002/jctb.2653]

[70] The Council of the European Communities. *Council Directive 88/344/EEC of 13 June 1988 on the approximation of the laws of the Member States on extraction solvents used in the production of foodstuffs and food ingredients,* **1988**.

[71] Bicas, J.L.; Fontanille, P.; Pastore, G.M.; Larroche, C. A bioprocess for the production of high concentrations of R-(+)-α-terpineol from R-(+)-limonene. *Process Biochem.,* **2010**, *45*, 481-48.
[http://dx.doi.org/10.1016/j.procbio.2009.11.007]

[72] Linares, D.; Fontanille, P.; Larroche, C. Protective effect of biomass components against interfacial inactivation of α-pinene oxide lyase from Pseudomonas rhodesiae CIP 107491. *Food Technol. Biotechnol.,* **2008**, *76*(2), 202-207.

[73] Déziel, E.; Comeau, Y.; Villemur, R. Two-liquid-phase bioreactors for enhanced degradation of hydrophobic/toxic compounds. *Biodegradation,* **1999**, *10*(3), 219-233.
[http://dx.doi.org/10.1023/A:1008311430525] [PMID: 10492889]

[74] Vaidya, A.M.; Halling, P.J.; Bell, G. Novel reactor for aqueous-organic two-phase biocatalysis: A packed bed hollow fiber membrane bioreactor. *Biotechnol. Tech.,* **1993**, *4*, 441.
[http://dx.doi.org/10.1007/BF00151881]

[75] Yang, W.; Cicek, N.; Ilg, J. State-of-the-art of membrane bioreactors: Worldwide research and commercial applications in North America. *J. Membr. Sci.,* **2005**, *270*, 201-211.
[http://dx.doi.org/10.1016/j.memsci.2005.07.010]

[76] Westgate, S.; Vaidya, A.M.; Bell, G.; Halling, P.J. High specific activity of whole cells in an aqueous-organic two-phase membrane bioreactor. *Enzyme Microb. Technol.,* **1998**, *22*(7), 575-577.
[http://dx.doi.org/10.1016/S0141-0229(97)00245-7]

[77] Boontawan, A.; Stuckey, D.C. A membrane bioreactor for the biotransformation of α-pinene oxide to

isonovalal by Pseudomonas fluorescens NCIMB 11671. *Appl. Microbiol. Biotechnol.,* **2006**, *69*(6), 643-649.
[http://dx.doi.org/10.1007/s00253-005-0025-7] [PMID: 16088347]

[78] Rios, G.M.; Belleville, M.P.; Paolucci-Jeanjean, D. Membrane engineering in biotechnology: quo vamus? *Trends Biotechnol.,* **2007**, *25*(6), 242-246.
[http://dx.doi.org/10.1016/j.tibtech.2007.04.003] [PMID: 17433844]

[79] Giorno, L.; Drioli, E. Biocatalytic membrane reactors: applications and perspectives. *Trends Biotechnol.,* **2000**, *18*(8), 339-349.
[http://dx.doi.org/10.1016/S0167-7799(00)01472-4] [PMID: 10899815]

[80] Zhang, S.; Sun, N.; He, X.; Lu, X.; Zhang, X. Physical Properties of Ionic Liquids: Database and Evaluation. *J. Phys. Chem. Ref. Data,* **2006**, *35*, 1475-1517.
[http://dx.doi.org/10.1063/1.2204959]

[81] Larroche, C.; Gros, J.B.; Fontanille, P. Microbial Processes. In: *Flavours and Fragrances. Chemistry, Bioprocessing and Sustainability*; Berger, R.G., Ed.; Springer, **2007**; pp. 573-595.
[http://dx.doi.org/10.1007/978-3-540-49339-6_24]

[82] Moniruzzamana, M.; Nakashimab, K.; Kamiyaa, N.; Goto, M. Recent advances of enzymatic reactions in ionic liquids. *Biochem. Eng. J.,* **2010**, *48*(3), 295-314.
[http://dx.doi.org/10.1016/j.bej.2009.10.002]

[83] Mai, N.L.; Koo, Y.M. Compatibility of Ionic Liquids with Enzymes. In: *Production of Biofuels and Chemicals with Ionic Liquids*; 257-273.

CHAPTER 7

Natural Colorants from Microorganisms

Júlio C. de Carvalho[a,*], **Juliano L. Bicas**[b], **Daniel E. R. Fernández**[a], **Adenise L. Woiciechowski**[a], **Adriane B.P. Medeiros**[a] and **Carlos R. Soccol**[a]

[a] *Bioprocess Engineering Laboratory, Federal University of Paraná, Curitiba, Brazil*

[b] *Department of Food Science, Faculty of Food Engineering, University of Campinas, Campinas, Brazil*

Abstract: Color additives are a necessity in the modern food industry. These additives are colored because of their peculiar molecular structure, which also frequently imparts other properties such as antioxidant activity. The natural diversity of microorganisms offers several opportunities for the development of "biopigments", which may be produced with high productivity and without seasonality concerns. While the number of artificial colors used foods is inescapably being reduced, that of permitted natural pigments is slowly growing. It is not possible to simply isolate a colored microorganism strain and use it as an additive: the color must be proven safe, and that is why only a dozen pigments from fungi, yeast, bacteria and microalgae are already permitted and are commercially produced. And yet, these few biopigments are paving the way for new developments, where the knowledge that involves microorganism isolation, bioactivity assays, biomass production and fractionation is used for the study of new alternatives. This chapter gives a general view of permitted natural colors, focusing on commercially relevant microbial biopigments and their production processes.

Keywords: *Arthrospira (Spirulina)*, *Ashbya gossypii*, Astaxanthin, Biopigments, *Blakeslea trispora*, β-carotene, *Chlorella*, *Dunaliella*, *Haematococcus*, Microalgae, Microorganisms, *Monascus*, Natural color, Phycobilin, *Porphyridium*, Riboflavin.

* **Address correspondence to Julio Carvalho:** Department of Bioprocess Engineering and Biotechnology, Universidade Federal do Paraná, Cx Postal 19011, ACF Centro Politécnico, 81531-990 Curitiba, Brazil; E-mail: jccarvalho@ufpr.br.

Juliano Lemos Bicas, Mário Roberto Maróstica, Glaucia Maria Pastore (Eds.)

INTRODUCTION

Color is the first sensation perceived by us when we come in contact with food. It is used for consumers to infer product properties: freshness, maturity, composition, and quality, for example. The consumer associates, by experience, cultural heritage and instinct, specific ranges of colors with specific classes of foods. In fact, certain matches such as green bananas or dark meat are immediately associated with unpleasant taste; and unnatural matches such as blue foods are usually repulsive.

As the first quality assessment of a food, color has impact in the other senses as well, influencing, for example, the perception of flavor identity [1]. In fact, color has a subjective effect so strong as to dominate even label information read by the consumer - as tested for drinks by [2] and [3].

Colorants must be added to most processed foods for several reasons, such as to standardize batches with natural variation, to compensate for color fading during processing or storage, to impart expected colors to products (*e.g.* red to an artificially strawberry-flavored candy), to intensify the colors naturally found in food and to protect other food components. However, it shall never be used to mask flaws due to bad processing or manipulation of food products [4].

To be used as food additives, these colorants must be chemically stable and nontoxic. The explosion of synthetic organic chemistry at the end of the 19th century led to the poorly regulated use of dozens of artificial color additives. These were gradually banned during the 20th century. Nowadays, there is intensification towards the use of natural color additives, which are perceived as safer and healthier than its artificial counterparts, as shown by market research [5].

In order to be used as a natural color additive, a colored molecule such as chlorophyll or a carotenoid must usually be extracted and concentrated from the raw material and processed into an additive, perhaps in solution or mixed with bulking agents. Several raw materials – vegetables, fruits, a few animals, and microorganisms - may be processed into color additives.

After a brief general discussion about natural color additives, this chapter emphasizes those produced by microorganisms, its production and uses.

DEFINITIONS, HISTORY AND REGULATION

Any substance may selectively absorb certain wavelengths of the electromagnetic spectrum; when this absorption occurs in the visible light range (wavelengths from 400 to 700nm), we observe color. The absorption of one color leads to the perception of its complementary color: a carotenoid which absorbs blue light, for example, will be seen as a yellow substance (Fig. **1**). Absorption over multiple wavelengths, which is usual for natural substances, creates different tones and shades.

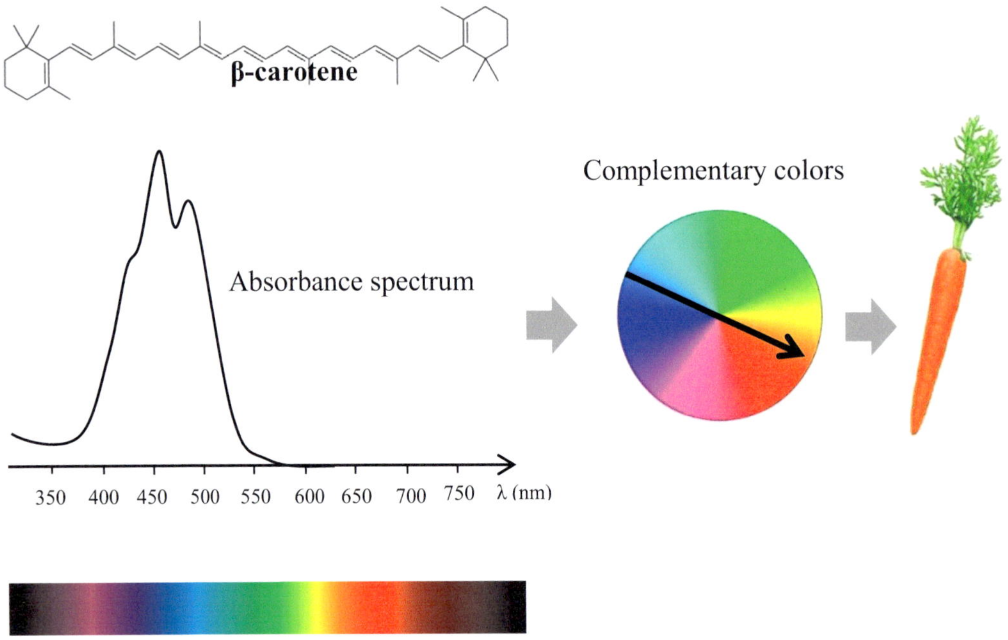

Fig. (1). Colored substance spectrum and color complementarity.

The absorption of radiant energy is central to life, as it is the basis for photosynthesis and vision. Nature offers a large variety of colors on the "warm" side of the spectrum, and substances responsible for the color may be extracted and used as "colorants" or "color additives". The main natural colorants present in foods may be grouped in different classes, based on their chemical structure:

porphyrins, which includes hemoglobin (red), myoglobin (red) and chlorophyll (green); betalains (yellow to red); flavonoids, specially anthocyanins (red to blue); carotenoids, such as lutein (yellow), β-carotene and lycopene (red-orange); tannins (pale yellow to light brown) and others (quinones, polyphenols *etc.*) [6].

The use of colored ingredients as a way of controlling prepared food appearance is probably as old as the use of spices. While it is difficult to precise when coloring substances were first used, there are reports of color modification of foods and beverages in ancient Greece and Rome [7], as well as to the coloring of candies in ancient Egypt [8] and the use of saffron in ancient Greece [9].

The use of color in processed foods became common with its commercialization. With this rise in use came the adulteration of foods, which in turn created the need for laws to control the practice [10]. In fact, the use of food color has always been pushed forward by the food processors and pushed back by regulation, and that became evident after the advent of synthetic organic colors. Early colors were limited to plant and animal extracts (usually safe or of moderate toxicity), and to inorganic additives (some innocuous, such as chalk, and others potentially dangerous, such as copper and lead salts). After 1850, with the synthesis of mauve by W. Perkins, there has been a flourishing of organic pigment industry and, as a complement, the weakening of traditional pigment agro industry such as indigo, in India [11]. There were around 700 new colors by the beginning of the 20th century, and over 80 were on the market [9]. Since then, active regulation has come to reduce the number of permitted colors, forbidding the use of most synthetic colors and defining the limits for the few permitted synthetic color additives. This regulation is done most notably by FDA (Food and Drug Administration) in USA, EFSA (European Food Safety Authority) in Europe, PFSB (the Pharmaceutical and Food Safety Bureau) in Japan, and ANVISA (the National Agency of Sanitary Vigilance) in Brazil.

There are around 20 permitted artificial food colors in the world, with the set of substances changing from country to country. The number of artificial colors permitted in food in the United States, Brazil, Japan and in most of Europe, is 7, 14, 12 and 15, respectively [12, 13]. A color additive may be banned not only due to its intrinsic toxicity, but in some cases due to the presence of toxic contaminants. That is the case for D&C Yellow #9 and #10, in USA: these additives cannot be produced and marketed with a guarantee that they *do not* contain traces of β-naphtylamine, and therefore are not permitted.

NATURAL COLORS – BACK TO PLANT EXTRACTS?

While permitted artificial color additives have limited dosage and, at least in the USA, must be certified (*i.e.*, each batch of color additive must be analyzed and in conformity with FDA standards), natural color additives may be used in an *ad quantum satis* manner and are usually exempt from certification. The most common permitted natural food colors are presented in Table **1**, with key properties for most of them: the color (although hues vary for different substances); its permitted usage levels in food and beverages; the main chemical substances responsible for the color – in natural extracts there are several substances, and a few are responsible for the color; the INS number, the maximum absorption wavelength and the correspondent extinction coefficient (color value), which are useful for analysis and quality control, giving an idea of how strong a pigment is. Some natural colors were omitted from the list: inorganic pigments, such as iron oxide; processing colors, such as caramel; organic iron salts; and fruit and vegetable juices whose coloring component fall into one of the categories found (Table **1**), such as anthocyanins. The table bases the permitted use as per FDA regulations, because USA has arguably one of the most restrictive legislations for food colors.

Table 1. Common natural colors used in food and feed. This list includes vegetal, animal and microbial sources.

Additive / Source	Color	Indicated use (in USA if not otherwise noted)	Main component	INS	Color value (*E1%1cm) and Maximum absorption wavelength (nm)
Algae meal, dried	Green to red	Chicken feed only: ≤0.3%	Mixture of carotenoids, xanthophyll and chlorophylls		Depends on specific components

(Table 1) contd.....

Additive / Source	Color	Indicated use (in USA if not otherwise noted)	Main component	INS	Color value (*E1%1cm) and Maximum absorption wavelength (nm)
Annatto extract (bixin, norbixin)	Red-brown to red - purple	GMP	Bixin	160b(i)	4200 @ 456nm in petroleum ether [14]
Anthocyanin extracts (fruits and green vegetables)	Purple to red	GMP	Mixture of anthocyanins	163i, 163ii (grape)	Variable with source and pH; typical 80-120 at 500-550nm in dilute HCl [15] and [16]
Astaxanthin and astaxanthin dimethyldisuccinate (several microorganisms, see text)	Orange-red	Salmonid fish feed only: ≤80 mg per kg of finished feed	Astaxanthin		2170@ 489nm in acetone
Beet juice or powder	Pink to dark red	GMP	Betanine	162	1120 @530nm
Beta-Apo-8'-carotenal (from carrot oil)	Reddish orange	General use: ≤15 mg/lb of solid or semisolid food or per pint of liquid food	all-trans- β-apo-8'-carotenal	160e, 160f	2640 @461nm
Beta carotene, natural and synthetic: from vegetables, *B. trispora* , and *D. salina* .	Orange	GMP	β-carotene	160a(ii), 160a(iii), 160a(iv)	2500 @455nm

(Table 1) contd.....

Additive / Source	**Color**	**Indicated use (in USA if not otherwise noted)**	**Main component**	**INS**	**Color value (*E1%1cm) and Maximum absorption wavelength (nm)**
Canthaxanthin; most of the pigment used in feeds is synthetic.	Orange pink	General use: ≤30 mg/lb of solid or semisolid food or per pint of liquid food; broiler chicken feed: ≤4.41 mg per kg (4 g/ton) of complete feed; salmonid fish feed: ≤80 mg per kg (72 gm/ton) of finished feed	ß-Carotene-4,4'-dione (canthaxanthin)	161g	2200 @470nm
Carmine, carminic acids, cochineal extract	Red	GMP	carminic acid	120	139 @494nm (in dilute HCl)
Carrot oil	Orange to yellow	GMP	carotenoids	160a (ii)	2500 @440-457nm
Gardenia red and yellow	Red, yellow	Not permitted in USA and EU. Regulated use in Japan	Crocin and crocetin		4320 @ 422nm in petroleum ether [14]

(Table 1) contd.....

Additive / Source	Color	Indicated use (in USA if not otherwise noted)	Main component	INS	Color value (*E1%1cm) and Maximum absorption wavelength (nm)
Grape skin extract (enocianina)	Mauve to blue	Still and carbonated drinks and ades, beverage bases and alcoholic beverages (in accordance with Parts 4 & 5 of 27 CFR)	Anthocyanins	163(ii)	500 @520nm, water at pH 1.5 [15]
Haematococcus algae meal	Orange red	Salmonid fish feed only. Total astaxanthin from all astaxanthin color additive sources listed in part 73, ≤80 mg per kg of finished feed	Astaxanthin		2170@ 489nm in acetone
Indigotine (indigo carmine): mostly synthetic, may be produced from *Indigofera sp.*	Blue	Not certified in USA and EU. 0.005-0.05 g/100g, depending on the use (in Brazil)	Disodium 3,3'-dioxo-[delta2,-'-biindoline]-5, 5'-disulfonate (main component)	132	416 @610nm in water [17]
Lac color, from *Laccifer lacca.*	Red	Not permitted in USA. Regulated use in Japan	Laccaic acid		123 @ 490nm (in water) [18]

(Table 1) contd.....

Additive / Source	Color	Indicated use (in USA if not otherwise noted)	Main component	INS	Color value (*E1%1cm) and Maximum absorption wavelength (nm)
Logwood color, from *Haematoxylum campechianum*	Red	Not permitted in USA. Regulated use in Japan	Hematin (oxidation product of Haematoxylin)		1133 @ 445nm (in water) [19]
Lycopene, tomato extract or concentrate (ii) or from *B. trispora(iii)*	Red	GMP	Lycopene	160d(ii), 160d(iii)	3460 @470nm, 5370 @503 nm in hexane [20]
Lutein, from marigold oleoresin		GMP (in Brazil)	Lutein	161b	2550 @ 445nm in ethanol [14]
Marigold color	Yellow	Not permitted in USA. Regulated use in Japan	Lutein		See lutein
Monascus color	Red (yellow to red if fractionated)	Not permitted in USA. Regulated use in Japan	Monascorubrin and ankaflavin		69 @505nm in ethanol**
Orchil dyes	Red	GMP (in Brazil)	Orcein		575 nm in ethanol [21]
Paprika & Paprika oleoresin	Red	GMP	Capsanthin, capsorubin	160c	2100 @462nm (in acetone)
Paracoccus pigment	Red	Salmonid fish feed only: ≤80 mg per kg of finished feed	Astaxanthin		2170@ 489nm in acetone

(Table 1) contd.....

Additive / Source	Color	Indicated use (in USA if not otherwise noted)	Main component	INS	Color value (*E1%1cm) and Maximum absorption wavelength (nm)
Phaffia yeast	Orange red	Salmonid fish feed only. Total astaxanthin from all astaxanthin color additive sources listed in part 73, ≤80 mg per kg of finished feed	Astaxanthin esters		2170@ 489nm in acetone
Riboflavin: from molds (see text)	Yellow	GMP	Riboflavin	101, 101a	332 @445 [22]
Saffron	Yellow to orange	GMP	α-Crocin	164	911 @443 in water [23]
Santalin (from sandalwood)	Red to violet	Not permitted in USA. Regulated use in Japan	Santalins A and B		100 @475nm in ethanol [24]
Chlorophyll: from leafy vegetables	Green	GMP	Chlorophylls (i) a and b, chlorophyllins (ii)	140(i), (ii)	700 @409nm (in chloroform)
copper chlorophylls and chlorophyllins:from leafy vegetables	Green	Citrus-based dry beverage mixes only: ≤0.2 percent by wt. of the dry mix	Copper chlorophylls and Chlorophyllins a and b.	141(i), (ii)	700 @405nm (in water, pH 9)

(Table 1) contd.....

Additive / Source	Color	Indicated use (in USA if not otherwise noted)	Main component	INS	Color value (*E1%1cm) and Maximum absorption wavelength (nm)
Tagetes (Aztec marigold) meal and extract	Yellow to orange	Chicken feed only: ethoxyquin content ≤0.3% by wt. of additive and ≤150 ppm by wt. of final feed	Lutein	161b	2520 @453nm in dioxane [25] in [26]
Turmeric & Turmeric oleoresin (EUA) Curcumin (EU)	Yellow to orange	GMP	1,7-Bis(4-hydroxy-3-methoxyphenyl)hepta-1, 6-diene-3,5-dione	100	1607 @426nm (in ethanol)

GMP: may be used as much as needed, provided that it is according to the good manufacturing practices defined for the specific food. Compiled from [27] , [28] for general information, and [29] for wavelengths, except where specified.

* The $E^{1\%}_{1cm}$ is the absorption of a solution of the pigment at 1% concentration

** estimated using spectral data from [30] .

BIOTECHNOLOGICALLY PRODUCED COLORANTS

Most of the natural colorants for food use are extracted from plants or animals. However, there are important cases in which whole or fractionated microbial biomass is the color additive, and these coloring substances will be treated in this section. Plant cell cultures, as well as modified plants, have potential use in the future but are not as important as microorganisms and, therefore, they will not be discussed here (for more information, we suggest the references [31] and [32]). The main motivations for the production of the so-called biopigments are their "natural" status, their independence on seasonal, climatic and geographical conditions and to the possible production under controlled conditions with predictable and optimized yields [33, 34].

There are thousands known of pigment-producing microorganisms (see examples in Table **2**), which use these colorants for protecting purposes (against ultraviolet radiation, oxidants, extremes of heat and cold or antimicrobials produced by other microorganisms), for competing with other microorganisms (antimicrobial activity), for the acquisition of nutrients or for participating in light capturing during photosynthesis (Liu and Nizet, 2009). However, in order to be used as a source of pigment, some criteria must be obeyed:

Table 2. Representative microorganisms for commercially relevant biopigment production. For their chemical structure, see Fig. (3).

Dominant Color	Microorganism	Coloring substance	References
Red	*Monascus* sp.	Rubropunctamine Monascorubramine	[11]
	Rocella tinctoria (a lichen)	Orceins Blue to pink	[21]
	Phaffia sp.	Astaxanthin	[36]
	Haematococcus sp.	Astaxanthin	[36]
	Paracoccus carotinifaciens	Astaxanthin	[37]
	Dunaliella sp.,	β-carotene	[38]
	Blakeslea sp.	β-carotene	[39]
Orange	*Paracoccus* sp.	Zeaxanthin	[40]
Yellow	*Monascus* sp.	Monascin, Ankaflavin	[41]
	Ashbya sp.	Riboflavin	[42]
	Several microalgae	Xanthophylls	[43]
Green	All microalgae	Chlorophyll	[44]
Blue	Spirulina	Phycocyanin	[45]

1. The microorganism must be easy to cultivate and produce the pigment in high concentration;
2. the pigment must be stable;
3. the pigment must not be toxic;
4. the microorganism should not produce toxins, and
5. the microorganism itself should not be harmful.

These assumptions may seem quite obvious, but must be carefully addressed when one tries to develop new biopigments. Several colored substances produced by fungi, for example, are actually bioactive, having antibiotic and cytotoxic activity [35]. Table **2** shows the most commercially important microorganisms used for color production.

Most of the microbial biopigments permitted for food use are carotenoids, some of which may be produced synthetically (but then will be termed "nature-identical", and not "natural"), or from plant material (but then, with lower yield than using microorganisms).

The general process for production of a microbial biopigment consist of the following steps: medium and inoculum preparation, biomass production (cultivation), biomass separation (harvesting and concentration), biomass processing (extraction), eventual concentration from the broth (for hydrosoluble color substances), eventual partial purification of the raw extract, and formulation – which may include encapsulation, emulsification, solubilization or drying (Fig. **2**)

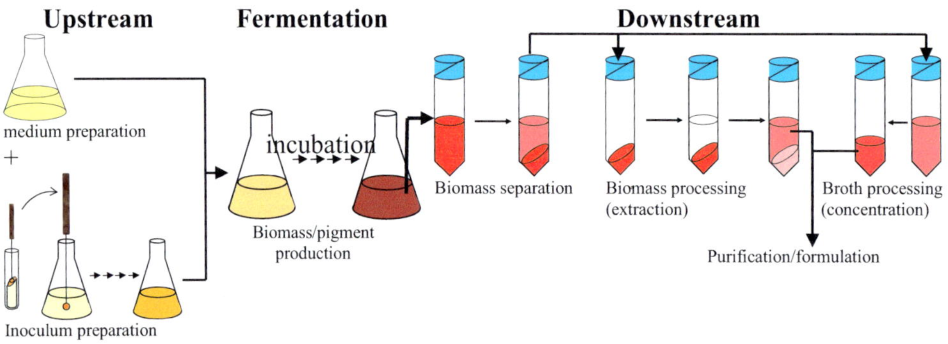

Fig. (2). General steps for producing microbial pigments, represented for a lab scale process.

The solvents permitted for processing of natural pigments are specified for each case by the regulatory agencies; the most commonly accepted are acetone, ethanol, methanol, propan-2-ol, hexane, and carbon dioxide; while less common solvents are vegetable oil, ethylene chloride, methylene chloride, dichloromethane, trichloroethylene and methyl ethyl ketone.

The structures of important biopigments are illustrated in Fig. (**3**), and their production process will be detailed in the next pages.

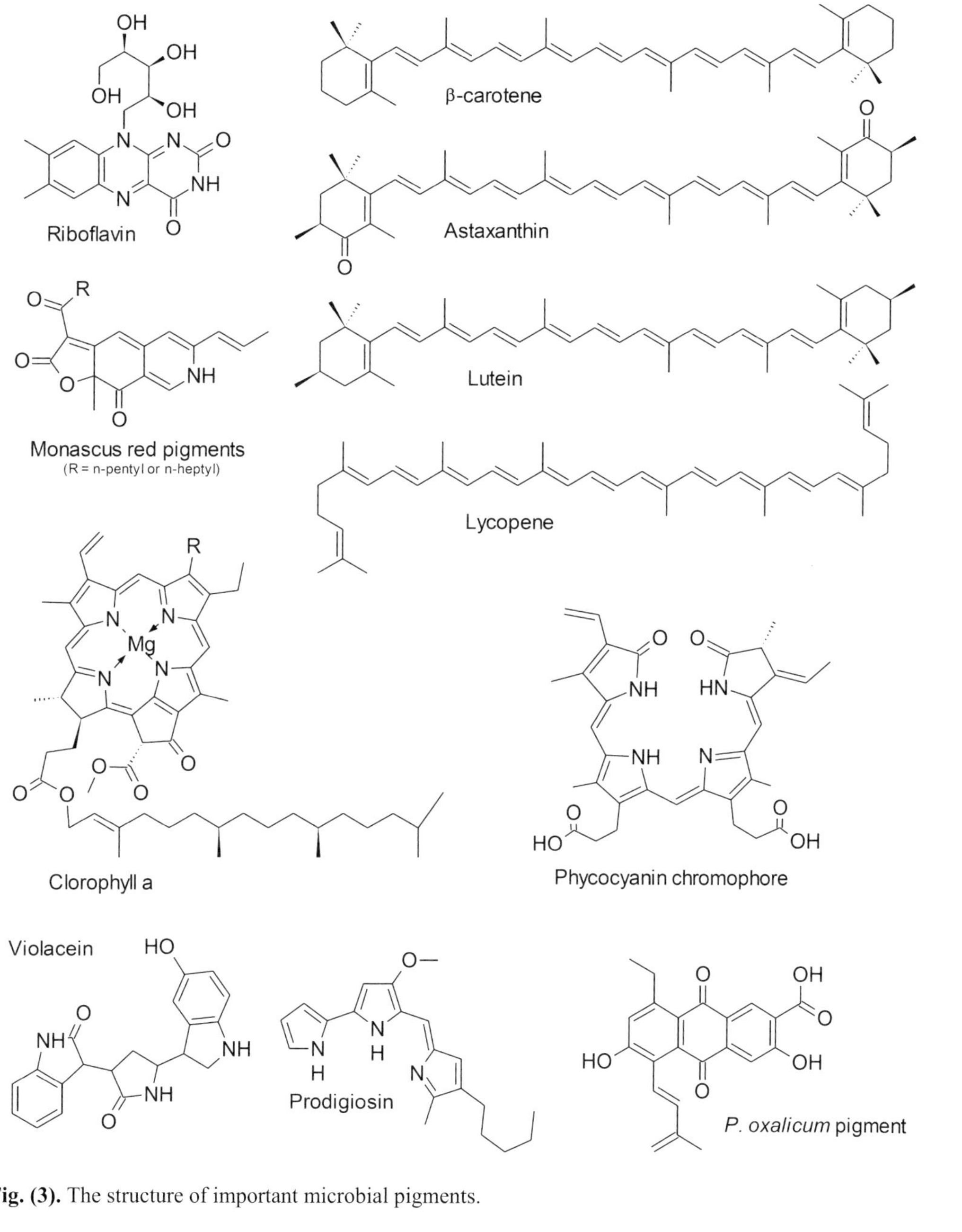

Fig. (3). The structure of important microbial pigments.

Monascus *Pigments*

These polyketide pigments are secondary metabolites produced by fungi from the genus *Monascus* as a mixture of substances whose color spans from yellow to purple red; the pigment is usually sold as a mixture. The six primary *Monascus* pigments are monascin and ankaflavin (yellow), rubropucntain and monascorubin (orange) and rubropunctamine and monascorubramine (red) [33, 46]. These pigments are not permitted in USA and Europe, but are traditionally used in oriental countries such as Japan, Thailand and China. The most important colored substances in *Monascus* extracts are bioactive, *i.e.* exhibiting antimicrobial, anti-inflammatory, and immunosuppressive activities [47]. There are also anti hyperlipidemic substances in the mix, and several strains produce the mycotoxin citrinin [46]. Therefore, it is unlikely that *Monascus* pigments will ever be permitted as a food color additive in (western) countries where it is not traditionally used, unless the production process is proven to eliminate specific molecules from the raw extract (citrinin and monacolin) and if the pigments' bioactivity is deemed low or beneficial to the products where they will be applied, *e.g.,* as a surface coloring for meat products.

The production of *Monascus* pigments may be done in Solid Substrate Fermentation (SSF) or in liquid media. There are several reports about the influence of cultivation conditions and media composition on the amount and proportion of the resulting pigment mix [41, 48 - 50]. One of the greatest challenges consists of producing the pigments free of citrinin, whose concentration varies for each strain and cultivation conditions. In this sense, [51] have employed mutagenesis and screening techniques to decrease the mycotoxins content. An example of a bioprocess for the production of such pigment will be illustrated in the sequence [11]:

- Biomass/pigment production: it may be done by SSF using steamed rice with 50-60% humidity as substrate, which is inoculated to 10^4-10^5 UFC with biomass from a previous solid or liquid fermentation, or with a spore suspension. The mixture is incubated at 30-32°C for 5-10 days. The fermented substrate may be dried and milled, or further processed for pigment extraction. In liquid media, *Monascus* may be cultivated in a medium composed by (g/L) starch (50), yeast extract (5.0), peptone (5.0), NH_4Cl (2.0), K_2HPO_4(1.5) and $MgSO_4.7H_2O$ (0.5). Dissolved oxygen should be kept around 60% of saturation and pH should be maintained at 6 for the whole production process (5-8 days).The broth containing part of the pigments is filtered before the pigment recovery, while the biomass, which also presents pigments, may be dried or processed for pigment extraction.The distribution of pigments on the broth or on the biomass depends on the strain and medium composition.

- Pigment extraction: For the process on SSF, the raw fermented substrate or the biomass may be extracted with methanol or 60% ethanol in water, followed by concentration by evaporation and filtering, giving a concentrated crude pigment. In case of submerged fermentation, the broth may be directly concentrated or extracted with a suitable solvent, such as butanol, which is then separated and concentrated.

Carotenoids

Carotenoids constitute the largest class of natural pigments, with more than 750 already described [52]. They are widely distributed in nature, conferring yellow-to-red colors to several microorganisms, vegetables (*e.g.,* papaya, guava, tomato, carrot, flowers *etc.*) and animals (*e.g.* birds, crustaceans and salmon). These are valuable ingredients to industry, since they are regarded as natural colorants to impart orange, yellow and red colors to food and beverages. Some of them (astaxanthin, lycopene, zeaxanthin and β-carotene) may be produced by microorganisms, as detailed in the next pages.

Astaxanthin Production

The ketocarotenoid astaxanthin is produced by several microorganisms and is part of the phytoplankton consumed by shrimps and salmonid fish, ultimately being responsible for the color of these animals. Astaxanthin supplementation for farmed fish became popular in the last decade, even for fish which usually do not have a strong color, such as farmed trouts [34]. Since astaxanthin is mainly used in aquaculture, most products consist of the dry biomass, eventually thermally processed to enhance digestibility. When the concentrated pigment is needed, the biomasses may be extracted with a solvent such as acetone, and the astaxanthin-rich extract is simply concentrated and dried.

Astaxanthin for natural additives is commercially produced using yeasts, microalgae or bacteria, as illustrated in the sequence:

- *Xanthophyllomyces dendrorhous*: this yeast may be grown on a culture medium composed of (in g/L): glucose (50), corn steep liquor (10), yeast extract (5.0), urea 1, KH_2PO_4 (0.5) and cultivated for 2-3 days at 22°C. The biomass may be separated by centrifugation or filtration, and then dried or the pigments may be extracted. The astaxanthin content in this yeast is typically around 0.03-0.1%, and current efforts are focused on enhancing its carotenoid content [53, 54]
- *Haematococcus pluvialis*: this microalga may be cultivated in an autotrophic medium composed of (in g/L): $NaNO_3$ (0.4), KH_2PO_4 (0.11) and 20 mL/L of soil extract, which is an organic matter-rich soil extracted with twice its weight of water. This culture medium is inoculated with 1.10^5 cells/mL (approximately 0.7g/L of biomass) and cultivated for 10-15 days under mild illumination (60 $\mu Mol.m^{-2}.s^{-1}$) and aeration (0.1-1 volumes of air/volume of culture medium). The air supplies CO_2 for the culture, although injection of supplementary CO_2 may be done. After a cell density above 1.10^6 cells/mL is reached, the illumination and salinity are increased (600 $\mu Mol.m^{-2}.s^{-1}$ and 0.5% salinity), which induces the carotenogenesis and the formation of astaxanthin-rich cysts. The biomass may be harvested by filtration or centrifugation after 5 days of the beginning of carotenogenesis [55]. The red cysts are used as a feed additive after drying, or are further processing for carotenoid extraction.
- *Paracoccus carotinifaciens*: the use of this bacteria for astaxanthin production has been recently authorized in EUA, EU and Norway. A typical culture medium is composed of (in g/L): sucrose (10), K_2HPO_4 (3.6), KH_2PO_4 (1.4), NaCl (8.8), $MgSO_4.7H_2O$ (0.7), peptone (2.0), yeast extract (1.0), $CaCl_2$ (0.07), $FeSO_4$ (0.015) [56]. The biomass is usually grown aerobically at pH 7, 28°C for 4 days. After harvesting the biomass (filtration), the biomass is dried or the pigments may be extracted.

β-Carotene Production

This carotenoid may be produced biotechnologically using several microorganisms. The two most important commercial sources of this pigment are the fungus *Blakeslea trispora* and the microalga *Dunaliella* sp., an halotolerant unicellular which grows in high salt concentration, can accumulate large amounts of β-carotene [57 - 59]. This biopigment is permitted by most regulating agencies. Currently, many companies are involved in the cultivation of *Dunaliella* for commercial production of this carotene for food, feed or pharmaceutical formulations [60]. The main steps for production of this carotenoid are described below.

- Production of biomass of *Blakeslea trispora*: spores of two mating types are mixed in the proportions 1:10 [58, 61], and the mixture is used as inoculum for submerged fermentation. A suitable culture medium usually contain (in g/L): glucose (50), vegetable oil (15), corn steep liquor (10), peptone (2.0), yeast extract (2.0), KH_2PO_4 (1.5), $MgSO_4.7H_2O$ (0.5) and sorbitan laurate (2.0). The microorganism is grown at 28°C and aeration of 0.5 to 1 vvm for 5 to 8 days. The composition of the carotenoid mixture and the productivity actually depend on the amount of oxygen transferred [62], therefore, agitation and aeration must be strictly controlled.
- Production of microalgal biomass: several species of *Dunaliella* sp. may be cultivated for β-carotene production. The ideal culture medium composition varies for each strain, but a suggested starting point would be seawater (artificial or natural) supplemented with 0.8 g/L $NaNO_3$, 0.22 g/L KH_2PO_4 and 30 mL/L of soil extract. Such a culture medium is inoculated with 1.10^5 cells/mL of *Dunaliella* (approximately 0.1g of biomass) and cultivated for 10-15 days under mild illumination (60 μMol.m^{-2}.s^{-1}) and aeration (0.1-1 volumes of air/volume of culture medium). As in *Haematococcus* cultures, the air supplies CO_2 for the culture and provides homogenization, but injection of supplementary CO_2 may be done. After a cell density reaches at least 3.10^6 cells/mL, nitrogen becomes limited and, under strong illumination (*e.g.,* 1700 μMol.m^{-2}.s^{-1} [63]), the carotenoid content of the cells will be increased. The main difficulties in this process are minimizing the contamination in open bioreactors and the control of illumination both for fast growth and carotenogenesis.
- Pigment extraction: after microbial growth, the resulting biomass is recovered from the broth (filtered or centrifuged), dried and extracted with a suitable solvent, such as ethyl acetate, isobutyl acetate, or carbon dioxide, which is then concentrated by evaporation. *Trans*-β-carotene can crystallize with variable amounts of *cis-* isomers and minor amounts of other carotenoids, such as xanthophylls, in the case of *Dunaliella*. Since carotenoids are prone to oxidation, an antioxidant (*e.g.* tocopherol) may be added to the biomass to be processed.

Photosynthesis-Associated Xanthophylls and Chlorophylls Production

In photosynthetic microorganisms, xanthophylls other than astaxanthin are essential for photoprotection [64]. Substances such as violaxanthin, anteraxanthin and zeaxanthin participate in the xanthophyll cycle, absorbing excess energy that would otherwise damage the cell [65]. These intracellular pigments may be industrially co-produced with other microalgal light harvesting molecules, such as phycocyanin. Their production is highly dependent of the microalgal species and the culture conditions, especially the light intensity. Its simplified production process for *Chlorella vulgaris* is described as the following steps:

- Biomass production: In a culture medium composed of (in g/L): $NaNO_3$ (1.0), KH_2PO_4 (0.4), $MgSO_4.7H_2O$ (0.1), $FeSO_4$ (0.01) and soil extract (30mL/L), an inoculum of $3.5.10^6$ cells/mL (approximately 0.1g/L) is added. The pH should be around 6.5. This microalga may be cultivated in open or closed bioreactors for 5-10 days at 25°C, with mild aeration (0.1-1vvm) and illumination (60-70 $\mu Mol.m^{-2}.s^{-1}$) [66]. The maximum biomass produced depends on the medium composition and the illumination used; higher irradiance may enhance xanthophyll production [67]. After cultivation, the biomass is harvested by flocculation followed by centrifugation [68].
- Pigment extraction: xanthophylls and chlorophylls are extracted from the dried biomass using an appropriated solvent, *e.g.* acetone, ethanol, diethyl ether, *etc.* Chlorophyll is soluble in moderately polar solvents, such as ethanol, while xanthophills are not – its solubility is much higher in low-polarity solvents [69]. Therefore, raw ethanolic extracts will contain low amounts of xanthophylls. The green pigment thus obtained may be converted to its cupric form, in which Mg^{2+} is substituted by Cu^{2+}, while the alkaline hydrolysis of the phytyl group by KOH or NaOH leads to green water-soluble chlorophyllins. On the other hand, the extraction microalgal biomass with a non-polar solvent, such as cyclohexane, yields a xanthophyll solution low in chlorophyll. The extracted pigments may be concentrated by solvent evaporation, or processed for further purification.

Riboflavin Production

Riboflavin is a vitamin with permitted use as a yellow pigment. It may be produced by bacteria and yeasts, but the most common sources are the fungi

Eremothecium ashbyii and *Ashbya gossypii*. The pigment is hydrosoluble and excreted by these fungi, being easily recovered from the broth. Some modified *B. subtilis* strains are also good sources of the biopigment [70]. The simplified production process requires the following steps:

- Cultivation: *A. gossypii* is inoculated into a fermentation medium composed of (in g/L): soy oil (50), gelatin (30), corn steep liquor (60), glycine (1.5), KH_2PO_4 (1.5), $CoCl_2.6H_2O$ (0.002), $MnCl_2.4H_2O$ (0.005), $ZnSO_4.7H_2O$ (0.01), MgS_2O_3 (0.001) [71]. The culture is maintained at 28°C for 7-8 days, at 200rpm in shakers.
- Pigment recovery: although riboflavin is hydrosoluble and excreted to the culture medium, part of the pigment remains bound to the fungal biomass. Heating the whole culture to 121°C helps to release the pigment [72]. The biomass is removed by filtration or centrifugation, and the pigment can be crystallized from the refrigerated broth. The crystals are acid-washed and dried. Since riboflavin is poorly soluble in water is low, part of the pigment – currently reaching concentration around 1g/L in the broth - may be already crystallized inside the biomass. Resolubilization, control of the crystallization and of the oxidized or reduced forms of riboflavin are the object of patents [72, 73]. The conversion of riboflavin (INS 101) to riboflavin phosphate (INS 101a) enhances its solubility and the use in aqueous products, such as juices.

Phycobilins

Besides chlorophyll and carotenoids, many types of microalgae produce other photosynthetic accessory pigments, such as phycoerythrin (red) and phycocyanin (blue), which are collectively known as phycobiliproteins [60, 74]. These are water soluble and proteins involved in light harvesting in cyanobacteria (blue-green algae, procaryotic), rhodophytes (red algae, eukaryotic), cryptomonads (biflagellate unicellular eukaryotic algae) and cyanelles (endosymbiotic plastid-like organelles) [75]. The main uses of such molecules seems to be as natural dyes/colorants and in clinical or research immunology, although an increasing number of investigations have shown on their health-promoting properties and broad range of pharmaceutical applications have been considered [74, 76].

The global market for phycobiliproteins was estimated at more than US$ 50 million in 1997, with prices ranging prices from US$ 3 to US$ 25/mg for native pigment [76]. Currently, many products containing phycobiliproteins are marketed by several companies [75]. In commercial scale, the cyanobacterium *Arthrospira* and the rhodophyte *Porphyridium* are the main microalgae used for the production of phycobiliproteins [75, 76]. Several articles describe processes for the production of these microalgae pigments: the production of (red) phycoerythrin and (blue) phycocyanin, mainly by *Porphyridium* sp. and by *Arthrospira* (*Spirulina*), respectively, has been reviewed in the past few years [60, 77, 78].

Phycocyanin Production

Phycocyanin is the blue antenna pigment of cyanobacteria. This pigment is part of an antenna complex which comprises other phycobilins and chlorophyll. It is the most important phycobilin, being commercially produced by strains of *Spirulina* sp., following a similar procedure to the one presented here:

- Biomass production: *Arthrospira (Spirulina)* sp. may be grow in several culture media, including agricultural residues such as cassava wastewater [79]. A classical synthetic medium, is the one described by [80], which is composed of (in g/L): $NaHCO_3$ (16.8), $NaNO_3$ (1.0), K_2HPO_4 (0.5), K_2SO_4 (1.0), NaCl (1.0), $CaCl_2$ (0.01), Na_2EDTA (0.08), $MgSO_4$ (0.04), $FeSO_4{\bullet}7H_2O$ (0.01), and 1.0 mL of each trace elements (TE) solutions (A and B) TE A (g/L): H_3BO_3 (2.86), $(NH_4)_6Mo_7O_{24}$ (0.02), $MnCl_2{\bullet}4H_2O$ (1.8), Cu_2SO_4 (0.08), $ZnSO_4{\bullet}7H_2O$ (0.22), $Co(NO_3)_2.6H_2O$ (0.05); TE B (g/L): NH_4NO_3 (0.23), $K_2Cr_2(SO_4)_4{\bullet}24H_2O$ (96), $NiSO_4{\bullet}7H_2O$ (47.8), $Na_2WO_4{\bullet}2H_2O$ (17.9), $Ti_2(SO_4)_3$ (40.0). The culture is cultivated under illumination of 25-40 $\mu Mol.m^{-2}.s^{-1}$ [81], and aeration of 0.1 to 0.5 vvm. Agitation through aeration is enough for this cyanobacteria, which is maintained at 35°C for 10-15 days. When the biomass reaches 2-3g/L, a partial harvesting by flocculation and filtration nets is carried out (the filtered medium, containing biomass, may be recycled) [45].
- Pigment extraction: as the pigments occur in the intracellular environment, the biomass must be permeabilized using a cell disruption equipment or simply by successive freeze-thawing (4 cycles). After such procedures a phycocyanin-rich aqueous solution may be separated from the cell debris. The solution may be

concentrated and partially purified by ammonium sulfate precipitation at 50% saturation, or directly through ultrafiltration, *e.g.* with a membrane with a 50kDa cutoff [82 - 84]. This process gives a food grade pigment; however, if the purified pigment is needed, processes such as chromatography and aqueous two-phases extraction may be used [85, 86] .

Other Pigments

Several other colored substances may be produced by microorganisms, and in fact the search for new biotechnologically derived color additives is very active [60, 87]. However, the regulatory process for approval of new pigments is complex: new additives must prove to be safe, and the bioactivity showed by several natural pigments is often undesirable. Several of the pigments from microalgae have only recently been approved for food use: for example, FDA has granted GRAS notices to dry biomass and C-phycocyanin extracts from *Arthrospira platensis* in 2012, for *Dunaliella bardawil* in 2011, and *Haematoccocus pluvialis* extracts in 2010 [88].

Already known pigments are easier to prove to be safe. Therefore, modifications of well stablished processes, already used for production of one industrial biopigment, are often considered as a strategy for the development of new products. For example, the red polyketide mix of *Monascus* pigments may be further fractionated into yellow and orange fractions [89]. Additionally, permitted color additives, such as lycopene or lutein, may be produced using novel strains(isolates) or modified organisms. Lycopene, for example, is traditionally extracted from tomato; but since this is an intermediary molecule in the biosynthesis of several other carotenoids, the use of suitable inhibitors in selected microorganism may stop the carotenoid synthesis cascade, leading to its accumulation – as is the case in a process using in *B. trispora,* traditionally used for β-carotene production [90]. In the same way, zeaxanthin, which occurs in several vegetable sources, may be produced by certain *Paracoccus* strains (microorganism already permitted for astaxanthin production) [40]. Also for zeaxanthin, a mutant *Dunaliella* strain (also traditionally used for β-carotene production) may be used in the future [91].

Some fungi are also reported as producers of commercial biopigments, such as

Arpink Red^{TM} produced by *Penicillium oxalicum* var. Armeniaca CCM 8242 by means of a patented process (Sardaryan, 2002). Other *Penicillium* strains are also able to generate yellow-to-bluish-green colored molecules, as reviewed in Mapari *et al.* (2005), as well as yellow, orange, and purple-red *Monascus*-like pigments free of citrinin (Mapari *et al.*, 2009).

Bikaverin is another fungal reddish pigment, which is claimed to present antitumor properties and antibiotic activity against a variety of organisms, including protozoa (especially *Leishmania brasiliensis*). It is produced by *Gibberella fujikuroi* (anamorph of *Fusarium fujikuroi*) or other *Fusarium* species using fluidized bed [92] or airlift bioreactors [93].

A variety of red, orange and yellow natural colors are already approved and in use, but natural blue and purple colorants are scarce (Bicas *et al.*, 2013), since the occurrence of blue colored compounds in the natural world is relatively rare [94]. Therefore, screening of new blue pigment-producing strains is important. Several authors report the great diversity of relatively unknown pigments [95, 96], and active search is necessary because besides finding microorganisms capable of excreting colored substances in high concentrations, it is essential that these substances do not present undesirable bioactivity for food applications. Recently studied pigments are strongly colored, but bioactive: Pyocianin (blue), for example, is cytotoxic [97] while prodigiosin (red) is immunosuppressant [98] and violacein (purple) presents antibiotic properties [99]. Thus, some of these pigments may never be approved for food use as color additives, although they may be suitable for coloring applications in cosmetics or textiles.

However, with the vast microorganism biodiversity and the high throughput methods developed such as in [100], it is likely that new pigments will keep being discovered, and among these some are likely to be safe alternatives for food use. It may be the case of a mixture of hydrosoluble blue pigments with an actinorhodin-like structure, produced by *Streptomyces coelicolor* at a concentration around 3g/L [101]. The pigment has a pH-dependent color and has shown low acute toxicity (LD_{50} in mice = 15000mg/kg).

FORMULATION OF PIGMENT ADDITIVES

The formulation of color additives depends on the properties of the pigment and on their expected application. In general, food industry demands hydrosoluble pigments (*e.g.* for soft drinks and candies), liposoluble pigments (*e.g.* for dairy products such as cheese and butter) and insoluble powdered pigments (for example, lakes for confectionery). The pure color substances may be predominantly liposoluble (*e.g.* carotenoids) or hydrosoluble (*e.g.* riboflavin); therefore, formulation strategies are necessary in order to modify the native solubility of the color additive. Table **3** shows some processing alternatives for formulation of such pigments into color additives suitable for food processing applications:

Table 3. Strategies for formulation of food pigments for different applications.

Desired form of dispersion	**Type of natural pigment molecule**	
	Hydrosoluble	**Liposoluble**
Hydrosoluble additive formulation	Aqueous concentrate, paste or powder. Example: copper chlorophyllin	Pigment dispersed in sugar and emulsifying agent, dried (powder). Example: β-carotene
Liposoluble additive formulation	Water in oil emulsion, with emulsifying agent. Example: Monascus pigments.	Pigment dissolved in vehicle, such as vegetable oil. Example: lycopene.
Lake pigments	Pigment dried over powdered vehicle, such as alumina. Carmine lake	

(Table **3**) indicates that pigments having solubility properties apparently incompatible with the product to be colored are not necessarily rejected. A liposoluble carotenoid, for example, may be used in the form of microemulsion in beverages (hydrosoluble matrix) [102, 103]. However, the stability of the formulation system must be verified throughout the development of the product to avoid, for example, the migration of pigments from its lakes or emulsions to the food matrix [104].

Most pigments have multiple insaturations and are therefore prone to oxidation. Their stability is improved by the use of antioxidants, most commonly tocopherol [105, 106]. Pigment stability is also affected by enzymes naturally present in the ingredients, processing and storage temperatures [107, 108], and careful process development is vital in order to define when color additives are to be added, so as to reduce color alteration.

PERSPECTIVES

The 20th century observed the rise and decline of the number of permitted synthetic colors in foods. The tendency for using natural food additives is likely to continue: the growing research on biopigments tends to broaden the palette of natural colors accessible for food and feed applications, with the development of new formulations for traditional natural pigments, and the discovery of new, safe color additives. In particular, it is probable that future developments of food colors will derive from a) strain selection, b) process development and c) genetic modification of known strains.

While strain selection is likely to assure new sources of known pigments: carotenoid producing bacteria, yeast and microalgae will remain important, but with a biodiversity comprising hundreds of thousands of microorganisms with unique characteristics, it is just a matter of time until the next "*Paracoccus xanthinifaciens*" or "*Dunaliella bardawil*" will be found and demonstrated to be a promising source of biopigments.

The process development for biopigment production is essential, especially for economic reasons (enhancing concentration, productivity and recovery) and product quality (separation of specific colors, warranting low contaminant content, and stabilization of formulation); therefore, active research is important not only in developing new colors, but also in developing industrial processes.

Finally, the genetic manipulation of known strains is already enabling the access to the whole palette of carotenoids produced by microorganisms and plant cultures (lycopene and zeaxanthin are the most important cases, as in [109, 110], and this approach may soon enable the production of plant pigments (selected anthocyanins, for example, such as in [111] in microbial systems, with high productivities and control.

CONFLICT OF INTEREST

The authors confirm that they have no conflict of interest to declare for this publication.

ACKNOWLEDGEMENTS

Decleared none.

REFERENCES

[1] Spence, C. Does food color influence taste and flavor perception in humans? *Chemosens. Percept.,* **2010**, *3*(1), 68-84. [http://dx.doi.org/10.1007/s12078-010-9067-z]

[2] Garber, L.L., Jr; Hyatt, E.M.; Starr, R.G., Jr The effects of food color on perceived flavor. *J. Mark. Theory Pract.,* **2000**, *8*(4), 59-72. [http://dx.doi.org/10.1080/10696679.2000.11501880]

[3] Zampini, M. The multisensory perception of flavor: Assessing the influence of color cues on flavor discrimination responses. *Food Qual. Prefer.,* **2007**, *18*(7), 975-984. [http://dx.doi.org/10.1016/j.foodqual.2007.04.001]

[4] Delgado-Vargas, F.; Paredes-López, O. *Natural colorants for food and nutraceutical uses*; CRC Press, **2002**. [http://dx.doi.org/10.1201/9781420031713]

[5] Colors, S.F. *Percepção de Cores.,* 2014 [[cited 2014 08/12/2014]]; Available from: http://www.sensientfoodcolors.com/color-insight/

[6] Bobbio, P.A.; Bobbio, F.O. *Química do processamento de alimentos*; Liv. Varela, **1992**.

[7] McKone, H.T. The history of food colorants before aniline dyes. *Bull. Hist. Chem.,* **1991**, *10*, 6.

[8] AACC. CHAPTER 1: Overview. In: *Colorants*; American Association of Cereal Chemists, **1999**; pp. 1-5. [http://dx.doi.org/10.1094/1891127004.001]

[9] Burrows, J. Palette of our palates: a brief history of food coloring and its regulation. *Comprehensive Reviews in Food Science and Food Safety,* **2009**, *8*(4), 394-408. [http://dx.doi.org/10.1111/j.1541-4337.2009.00089.x]

[10] Hart, F.L. Adulteration of food before 1906. *Food Drug Cosmetic Law,* **1952**, *7*(1)

[11] Carvalho, J. Production of Pigments. In: *Current Developments in Solid-state Fermentation*; Pandey, A.; Soccol, C.; Larroche, C., Eds.; Springer New York, **2008**; pp. 337-355. [http://dx.doi.org/10.1007/978-0-387-75213-6_15]

[12] Prado, M.A.; Godoy, H.T. Corantes artificiais em alimentos. *Alimentos e Nutricao Araraquara.,* **2009**, *14*(2)

[13] Sharma, V.; McKone, H.T.; Markow, P.G. A Global Perspective on the History, Use, and Identification of Synthetic Food Dyes. *J. Chem. Educ.,* **2010**, *88*(1), 24-28. [http://dx.doi.org/10.1021/ed100545v]

[14] Rodriguez-Amaya, D.B. *A guide to carotenoid analysis in foods*; C DC: ILSI press Washington^ eD,

2001.

[15] Markakis, P. *Anthocyanins as food colors*; Elsevier, **2012**.

[16] Harborne, J.B. Spectral methods of characterizing anthocyanins. *Biochem. J.,* **1958**, *70*(1), 22-28. [http://dx.doi.org/10.1042/bj0700022] [PMID: 13584295]

[17] Sousa, M.M.; Miguel, C.; Rodrigues, I.; Parola, A.J.; Pina, F.; Seixas de Melo, J.S.; Melo, M.J. A photochemical study on the blue dye indigo: from solution to ancient Andean textiles. *Photochem. Photobiol. Sci.,* **2008**, *7*(11), 1353-1359. [http://dx.doi.org/10.1039/b809578g] [PMID: 18958322]

[18] GOSWAMI. Spectrophotometric studies on the interaction between laccaic acid and DNA , **1988**; pp. 30-32. Current science assn cv raman avenue, PO Box 8005, Bangalore 560 080, India.

[19] Bettinger, C.; Zimmermann, H.W. New investigations on hematoxylin, hematein, and hematein-aluminium complexes. II. Hematein-aluminium complexes and hemalum staining. *Histochemistry,* **1991**, *96*(3), 215-228. [http://dx.doi.org/10.1007/BF00271540] [PMID: 1717413]

[20] Ranganna, S. *Handbook of analysis and quality control for fruit and vegetable products*; Tata McGraw-Hill Education, **1986**.

[21] Beecken, H.; Gottschalk, E.M.; v Gizycki, U.; Krämer, H.; Maassen, D.; Matthies, H.G.; Musso, H.; Rathjen, C.; Zdhorszky, U. Orcein and litmus. *Biotech. Histochem.,* **2003**, *78*(6), 289-302. [http://dx.doi.org/10.1080/10520290410001671362] [PMID: 15473576]

[22] Cha, G.S.; Meyerhoff, M.E. Solid phase enzyme-linked competitive binding assay for riboflavin. *Anal. Biochem.,* **1988**, *168*(1), 216-227. [http://dx.doi.org/10.1016/0003-2697(88)90031-0] [PMID: 3364714]

[23] Escribano, J.; Alonso, G.L.; Coca-Prados, M.; Fernandez, J.A. Crocin, safranal and picrocrocin from saffron (Crocus sativus L.) inhibit the growth of human cancer cells in vitro. *Cancer Lett.,* **1996**, *100*(1-2), 23-30. [http://dx.doi.org/10.1016/0304-3835(95)04067-6] [PMID: 8620447]

[24] Attokaran, M. *Natural food flavors and colorants*; John Wiley & Sons, **2011**. [http://dx.doi.org/10.1002/9780470959152]

[25] Mayer, H.; Rüttimann, A. Synthese von optisch aktiven, natürlichen Carotinoiden und strukturell verwandten Naturprodukten. IV. Synthese von (3R, 3′ R, 6′ R)□Lutein. Vorläufige Mitteilung. *Helv. Chim. Acta,* **1980**, *63*(6), 1451-1455. [http://dx.doi.org/10.1002/hlca.19800630614]

[26] Roy, S. *Phytoplankton pigments: characterization, chemotaxonomy and applications in oceanography*; Cambridge University Press, **2011**. [http://dx.doi.org/10.1017/CBO9780511732263]

[27] FDA. U.S.F.a.D.A *Color Additive Status List,* **2009**.

[28] JECFA. J.F.W.E.C.o.F.A. *Combined Compendium of Food Additive Specifications,* **2014**. FAO - Food and Agriculture Organization of the United Nations

[29] O'Neil, M.J. *The Merck index: an encyclopedia of chemicals, drugs, and biologicals*; RSC Publishing,

2013.

[30] Teng, S.; Feldheim, W. Analysis of anka pigments by liquid chromatography with diode array detection and tandem mass spectrometry. *Chromatographia,* **1998**, *47*(9-10), 529-536. [http://dx.doi.org/10.1007/BF02467490]

[31] Cormier, F. Food Colorants from Plant Cell Cultures. In: *Functionality of Food Phytochemicals*; Johns, T.; Romeo, J., Eds.; Springer, US, **1997**; pp. 201-222. [http://dx.doi.org/10.1007/978-1-4615-5919-1_9]

[32] Dörnenburg, H.; Knorr, D. Generation of colors and flavors in plant cell and tissue cultures. *Crit. Rev. Plant Sci.,* **1996**, *15*(2), 141-168. [http://dx.doi.org/10.1080/07352689.1996.10393184]

[33] Babitha, S. Microbial pigments. In: *Biotechnology for agro-industrial residues utilisation*; Springer, **2009**; pp. 147-162. [http://dx.doi.org/10.1007/978-1-4020-9942-7_8]

[34] De Carvalho, J.C. Microbial Pigments. In: *Biotransformation of Waste Biomass into High Value Biochemicals*; Springer, **2014**; pp. 73-97. [http://dx.doi.org/10.1007/978-1-4614-8005-1_4]

[35] Limón, M.C.; Rodríguez-Ortiz, R.; Avalos, J. Bikaverin production and applications. *Appl. Microbiol. Biotechnol.,* **2010**, *87*(1), 21-29. [http://dx.doi.org/10.1007/s00253-010-2551-1] [PMID: 20376635]

[36] Domínguez-Bocanegra, A.R.; Ponce-Noyola, T.; Torres-Muñoz, J.A. Astaxanthin production by Phaffia rhodozyma and Haematococcus pluvialis: a comparative study. *Appl. Microbiol. Biotechnol.,* **2007**, *75*(4), 783-791. [http://dx.doi.org/10.1007/s00253-007-0889-9] [PMID: 17333170]

[37] Tsubokura, A.; Yoneda, H.; Mizuta, H. Paracoccus carotinifaciens sp. nov., a new aerobic gram-negative astaxanthin-producing bacterium. *Int. J. Syst. Bacteriol.,* **1999**, *49*(Pt 1), 277-282. [http://dx.doi.org/10.1099/00207713-49-1-277] [PMID: 10028273]

[38] Ben-Amotz, A.; Shaish, A.; Avron, M. Mode of action of the massively accumulated β-carotene of Dunaliella bardawil in protecting the alga against damage by excess irradiation. *Plant Physiol.,* **1989**, *91*(3), 1040-1043. [http://dx.doi.org/10.1104/pp.91.3.1040] [PMID: 16667108]

[39] Kim, S-W.; Seo, W-T.; Park, Y-H. Enhanced production of b-carotene from Blakeslea trispora with Span 20. *Biotechnol. Lett.,* **1997**, *19*(6), 557-560. [http://dx.doi.org/10.1023/A:1018397606101]

[40] Berry, A.; Janssens, D.; Hümbelin, M.; Jore, J.P.; Hoste, B.; Cleenwerck, I.; Vancanneyt, M.; Bretzel, W.; Mayer, A.F.; Lopez-Ulibarri, R.; Shanmugam, B.; Swings, J.; Pasamontes, L. Paracoccus zeaxanthinifaciens sp. nov., a zeaxanthin-producing bacterium. *Int. J. Syst. Evol. Microbiol.,* **2003**, *53*(Pt 1), 231-238. [http://dx.doi.org/10.1099/ijs.0.02368-0] [PMID: 12656178]

[41] Carvalho, J.C. Effect of substrates on the production of Monascus biopigments by solid-state fermentation and pigment extraction using different solvents. *Indian J. Biotechnol.,* **2007**, *6*(2), 194.

[42] Stahmann, K-P.; Revuelta, J.L.; Seulberger, H. Three biotechnical processes using *Ashbya gossypii*, Candida famata, or Bacillus subtilis compete with chemical riboflavin production. *Appl. Microbiol. Biotechnol.,* **2000**, *53*(5), 509-516.
[http://dx.doi.org/10.1007/s002530051649] [PMID: 10855708]

[43] Jin, E. Xanthophylls in microalgae: from biosynthesis to biotechnological mass production and application. *J. Microbiol. Biotechnol.,* **2003**, *13*(2), 165-174.

[44] Hosikian, A. Chlorophyll extraction from microalgae: A review on the process engineering aspects. *Int. J. Chem. Eng.,* **2010**.
[http://dx.doi.org/10.1155/2010/391632]

[45] Walter, A. Study of phycocyanin production from Spirulina platensis under different light spectra. *Braz. Arch. Biol. Technol.,* **2011**, *54*, 675-682.
[http://dx.doi.org/10.1590/S1516-89132011000400005]

[46] Carvalho, J.C. Biopigments from Monascus: strains selection, citrinin production and color stability. *Braz. Arch. Biol. Technol.,* **2005**, *48*(6), 885-894.
[http://dx.doi.org/10.1590/S1516-89132005000800004]

[47] Lin, Y-L.; Wang, T.H.; Lee, M.H.; Su, N.W. Biologically active components and nutraceuticals in the Monascus-fermented rice: a review. *Appl. Microbiol. Biotechnol.,* **2008**, *77*(5), 965-973.
[http://dx.doi.org/10.1007/s00253-007-1256-6] [PMID: 18038131]

[48] Babitha, S. Effect of light on growth, pigment production and culture morphology of Monascus purpureus in solid-state fermentation. *World J. Microbiol. Biotechnol.,* **2008**, *24*(11), 2671-2675.
[http://dx.doi.org/10.1007/s11274-008-9794-3]

[49] Hajjaj, H.; François, J.M.; Goma, G.; Blanc, P.J. Effect of amino acids on red pigments and citrinin production in Monascus ruber. *J. Food Sci.,* **2012**, *77*(3), M156-M159.
[http://dx.doi.org/10.1111/j.1750-3841.2011.02579.x] [PMID: 22384962]

[50] Feng, Y.; Shao, Y.; Chen, F. Monascus pigments. *Appl. Microbiol. Biotechnol.,* **2012**, *96*(6), 1421-1440.
[http://dx.doi.org/10.1007/s00253-012-4504-3] [PMID: 23104643]

[51] Wang, J-J.; Lee, C-L.; Pan, T-M. Modified mutation method for screening low citrinin-producing strains of *Monascus purpureus* on rice culture. *J. Agric. Food Chem.,* **2004**, *52*(23), 6977-6982.
[http://dx.doi.org/10.1021/jf049783o] [PMID: 15537306]

[52] Britton, G.; Liaaen-Jensen, S.; Pfander, H. *Carotenoids: handbook*; Springer, **2004**.
[http://dx.doi.org/10.1007/978-3-0348-7836-4]

[53] Schmidt, I.; Schewe, H.; Gassel, S.; Jin, C.; Buckingham, J.; Hümbelin, M.; Sandmann, G.; Schrader, J. Biotechnological production of astaxanthin with Phaffia rhodozyma/Xanthophyllomyces dendrorhous. *Appl. Microbiol. Biotechnol.,* **2011**, *89*(3), 555-571.
[http://dx.doi.org/10.1007/s00253-010-2976-6] [PMID: 21046372]

[54] Lin, G. Genetic Manipulation of Xanthophyllomyces dendrorhous and Phaffia rhodozyma. In: *Microbial Carotenoids From Fungi*; Springer, **2012**; pp. 235-249.
[http://dx.doi.org/10.1007/978-1-61779-918-1_16]

[55] Ghiggi, V. *Estudo do crescimento e inducao da producao do pigmento staxantantina por Haematococcus pluvialis*; Dissertacao (Mestrado)-Programa de Pos-Graduacao em Engenharia de Bioprocessos e Biotecnologia, Universidade Federal do Parana, Curitiba, **2007**.

[56] Tanaka, T.; Ide, T. *Novel microorganism and method for producing carotenoid using the same*; Google Patents, **2006**.

[57] Ribeiro, B.D.; Barreto, D.W.; Coelho, M.A. Technological aspects of β-carotene production. *Food and Bioprocess Technology,* **2011**, *4*(5), 693-701. [http://dx.doi.org/10.1007/s11947-011-0545-3]

[58] Varzakakou, M.; Roukas, T.; Kotzekidou, P. Effect of the ratio of (+) and (−) mating type of Blakeslea trispora on carotene production from cheese whey in submerged fermentation. *World J. Microbiol. Biotechnol.,* **2010**, *26*(12), 2151-2156. [http://dx.doi.org/10.1007/s11274-010-0398-3]

[59] Kleinegris, D.M.; Janssen, M.; Brandenburg, W.A.; Wijffels, R.H. Continuous production of carotenoids from Dunaliella salina. *Enzyme Microb. Technol.,* **2011**, *48*(3), 253-259. [http://dx.doi.org/10.1016/j.enzmictec.2010.11.005] [PMID: 22112908]

[60] Dufossé, L. Microorganisms and microalgae as sources of pigments for food use: a scientific oddity or an industrial reality? *Trends Food Sci. Technol.,* **2005**, *16*(9), 389-406. [http://dx.doi.org/10.1016/j.tifs.2005.02.006]

[61] López-Nieto, M.J.; Costa, J.; Peiro, E.; Méndez, E.; Rodríguez-Sáiz, M.; de la Fuente, J.L.; Cabri, W.; Barredo, J.L. Biotechnological lycopene production by mated fermentation of Blakeslea trispora. *Appl. Microbiol. Biotechnol.,* **2004**, *66*(2), 153-159. [http://dx.doi.org/10.1007/s00253-004-1669-4] [PMID: 15248041]

[62] Mantzouridou, F.; Roukas, T.; Achatz, B. Effect of oxygen transfer rate on β-carotene production from synthetic medium by< i> Blakeslea trispora</i> in shake flask culture. *Enzyme Microb. Technol.,* **2005**, *37*(7), 687-694. [http://dx.doi.org/10.1016/j.enzmictec.2005.02.020]

[63] Kleinegris, D.M.; Janssen, M.; Brandenburg, W.A.; Wijffels, R.H. The selectivity of milking of Dunaliella salina. *Mar. Biotechnol. (NY),* **2010**, *12*(1), 14-23. [http://dx.doi.org/10.1007/s10126-009-9195-0] [PMID: 19475448]

[64] Demmig-Adams, B.; Adams, W.W., III The role of xanthophyll cycle carotenoids in the protection of photosynthesis. *Trends Plant Sci.,* **1996**, *1*(1), 21-26. [http://dx.doi.org/10.1016/S1360-1385(96)80019-7]

[65] Goss, R.; Jakob, T. Regulation and function of xanthophyll cycle-dependent photoprotection in algae. *Photosynth. Res.,* **2010**, *106*(1-2), 103-122. [http://dx.doi.org/10.1007/s11120-010-9536-x] [PMID: 20224940]

[66] Sorokin, C.; Krauss, R.W. The effects of light intensity on the growth rates of green algae. *Plant Physiol.,* **1958**, *33*(2), 109-113. [http://dx.doi.org/10.1104/pp.33.2.109] [PMID: 16655087]

[67] Buma, A.G. PAR acclimation and UVBR-induced DNA damage in Antarctic marine microalgae. *Mar. Ecol. Prog. Ser.,* **2006**, *315*, 33.

[http://dx.doi.org/10.3354/meps315033]

[68] Salim, S.; Bosma, R.; Vermuë, M.H.; Wijffels, R.H. Harvesting of microalgae by bio-flocculation. *J. Appl. Phycol.,* **2011**, *23*(5), 849-855. [http://dx.doi.org/10.1007/s10811-010-9591-x] [PMID: 21957329]

[69] Budavari, S. *The Merck Index*; Merck Research Laboratories Division of Merck & Co. Inc.: Whitehouse Station, NJ, **1996**, p. 1172.

[70] Wu, Q-L.; Chen, T.; Gan, Y.; Chen, X.; Zhao, X.M. Optimization of riboflavin production by recombinant Bacillus subtilis RH44 using statistical designs. *Appl. Microbiol. Biotechnol.,* **2007**, *76*(4), 783-794. [http://dx.doi.org/10.1007/s00253-007-1049-y] [PMID: 17576552]

[71] Lim, S.H. Improvement of riboflavin production using mineral support in the culture of Ashbya gossypii. *Food Technol. Biotechnol.,* **2003**, *41*(2), 137-144.

[72] Gloor, A. *Process for the purification of riboflavin*; Google Patents, **2010**.

[73] Wagner, G. *Purification and crystallization of riboflavin*; Google Patents, **2000**.

[74] Milledge, J.J. Commercial application of microalgae other than as biofuels: a brief review. *Rev. Environ. Sci. Biotechnol.,* **2011**, *10*(1), 31-41. [http://dx.doi.org/10.1007/s11157-010-9214-7]

[75] Sekar, S.; Chandramohan, M. Phycobiliproteins as a commodity: trends in applied research, patents and commercialization. *J. Appl. Phycol.,* **2008**, *20*(2), 113-136. [http://dx.doi.org/10.1007/s10811-007-9188-1]

[76] Spolaore, P.; Joannis-Cassan, C.; Duran, E.; Isambert, A. Commercial applications of microalgae. *J. Biosci. Bioeng.,* **2006**, *101*(2), 87-96. [http://dx.doi.org/10.1263/jbb.101.87] [PMID: 16569602]

[77] Eriksen, N.T. Production of phycocyanin--a pigment with applications in biology, biotechnology, foods and medicine. *Appl. Microbiol. Biotechnol.,* **2008**, *80*(1), 1-14. [http://dx.doi.org/10.1007/s00253-008-1542-y] [PMID: 18563408]

[78] Kaur, G. *Microalgae: A Source of Natural Colours*; Algal Biology and Biotechnology, **2009**.

[79] Walter, A. *Estudo do processo biotecnológico para obtenção de ficocianina a partir da microalga Spirulina Platensis sob diferentes condições de cultivo .,* **2012**.

[80] Zarrouk, C. *Contribution à l'étude d'une cyanophycée - influence de divers facteurs physique et chimiques sur la croissance et la photosynthèse de Spirulina maxima*; Université de Paris, **1966**.

[81] Kumar, M.; Kulshreshtha, J.; Singh, G.P. Growth and biopigment accumulation of *cyanobacterium Spirulina* platensis at different light intensities and temperature. *Braz. J. Microbiol.,* **2011**, *42*(3), 1128-1135. [http://dx.doi.org/10.1590/S1517-83822011000300034] [PMID: 24031731]

[82] Silva, L.A. *Estudo do processo biotecnologico de producao, extracao e recuperacao do pigmento ficocianina da Spirulina platensis,* **2009**.

[83] Herrera, A. Recovery of c-phycocyanin from the cyanobacteriumSpirulina maxima. *J. Appl. Phycol.,* **1989**, *1*(4), 325-331.

[http://dx.doi.org/10.1007/BF00003469]

[84] Chaiklahan, R.; Chirasuwan, N.; Loha, V.; Tia, S.; Bunnag, B. Separation and purification of phycocyanin from Spirulina sp. using a membrane process. *Bioresour. Technol.,* **2011**, *102*(14), 7159-7164.
[http://dx.doi.org/10.1016/j.biortech.2011.04.067] [PMID: 21570281]

[85] Liao, X. Purification of C-Phycocyanin from Spirulina platensis by Single-Step Ion-Exchange Chromatography. *Chromatographia,* **2011**, *73*(3-4), 291-296.
[http://dx.doi.org/10.1007/s10337-010-1874-5]

[86] Patil, G.; Raghavarao, K. Aqueous two phase extraction for purification of C-phycocyanin. *Biochem. Eng. J.,* **2007**, *34*(2), 156-164.
[http://dx.doi.org/10.1016/j.bej.2006.11.026]

[87] Mapari, S.A.; Thrane, U.; Meyer, A.S. Fungal polyketide azaphilone pigments as future natural food colorants? *Trends Biotechnol.,* **2010**, *28*(6), 300-307.
[http://dx.doi.org/10.1016/j.tibtech.2010.03.004] [PMID: 20452692]

[88] FDA. U.S.F.a.D.A. *GRAS Notice Inventory.,* **2014**.

[89] Patakova, P. Monascus secondary metabolites: production and biological activity. *J. Ind. Microbiol. Biotechnol.,* **2013**, *40*(2), 169-181.
[http://dx.doi.org/10.1007/s10295-012-1216-8] [PMID: 23179468]

[90] Choudhari, S.M.; Ananthanarayan, L.; Singhal, R.S. Use of metabolic stimulators and inhibitors for enhanced production of β-carotene and lycopene by Blakeslea trispora NRRL 2895 and 2896. *Bioresour. Technol.,* **2008**, *99*(8), 3166-3173.
[http://dx.doi.org/10.1016/j.biortech.2007.05.051] [PMID: 17637505]

[91] Jin, E.; Feth, B.; Melis, A. A mutant of the green alga Dunaliella salina constitutively accumulates zeaxanthin under all growth conditions. *Biotechnol. Bioeng.,* **2003**, *81*(1), 115-124.
[http://dx.doi.org/10.1002/bit.10459] [PMID: 12432587]

[92] Chávez-Parga, M.C. Mathematical description of bikaverin production in a fluidized bed bioreactor. *World J. Microbiol. Biotechnol.,* **2005**, *21*(5), 683-688.
[http://dx.doi.org/10.1007/s11274-004-3854-0]

[93] Chavez-Parga, M.C. Kinetic of the gibberellic acid and bikaverin production in an airlift bioreactor. *Process Biochem.,* **2008**, *43*(8), 855-860.
[http://dx.doi.org/10.1016/j.procbio.2008.04.007]

[94] Newsome, A.G.; Murphy, B.T.; Breemen, R.B. Isolation and Characterization of Natural Blue Pigments from Underexplored Sources. In: *Physical Methods in Food Analysis*; American Chemical Society, **2013**; pp. 105-125.
[http://dx.doi.org/10.1021/bk-2013-1138.ch008]

[95] Sajilata, M.; Singhal, R. Isolation and stabilisation of natural pigments for food applications. *Stewart Postharvest Review,* **2006**, *2*(5), 1-29.
[http://dx.doi.org/10.2212/spr.2006.5.11]

[96] Venil, C.K.; Zakaria, Z.A.; Ahmad, W.A. Bacterial pigments and their applications. *Process Biochem.,* **2013**, *48*(7), 1065-1079.

[http://dx.doi.org/10.1016/j.procbio.2013.06.006]

[97] O'Malley, Y.Q.; Abdalla, M.Y.; McCormick, M.L.; Reszka, K.J.; Denning, G.M.; Britigan, B.E. Subcellular localization of Pseudomonas pyocyanin cytotoxicity in human lung epithelial cells. *Am. J. Physiol. Lung Cell. Mol. Physiol.,* **2003**, *284*(2), L420-L430. [http://dx.doi.org/10.1152/ajplung.00316.2002] [PMID: 12414438]

[98] Han, S.B.; Kim, H.M.; Kim, Y.H.; Lee, C.W.; Jang, E.S.; Son, K.H.; Kim, S.U.; Kim, Y.K. T-cell specific immunosuppression by prodigiosin isolated from Serratia marcescens. *Int. J. Immunopharmacol.,* **1998**, *20*(1-3), 1-13. [http://dx.doi.org/10.1016/S0192-0561(97)00062-3] [PMID: 9717078]

[99] Liu, G.Y.; Nizet, V. Color me bad: microbial pigments as virulence factors. *Trends Microbiol.,* **2009**, *17*(9), 406-413. [http://dx.doi.org/10.1016/j.tim.2009.06.006] [PMID: 19726196]

[100] Stafsnes, M.H.; Josefsen, K.D.; Kildahl-Andersen, G.; Valla, S.; Ellingsen, T.E.; Bruheim, P. Isolation and characterization of marine pigmented bacteria from Norwegian coastal waters and screening for carotenoids with UVA-blue light absorbing properties. *J. Microbiol.,* **2010**, *48*(1), 16-23. [http://dx.doi.org/10.1007/s12275-009-0118-6] [PMID: 20221724]

[101] Zhang, H. A kind of potential food additive produced by Streptomyces coelicolor: Characteristics of blue pigment and identification of a novel compound, λ-actinorhodin. *Food Chem.,* **2006**, *95*(2), 186-192. [http://dx.doi.org/10.1016/j.foodchem.2004.12.028]

[102] Chanamai, R. *Microemulsions for use in food and beverage products*; Google Patents, **2006**.

[103] Mortensen, A. Carotenoids and other pigments as natural colorants. *Pure Appl. Chem.,* **2006**, *78*(8), 1477-1491. [http://dx.doi.org/10.1351/pac200678081477]

[104] Knehr, E. *Maintaining color stability*; Food Product Design, **2006**.

[105] Heinonen, M. Inhibition of oxidation in 10% oil-in-water emulsions by β-carotene with α-and γ-tocopherols. *J. Am. Oil Chem. Soc.,* **1997**, *74*(9), 1047-1052. [http://dx.doi.org/10.1007/s11746-997-0023-2]

[106] Ribeiro, H.; Ax, K.; Schubert, H. Stability of lycopene emulsions in food systems. *J. Food Sci.,* **2003**, *68*(9), 2730-2734. [http://dx.doi.org/10.1111/j.1365-2621.2003.tb05796.x]

[107] Crinó, M.A.; Heenan, C.N.; Nguyen, M.H.; Stathopoulos, C.E. The stability of natural red/pink food colours in ultrahigh-temperature (UHT) products. *J. Sci. Food Agric.,* **2013**, *93*(8), 2022-2027. [http://dx.doi.org/10.1002/jsfa.6009] [PMID: 23254477]

[108] Xianquan, S.; Shi, J.; Kakuda, Y.; Yueming, J. Stability of lycopene during food processing and storage. *J. Med. Food,* **2005**, *8*(4), 413-422. [http://dx.doi.org/10.1089/jmf.2005.8.413] [PMID: 16379550]

[109] Breitenbach, J.; Visser, H.; Verdoes, J.C.; van Ooyen, A.J.; Sandmann, G. Engineering of geranylgeranyl pyrophosphate synthase levels and physiological conditions for enhanced carotenoid and astaxanthin synthesis in Xanthophyllomyces dendrorhous. *Biotechnol. Lett.,* **2011**, *33*(4), 755-

761.
[http://dx.doi.org/10.1007/s10529-010-0495-2] [PMID: 21165672]

[110] Ide, T. Enhanced production of astaxanthin in *Paracoccus* sp. strain N-81106 by using random mutagenesis and genetic engineering. *Biochem. Eng. J.,* **2012**, *65*, 37-43.
[http://dx.doi.org/10.1016/j.bej.2012.03.015]

[111] Yan, Y.; Chemler, J.; Huang, L.; Martens, S.; Koffas, M.A. Metabolic engineering of anthocyanin biosynthesis in *Escherichia coli*. *Appl. Environ. Microbiol.,* **2005**, *71*(7), 3617-3623.
[http://dx.doi.org/10.1128/AEM.71.7.3617-3623.2005] [PMID: 16000769]

CHAPTER 8

Microbial Single-Cell Oils: Precursors of Biofuels and Dietary Supplements

Vanessa Béligon[a,b], Gustavo Molina[a,d], Gwendoline Christophe[a,b,c], Christian Larroche[a,b,c], André Lebert[a,b,c] and Pierre Fontanille[a,b,c,*]

[a] *Clermont Université, Université Blaise Pascal, Institut Pascal, BP20206, F-63174 aubiere cedex, France*

[b] *CNRS, UMR 6602, Institut Pascal, F-63171 AubiÉre, France*

[c] *Clermont Université, Université Blaise Pascal, LABEX IMobS3, BP80026, F-63171 aubiere cedex, France*

[d] *Institute of Science and Technology, Food Engineering, UFVJM, Diamantina, MG, Brazil*

Abstract: For centuries, men have used microorganisms for their activities and abilities to produce metabolites of interest such as antibiotics or pigments. Lipids are now under the spotlight as applications can be found in several domains. With the growing awareness of climate change and the depletion of petroleum resources, microbial lipids, which share similar fatty acids profiles with those of vegetable oils currently used in biofuels, compete as potential candidates for the development of green biodiesel. Oleaginous microorganisms can assure this production with substantial productivity, using various low-cost types of substrates. Lipids formed by microorganisms can also be interesting from a dietary point of view, as some microorganisms are able to produce polyunsaturated fatty acids. These PUFAs, such as those belonging to the omega-3 and omega-6 series, are known for their benefits to human health. The use of microorganisms represents a promising way to produce PUFAs at lower cost and with a higher yield. This chapter discusses various potent microorganisms, especially bacteria and fungi, for single-cell oils production designed either for the energy field or the dietary domain, the metabolic ways involved, the culture conditions and the downstream processes of manufacturing.

[*] **Address correspondence to Pierre Fontanille:** Axe Génie des Procédés, Energétique et Biosystémes, Institut Pascal, UMR 6602 UBP/CNRS/IFMA. 24, avenue des Landais, BP 20206, 63174 Aubiére cedex, France; Tel: +33 (0)4 73 40 74 27; Fax: +33 (0)4 73 40 78 29; E-mail: pierre.fontanille@univ-bpclermont.fr.

Juliano Lemos Bicas, Mário Roberto Maróstica, Glaucia Maria Pastore (Eds.)

Keywords: Alpha-linolenic acid, Arachidonic acid, Biodiesel, Biofuel, Dietary supplement, Docosahexaenoic acid, Edible oil, Eicosapentaenoic acid, Essential fatty acid, Fatty acid, Fungi, Gamma-linolenic acid, Linoleic acid, Lipid, Oleaginous yeast, Omega-3 series, Omega-6 series, Polyunsaturated Fatty Acid, Transesterification, Vegetable oil.

INTRODUCTION

With an ongoing growth of the world's population, planned to reach 9.6 billion in 2050 according to experts, numerous challenges concerning food and energetic demands are expected. Currently, the daily world consumption of petrol reaches 84 millions of barrels, which amounts to 139,000 litres used per second. More than half of this quantity is used to feed vehicle motors.

One tenth is consumed in thermal power stations producing electricity whereas one twentieth is used for heating. However, the heavy exploitation of fossil fuels has led to a decrease in the world reserves. Consequently, a constant increase in the price of oil is observed reaching $108 per barrel in September 2013.

Along with the drop of the fossil fuels reserves come several concerns about climate change. Combustion of fossil fuels causes the production and the release of greenhouse gases (GHGs) into the environment. Carbon dioxide (CO_2), methane (CH_4) and nitrous oxide (N_2O) are the major GHGs produced. These gases absorb part of the solar rays and redistribute them as radiations in the atmosphere, leading to a rise in the global temperature. That is why the question of petrol replacement by a green substitute is more than ever a burning issue. One of the possible alternatives to traditional fuels is the use of biofuels. Biodiesel is defined as a mixture of fatty acid alkyl esters which can be produced from renewable sources and can be solid or liquid [1]. The technical properties of biodiesel must be the availability [2], lack of negative effect upon the environment, biodegradability [3], low potential risk (no explosive vapour) [4] and easy transportability to existent storage places [3, 5, 6]. With characteristics similar to petrodiesel [7], biodiesel is an attractive alternative in the long run for its biodegradable, nontoxic and clean renewable qualities. So far, the raw materials tested for the production of biodiesel are vegetable and animal oils or fats and waste from cooking oils [8] but industrial biodiesel is mainly produced from vegetable oils (such as wheat, soy, rapeseed). Despite its efficiency, this type of biodiesel cannot be used on a global scale. The use of oleaginous plants to produce biodiesel is a threat to the human population because it diverts edible elements from their food role, creating a food security problem. The SCO (single cell oils) produced by oleaginous microorganisms and similar to vegetable oils appear as a response. They share a similar fatty acids profile as that of vegetable oils and can thus be substituted for petrodiesel after a transesterification step [9].

The other emerging problem will be to provide enough food resources to an increasing population, while these resources are diminishing due to an over-exploitation of the planet: modern agriculture causes the impoverishment of soils, water exhaustion and chemical pollution, putting a pressure on the environment and the ecosystems. Searches for new food sources should particularly insist on finding alternatives to essential nutrients (carbohydrates, lipids, proteins, mineral salts, vitamins) present in "traditional" foodstuff (cereals, meat, milk, fish, vegetables and fruit). Among these fundamental elements, the essential fatty acids are of major importance. The essential fatty acids are polyunsatured fatty acids (PUFAs) which cannot be synthesized by mammals and must be therefore brought by food [10]. Two families of PUFAs exist: the omega-3 family and the omega-6 family. Their health benefits have been described which make them indispensable to a proper development [11]. The fatty acids composing the omega-3 family can be obtained by consuming oily fish such as salmon, white tuna and sardine. The omega-6 family is present in borage oil, evening primrose oil, blackcurrant seeds and also in mother's milk. However, while mammals cannot produce these fatty acids by themselves, some microorganisms possess the ability to synthesize them. This competency creates a new prospect in the domain of food supplements, as we can therefore contemplate the option of producing and administering PUFAs to a larger number of people.

This chapter will discuss the single-cell oils produced by oleaginous microorganisms in particular fungi and bacteria which can be used either in the food industry or in the energy domain. It will present the different types of microorganisms involved in the production of the lipids of interest, their culture conditions and the production strategies of these oils. The downstream process of oil production will also be treated.

SINGLE-CELL OILS AS PRECURSOR OF BIODIESEL

Bio-oils as Biodiesel

The rarefaction of fossil oils together with growing concerns about climate changes cause scientists to search for alternatives to petrol. Due to its omnipresence in the energy field, notably as fuels for vehicles, its replacement by green biofuels has been one of the aims of many studies. A classification of biofuels is presented in Fig. (**1**). Biofuels can be grouped in two categories: the primary one which is used in an unprocessed form contains fuel wood, organic waste, forest litter and landfill gases and the secondary one, which is itself divided into three generations. The first-generation biofuels use seeds or sugar as a substrate. The final products are bioethanol and butanol from starch and sugar fermentation, or biodiesel extracted from seeds oil. The major disadvantage of using this first-generation biofuel is the direct competition with human and animal food. Diversion of agricultural land for the production of biofuels causes a rise of the raw materials price such as corn. Furthermore, the high interest for the first-generation biofuels induces environmental problems in countries such as Brazil and Indonesia, which observed a massive deforestation of their primary forests in order to increase agricultural lands of sugarcane and palm oil cultures.

The second-generation biofuels use lignocellulosic biomass as substrate and currently are in a development stage. The production of these fuels will not interfere with normal food production, and as such could help to resolve the biofuels *versus* food opposition: by using only non-edible parts of plants, biofuels produced from lignin and cellulose could allow the simultaneous production of food and biofuels in the same agricultural area.

The third-generation biofuels involve the use of microorganisms such as microalgae and yeasts. Many substrates can be used by these microorganisms which turn carbon sources into lipids (and subsequently biodiesel). Biodiesel is constituted of fatty acids methyl esters. Just like crude oil, single-cell oil needs pretreatment before being used as a suitable alternative to diesel fuel. Four crude oil modifications are used: dilution or blending, microemulsion, pyrolysis and transesterification, of which the latter is the most commonly preferred treatment. The aim of transesterification is to convert free fatty acids, mono-, di- and triglycerides into alkyl esters having properties similar to the conventional diesel [2].

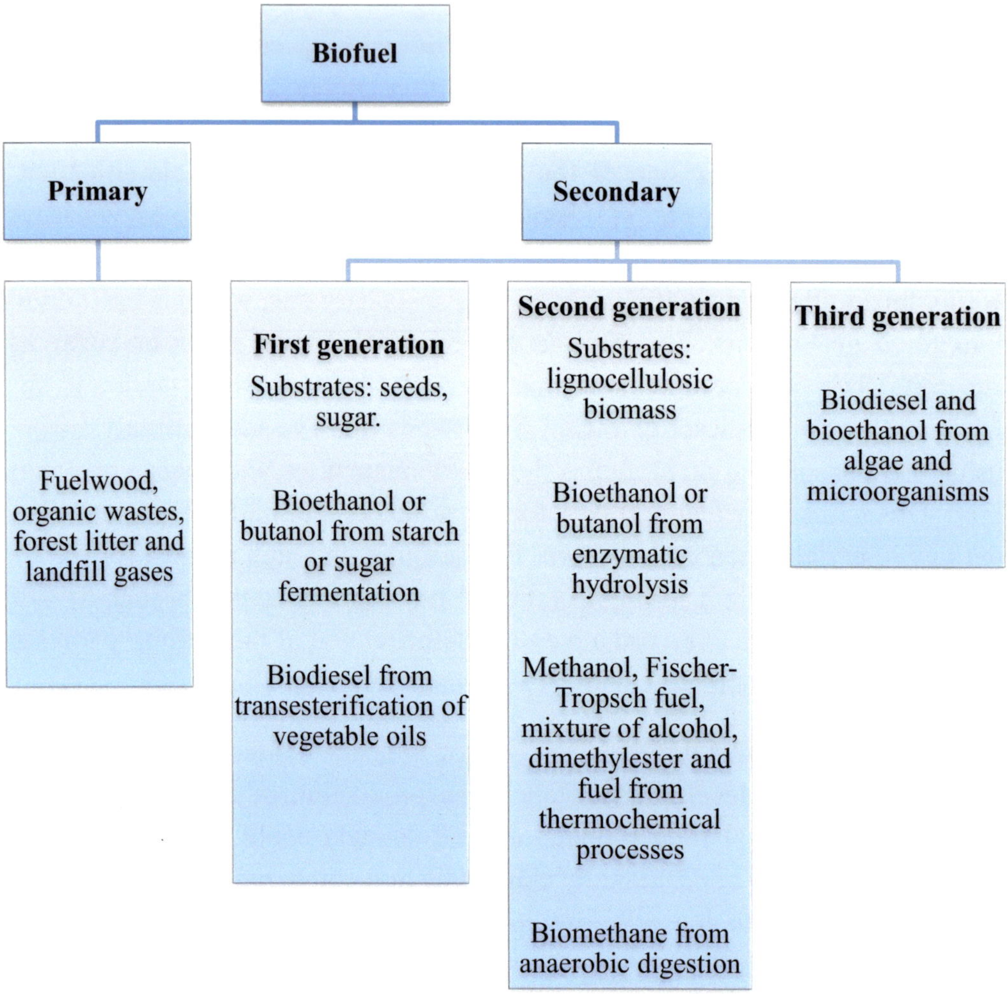

Fig. (1). Classification of biofuels [13].

Therefore, biodiesel production using microbial oils could represent an alternative to producing bio-oils for biodiesel production. The progress of research in this field shows that the synthesis of lipids in organisms like fungi and bacteria is now known and controlled. A large number of strains have been described and selected for their great ability to produce lipids and the methods of assay and extraction of these molecules are now established [12].

Potential Microorganisms for The Production of Lipids

All microorganisms possess the ability to produce lipids to maintain their membrane system. However, some of them can accumulate lipids more than 20% of their dry weights. These particular microorganisms are qualified as oleaginous and include various types of species such as yeasts, fungi, moulds and microalgae. Table **1** summarizes the strains cited in literature. Among 600 known strains of yeasts, around 30 of them have been identified as oleaginous [14]. The most productive yeasts are *Lipomyces starkeyi*, *Rhodosporidium toruloides*, *Cryptococcus curvatus* and *Yarrowia lipolytica*.

Table 1. Example of cultivation of oleaginous microorganisms in different culture conditions and lipid production.

Strains	Reference	Lipids (wt. %)	Culture conditions		
			Substrate	Culture mode	Experimental conditions
Aspergillus oryzae	[17]	*ca.* 40	Potato processing wastewater	Flask	N and P limitations
Candida sp.	[18]	55.02	Crude glycerol	Flask	Initial pH 6.5
Cryptococcus curvatus	[19]	23	Waste active sludge supernatant	Flask	
Cryptococcus curvatus	[20]	46	Waste from brewery industry + glycerol	Flask	C/N 35
Cryptococcus sp.	[21]	62.2	Glucose	Flask fed-batch culture	
		60.9	Corncob hydrolysate	Flask fed-batch culture	
Lipomyces starkeyi	[22]	50	Cellobiose	Flask	
		53	Glucose	Flask	
		57	Xylose	Flask	
Rhodosporidium toruloides	[23]	47 ± 11.95	Non-sterile distillery water	Flask	COD of effluent: 2419 ± 152 mg/L

(Table 1) contd.....

Strains	Reference	Lipids (wt. %)	Culture conditions		
			Substrate	Culture mode	Experimental conditions
Rhodosporidium toruloides	[24]	61.8 ± 0.9	Glucose	Continuous culture	Consumed C/N 61.6 Dilution rate 0.02 h-1
Rhodosporidium toruloides	[25]	40.9	Biodiesel by-products	Fed-batch culture in bioreactor	
Rhodosporidium toruloides	[26]	36.9	Glucose	Flask	C/N 27
Rhodosporidium toruloides		34.5	Levoglucosan	Flask	C/N 27
Rhodotorula glutinis		35.8	Glucose	Flask	C/N 27
Rhodotorula glutinis		39.7	Levoglucosan	Flask	C/N 27
Trichosporon coremiiforme	[27]	19.5	Xylose, arabinose, glucose	Flask	Sugar composition similar to corncob hydrolysate
		37.8	Corncob acid hydrolysate	Flask	
Trichosporon cutaneum	[28]	47.1	Synthetic corncob acid hydrolysate	Flask	C/N 180
		52	Glucose	Flask	
		37.5	Xylose	Flask	
		43.7	cellobiose	Flask	
		45.4	Corncob acid hydrolysate	Flask	C/N 100
Trichosporon dermatis	[29]	13.5	Wastewater from butanol production	Flask	
Trichosporon fermentans	[30]	35.6	Waste sweetpotato vines hydrolysates	Flask	
		64.48	Sweetpotato hydrolysate + wheat straw hydrolysate	Flask	
Yarrowia lipolytica	[31]	64	Crude glycerol	Flask	4% glycerol
		8-13	Serum latex	Flask	C/N 121
		14-20	Serum latex + molasses	Flask	C/N 327
		ca. 70	Palm oil mill effluent + crude glycerol 4%	Flask	

(Table 1) contd.....

Strains	Reference	Lipids (wt. %)	Culture conditions		
			Substrate	Culture mode	Experimental conditions
Yarrowia lipolytica	[32]	48.7	Glucose	Flask	Disruption of the MIG1 gene
Yarrowia lipolytica	[33]	40.69	Glucose + acetic acid	Fed-batch culture in bioreactor	C/N 50
		38.06	Glucose + propionic acid	Fed-batch culture in bioreactor	C/N 50
		24	Glucose + butyric acid	Fed-batch culture in bioreactor	C/N 50
Aspergillus awamori	[34]	28.3	Modified sabourauds dextrose broth + glucose	Flask	pH 6, 28 °C
Aspergillus terreus	[35]	*ca.* 8.6	Sugarcane bagasse	Flask	
Mortierella isabellina	[36]	71 ± 1.6	Xylose	Batch culture in flask	Molar ratio of N/yeast extract/urea : ⅔/⅓/0
Mortierella isabellina	[37]	66.68 ± 2.22	Xylose	Flask	
		66.50 ± 2.12	Glucose	Flask	
		4.84 ± 1.26	CMC	Flask	
		53	Wheat straw hydrolysate	Flask	
Mortierella isabellina	[38]	54.8	Corn fiber hydrolysates	Batch culture in flask	
Mortierella isabellina	[39]	43.9	Glucose	Flask	C/N 70.3
		38.8	Xylose	Flask	C/N 66.8
Mucor circinelloides	[40]	46	Corn-ethanol stillage	Bioreactor	
Cunninghamella echinulata	[41]	72	glucose	Batch	C/N 264
Cunninghamella echinulata	[42]	24	glycerol	flask	C/N 235
		58	xylose	flask	C/N 285
Thamnidium elegans	[43]	70	Glucose, fructose, xylose	flask	C/N 240

(Table 1) contd.....

Strains	Reference	Lipids (wt. %)	Culture conditions		
			Substrate	Culture mode	Experimental conditions
Thamnidium elegans	[44]	71	Glycerol	Flask	C/N 264
Colletotrichum sp.	[45]	36.4 ± 0.4	Glucose	Flask	C/N 40
		40.1 ± 0.3	Glucose	Flask	C/N 120
		46.8 ± 0.3	Glucose	Flask	C/N 240
		44.5 ± 0.4	Glucose	Flask	C/P 86
Alternaria sp.		40.1 ± 0.6	Glucose	Flask	C/N 40
		49.6 ± 0.4	Glucose	Flask	C/N 120
		55.1 ± 0.2	Glucose	Flask	C/N 240
		51.4 ± 0.3	Glucose	Flask	C/P 86

While non-oleaginous yeasts contain 5 to 10% of their dry weight as lipids implied in the membrane structure, oleaginous yeasts are known to accumulate 40 to 70% of their biomass as lipids mostly stored in lipid bodies as triglycerides. Some moulds such as Zygomycetes (*Mortierella* and *Cunninghamella*) also possess the ability of lipid accumulation [15].

Table 2. FAME composition of vegetable and microbial oils [15, 46 - 50].

Species	Palmitic acid C16:0	Palmitoleic acid C16:1*n*-9	Stearic acid C18:0	Oleic acid C18:1*n*-9	Linoleic acid C18:2*n*-6	Linolenic acid C18:3*n*-3
Lipomyces starkeyi	33-56	2-6	5-14	26-55	0.1-3	0.12
Lipomyces lipofera	37	4	7	48	3	-
Rhodosporidium toruloides	18-37	1	3-36	19-60	2-13	0-3.5
Trichosporon pullulans	15	Traces	2-10	57	7-24	1
Cryptococcus curvatus	17-25	-	12	55	8	2
Cunninghamella echinulata	16-19	-	12-14	40-48	4	3
Mortierella isabellina	20-27	1-4	2-6	44-54	4-18	3-8
Yarrowia lipolytica	11	6	1	28	51	1
Oil palm	32-59	-	1-8	27-52	5	Traces
Sunflower oil	3-10	-	1-10	14-65	20-75	Traces
Soya oil	7-14	-	1-6	19-30	44-62	4-11

(Table 2) contd.....

Species	Palmitic acid C16:0	Palmitoleic acid C16:1*n*-9	Stearic acid C18:0	Oleic acid C18:1*n*-9	Linoleic acid C18:2*n*-6	Linolenic acid C18:3*n*-3
Peanut oil	6-13	Traces	3-6	37-61	13-41	1
Cottonseed oil	27-28	-	1	13-18	51-57	Traces
Soybean oil	11	-	4	22	53	8
Rapeseed oil	3-14	-	1-2	56-64	22-26	8-10
Corn oil	12	-	1	25	61	1

The most common fatty acids in the lipid bodies are oleic acid (*cis*-9-octadecenoic acid or C18:1*n*9), linoleic acid (*cis*-9,12-octadecadienoic acid or C18:2*n*-6), palmitic acid (hexadecanoic acid or C:16), palmitoleic acid (*cis*-9-hexadecenoic acid or C16:1*n*-9) and stearic acid (octadecanoic acid or C18:0) [14]. The similarity of the fatty acids profiles produced by oleaginous microorganisms and those of vegetable oils currently used in the biofuels domain indicates that oleaginous microorganisms can represent interesting tools for the production of biodiesel (Table **2**). After esterification, the fatty acids methyl esters (FAME) contain high concentration of saturated fatty acids and good characteristics for combustion such as a high cetane number [16]. For example, the characteristics of the lipids produced by *Cunninghamella japonica* and *Cunninghamella echinulata* are in agreement with the European standards (EN 14214) for biodiesel production.

Lipogenesis

The process of lipid accumulation can be divided into two parts: the first one is the production of acetyl-CoA and the second one involves the conversion of acetyl-CoA into lipids. The scheme for lipid production in yeast is shown in Fig. (**2**). Glucose *via* glycolysis leads to proton-linked pyruvate transported into the mitochondrion. In the mitochondrion, the enzyme pyruvate decarboxylative dehydrogenase converts pyruvate into acetyl-CoA which reacts with oxaloacetate to give citrate. Citrate, under aerobic conditions enters the Krebs cycle, leading to the synthesis of ATP *via* an electron transport chain in the mitochondrial membrane. Under lipid accumulating conditions (usually limitation of nitrogen source), citrate is exchanged *via* a citrate/malate antiport (citrate/malate translocase) with intracellular malate and is cleaved by the ATP: citrate lyase into acetyl-CoA and oxaloacetate according to reaction 1.

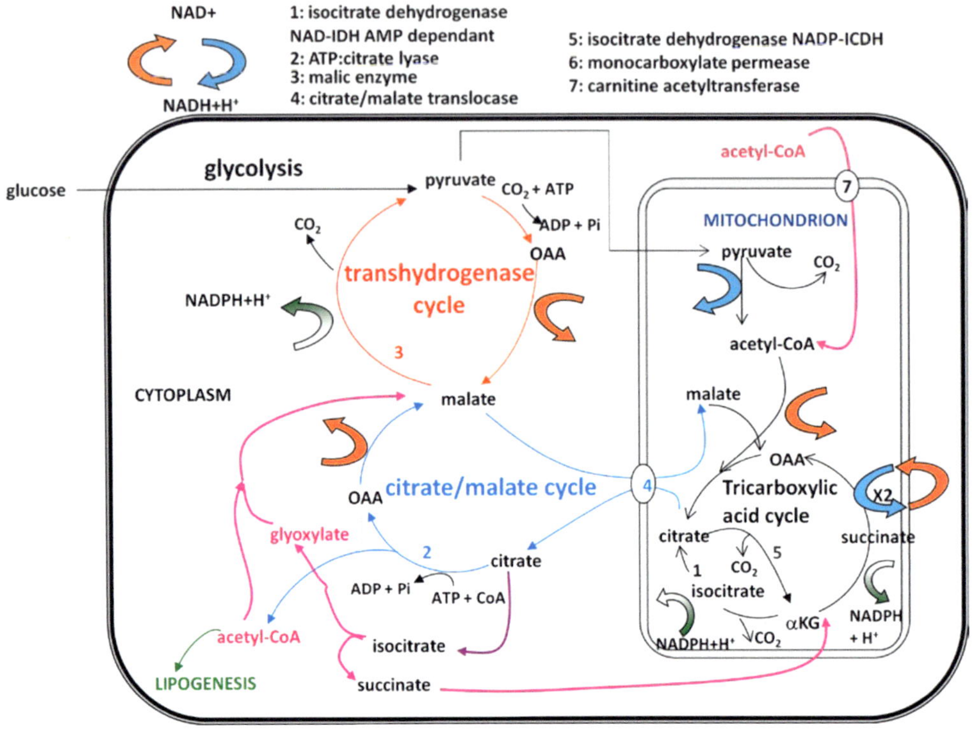

Fig. (2). Lipogenesis in yeast [9]. Glucose *via* glycolysis leads to pyruvate. In the mitochondrion, the pyruvate decarboxylative dehydrogenase converts pyruvate into acetyl-CoA which reacts with oxaloacetate to give citrate. Citrate enters the Krebs cycle, leading to the synthesis of ATP *via* an electron transport chain in the mitochondrial membrane. Citrate is exchanged *via* a citrate/malate antiport (citrate/malate translocase) with intracellular malate and is cleaved by the ATP: citrate lyase into acetyl-CoA and oxaloacetate.
Acetyl-CoA is converted into palmitic acid by lipogenesis pathway, which is the primer for longer-chained saturated or unsaturated fatty acids. Acetyl-CoA also acts as the primer for synthesis of malonyl-Acyl carrier protein and acetyl-ACP.

1: citrate + ATP + CoA → acetyl-CoA + oxaloacetate + ADP + Pi

Acetyl-CoA obtained from the above reaction is converted into palmitic acid by lipogenesis pathway, which is the primer for longer-chained saturated or unsaturated fatty acids.

Acetyl-CoA acts as the primer for the synthesis of malonyl-Acyl Carrier Protein (ACP) and acetyl-ACP.

Malonyl-CoA is obtained by the carboxylation of acetyl-CoA by an acetyl-CoA transacetylase or carboxylase and then transferred to an Acyl Carrier Protein (ACP) by malonyl transacetylase, with release of CoA. Malonyl-ACP allows the addition of g C_2 units to the primer acetyl-ACP obtained by the transfer of the acetyl group of acetyl-CoA on an ACP with release of CoA. The next steps are operated by the fatty acid synthase (FAS). FAS in an enzymatic complex pretty much conserved in many species [51 - 53]. First, a β-ketoacyl-ACP synthase (KAS) adds the C_2-unit of malonyl-ACP on an acetyl ACP, decarboxylating the malonyl group and causing release of an ACP to give acetoacetyl-ACP. Acetoacyl-ACP reductase (KAR) leads to β-hydroxybutyryl-ACP by regeneration of $NADP^+$ from NADPH and H^+. The dehydration of β-hydroxybutyryl-ACP by a β-hydroxyacyl-ACP dehydratase (HAD) produces a crotonyl-ACP which is reduced by an enoyl-ACP reductase (EAR) into butyryl-ACP with regeneration of a NAD+. Elongation of the fatty acid chain is continued by the cycling addition of C2 unit in the form of the acyl group from malonyl-ACP [14, 54].

The overall equation is:

2: acetyl-CoA + 7 malonyl-CoA + 14 NADPH + 14 $H^+ \rightarrow$ palmitic acid + 7 CO_2 + 8 CoA + 14 $NADP^+$ + 6 H_2O

Then specific enzymes, desaturases and/or elongases convert palmitic acid into unsaturated or polyunsaturated fatty acids (PUFAs) or into longer fatty acyl chains. Desaturase catalyzes the introduction of double bond(s) into the fatty acid chain and elongase in a sequence similar to that of C2-unit cycling addition. Biosynthesis of PUFAs is detailed in the second part of the chapter.

Lipid Accumulation Conditions

In mainly oleaginous yeasts and fungi, lipid accumulation is triggered when the carbon source is present in excess and a nutrient becomes limiting in the growth medium. The limitation of many elements can result in lipid accumulation but usually it is nitrogen limitation which is used for this purpose because it is the most efficient one for the induction of lipid accumulation [55 - 58]. Under nitrogen starvation, many microorganisms continue to assimilate the carbon source but the cell proliferation stops as nitrogen is required for protein and nucleic acid synthesis.

Under these conditions, the carbon flux is diverted towards lipid synthesis. Triacylglycerols begin to accumulate within discrete lipid bodies in the cells. Lipids formed have to be stored within the existing cells which can no longer divide. If non-oleaginous microorganisms are placed under the same conditions, then they either stop to proliferate or accumulate polysaccharides including starch, glycogen and various other glucans and mannans [49].

In oleaginous fungi, nitrogen exhaustion induces cascades of biochemical events leading to an accumulation of lipids, the first of them being an increase of AMP desaminase activity to compensate for the lack of nitrogen feeding [49]. The first identifiable biochemical event following exhaustion of nitrogen from the growth medium is the activation of an AMP deaminase, which catalyses the following reaction:

3: AMP $\rightarrow$ IMP + NH_4^+

The activation of AMP deaminase results in the reduction of mitochondrial AMP concentration and the increase in cellular ammonium concentration. This reaction can only be a short-term solution to relieve the deficiency of nitrogen within the cells [14]. The enzyme can be regarded as an ammonium scavenging enzyme and its enhanced activity during nitrogen limited conditions could be viewed as a means of garnering further nitrogen for protein and nucleic acid biosynthesis. This decrease in AMP concentration is not correlated with a decrease in ATP or ADP concentration. This decrease of AMP concentration happens when nitrogen is exhausted and just before lipid accumulates [54].

Key Enzymes of Lipid Accumulation

Three enzymes are key enzymes for lipid accumulation: isocitrate dehydrogenase, ATP: citrate lyase and malic enzyme.

Isocitrate Dehydrogenase

The decrease in AMP concentration inhibits the activity of isocitrate (NAD^+-dependent) dehydrogenase (IDH), which is one of the key enzymes for lipid accumulation [54]. The enzyme catalyses the following reaction:

4: Isocitrate + $NAD^+ \rightarrow$ 2-oxoglutarate + NADH + H^+

In oleaginous yeasts, the enzyme IDH has an absolute requirement for AMP directed lipid accumulation, where AMP acts as a positive effector *via* allosteric regulation. In non-oleaginous microorganisms, the enzyme has no such dependency [14]. Cessation of IDH activity results in an accumulation of isocitrate, which is no longer oxidized by citric acid cycle. This isocitrate accumulation is rapidly equilibrated with citrate *via* enzyme aconitase. Mitochondrial citrate is then transferred out *via* malate/citrate translocase system [14].

In the cytoplasm, the excess of citrate inhibits the glycolytic enzyme phosphofructokinase (PFK) that would stop the glycolytic flux. This inhibition is released by NH_4^+ ions and probably warrant pyruvate supply for lipid synthesis because NH_4^+ fixation on PFK makes a stable complex before the exhaustion of nitrogen [54] even when the NH_4^+ ions concentration is not sufficient to ensure the synthesis of amino acids. Furthermore the release of NH_4^+ ion by AMP desaminase (reaction 3) allows the activity of PFK. When glutamate was used in place of ammoniun ion for nitrogen feeding, for lipid accumulation in *Rhodosporidium toruloides* culture [54], the intracellular concentration of NH_4^+ ions was higher than it was with ammonia. Therefore the activity of enzyme PFK was not inhibited, thus, it may be an interesting way of keeping a high PFK activity and consequently a higher synthesis of lipids. Pyruvate availability for the lipid synthesis means that PFK as hypothesized by Ratledge and Wynn [54] remains active or that another pathway leads to pyruvate supply from G3P. It may be possible that the pentose-phosphates pathway represents an alternative way to the Embder-Meyeroff pathway. This pathway does not need ATP for uptake of glucose, at the time when the electron transport chain could not be active enough due to inhibition of the TCA cycle at the level of NAD-IDH. In yeast and fungi, the mitochondrial isocitrate dehydrogenase (IDH) depends on both, NAD^+ and AMP. Two isoenzymatic forms of $NADP^+$ dependent isocitrate dehydrogenase (ICDH) are present, one of them is cytoplasmic and the other is the mitochondrial [59]. They ensure the supply of α-ketoglutarate (precursor of amino acids) in the mitochondrion when nitrogen is not limited, and in the cytoplasm to produce the NADPH required for lipogenesis. Even if their activities are very low compared to NAD^+ dependent ICDH (in acetate grown culture, the IDH/ICDH is equal to 4), the mitochondrial ICDH provides α-ketoglutarate to the citric acid cycle, that allows electron transport chain ATP synthesis required to maintain cellular energetics.

ATP:Citrate Lyase

The citrate transported out of mitochondrion into cytoplasm is cleaved by an ATP-citrate lyase (ACL) to produce oxaloacetate and acetyl-CoA (reaction 5). This enzyme does not exist in non-oleaginous yeasts, such as *Saccharomyces cerevisiae* [49].

5: citrate + ATP + CoA → acetyl-CoA + oxaloacetate + ADP + Pi

This reaction furnishes large amount of acetyl-CoA for fatty acid synthesis. The presence of ACL activity is strongly linked to the ability to store lipids. This particularity has been observed in yeasts, filamentous fungi and other oleaginous microorganisms [10, 14, 60]. Only organisms able to produce more than 20% of triacylglycerols possess an ACL activity. Though the enzyme has been found in some non-oleaginous yeasts, they are not able to accumulate lipids beyond a little percent of their biomass [14, 54]. Thus the enzyme is mandatory for lipid accumulation but its presence does not strictly imply that the microorganism is an oleaginous species. Hence the presence of ACL cannot be the sole criterion for an organism to be oleaginous.

Malic Enzyme

In addition to continuous supply of acetyl-CoA, there must also be a supply of NADPH for fatty acid synthesis to occur. Fatty acid synthase (FAS) requires two moles of NADPH for every acetyl (or malonyl) unity added to the growing fatty acyl chain during lipid synthesis. Each of the two reductive steps in FAS: 3-ketoacyl reductase and 2,3-enoyl reductase, require one mole of NADPH [14]. The NADPH for lipogenesis is provided by malic enzyme which catalyzes the following reaction (6) [61].

6: L-malate + $NADP^+ \rightarrow$ Pyruvate + CO2 + NADPH + H^+

Though malic enzyme produces less than 15% of the total NADPH (the others being glucose-6-phosphate dehydrogenase, 6-phosphogluconate dehydrogenase, and cytosolic NADPH-isocitrate dehydrogenase) the NADPH produced is essential for lipid accumulation [10]. Malic enzyme would assume prime importance if a substrate such as pyruvate or even acetate is used in place of glucose. If the activity of the enzyme vanishes, either by inhibiting or mutation, then lipid accumulation stops. Thus malic enzyme is vital for the process of lipid accumulation [54]. Wynn *et al.* [56] have also shown a strong correlation between the malic enzyme activity and the extent of lipid accumulation.

The lipid metabolism in many oleaginous yeasts is well-known today. Thanks to bioinformatic data, many genes implied in fatty acid synthesis in *Y. lipolytica* and *S. cerevisiae* are identified and lead to novel approaches to the engineering of unicellular microorganisms [49]. This knowledge allows also a rapid screening of efficient strains for lipids production [62].

Cultivation Modes for Lipid Accumulation

For lipid accumulation in oleaginous microorganisms, different modes of biomass production can be considered: batch, fed-batch and continuous culture modes. The mode of culture is an important parameter that will be crucial for optimizing lipid production and will therefore have a direct economic impact on the process. The tools of chemical engineering will have to be mastered to develop the production process on an industrial scale.

Batch Operation

Batch cultivation refers to the culture of cells in a fixed volume of culture medium and under particular environmental conditions (*e.g.* temperature, nutrient type, pressure, agitation, aeration, *etc.*). Most of the studies in literature have been done using this mode of culture (Table **1**) which simply allows to determine the influence of some parameters on lipid accumulation. For example, lipid production by oleaginous organisms can be triggered by a nutrient limitation (usually nitrogen) [48]. Therefore, batch cultivations of oleaginous microorganisms are performed with a high C/N ratio to channel the extra carbon into lipid accumulation after the establishment of nitrogen limiting conditions. As long as the nitrogen is not depleted, the culture remains in the exponential phase and the biomass keeps on increasing. When the nitrogen concentration becomes very low, the growth almost stops and the culture enters the lipid accumulation phase.

The batch culture is often used in flask culture. However, a high initial glucose concentration can inhibit the microorganism growth and make it difficult to obtain a high cell density biomass. For example, for *Cryptococcus sp.* grown in flask containing 4% of glucose, the biomass reaches 9.4 g/L whereas an increased glucose concentration from 6 to 10% leads to a drop of biomass production to 6.6 g/L. This may be due to an inefficient consumption of the carbon source and the decrease of the specific growth rate [21].

In batch cultures, carbon substrates and minerals are initially added in the bioreactor with a high initial C/N ratio to induce lipid accumulation. Usually this ratio is 50:1 [57]. Nitrogen consumption starts right from the beginning of the culture. The rC/rN (residual carbon to residual nitrogen) ratio constantly increases, tending to infinity. But, after an active phase of lipids production, citric acid production is generally observed as a function of rC/rN ratio. This citric acid production results in a change of microbial metabolism and in the decrease the lipid yield from the carbon substrate. The consumed carbon to consumed nitrogen ratio has to be controlled to prevent citric acid secretion, hence the importance of monitoring rC/rN in fed-batch and continuous cultures.

Continuous Operation

In a continuous culture, fresh nutrient medium is continuously supplied to a well-stirred culture and products and cells are simultaneously withdrawn. The C/N and rC/rN ratios in the culture medium are constant for a given dilution rate when the steady state is reached. In these conditions, lipid accumulation strongly depends on the dilution rate and molar C/N ratio of the growth medium [63, 64]. The dilution rate (D) is defined as the rate of the flow of the medium over the volume of culture in the bioreactor and is expressed in h^{-1}. Low dilution rates favours lipid accumulation, as shown by Shen *et al.* [24]. The yeast *R. toruloides* grown on glucose and under nitrogen limitation in a chemostat produces highest lipid concentration (5.36 g/L) and non-lipid cell mass concentration (3.32 g/L) at a dilution rate of 0.02 h^{-1}. By contrast, a dilution rate of 0.2 h^{-1} induces a drop of lipid concentration (0.21 g/L) and non-lipid cell mass concentration (1.42 g/L) indicating a change in production yields of biomass and lipids between the two dilution rates. The same case has been observed by Papanikolaou and Aggelis [65] who determined the best dilution rate for *Y. lipolytica*'s growth. From D = 0.02 h^{-1} to D = 0.03 h^{-1}, a significant growth is observed. However, high dilution rates resulted in significant amount of glycerol detected in the culture medium, meaning a lower usage of the carbon source. As for lipid yield, it reaches its highest value (0.1 g/g) for a dilution rate of 0.04 h^{-1}.

C/N ratios from 40:1 to 60:1 are the most cited in literature for being the ratios which promote lipid accumulation [24, 49, 57]. The optimization of the process involves determining the optimal C/N ratio in addition to the optimal dilution rate in continuous culture.

Fed-Batch Operation

The fed-batch operation is characterized by intermittent or continuous feedings of nutrients. This mode allows a prolongation of a traditional batch and a more precise control of the concentration of the substrate or of any other nutrients studied. Consequently, a fed-batch operation is usually used to overcome catabolite repression or substrate inhibition by sequential feedings of the substrate. This feeding improves the productivity of the fermentation by regulating the environmental variables to maximize the stability of the metabolic state. Nitrogen and carbon flow monitoring allows controlling the specific growth rate and the rC/rN ratio [49]. Lipid production is generally performed in two main phases. The first phase promotes growth, where all nutrients are present in excess. The C/N flux ratio is equal to the catalytic biomass production requirements. Once the nutrients limitation occurs (generally nitrogen limitation), the C/N ratio is then controlled to promote lipid accumulation.

Among the numerous studies dealing with the fed-batch mode, Li *et al.* [48] led a high density culture of *R. toruloides* grown in a fed-batch bioreactor. The final biomass concentration reaches 106.5 g/L, the lipid content 67.5% with a lipid productivity of 0.54 g/L/h. Fontanille *et al.* [33] reported a biomass concentration of 30.83 g/L and a lipid content of 40.69% through fed-batch cultivation of *Y. lipolytica*. The cultivation of *C. curvatus* on glucose allows to obtain biomass concentration, lipid content and lipid productivity of respectively 104.1 g/L, 82.7% and 0.47 g/L/h. However a precise control of the substrate input should be respected in order to avoid a carbon limitation and the reserve lipids consumption which leads to a decrease of the cellular lipid concentration [66].

Extraction, Quantification and Identification of Lipids

Extraction Methods

One of the most important parameters for lipid production is the lipid content of the biomass. This piece of information allows the operator to control, to validate and to optimize the lipid production processes. As a consequence, the extraction method and the lipid quantification must be fast and efficient. This method should also be applicable to an industrial scale-up. Several extraction methods have been referenced in literature. Among these methods, the Soxhlet method (1879) is the oldest. This method allows extraction using solvents and consists of a glass apparatus designed for solid/liquid extraction. It allows continuous lipid extraction by cycling washes of the sample by the solvent. More recently, the Folch method [67] allowed the extraction of lipids contained in animal tissues. Despite its fastness and its easiness to operate, this method is not completely adapted for the extraction of lipids contained into micro-organisms. Since then, the Folch method has been modified, in particular by Cot *et al.* [68], making it more efficient for microbial lipids extraction but also longer. However, these modifications, consisting in a pretreatment step of lyophilisation and several washing cycles of 24 hours, take more than one week.

A well-known method from 1959 is the Bligh and Dyer method [69] which is also a derived version of the Folch method. This method has the advantage of reducing the solvent volume used and warrants recovery of more than 95 % of lipid. This method is, originally, meant for lipid extraction from fish tissues. It employs classical organic solvents such as chloroform and methanol. This method is the starting point of a new extraction system used by Cescut *et al.* [70]. The pressurized liquid extraction system consists of a controlled device meant for compounds extraction. Parameters of pressure, temperature, volume of extraction solvent and extracting time are tightly regulated. Lyophilized samples are automatically treated with a modified version of the Bligh and Dyer method.

Some studies focused on the use of green solvents in lipids extraction. Among the alternative solvents, terpenes such as alpha-pinene, limonene and cymene are used for lipid extraction of microalgae [71]. Emergent methods using supercritical fluids show that CO2 can be used as an extraction solvent for lipid recovery and allows the recovery of 38,73% of lipids from *Mortierella alpina* [72]. Organic salts, associated with an ultrasonication step, have also been used for recovering microalgal lipids. Ionic liquids have also been tested due to their different solubilities and their high dissolving power for algal biomass. They are also classified as a green solvent [73].

If many pieces of research focus on lipids extraction methods, studies of biomass pretreatment are also on the go. Lipid reserves from oleaginous microorganisms are usually stored inside the cell as lipid globules. The presence of a cell wall makes their extraction difficult and slows down the solvents entry. Classically, the most employed pretreatment is the drying of the samples, by oven-drying or freeze-drying [70, 71]. However this pretreatment is not applicable to an industrial scale-up, due to its cost and its duration. Other techniques consist in a cell disruption, using mechanical or non-mechanical methods. Mechanical methods are ultrasonication, high pressure homogenization, French pressing and bead milling. Non mechanical methods include chemical lysis (using acids or enzymes), osmotic shocks [72] and microwave irradiation. Lee and *et al.* [74] analysed the effect of different pretreatments of microalgae biomass such as an autoclaving at 125 °C, a bead beating, a microwave irradiation, a sonication, an osmotic shock using a 10% NaCl solution. A better efficiency of a microwave irradiation pretreatment (28.6 % of lipid content obtained) is demonstrated.

Lipid Quantification

Gravimetric Assay

In the gravimetric method, after being dried, the total lipids are weighed and lipid content is calculated with respect to the dry weight of the extracted sample. In order to have an accurate precision, this type of methods generally requires high weight samples, *i.e.* more than 1 g dry weight, which is unusual with microbial cultures on a laboratory scale and therefore requires great culture volumes.

Triphenyltetrazolium Chloride Assay (TTC)

Reduction of TTC into red coloured triphenylformasan (TF) is the basis of this method for the screening of fatty acids producing strains but can also be used to quantify lipids. Usually, TTC is used to test the viability of cells. This compound is taken up by the living cells and reduced by cellular dehydrogenases. This reduction reaction is enhanced by the addition of phenazin methosulfate (PMS) which competes with the flavoprotein of respiratory electron transport. This method was employed by Zhu *et al.* [75] to detect arachidonic acid production in *M. alpina*. The hyphae of *M. alpina* become red coloured due to the activity of Δ5 fatty acid dehydrogenase-desaturase while hyphae of non-lipid producing strains remain uncoloured.

Nile Red Photofluorescent Assay

Neutral lipids are assayed in moulds, yeasts and microalgae extracts by Nile Red coloration [8, 76, 77]. This method is considered as a sensitive and quantitative method and may also be used for the screening of oleaginous strains. In this method, lipid extracts are suspended in a KH_2PO_4/KCl buffer of pH 7 and a solution of Nile Red in acetone (0.26 μM Nile Red as the final concentration) is added to it. The microbial extracts are excited at 488 nm and reemission is measured at 570 nm before and after the addition of Nile Red. Reemission spectra are recorded between 500 and 700 nm. Lipid concentration may be correlated with the difference between the emission intensity at 570 nm. Lee *et al.* [78] modified this method to assay the lipids produced by *Botryococcus braunii* without involving extraction. After the addition of Nile Red, the cells were excited at 490 nm and the reemission measured at 585 nm. The difference between the peak intensities with and without Nile Red was found proportional to the lipid concentration.

Another lipophilic bright green fluorescent dye, BODIPY (boron-dipyrromethene), is presented as an alternative for Nile Red [79]. Indeed, it has specificity for intracellular lipid bodies, a high oil/water partition coefficient, allowing to cross the cell membranes efficiently.

Vanillin Assay

Under acidic conditions, lipids make a coloured complex with vanillin. Lipids extracts are heated in H_2SO_4 at 100 °C followed by addition of 20 volumes of vanillin reactive (3 g vanillin, 0.5 L deionised H2O and 2 L phosphoric acid). After 30 minutes of incubation at room temperature, the absorbance of the mixture is recorded at 525 or 528 nm. A standard curve is prepared using olive oil at 525 nm [80] or cholesterol at 528 nm [81] which is used for extrapolating the concentration of lipids.

Fatty Acids Identification Methods

Many methods to determine the fatty acid profile which is decisive for future use such as biofuel or for the production of a specific fatty acid exist in the literature. Lipids are generally transesterified before analysis, even on crude samples. Gas chromatography remains the principal method used for separation and identification of esterified fatty acids [61]. Detectors such as mass spectrometry (MS), infra-red spectrophotometry (IR) [82] or nuclear magnetic resonance (NMR) [83] have also been used for the identification of lipids. All these methods allow relative quantification of different fatty acids. Table **3** presents the main techniques found in the literature.

Table 3. Lipid identification methods.

References	Lipid fractionation	Transesterification on sample	Transesterification on extracted lipids	Identification methods
[84]	+	-	+	GC
[85]	+	-	+	GC

(Table 3) contd.....

References	Lipid fractionation	Transesterification on sample	Transesterification on extracted lipids	Identification methods
[86]	-	-	+	GC
[10]	-	-	+	GC
[87]	-	-	+	GC
[88]	+	-	+	GCMS
[65, 89]	+	-	+	GC-GCMS
[82]	+	-	+	GC-IR
[90]	-	+	-	GCMS
[47]	-	+	-	GC
[48]	-	+	-	GC
[91]	-	+	-	
[16]	-	+	-	GC

Single-Cell Oil as Future Biodiesel?

Commercial production of lipids is slowed down by the high cost of the substrates employed. Although glucose is an efficient carbon source for biomass production and lipid accumulation, its high price might interfere with the development of the third-generation biofuels. It has been shown that the raw materials cost represent roughly 88 % of the total estimated biodiesel production cost [92]. Since carbon sources constitute over 60 % of the total production cost in typical fermentation processes, low-cost carbon sources should therefore be considered. Moreover, the strains should also be able to use these substrates and produce lipids from them.

Various low-cost substrates have already been evaluated: molasses, raw materials from the food industry [20, 27, 30]; wastewaters [17, 29], crude glycerol; whey [63] and they turn out to be appropriate substrates for lipid production. The use of these cheap carbon sources will contribute to lower the final cost of biodiesel. Fei *et al.* [93] show the cost of volatile fatty acids produced from anaerobic-digested foodwaste at around $30/ton whereas glucose reaches $500/ton. So although low-cost substrates are not favoured for biomass production and lipid accumulation, they could constitute an economically potential source for lipid and high-value biodiesel production.

An interesting way to reduce biodiesel cost is to add value to the by-products of the process, in particular biomass and crude glycerol, in a biorefinery concept. The residual biomass from biodiesel production can be potentially re-used as animal food as well as for methane production by anaerobic digestion. The waste glycerol produced during the transesterification reaction can also be used as a substrate for lipid production. Kiran *et al.* [25] report a production of 40.9% of lipid by *R. toruloides* grown on crude glycerol. Unrefined glycerol is also one of the constituents of many chemical products such as 1,2-propanediol, 1,3-propanediol, polyesters, among others [94]. The production of hydrogen from unrefined glycerol is another interesting application. Crude glycerol has been tested for the production of syngas, after an air-blown gasification step [95]. The final syngas composition was about 24%, 19%, 6% and 6% for H_2, CO, CO_2 and CH_4 respectively.

Although the production of biodiesel from yeast or fungi is still at laboratory level, this technology represents an interesting alternative for the production of biofuels in the future. The variety of existing microorganisms, the possibility of using low cost substrates and of recycling by-products of the reaction along with a good knowledge of the process are important arguments for the development of single-cell oils production as a renewable energy source.

MICROBIAL PRODUCTION OF OMEGA-3 AND OMEGA-6

An Original Way to Produce PUFAs

Polyunsaturated fatty acids (PUFAs) are vital important structural elements of cell membranes. They particularly confer fluidity, flexibility and selective permeability to membranes, and, therefore their administration is essential to ensure the development of organisms [96]. Among the existing PUFAs, the omega-3 and omega-6 families have benefited from a growing attention due to their impact on human health. The omega-3 family (also referred to as ω-3 or *n*-3) includes α-linolenic acid (ALA or all-*cis*-9,12,15-octadecatrienoic acid or C18:3*n*-3), eicosapentaenoic acid (EPA or all-*cis*-5,8,11,14,17-eicosapentaenoic acid or C20:5*n*-3) and docosahexaenoic acid (DHA or all-*cis*-4,7,10,13,16,1--docosahexaenoic acid or C22:6*n*-3), while the omega-6 family (ω-6 or *n*-6) gathers γ-linolenic acid (GLA or all-*cis*-6,9,12-octadecatrienoic acid or C18:3*n*-6), linoleic acid (LA), arachidonic acid (ARA or all-*cis*-5,8,11,1--eicosatetraenoic acid or C20:4*n*-6) and conjugated linolenic acid (CLA). ALA and LA are, for each family, the precursors of the other family members. They are qualified as "essential" because humans, and in general mammals, cannot synthesize these acids due to their lack of specific desaturase enzymes. Consequently, they need to find these acids through their diet.

Omega-3 Fatty Acids

The omega-3 fatty acids family is defined by a double bond beginning at the third carbon from the methyl end. Mammals cannot insert double bonds more proximal to the methyl end than the ninth carbon atom. As a consequence, omega-3 fatty acids cannot be synthesized *de novo*. They must be present in the diet. ALA is the simplest member of the omega-3 series. It can be found in vegetable oils such as flaxseed oil and camelina oil and also in fish oils. ALA is the essential precursor of all longer-chained fatty acids because it is the metabolic precursor from which longer chain *n*-3 fatty acids are synthesized through a series of elongation and desaturation reactions.

EPA and DHA are the PUFAs derived from ALA. Their actions against cardiovascular diseases have been proven as they reduce arterial LDL-cholesterol delivery [97, 98], participate in the inflammatory response through resolvins and neuroprotectins generation [98, 99] and suppress vascular smooth muscle cell proliferation [98, 100]. Patients awaiting carotid endarterectomy who were given fish oil capsules enriched in ω-3 showed a more stable morphology of their carotid plaques and a decreased macrophage infiltration [101]. DHA is also a major structural component of the gray matter of the brain and the eye retina. Therefore DHA is essential to ensure optimum neural and visual functions [102]. However, using fish oils as nutritional supplements, and notably for infants, is debatable because of the presence of environmental pollutants such as dioxins, heavy metals and PCB which might be absorbed by fish and concentrated in their livers and other organs.

The other important source of DHA is mother's milk. Also rich in ARA, mother's milk brings the indispensable fatty acids for neural development and visual acuity to new-borns. DHA and ARA are absent from cow's milk used in place of mother's milk. Therefore, they should be added to the diet of babies to ensure a normal development.

Omega-6 Fatty Acids

N-6 fatty acids present a double bond starting at the sixth carbon from the methyl end. LA acts as a precursor for the synthesis of more highly unsaturated and longer-chained omega-6 family. LA is present in every food we eat. It is the predominant PUFA in dairy, vegetables, land-based meats, cereals, vegetable oils, nuts, seeds, fruits, legumes and breads [103]. GLA is produced in the body as a product of LA metabolism but can also be found in some plants oils such as evening primrose and borage oils. GLA has been incorporated into infant formula. It is also implied in treatments of several pathologies such as rheumatoid arthritis, atopic eczema, premenstrual tension and multiple sclerosis [104]. Its conversion to ARA is catalyzed by the Δ-5 desaturase which is widely expressed in human tissues with the highest level in liver [105]. The human organs, muscle, blood tissues and brain are particularly rich in ARA. This omega-6 fatty acid has a major role as a structural lipid associated predominantly with phospholipids. Together with DHA, ARA is of crucial importance for the brain development of children. ARA is a direct precursor of a number of eicosanoids regulating lipoprotein metabolism, blood rheology, leucocyte function and platelet activation. This PUFA is found in animal liver and egg yolks, but also in human breast milk. CLA refers to a group of PUFAs existing as positional and stereo-isomers of LA with a conjugated double bond. Many forms are possible [106] but the major one naturally occurring is *cis*-9, *trans*-11 octadecadienoic acid found in ruminant milks, dairy products and ruminant meats.

Currently, the omega-6 to omega-3 ratio brought by the Western diet reaches 15:1 to 16.7:1 [107]. Ideally, this ratio should be around 1/1. Over-consumption of omega-6 enhances pathogenesis of many diseases such as cardiovascular diseases, cancer, autoimmune and inflammatory diseases [108, 109]. On the other hand, an enriched diet in omega-3 reduces the risk of heart disease [96]. Many countries, as well as WHO, recommend a daily consumption of 0.8 to 1.1 g for ALA and 0.3 to

0.5 g for DHA+EPA. Daily dietary intake of ALA should vary from 0.5 to 2% of total dietary energetic input and omega-6 intake from 2.5 to 9%. ARA intake is not essential for healthy adults whose habitual dietary LA input represents 2.5% of total energetic input. For infants from 0 to 6 months, ARA intake should be supplied in the diet within the range of 0.2 to 0.3% based on human milk composition as a criterion [110].

An Original and Future Way to Produce PUFAs

With the ongoing interest in PUFAs, the world wholesale market for infant formula has increased tremendously. PUFAs naturally brought by human milk have to be supplied in infant formulas in order to warrant a normal development of babies. Traditional sources of PUFAs are well-known (Table **4)**, however, the search for cheaper, more accessible alternative sources of production began a few decades ago. Oleaginous organisms do not only produce C16 to C18 fatty acids, they also can manufacture longer chained fatty acids such as PUFAs. It is well established that some fungi such as *Mortierella isabellina* or *Mortierella alpina* and microalgae (*Crypthecodinium cohnii*) are able to produce these high-value oils at levels of over 20% of the total fatty acids [111]. Therefore, using these oleaginous microorganisms represents a promising way to produce PUFAs.

Table 4. Dietary sources of functional lipids.

Family	PUFA	Dietary sources
Omega-3	ALA	Vegetable oil (flaxseed oil, chia seed oil, canola oil, egg, meat, walnut, hazelnut) Fatty fish oil
	EPA	Fatty fish oil
	DHA	Fatty fish oil Breast milk
Omega-6	LA	Vegetable oil Nuts
	GLA	Evening primrose oil Borage oil
	ARA	Egg yolk Animal fat Breast milk

Currently, a growing number of companies works on research, development, manufacturing and selling of SCO-PUFAs or PUFAs containing products: DSM (ex-Martek Bioscience Inc.), Nestle S.A., Nutricia, to name but a few. However, before being offered for sale on the market, the product has to overcome tests ensuring its safety. The Food and Drug Administration (FDA) in particular has developed a program called the "Generally Recognized as Safe (GRAS) Notification Program. This program stipulates that "any substance that is intentionally added to food is a food additive and is subject to premarket review and approval by FDA, unless the substance is generally recognized, among qualified experts, as having been adequately shown to be safe under the conditions of its intended use, or unless the use of the substance is otherwise excluded from the definition of a food additive" [112]. In 2001, Martek Bioscience Inc. obtained by the FDA the GRAS status for its ARA fungal oil (ARASCO) and DHA algal oil (DHASCO) infant formula ingredients.

Biosynthesis of PUFAs

Elongation/Desaturation Pathway

Understanding the metabolic pathway implied in the biosynthesis of PUFAs and how microorganisms are able to accumulate lipids of interest is of crucial importance in order to achieve the highest yield and productivity possible. Two distinct schemes for the production of PUFAs have been identified: the elongation/desaturation pathway which involves elongase and desaturase and the anaerobic polyketide pathway which occurs mainly in thraustochytrids and are large-celled marine heterokonts.

In eukaryotic and prokaryotic cells, lipogenesis leads to the final production of palmitic acid from malonyl- and acetyl-CoA. The fatty acid synthase is the complex of enzymes implied in this process. The palmitic acid obtained is modified by series of desaturation and elongation to produce longer-chained fatty acids. In oleaginous yeasts, the type of fatty acids is limited: the most frequently found are oleic, linoleic, palmitic and palmitoleic acids. PUFAs appear at levels of over 20% of the total fatty acids only in fungi and microalgae [111]. The succession of reactions leading to PUFAs are realised by desaturase enzyme which inserts double bonds at specific carbon atoms in the fatty acid chain while the elongase enzyme adds two-carbon increments to the molecules. In fungal producers of PUFAs, stearic acid is desaturated by Δ-9 desaturase. The forming of oleic acid is followed by a desaturation step, catalyzed by Δ-12 desaturase and leading to the production of LA (Fig. **3**). LA is the substrate of a second desaturation step which implies Δ-15 desaturase and ensures the production of ALA. Then, a series of reactions described in Fig. (**3**) leads to the formation of others PUFAs.

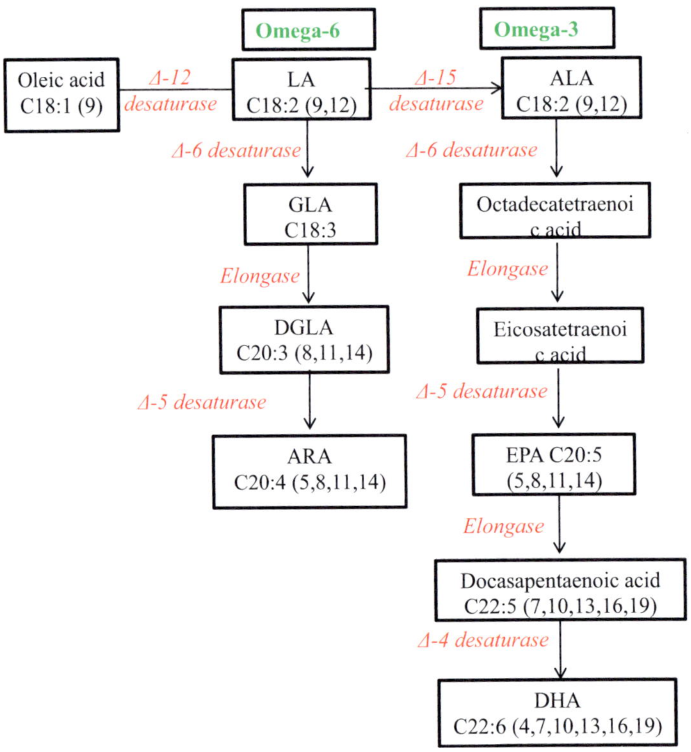

Fig. (3). Route of polyunsaturated fatty acids biosynthesis. Oleic acid is desaturated by the Δ-12 desaturase and leading to the production of LA. LA is the substrate of a second desaturation step implying the Δ-15 desaturase which ensures the production of ALA.
From ALA, the Δ-6 desaturase inserts a double bond to produce the C18:4(6,9,12,15) intermediate which is elongated to form C20:4(8,11,14,17) molecule. A second desaturation mediated by the Δ-5 desaturase leads to the synthesis of EPA (C20:5) which is at last elongated and desaturated by the Δ-4 desaturase, allowing the final production of DHA (C22:6). LA is the precursor of the omega-6 series. First LA is desaturated by the Δ-6 desaturase for the production of GLA (C18:3). GLA is then elongated and desaturated by the Δ-5 desaturase to form ARA (C20:4). ARA can be elongated as well and submitted to a desaturation step mediated by the Δ-4 desaturase to produce docosapentaenoic acid (DPA, C22:5).

It should be noticed that Δ-5 and Δ-6 desaturases can potentially operate in either the n-3 or the n-6 fatty acid pathway, converting either DGLA to ARA or 20:4*n*-3 intermediate to EPA. They belong to a subgroup of desaturases that have been designated the "front-end" desaturases. The front-end desaturases have similar sequence homology and share distinct features that separate them from the desaturases that insert double bonds to the methyl end of the fatty acid relative to the initial Δ-9 desaturation [54].

In mammals, Δ-12 and Δ-15 desaturases are lacking. Therefore, ALA and LA cannot be produced by humans and they require these EFAs in their diet [113].

Anaerobic Polyketide Synthase Pathway

Recently, an alternative pathway for DHA biosynthesis has been discovered in *Schizochytrium*, a genus of unicellular micro-organisms found in coastal marine habitats belonging to the family of Thraustochytriaceae [114, 115]. The characteristic of this new pathway is that double bonds are introduced in the acyl chain during fatty acid biosynthesis, without the requirement of aerobic desaturation. Thus it is also called the anaerobic pathway for DHA synthesis. The term "anaerobic" emphasizes that oxygen is not implied in the synthesis of double bonds. In bacteria and fungi, the PKS pathway is implied in the synthesis of a large collection of antibiotics. This pathway has been discovered in particular in marine bacteria such as *Shewanella* and *Moritella marina*, known to produce EPA and DHA, respectively.

The polyketide synthase (PKS) is an enzymatic complex similar to the fatty acid synthase found in bacteria. It uses an acyl carrier protein as a covalent attachment for chain synthesis, proceeding with reiterative cycles (Fig. **4**). An acyl-ACP and a malonyl-ACP first condense thanks to the 3-ketoacyl-ACP synthase to form 3-ketoacyl-ACP. The latter is reduced by a 3-hydroxyacyl-ACP reductase and the reaction produces 3-hydroxyacyl-ACP. A dehydrase/isomerase removes a water molecule and induces the production of enoyl-ACP. Finally, enoyl-ACP is reduced to form a saturated acyl-chain [116]. However, contrary to fatty acids synthesis, some steps such as dehydration and reduction do not occur during the polyketide synthesis.

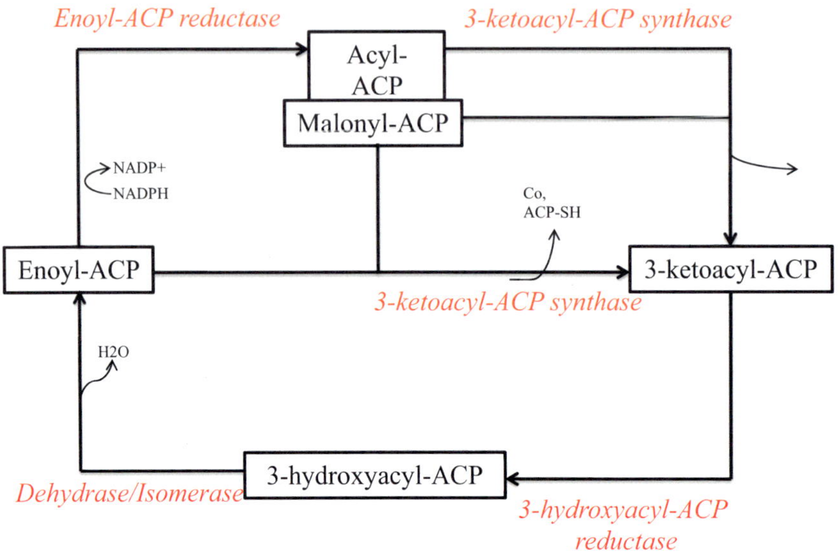

Fig. (4). The anaerobic polyketide synthase pathway for DHA biosynthesis [116]. The PKS system uses acyl carrier protein as a covalent attachment for chain synthesis. An acyl-ACP and a malonyl-ACP first condense thanks to the 3-ketoacyl-ACP synthase to form 3-ketoacyl-ACP. The latter is reduced by a 3-hydroxyacy--ACP reductase and the reaction produces 3-hydroxyacyl-ACP. A dehydrase/isomerase removes a water molecule and induces the production of enoyl-ACP. Finally, enoyl-ACP is reduced to form a saturated acyl-chain.

The exact mechanism underlying DHA synthesis with this new pathway remains to be defined but the key features likely include condensation and isomerisation steps. Ketosynthase is known as the key enzyme of this pathway. This enzyme can catalyze the condensation of a wide range of substrates, including unsaturated fatty acids. Enoyl-ACP with malonyl-ACP can be condensed by ketosynthase, leading to the production of a fatty acid with double bonds. The isomerase enzyme is also of crucial importance for DHA synthesis. This enzyme, occurring either with dehydratase or separately, would warrant precise location and stereospecific configuration of double bonds in the PUFAs [116].

Omega-3 and Omega-6 Produced by Micro-Organisms

Microbial Production of GLA

One of the most important and traditional sources of GLA is the evening primrose. Its seed in particular can store lipids up to 20-30% of its dry weight and 7-10% GLA [117]. The impact of GLA on human health has been previously described. According to Life Sciences Alliance, up to 2000 tons of GLA, mostly from China, are produced per year at a bulk price of around $15 per kilogram. The expensive price is due to high fluctuations in availability and the high costs of production and purification.

Among alternative sources of production, Mucorales, including *Mortierella* and *Mucor* genus, are well-known for their tendency to produce GLA. It is worth noticing that already in 1985, *Mucor (javanicus) circinelloides* had been selected for its ability to produce high amount of GLA (18-22%) over twice that of evening primrose oil. The fungus has been cultivated in large-scale fermenters (220 m^3) by J. & E. Sturge Ltd. (England), allowing a GLA content of 20-25% produced in 72-96 hours. The oil was sold under the name of Oil of Javanicus or it was mixed with evening primrose oil. However, the production stopped in 1990 when the company was sold to Rhone-Poulenc [14].

Since then, GLA production by fungi has not stopped and research is still in progress. Table **5** references the work undertaken in this field. Strains such as *Mortierella isabellina* and *Cunnighamella echinulata* represent promising ways for the development of microbial production of GLA.

Table 5. Strains involved in GLA microbial production. TFA = Total Fatty Acids.

Strains	**Ref.**	**GLA**	**Culture conditions**		
		% TFA	**Substrate**	**Culture mode**	**Experimental conditions**
Mortierella ramanniana	[127]	13.3[a]	Dextrose + yeast extract	Flask	Supplementation of 5 mg/L Mn2+
Cunninghamella echinulata	[128]	11.2 - 13.4[a]	Soluble starch	Flask	
Mucor circinelloides	[129]	0.8-2[b]	Cereal substrates	Solid state fermentation	
Mortierella isabellina	[130]	5.48	Deproteinized whey + 16% lactose	Flask	

(Table 5) contd.....

Strains	Ref.	GLA	Culture conditions		
		% TFA	Substrate	Culture mode	Experimental conditions
Mortierella isabellina	[37]	20.29	CMC	Flask	
		3.74	Xylose	Flask	Presence of a lignin derivative (vanillin)
		4.78	Xylose	Flask	Presence of a weak acid (levulinic acid)
Mortierella isabellina	[38]	3.85	Corn fiber hydrolysates	Flask	
Cunninghamella echinulata	[42]	16.7	Xylose	Flask	C/N 78
		16.6	Xylose	Flask	C/N 117
		21	Xylose	Flask	C/N 157
		19.5	Xylose	Flask	C/N 235
		22	Xylose	Flask	C/N 285
		18	Raw glycerol	Flask	C/N 78
		10	Raw glycerol	Flask	C/N 117
		13	Raw glycerol	Flask	C/N 157
		16.2	Raw glycerol	Flask	C/N 235
		15.5	Raw glycerol	Flask	C/N 285
Mortierella isabellina		8.7	Raw glycerol	Flask	C/N 78
		11.6	Raw glycerol	Flask	C/N 117
		12.8	Raw glycerol	Flask	C/N 157
		5.2	Raw glycerol	Flask	C/N 235
		9.7	Raw glycerol	Flask	C/N 285

[a] % wt.; [b] Yield (mg/g).

Microbial Production of DHA

Currently, two strains are commercially used for the production of DHA: *Crypthecodinium cohnii* and strains from the traustochytrid marine protists (Table **6**). *C. cohnii* is a heterotrophic dinoflagellate which can produce DHA content of more than 50% of its total fatty acids produced [118]. This dinoflagellate has been chosen by Kyle in 1992 [119] to produce DHA used in DHASCO, a commercial mixture of DHA (40-45%) and high oleic sunflower oil incorporated in infant formulae and other foods. The process is done in the United

States and is exploited by Martek Biosciences Corporation (now acquired by DSM). It uses stirred tank fermenters with glucose as the principal feedstock. In 1999, De Swaaf *et al.* [120] developed a high density fed-batch culture system for *C. cohnii* grown on ethanol and reported a final biomass concentration of 83 g/L, lipid concentration of 35 g/L and DHA concentration of 11.7 g/L in a 220 hours fermentation. In 2003, De Swaaf *et al.* [121] tested a fed-batch culture of *C. cohnii* using glucose as initial substrate and acetic acid as feeding medium and this system allows a final biomass concentration of 109 g/L, lipid concentration of 61 g/L and DHA concentration of 19 g/L in a 400-hour fermentation. The authors reported these results as being the highest biomass, lipid and DHA yields reported in heterotrophic dynoflagellate.

Table 6. Strains involved in DHA microbial production. TFA = Total Fatty Acids.

Strains	Reference	DHA		Culture conditions		
		% wt.	% TFA	Substrate	Culture mode	Experimental conditions
Thraustochytrids	[131]		28	Glucose + yeast extract	Bioreactor	
Thraustochytrium aureum	[132]		40	Glucose	Flask	Medium supplemented with glutamate
Schizochytrium sp.	[133]	5.23	13.22	Glucose	Flask	pH 7
		4.79	14.24	Fructose	Flask	
		5.57	16.08	Glucose	Flask	pH 7 + yeast extract 0.4 %
Aurantiochytrium limacinum	[134]		49.5-55.2	Glycerol	Fed-batch culture in bioreactor (10L)	C/N 0.5
			51.4-58.2	Glycerol	Fed-batch culture in bioreactor (10 L)	C/N 1.25
			57.3-66.6	Glycerol	Fed-batch culture in bioreactor (10 L)	C/N 1.875
			21.36	Glycerol	Flask	

(Table 6) contd.....

Strains	Reference	DHA		Culture conditions		
		% wt.	% TFA	Substrate	Culture mode	Experimental conditions
Crypthecodinium cohnii	[118]		51.12	Glucose	Flask	
Schizochytrium mangrovei	[135]		*ca.* 40	Glucose	Flask	Medium supplemented with coconut water

Thraustochytrids form a group of common marine microheterotrophs, originally classified as marine fungi and now taxonomically aligned with heterokont algae [122]. *Thraustochytrium sp.* and *Schizochytrium sp.* are the main organisms investigated for DHA production (Table **6**). Thanks to their efficiency to produce DHA, thraustochytrids participate actively in the transfer of this omega-3 into marine ecosystem and consequently to land [123]. Currently, DSM commercialises life's. DHA which is a nutraceutical containing DHA produced by *Schizochytrium sp.* or *C. cohnii.*

Microbial Production of EPA

Fungi belonging to Mucorales produce significant quantity of EPA (Table **7**). Usually this fatty acid occurs in lower quantities than the other fatty acids produced in fungi. Thraustochytrids have been envisaged as a novel source of production of EPA, together with DHA [124]. Algae are also potential candidate for an industrial production of EPA. The dietary supplement, life'sOMEGA produced by DSM, containing a mixture of EPA and DHA has an algal origin.

Table 7. Strains involved in EPA microbial production. TFA = Total Fatty Acids.

Strains	Reference	EPA		Culture conditions		
		% TFA	Yield mg/g	Substrate	Culture mode	Experimental conditions
Mortierella elongata	[136]	13		Soybean oil 3%	Flask	
Pythium irregulare		4.6		Soybean oil 3%	Flask	

(Table 7) contd.....

Strains	Reference	EPA		Culture conditions		
		% TFA	Yield mg/g	Substrate	Culture mode	Experimental conditions
Pythium ultimum	[87]	4.4	1.8	Coconut oil + spent malt grains	Solid-state cultivation	
		5.6	2.7	Peanut oil + spent malt grains	Solid-state cultivation	
		5.5	2.6	Sunflower oil + spent malt grains	Solid-state cultivation	
		8.2	3.4	Linseed oil + barley + spent malt grains	Solid-state cultivation	
Pythium irregulare	[137]		9.70 ± 1.73	Thin stillage	Flask	
			7.93 ± 2.40	Glycerol + thin stillage	Flask	
			8.40 ± 1.36	Glycerol + yeast extract	Flask	
			9.65 ± 1.24	Glucose + yeast extract	Flask	
Mortierella alpina	[138]	16.9 ± 0.7		Rice bran	Solid-state fermentation	Nitrogen supplementation

Microbial Production of ARA

ARA was the second PUFA produced by microorganisms to be developed after GLA production. Work on ARA production by oleaginous organisms is presented on Table **8**. Genus *Mortierella* is the most important producer of ARA representing up to 50% of the total fatty acids produced in wild strains. *Mortierella alpina* is the fungi currently used for the production of dietary supplements enriched in ARA commercialised by DSM. Takeno *et al.* [125] used transformants of *M. alpina* which showed higher percentages of ARA content compared to the wild strain. The transformant *M. alpina* pDZeoGLELO 63 presented an ARA content of over 70% of the total fatty acids, whereas wild *M. alpina* produced an ARA content of *ca.* 50%. More and more studies focus on mutants of fungi of interest to improve their ability to produce PUFAs [125] [126]. However, if metabolic engineering can lead to better results concerning the production of PUFAs, it is unsure that oils coming from genetically modified organisms will be accepted by the population, especially if it concerns infant formula and health.

Table 8. Strains involved in ARA microbial production. TFA = Total Fatty Acids.

Strains	Reference	ARA			Culture conditions		
		% wt.	% TFA	Yield mg/g	Substrate	Culture mode	Experimental conditions
Mortierella alpina	[125]		53.4 ± 0.98		Glucose + yeast extract	Flask	Wild strain
			62.2 ± 1.12		Glucose + yeast extract	Flask	Tranformant strain
			71.2 ± 1.34		Glucose + yeast extract	Flask	Transformant strain
Mortierella alpina	[139]	5.77 ± 0.42	15.29 ± 2.1		Glucose	Flask	Supplementation with glutamate 0.8 g/L
		5.02 ± 0.21	14.78 ± 0.97		Glucose	Fermenter 6L	Supplementation with glutamate 0.8 g/L
Mortierella elongata	[136]		58		Soybean oil 3%	Flask	
Pythium irregulare			8.3		Soybean oil 3%	Flask	
Pythium ultimum	[87]		5.2	2.2	Coconut oil + spent malt grains	Solid-state cultivation	
			4.4	2.1	Peanut oil + spent malt grains	Solid-state cultivation	
			4.4	2.0	Sunflower oil + spent malt grains	Solid-state cultivation	
			6.1	2.5	Linseed oil + barley + spent malt grains	Solid-state cultivation	

Processes for Microbial PUFAs Production

Submerged Fermentation

Submerged fermentation processes are similar to those presented in the Cultivation modes for lipid accumulation part. On a laboratory-scale, studies are mainly led in flasks with a batch process, as shown in Tables **5-8**. Fungi are also cultivated in bioreactors, using a batch, fed-batch or continuous strategy. This scale-up improves final biomass and lipid concentrations, since parameters are more efficiently regulated. However, several barriers have been identified for the growth of fungi in bioreactors.

To achieve a bioconversion in a fermenter, mass and heat transfer, gas dispersion and mixing of the medium are required [140]. In the case of fungal fermentation, the agitation fulfils the above functions but may also influence the morphology of fungi [140]. The fungal organization is often considered as one of the key parameters in industrial production in large fermenters. Two forms exist: dispersed mycelia or pellets. It has a strong effect on the physical properties of the fermentation broth. The fungal morphology is also a determinant component in the production of lipids of interest because the lipid concentration depends on the morphology. It has been demonstrated by Higashiyama *et al.* [141] that the morphology of *M. alpina* affects the productivity of ARA, with small pellets showing the highest productivity.

Microalgae are also producers of PUFAs of interest. Heterotrophic growth seems to be the preferential mode of culture among the others (mixotrophic, photoheterotrophic and phototrophic) which offers better oil productivity. However, heterotrophic culture can be contaminated very easily, particularly in open-culture systems. This risk of contamination hampers large-scale production. Phototrophic cultivation is mainly attractive due to the fact that CO_2 can be used as the sole carbon source and that no additional energy source is necessary besides light. The oil productivity of this approach is usually lower than that of heterotrophic cultivation. This is mainly due to slow cell growth and low biomass productivity. But phototrophic cultivation remains attractive [142].

Solid-state Fermentation

Solid-state fermentation (SSF) requires solid matrix and is carried out in absence or near absence of free water. However, the substrate should be moist enough to support microbial growth and metabolism. The solid matrix could be either the source of nutrients or simply a support impregnated by the nutrients [143]. In general, SSF is simple in terms of equipment and process control and less expensive than submerged fermentation. The elimination of water and mineral salts traditionally used for the constitution of a growth medium makes the process economically interesting. The substrates used in this fermentation mode are usually waste from raw agromaterials or by-products from the food industry such as cereal, rice bran and spent malt grain [87, 129, 138]. The feedstock used can be valorized by this approach as they are enriched in PUFAs. As a result, the product obtained from this fermentation consists in a cereal-derived product with a high content of valuable metabolites, which may be envisaged as a feed supplement in the future. Certik *et al.* [129] have demonstrated the efficiency of *M. circinelloides* to enrich cereal substrates naturally poor in PUFAs with GLA (24.2 g/kg solid substrate) and β-carotene. However, SSF remains actually well controlled in laboratory but is not yet used on an industrial scale for oil production.

Downstream Processing of PUFAs

Recovery of Oils in Fungi

The main typical processes for microbial lipids extraction have been described previously. However, resistance of the fungal wall becomes the rate limiting step for the recovery of fungal oils [72]. The fungal wall is an intricate structure composed of chitin, 1, 3-β- and 1, 6-β-glucan, mannan and proteins with a higher rigidity than that of plant cells. Consequently a pretreatment step has to be done before the extraction. Nisha *et al.* [72] have reported a better lipid yield when samples undergo enzymatic or HCl pretreatment than when the extraction was realized directly on control samples.

The same authors also used supercritical fluid to prevent lipids from auto-oxidative degradation. Used to extract lipids of a culture of *M. alpina*, supercritical CO_2, associated to an HCl pretreatment of the biomass, allows the recovery of 26.78% lipids and an ARA content of 33.28% [72].

Refining and Modification of Microbial oils

The final processing of microbial oils involves purification or refining and modification steps. Contaminants of the crude oil, which may affect the quality of the final product, should be removed or reduced by refining treatments. Antioxydants such as tocotrienols or tocopherol, which are present in the crude oil, have a beneficial effect on product quality. They also ensure a protection against oxidation. Purification and concentration of microbial oils require the use of several processes depending on the extracted fatty acid and the operation scale [10].

A summary of the general PUFA enrichment processes is presented in Table **9**. Solvent winterization applied on *Borago officinalis* oil allows a GLA enrichment of nearly 90% when hexane was used and when the temperature was reduced to 70 °C [144]. The efficiency of PUFA enrichment using solvent winterization is strongly temperature-dependant [145].

Table 9. Summary of general PUFA enrichment processes [96].

Method	**Procedure**
Winterization	Removal of saturated fats from oils by gentle cooling.
Triglyceride hydrolysis	Production of glycerol and free fatty acids using saponification or enzymes such as lipases
Urea complexation	Addition of urea and aqueous ethanol to the fatty acids mixture at high temperature. Formation of urea-saturated complexes by cooling down the mixture. Removal of complexes by filtration or centrifugation. Addition of hexane to the supernatant Recovery of PUFAs as free fatty acids from hexane by evaporation
Enzyme splitting	Selective esterification of saturated fatty acids by lipases Separation of PUFAs from the esterified saturated fatty acids fraction Repeat esterification for further PUFA enrichment
PUFA transformations	Esterification of PUFAs Interesterifications to enrich non-highly unsaturated fatty acids oils with PUFAs

Urea inclusion is also an enrichment method mainly based on the degree of unsaturation of the fatty acids. Saturated and mono-unsaturated fatty acids easily form urea compounds to be removed from the urea concentrate. When used for an ARA enrichment from *M. alpina* lipids, this technique allows to obtain an ARA content in the urea concentrate of 57.1% and a final ARA recovery of 81.9% [47].

Enzyme-based procedures are especially advantageous because enzymes such as lipases act under mild reaction conditions. Long chain PUFAs are very labile.

Methods implying extreme pH or temperature conditions can destroy the all *cis* nature of omega-3 fatty acids (such as EPA or DHA), by *cis-trans* isomerization, oxidation or migration of double bonds [96].

Lipase enzymes could be used to enrich PUFAs in oils. It can also produce different forms and compositions of PUFAs such as triglycerides, phospholipids, other fatty acids esters and free fatty acids [96]. The natural function of lipases is the hydrolysis of oils. Hydrolysis of triglycerides is an essential reaction for the production of fatty acids. Kahveci and Xu have reported that a lipase-catalyzed hydrolysis can contribute to enrich salmon oil with omega-3 PUFAs to more than the double of the initial level [146].

Future Outlook

The production of PUFAs by microbial strains is an industrial reality. The infant formula market is now around $11.5 billion and is still growing. The health benefits toward newborns, adults and elderly people brought by omega-3 and omega-6 are now widely known.

A better knowledge of the microbial capacities of production allows the synthesis of these important PUFAs on a large-scale level, provided by oleaginous organisms. Among them, fungi belonging to the Mucorales and thraustochytrids are efficient candidates to realize this production. *Mortierella alpina* in particular is widely used in large-scale production as it produces important quantities of different PUFAs with interesting growth rate and lipid productivity. There is also a growing interest toward the development of genetically modified organisms. Pieces of work focusing on recombinant strains which possess the ability to produce higher quantity of selective PUFA are increasing. The knowledge of metabolic pathways implied in the lipid production, together with the progress of molecular biology techniques allows to demonstrate the feasibility of this approach. However, as for transgenic plants, time, extensive investigations and testing procedures are necessary before the public opinion accepts this industrial mode of production.

Currently, the industrial production of PUFAs takes place using traditional fermenters and submerged fermentation. The development of solid-state fermentation which uses substrates such as agroindustrial and lignocellulosic waste could lead to the development of a new type of animal feed, as the final products could be enriched with PUFAs. The production of enriched-cereal with PUFAs can also be considered for the human diet.

The use of low-cost and renewable substrates should be considered in an ecological approach. Organic waste is perfectly suitable for the growth of microorganisms and studies have proved it can be bioconverted into PUFAs by particular microorganisms. In addition to be easily available, waste and byproducts contribute also to the decreasing of the final price of the product.

CONFLICT OF INTEREST

The authors confirm that they have no conflict of interest to declare for this publication.

ACKNOWLEDGEMENTS

Declared none.

REFERENCES

[1] Ghaly, A.E.; Dave, D.; Brooks, M.S.; Budge, S. Production of biodiesel by enzymatic transesterification. *Am. J. Biochem. Biotechnol.,* **2010**, *6*, 54-76. [http://dx.doi.org/10.3844/ajbbsp.2010.54.76]

[2] Demirbas, A. Biofuels sources, biofuel policy, biofuel economy and global biofuel projections. *Energy Convers. Manage.,* **2008**, *49*, 2106-2116. [http://dx.doi.org/10.1016/j.enconman.2008.02.020]

[3] Zhang, Y.; Dubé, M.A.; McLean, D.D.; Kates, M. Biodiesel production from waste cooking oil: 2. Economic assessment and sensitivity analysis. *Bioresour. Technol.,* **2003**, *90*(3), 229-240. [http://dx.doi.org/10.1016/S0960-8524(03)00150-0] [PMID: 14575945]

[4] Muniyappa, P.R.; Brammer, S.C.; Noureddini, H. Improved conversion of plant oils and animal fats into biodiesel and co-product. *Bioresour. Technol.,* **1996**, *56*, 19-24. [http://dx.doi.org/10.1016/0960-8524(95)00178-6]

[5] Demirbas, A. Biodiesel production *via* non-catalytic SCF method and biodiesel fuel characteristics. *Energy Convers. Manage.,* **2006**, *47*, 2271-2282. [http://dx.doi.org/10.1016/j.enconman.2005.11.019]

[6] Fischer, C.R.; Klein-Marcuschamer, D.; Stephanopoulos, G. Selection and optimization of microbial hosts for biofuels production. *Metab. Eng.,* **2008**, *10*(6), 295-304. [http://dx.doi.org/10.1016/j.ymben.2008.06.009] [PMID: 18655844]

[7] Sharma, Y.C.; Singh, B. Development of biodiesel: Current scenario. *Renew. Sustain. Energy Rev.*, **2009**, *13*, 1646-1651.
[http://dx.doi.org/10.1016/j.rser.2008.08.009]

[8] Liu, Z.Y.; Wang, G.C.; Zhou, B.C. Effect of iron on growth and lipid accumulation in *Chlorella vulgaris*. *Bioresour. Technol.*, **2008**, *99*(11), 4717-4722.
[http://dx.doi.org/10.1016/j.biortech.2007.09.073] [PMID: 17993270]

[9] Christophe, G.; Kumar, V.; Nouaille, R.; Gaudet, G.; Fontanille, P.; Pandey, A.; Soccol, C.R.; Larroche, C. Recent developments in microbial oils production: a possible alternative to vegetable oils for biodiesel without competition with human food? *Braz. Arch. Biol. Technol.*, **2012**, *55*, 29-46.
[http://dx.doi.org/10.1590/S1516-89132012000100004]

[10] Certik, M.; Shimizu, S. Biosynthesis and regulation of microbial polyunsaturated fatty acid production. *J. Biosci. Bioeng.*, **1999**, *87*(1), 1-14.
[http://dx.doi.org/10.1016/S1389-1723(99)80001-2] [PMID: 16232418]

[11] Alabdulkarim, B.; Bakeet, Z.A.N.; Arzoo, S. Role of some functional lipids in preventing diseases and promoting health. *J. King Saud Univ. Sci.*, **2012**, *24*, 319-329.

[12] Subramaniam, R.; Dufreche, S.; Zappi, M.; Bajpai, R. Microbial lipids from renewable resources: production and characterization. *J. Ind. Microbiol. Biotechnol.*, **2010**, *37*(12), 1271-1287.
[http://dx.doi.org/10.1007/s10295-010-0884-5] [PMID: 21086103]

[13] Nigam, P.S.; Singh, A. Production of liquid biofuels from renewable resources. *Pror. Energy Combust. Sci.*, **2011**, *37*, 52-68.
[http://dx.doi.org/10.1016/j.pecs.2010.01.003]

[14] Ratledge, C. Regulation of lipid accumulation in oleaginous micro-organisms. *Biochem. Soc. Trans.*, **2002**, *30*(Pt 6), 1047-1050.
[http://dx.doi.org/10.1042/bst0301047] [PMID: 12440969]

[15] Papanikolaou, S.; Galiotou-Panayotou, M.; Fakas, S.; Aggelis, G. Lipid production by oleaginous *Mucorales* cultivated on renewable carbon sources. *Eur. J. Lipid Sci. Technol.*, **2007**, *109*, 1060-1070.
[http://dx.doi.org/10.1002/ejlt.200700169]

[16] Angerbauer, C.; Siebenhofer, M.; Mittelbach, M.; Guebitz, G.M. Conversion of sewage sludge into lipids by *Lipomyces starkeyi* for biodiesel production. *Bioresour. Technol.*, **2008**, *99*(8), 3051-3056.
[http://dx.doi.org/10.1016/j.biortech.2007.06.045] [PMID: 17719773]

[17] Muniraj, I.K.; Xiao, L.; Hu, Z.; Zhan, X.; Shi, J. Microbial lipid production from potato processing wastewater using oleaginous filamentous fungi *Aspergillus oryzae*. *Water Res.*, **2013**, *47*(10), 3477-3483.
[http://dx.doi.org/10.1016/j.watres.2013.03.046] [PMID: 23597680]

[18] Hartwig Duarte, S.; Ghiselli, G.; Maugeri, F. Influence of lipid production by *Candida* sp. LEB-M3 using glycerol from biodiesel synthesis. *Biocatal. Agricult. Biotechnol*, **2013**, *2*, 339-343.

[19] Seo, Yh.; Lee, Ig.; Han, Ji. Cultivation and lipid production of yeast *Cryptococcus curvatus* using pretreated waste active sludge supernatant. *Bioresour. Technol.*, **2013**, *135*, 304-308.
[http://dx.doi.org/10.1016/j.biortech.2012.10.024] [PMID: 23177208]

[20] Ryu, B.G.; Kim, J.; Kim, K.; Choi, Y.E.; Han, J.I.; Yang, J.W. High-cell-density cultivation of oleaginous yeast *Cryptococcus curvatus* for biodiesel production using organic waste from the brewery industry. *Bioresour. Technol.,* **2013**, *135*, 357-364. [http://dx.doi.org/10.1016/j.biortech.2012.09.054] [PMID: 23177209]

[21] Chang, Y.H.; Chang, K.S.; Hsu, C.L.; Chuang, L.T.; Chen, C.Y.; Huang, F.Y.; Jang, H.D. A comparative study on batch and fed-batch cultures of oleaginous yeast *Cryptococcus* sp. in glucose-based media and corncob hydrolysate for microbial oil production. *Fuel,* **2013**, *105*, 711-717. [http://dx.doi.org/10.1016/j.fuel.2012.10.033]

[22] Gong, Z.; Wang, Q.; Shen, H.; Hu, C.; Jin, G.; Zhao, Z.K. Co-fermentation of cellobiose and xylose by *Lipomyces starkeyi* for lipid production. *Bioresour. Technol.,* **2012**, *117*, 20-24. [http://dx.doi.org/10.1016/j.biortech.2012.04.063] [PMID: 22609709]

[23] Ling, J.; Nip, S.; Shim, H. Enhancement of lipid productivity of *Rhodosporidium toruloides* in distillery wastewater by increasing cell density. *Bioresour. Technol.,* **2013**, *146*, 301-309. [http://dx.doi.org/10.1016/j.biortech.2013.07.023] [PMID: 23948267]

[24] Shen, H.; Gong, Z.; Yang, X.; Jin, G.; Bai, F.; Zhao, Z.K. Kinetics of continuous cultivation of the oleaginous yeast *Rhodosporidium toruloides. J. Biotechnol.,* **2013**, *168*(1), 85-89. [http://dx.doi.org/10.1016/j.jbiotec.2013.08.010] [PMID: 23965273]

[25] Uçkun Kiran, E.; Trzcinski, A.; Webb, C. Microbial oil produced from biodiesel by-products could enhance overall production. *Bioresour. Technol.,* **2013**, *129*, 650-654. [http://dx.doi.org/10.1016/j.biortech.2012.11.152] [PMID: 23298770]

[26] Lian, J.; Garcia-Perez, M.; Chen, S. Fermentation of levoglucosan with oleaginous yeasts for lipid production. *Bioresour. Technol.,* **2013**, *133*, 183-189. [http://dx.doi.org/10.1016/j.biortech.2013.01.031] [PMID: 23425586]

[27] Huang, C.; Chen, X.F.; Xiong, L.; Yang, X.Y.; Chen, X.D.; Ma, L.L.; Chen, Y. Microbial oil production from corncob acid hydrolysate by oleaginous yeast *Trichosporon coremiiforme. Biomass Bioenergy,* **2013**, *49*, 273-278. [http://dx.doi.org/10.1016/j.biombioe.2012.12.023]

[28] Chen, X.F.; Huang, C.; Yang, X.Y.; Xiong, L.; Chen, X.D.; Ma, L.L. Evaluating the effect of medium composition and fermentation condition on the microbial oil production by *Trichosporon cutaneum* on corncob acid hydrolysate. *Bioresour. Technol.,* **2013**, *143*, 18-24. [http://dx.doi.org/10.1016/j.biortech.2013.05.102] [PMID: 23774292]

[29] Peng, W.F.; Huang, C.; Chen, X.F.; Xiong, L.; Chen, X.D.; Chen, Y.; Ma, L.L. Microbial conversion of wastewater from butanol fermentation to microbial oil by oleaginous yeast *Trichosporon dermatis. Renew. Energy,* **2013**, *55*, 31-34. [http://dx.doi.org/10.1016/j.renene.2012.12.017]

[30] Zhan, J.; Lin, H.; Shen, Q.; Zhou, Q.; Zhao, Y. Potential utilization of waste sweetpotato vines hydrolysate as a new source for single cell oils production by *Trichosporon fermentans. Bioresour. Technol.,* **2013**, *135*, 622-629. [http://dx.doi.org/10.1016/j.biortech.2012.08.068] [PMID: 22985824]

[31] Cheirsilp, B.; Louhasakul, Y. Industrial wastes as a promising renewable source for production of

microbial lipid and direct transesterification of the lipid into biodiesel. *Bioresour. Technol.,* **2013**, *142*, 329-337.
[http://dx.doi.org/10.1016/j.biortech.2013.05.012] [PMID: 23747444]

[32] Wang, Z.P.; Xu, H.M.; Wang, G.Y.; Chi, Z.; Chi, Z.M. Disruption of the MIG1 gene enhances lipid biosynthesis in the oleaginous yeast *Yarrowia lipolytica* ACA-DC 50109. *Biochim. Biophys. Acta, Mol. Cell Biol. Lipids,* **1831**, *2013*, 675-682.

[33] Fontanille, P.; Kumar, V.; Christophe, G.; Nouaille, R.; Larroche, C. Bioconversion of volatile fatty acids into lipids by the oleaginous yeast *Yarrowia lipolytica. Bioresour. Technol.,* **2012**, *114*, 443-449.
[http://dx.doi.org/10.1016/j.biortech.2012.02.091] [PMID: 22464419]

[34] Venkata Subhash, G.; Venkata Mohan, S. Lipid accumulation of biodiesel production by oleaginous fungus *Aspergillus awamori*: Influence of critical factors. *Fuel,* **2014**, *116*, 509-515.
[http://dx.doi.org/10.1016/j.fuel.2013.08.035]

[35] Kamat, S.; Khot, M.; Zinjarde, S.; RaviKumar, A.; Gade, W.N. Coupled production of single cell oil as biodiesel feedstock, xylitol and xylanase from sugarcane bagasse in a biorefinery concept using fungi from the tropical mangrove wetlands. *Bioresour. Technol.,* **2013**, *135*, 246-253.
[http://dx.doi.org/10.1016/j.biortech.2012.11.059] [PMID: 23260270]

[36] Gao, D.; Zeng, J.; Zheng, Y.; Yu, X.; Chen, S. Microbial lipid production from xylose by *Mortierella isabellina. Bioresour. Technol.,* **2013**, *133*, 315-321.
[http://dx.doi.org/10.1016/j.biortech.2013.01.132] [PMID: 23434808]

[37] Zeng, J.; Zheng, Y.; Yu, X.; Yu, L.; Gao, D.; Chen, S. Lignocellulosic biomass as a carbohydrate source for lipid production by *Mortierella isabellina. Bioresour. Technol.,* **2013**, *128*, 385-391.
[http://dx.doi.org/10.1016/j.biortech.2012.10.079] [PMID: 23201519]

[38] Xing, D.; Wang, H.; Pan, A.; Wang, J.; Xue, D. Assimilation of corn fiber hydrolysates and lipid accumulation by *Mortierella isabellina. Biomass Bioenergy,* **2012**, *39*, 494-501.
[http://dx.doi.org/10.1016/j.biombioe.2012.01.024]

[39] Ruan, Z.; Zanotti, M.; Wang, X.; Ducey, C.; Liu, Y. Evaluation of lipid accumulation from lignocellulosic sugars by *Mortierella isabellina* for biodiesel production. *Bioresour. Technol.,* **2012**, *110*, 198-205.
[http://dx.doi.org/10.1016/j.biortech.2012.01.053] [PMID: 22330588]

[40] Mitra, D.; Rasmussen, M.L.; Chand, P.; Chintareddy, V.R.; Yao, L.; Grewell, D.; Verkade, J.G.; Wang, T.; van Leeuwen, J.H. Value-added oil and animal feed production from corn-ethanol stillage using the oleaginous fungus Mucor circinelloides. *Bioresour. Technol.,* **2012**, *107*, 368-375.
[http://dx.doi.org/10.1016/j.biortech.2011.12.031] [PMID: 22237170]

[41] Chatzifragkou, A.; Fakas, S.; Galiotou-Panayotou, M.; Aggelis, G.; Papanikolaou, S. Commercial sugars as substrates for lipid accumulation in *Cunninghamella echinulata* and *Mortierella isabellina* fungi. *Eur. J. Lipid Sci. Technol.,* **2010**, *112*, 1048-1057.
[http://dx.doi.org/10.1002/ejlt.201000027]

[42] Fakas, S.; Papanikolaou, S.; Batsos, A.; Galiotou-Panayotou, M.; Mallouchos, A.; Aggelis, G. Evaluating renewable carbon sources as substrates for single cell oil production by *Cunninghamella echinulata* and *Mortierella isabellina. Biomass Bioenergy,* **2009**, *33*, 573-580.
[http://dx.doi.org/10.1016/j.biombioe.2008.09.006]

[43] Papanikolaou, S.; Diamantopoulou, P.; Chatzifragkou, A.; Philippoussis, A.; Aggelis, G. Suitability of low-cost sugars as substrates for lipid production by the *fungus Thamnidium elegans*. *Energy Fuels,* **2010**, *24*, 4078-4086. [http://dx.doi.org/10.1021/ef1004804]

[44] Chatzifragkou, A.; Makri, A.; Belka, A.; Bellou, S.; Mavrou, M.; Mastoridou, M.; Mystrioti, P.; Onjaro, G.; Aggelis, G.; Papanikolaou, S. Biotechnological conversions of biodiesel derived waste glycerol by yeast and fungal species. *Energy,* **2011**, *36*, 1097-1108. [http://dx.doi.org/10.1016/j.energy.2010.11.040]

[45] Dey, P.; Banerjee, J.; Maiti, M.K. Comparative lipid profiling of two endophytic fungal isolates--*Colletotrichum* sp. and *Alternaria* sp. having potential utilities as biodiesel feedstock. *Bioresour. Technol.,* **2011**, *102*(10), 5815-5823. [http://dx.doi.org/10.1016/j.biortech.2011.02.064] [PMID: 21414775]

[46] Ratledge, C. Single cell oils--have they a biotechnological future? *Trends Biotechnol.,* **1993**, *11*(7), 278-284. [http://dx.doi.org/10.1016/0167-7799(93)90015-2] [PMID: 7763950]

[47] Zhu, M.; Zhou, P.P.; Yu, L.J. Extraction of lipids from *Mortierella alpina* and enrichment of arachidonic acid from the fungal lipids. *Bioresour. Technol.,* **2002**, *84*(1), 93-95. [http://dx.doi.org/10.1016/S0960-8524(02)00028-7] [PMID: 12137275]

[48] Li, Y.; Zhao, Z.K.; Bai, F. High density cultivation of oleaginous yeast *Rhodosporidium toruloides* Y4 in fedbatch culture. *Enzyme Microb. Technol.,* **2007**, *41*, 312-317. [http://dx.doi.org/10.1016/j.enzmictec.2007.02.008]

[49] Beopoulos, A.; Cescut, J.; Haddouche, R.; Uribelarrea, J.L.; Molina-Jouve, C.; Nicaud, J.M. *Yarrowia lipolytica* as a model for bio-oil production. *Prog. Lipid Res.,* **2009**, *48*(6), 375-387. [http://dx.doi.org/10.1016/j.plipres.2009.08.005] [PMID: 19720081]

[50] Feofilova, E.P.; Sergeeva, IaE.; Ivashechkin, A.A. [Biodiesel-fuel: content, production, producers, contemporary biotechnology (review)]. *Prikl. Biokhim. Mikrobiol.,* **2010**, *46*(4), 405-415. [Review]. [PMID: 20873163]

[51] Schujman, G.E.; de Mendoza, D. Regulation of type II fatty acid synthase in Gram-positive bacteria. *Curr. Opin. Microbiol.,* **2008**, *11*(2), 148-152. [http://dx.doi.org/10.1016/j.mib.2008.02.002] [PMID: 18372209]

[52] Wen, J.; Kerr, R.G. Purification and characterization of the fatty acid synthase from Bugula neritina. Comp. Biochem. Physiol., B. *Comp. Biochem,* **2001**, *128*, 445-450. [http://dx.doi.org/10.1016/S1096-4959(00)00343-2]

[53] Lomakin, I.B.; Xiong, Y.; Steitz, T.A. The crystal structure of yeast fatty acid synthase, a cellular machine with eight active sites working together. *Cell,* **2007**, *129*(2), 319-332. [http://dx.doi.org/10.1016/j.cell.2007.03.013] [PMID: 17448991]

[54] Ratledge, C.; Wynn, J.P. The biochemistry and molecular biology of lipid accumulation in oleaginous microorganisms. *Adv. Appl. Microbiol.,* **2002**, *51*, 1-51. [http://dx.doi.org/10.1016/S0065-2164(02)51000-5] [PMID: 12236054]

[55] Rattray, J.B.; Schibeci, A.; Kidby, D.K. Lipids of yeasts. *Bacteriol. Rev.,* **1975**, *39*(3), 197-231.

[PMID: 240350]

[56] Wynn, J.P.; bin Abdul Hamid, A.; Ratledge, C. The role of malic enzyme in the regulation of lipid accumulation in filamentous fungi. *Microbiology,* **1999**, *145*(Pt 8), 1911-1917. [http://dx.doi.org/10.1099/13500872-145-8-1911] [PMID: 10463157]

[57] Nigam, P. Production of oils and fatty acids. In: *Encyclopedia of Food Microbiology,* 2nd ed; Robinson, C.A.; Batt, P.D., Eds.; , **2000**; pp. 18-729.

[58] Beopoulos, A.; Chardot, T.; Nicaud, J.M. *Yarrowia lipolytica*: A model and a tool to understand the mechanisms implicated in lipid accumulation. *Biochimie,* **2009**, *91*(6), 692-696. [http://dx.doi.org/10.1016/j.biochi.2009.02.004] [PMID: 19248816]

[59] Gálvez, S.; Gadal, P. On the function of the NADP dependent isocitrate dehydrogenase isoenzymes in living organisms. *Plant Sci.,* **1995**, *105*, 1-14. [http://dx.doi.org/10.1016/0168-9452(94)04041-E]

[60] Meng, X.; Yang, J.; Xu, X.; Zhang, L.; Nie, Q.; Xian, M. Biodiesel production from oleaginous microorganisms. *Renew. Energy,* **2009**, *34*, 1-5. [http://dx.doi.org/10.1016/j.renene.2008.04.014]

[61] Wynn, J.P.; Ratledge, C. Malic enzyme is the major source of NADPH for lipid accumulation by *Aspergillus nidulans. Microbiology,* **1997**, *143*, 253-257. [http://dx.doi.org/10.1099/00221287-143-1-253]

[62] Peralta-Yahya, P.P.; Keasling, J.D. Advanced biofuel production in microbes. *Biotechnol. J.,* **2010**, *5*(2), 147-162. [http://dx.doi.org/10.1002/biot.200900220] [PMID: 20084640]

[63] Ykema, A.; Verbree, E.C.; van Verseveld, H.W.; Smit, H. Mathematical modelling of lipid production by oleaginous yeasts in continuous cultures. *Antonie van Leeuwenhoek,* **1986**, *52*(6), 491-506. [http://dx.doi.org/10.1007/BF00423410] [PMID: 3813522]

[64] Ratledge, C. Yeasts, Moulds, Algae and Bacteria as Sources of Lipids. In: *Technological Advances in Improves and Alternative Sources of Lipids*; Kamel, S.; Kakuda, Y., Eds.; , **1994**; pp. 235-291. [http://dx.doi.org/10.1007/978-1-4615-2109-9_9]

[65] Papanikolaou, S.; Aggelis, G. Lipid production by *Yarrowia lipolytica* growing on industrial glycerol in a single-stage continuous culture. *Bioresour. Technol.,* **2002**, *82*(1), 43-49. [http://dx.doi.org/10.1016/S0960-8524(01)00149-3] [PMID: 11848376]

[66] Zhang, J.; Fang, X.; Zhu, X.L.; Li, Y.; Xu, H.P.; Zhao, B.F.; Chen, L.; Zhang, X.D. Microbial lipid production by the oleaginous yeast *Cryptococcus curvatus* O3 grown in fed-batch culture. *Biomass Bioenergy,* **2011**, *35*, 1906-1911. [http://dx.doi.org/10.1016/j.biombioe.2011.01.024]

[67] Folch, J.; Lees, M.; Sloane Stanley, G.H. A simple method for the isolation and purification of total lipides from animal tissues. *J. Biol. Chem.,* **1957**, *226*(1), 497-509. [PMID: 13428781]

[68] Cot, M.; Alfenore, S.; Cameleyre, X.; Le Berre, V.; Sokil, S.; Jasson, S.; Uribelarrea, J.L.; Molina-Jouve, C.; Goma, G.; Francois, J.; Guillouet, S.E.; Benbadis, L. A genomic analysis of the dynamic adaptation of the yeast *Saccharomyces cerevisiae* to high concentration of produced ethanol. *26th*

Symposium on Biotechnology for Biofuels and Chemicals, **2004**.Chattanooga, USA

[69] Bligh, E.G.; Dyer, W.J. A rapid method of total lipid extraction and purification. *Can. J. Biochem. Physiol.,* **1959**, *37*(8), 911-917.
[http://dx.doi.org/10.1139/o59-099] [PMID: 13671378]

[70] Cescut, J.; Severac, E.; Molina-Jouve, C.; Uribelarrea, J.L. Optimizing pressurized liquid extraction of microbial lipids using the response surface method. *J. Chromatogr. A,* **2011**, *1218*(3), 373-379.
[http://dx.doi.org/10.1016/j.chroma.2010.12.003] [PMID: 21185025]

[71] Dejoye Tanzi, C.; Abert Vian, M.; Chemat, F. New procedure for extraction of algal lipids from wet biomass: a green clean and scalable process. *Bioresour. Technol.,* **2013**, *134*, 271-275.
[http://dx.doi.org/10.1016/j.biortech.2013.01.168] [PMID: 23500584]

[72] Nisha, A.; Udaya Sankar, K. Venkateswaran, G. Supercritical CO2 extraction of *Mortierella alpina* single cell oil: Comparison with organic solvent extraction. *Food Chem.,* **2012**, *133*, 220-226.
[http://dx.doi.org/10.1016/j.foodchem.2011.12.081]

[73] Kim, Y.H.; Choi, Y.K.; Park, J.; Lee, S.; Yang, Y.H.; Kim, H.J.; Park, T.J.; Hwan Kim, Y.; Lee, S.H. Ionic liquid-mediated extraction of lipids from algal biomass. *Bioresour. Technol.,* **2012**, *109*, 312-315.
[http://dx.doi.org/10.1016/j.biortech.2011.04.064] [PMID: 21601447]

[74] Lee, J.Y.; Yoo, C.; Jun, S.Y.; Ahn, C.Y.; Oh, H.M. Comparison of several methods for effective lipid extraction from microalgae. *Bioresour. Technol.,* **2010**, *101* Suppl. 1, S75-S77.
[http://dx.doi.org/10.1016/j.biortech.2009.03.058] [PMID: 19386486]

[75] Zhu, M.; Yu, L.J.; Liu, Z.; Xu, H.B. Isolating *Mortierella alpina* strains of high yield of arachidonic acid. *Lett. Appl. Microbiol.,* **2004**, *39*(4), 332-335.
[http://dx.doi.org/10.1111/j.1472-765X.2004.01581.x] [PMID: 15355534]

[76] Chen, W.; Zhang, C.; Song, L.; Sommerfeld, M.; Hu, Q. A high throughput Nile red method for quantitative measurement of neutral lipids in microalgae. *J. Microbiol. Methods,* **2009**, *77*(1), 41-47.
[http://dx.doi.org/10.1016/j.mimet.2009.01.001] [PMID: 19162091]

[77] Huang, G.H.; Chen, G.; Chen, F. Rapid screening method for lipid production in alga based on Nile red fluorescence. *Biomass Bioenergy,* **2009**, *33*, 1386-1392.
[http://dx.doi.org/10.1016/j.biombioe.2009.05.022]

[78] Lee, S.J.; Yoon, B.D.; Oh, H.M. Rapid method for the determination of lipid from the green alga *Botryococcus braunii. Biotechnol. Tech.,* **1998**, *12*, 553-556.
[http://dx.doi.org/10.1023/A:1008811716448]

[79] Govender, T.; Ramanna, L.; Rawat, I.; Bux, F. BODIPY staining, an alternative to the Nile Red fluorescence method for the evaluation of intracellular lipids in microalgae. *Bioresour. Technol.,* **2012**, *114*, 507-511.
[http://dx.doi.org/10.1016/j.biortech.2012.03.024] [PMID: 22464420]

[80] Lu, Y.; Ludsin, S.A.; Fanslow, D.L.; Pothoven, S. Comparison of three microquantity techniques for measuring total lipids in fish. *Can. J. Fish. Aquat. Sci.,* **2008**, *65*, 2233-2241.
[http://dx.doi.org/10.1139/F08-135]

[81] Gessner, M.O.; Neumann, P.T. Total lipids. In: *Methods to study litter decomposition: A practical*

guide; MAS, Graça; F, Bärlocher; MO, Gessner, Eds.; Berlin : Springer, **2005**; pp. 91-96. [http://dx.doi.org/10.1007/1-4020-3466-0_13]

[82] Peng, X.; Chen, H. Rapid estimation of single cell oil content of solid-state fermented mass using near-infrared spectroscopy. *Bioresour. Technol.,* **2008**, *99*(18), 8869-8872. [http://dx.doi.org/10.1016/j.biortech.2008.04.055] [PMID: 18534844]

[83] Horst, M.; Urbin, S.; Burton, R.; McMillan, C. Using proton nuclear magnetic resonance as a rapid response tool for methyl ester characterization in biodiesel. *Lipid Tech,* **2009**, *21*, 39-41. [http://dx.doi.org/10.1002/lite.200900004]

[84] Gill, C.O.; Hall, M.J.; Ratledge, C. Lipid accumulation in an oleaginous yeast (*Candida* 107) growing on glucose in single-stage continuous culture. *Appl. Environ. Microbiol.,* **1977**, *33*(2), 231-239. [PMID: 848947]

[85] Kendrick, A.; Ratledge, C. Lipid formation in the oleaginous mould *Entomophthora exitalis* grown in continuous culture: effects of growth rate, temperature and dissolved oxygen tension on polyunsaturated fatty acids. *Appl. Microbiol. Biotechnol.,* **1992**, *37*, 18-22. [http://dx.doi.org/10.1007/BF00174196]

[86] Shimada, Y.; Sugihara, A.; Maruyama, K.; Nagao, T.; Nakayama, S.; Nakano, H.; Tominaga, Y. Enrichment of arachidonic acid: Selective hydrolysis of a single-cell oil from *Mortierella* with *Candida cylindracea* lipase. *J. Am. Oil Chem. Soc.,* **1995**, *72*, 1323-1327. [http://dx.doi.org/10.1007/BF02546206]

[87] Stredansky, M.; Conti, E.; Salaris, A. Production of polyunsaturated fatty acids by *Pythium ultimum* in solid-state cultivation. *Enzyme Microb. Technol.,* **2000**, *26*(2-4), 304-307. [http://dx.doi.org/10.1016/S0141-0229(99)00146-5] [PMID: 10689092]

[88] Kavadia, A.; Komaitis, M.; Chevalot, I.; Blanchard, F.; Marc, I.; Aggelis, G. Lipid and γ-linolenic acid accumulation in strains of Zygomycetes growing on glucose. *J. Am. Oil Chem. Soc.,* **2001**, *78*, 341-346. [http://dx.doi.org/10.1007/s11746-001-0266-3]

[89] Papanikolaou, S.; Komaitis, M.; Aggelis, G. Single cell oil (SCO) production by *Mortierella isabellina* grown on high-sugar content media. *Bioresour. Technol.,* **2004**, *95*(3), 287-291. [http://dx.doi.org/10.1016/j.biortech.2004.02.016] [PMID: 15288271]

[90] El Menyawi, I.; Wögerbauer, M.; Sigmund, H.; Burgmann, H.; Graninger, W. Identification of yeast species by fatty acid profiling as measured by gas-liquid chromatography. *J. Chromatogr. B Biomed. Sci. Appl.,* **2000**, *742*(1), 13-24. [http://dx.doi.org/10.1016/S0378-4347(00)00044-X] [PMID: 10892580]

[91] Xue, F.; Miao, J.; Zhang, X.; Luo, H.; Tan, T. Studies on lipid production by *Rhodotorula glutinis* fermentation using monosodium glutamate wastewater as culture medium. *Bioresour. Technol.,* **2008**, *99*(13), 5923-5927. [http://dx.doi.org/10.1016/j.biortech.2007.04.046] [PMID: 18420404]

[92] Haas, M.J.; McAloon, A.J.; Yee, W.C.; Foglia, T.A. A process model to estimate biodiesel production costs. *Bioresour. Technol.,* **2006**, *97*(4), 671-678. [http://dx.doi.org/10.1016/j.biortech.2005.03.039] [PMID: 15935657]

[93] Fei, Q.; Chang, H.N.; Shang, L.; Choi, J.D.; Kim, N.; Kang, J. The effect of volatile fatty acids as a sole carbon source on lipid accumulation by *Cryptococcus albidus* for biodiesel production. *Bioresour. Technol.,* **2011**, *102*(3), 2695-2701.
[http://dx.doi.org/10.1016/j.biortech.2010.10.141] [PMID: 21134744]

[94] Fan, X.; Burton, R.; Zhou, Y. Glycerol (bioproduct of biodiesel production) as a source for fuels and chemicals – Mini review. *TOEFJ,* **2010**, *3*, 17-22.
[http://dx.doi.org/10.2174/1876973X01003010017]

[95] Yoon, S.J.; Choi, Y.C.; Son, Y.I.; Lee, S.H.; Lee, J.G. Gasification of biodiesel by-product with air or oxygen to make syngas. *Bioresour. Technol.,* **2010**, *101*(4), 1227-1232.
[http://dx.doi.org/10.1016/j.biortech.2009.09.039] [PMID: 19819133]

[96] Ward, O.P.; Singh, A. Omega-3/6 fatty acids: Alternative sources of production. *Process Biochem.,* **2005**, *40*, 3627-3652.
[http://dx.doi.org/10.1016/j.procbio.2005.02.020]

[97] Chang, C.L.; Seo, T.; Matsuzaki, M.; Worgall, T.S.; Deckelbaum, R.J. n-3 fatty acids reduce arterial LDL-cholesterol delivery and arterial lipoprotein lipase levels and lipase distribution. *Arterioscler. Thromb. Vasc. Biol.,* **2009**, *29*(4), 555-561.
[http://dx.doi.org/10.1161/ATVBAHA.108.182287] [PMID: 19201689]

[98] Sudheendran, S.; Chang, C.C.; Deckelbaum, R.J. N-3 *vs.* saturated fatty acids: effects on the arterial wall. *Prostaglandins Leukot. Essent. Fatty Acids,* **2010**, *82*(4-6), 205-209.
[http://dx.doi.org/10.1016/j.plefa.2010.02.020] [PMID: 20207121]

[99] Kohli, P.; Levy, B.D. Resolvins and protectins: mediating solutions to inflammation. *Br. J. Pharmacol.,* **2009**, *158*(4), 960-971.
[http://dx.doi.org/10.1111/j.1476-5381.2009.00290.x] [PMID: 19594757]

[100] Terano, T.; Tanaka, T.; Tamura, Y.; Kitagawa, M.; Higashi, H.; Saito, Y.; Hirai, A. Eicosapentaenoic acid and docosahexaenoic acid inhibit vascular smooth muscle cell proliferation by inhibiting phosphorylation of Cdk2-cyclinE complex. *Biochem. Biophys. Res. Commun.,* **1999**, *254*(2), 502-506.
[http://dx.doi.org/10.1006/bbrc.1998.9976] [PMID: 9918868]

[101] Thies, F.; Garry, J.M.; Yaqoob, P.; Rerkasem, K.; Williams, J.; Shearman, C.P.; Gallagher, P.J.; Calder, P.C.; Grimble, R.F. Association of n-3 polyunsaturated fatty acids with stability of atherosclerotic plaques: a randomised controlled trial. *Lancet,* **2003**, *361*(9356), 477-485.
[http://dx.doi.org/10.1016/S0140-6736(03)12468-3] [PMID: 12583947]

[102] Birch, E.E.; Garfield, S.; Castañeda, Y.; Hughbanks-Wheaton, D.; Uauy, R.; Hoffman, D. Visual acuity and cognitive outcomes at 4 years of age in a double-blind, randomized trial of long-chain polyunsaturated fatty acid-supplemented infant formula. *Early Hum. Dev.,* **2007**, *83*(5), 279-284.
[http://dx.doi.org/10.1016/j.earlhumdev.2006.11.003] [PMID: 17240089]

[103] Whelan, J. The health implications of changing linoleic acid intakes. *Prostaglandins Leukot. Essent. Fatty Acids,* **2008**, *79*(3-5), 165-167.
[http://dx.doi.org/10.1016/j.plefa.2008.09.013] [PMID: 18990554]

[104] Horrobin, D.F. Nutritional and medical importance of gamma-linolenic acid. *Prog. Lipid Res.,* **1992**, *31*(2), 163-194.

[http://dx.doi.org/10.1016/0163-7827(92)90008-7] [PMID: 1334266]

[105] Cho, H.P.; Nakamura, M.T.; Clarke, S.D. Cloning, expression, and nutritional regulation of the mammalian Δ-6 desaturase. *J. Biol. Chem.,* **1999**, *274*(1), 471-477. [http://dx.doi.org/10.1074/jbc.274.1.471] [PMID: 9867867]

[106] Sehat, N.; Kramer, J.K.; Mossoba, M.M.; Yurawecz, M.P.; Roach, J.A.; Eulitz, K.; Morehouse, K.M.; Ku, Y. Identification of conjugated linoleic acid isomers in cheese by gas chromatography, silver ion high performance liquid chromatography and mass spectral reconstructed ion profiles. Comparison of chromatographic elution sequences. *Lipids,* **1998**, *33*(10), 963-971. [http://dx.doi.org/10.1007/s11745-998-0293-8] [PMID: 9832075]

[107] Simopoulos, A.P. The importance of the ratio of omega-6/omega-3 essential fatty acids. *Biomed. Pharmacother.,* **2002**, *56*(8), 365-379. [http://dx.doi.org/10.1016/S0753-3322(02)00253-6] [PMID: 12442909]

[108] Momiyama, Y. Association between serum omega-3 to omega-6 polyunsaturated fatty acid ratio and cardiovascular events in a general Japanese population. *Atherosclerosis,* **2013**, *231*(2), 281-282. [http://dx.doi.org/10.1016/j.atherosclerosis.2013.09.032] [PMID: 24267240]

[109] Simopoulos, A.P. Evolutionary aspects of diet, the omega-6/omega-3 ratio and genetic variation: nutritional implications for chronic diseases. *Biomed. Pharmacother.,* **2006**, *60*(9), 502-507. [http://dx.doi.org/10.1016/j.biopha.2006.07.080] [PMID: 17045449]

[110] FAO. Fats and fatty acids in human nutrition. Report of an expert consultation. *FAO Food Nutr. Pap.,* **2010**, *91*, 1-166. [PMID: 21812367]

[111] Ratledge, C. Fatty acid biosynthesis in microorganisms being used for Single Cell Oil production. *Biochimie,* **2004**, *86*(11), 807-815. [http://dx.doi.org/10.1016/j.biochi.2004.09.017] [PMID: 15589690]

[112] FDA. *Generally Recognized as Safe (GRAS). [ONLINE],* [Accessed 02 October 2013]; Available at: http://www.fda.gov/Food/IngredientsPackagingLabeling/GRAS/

[113] Wallis, J.G.; Watts, J.L.; Browse, J. Polyunsaturated fatty acid synthesis: what will they think of next? *Trends Biochem. Sci.,* **2002**, *27*(9), 467-473. [http://dx.doi.org/10.1016/S0968-0004(02)02168-0] [PMID: 12217522]

[114] Metz, J.G.; Kuner, J.; Rosenzweig, B.; Lippmeier, J.C.; Roessler, P.; Zirkle, R. Biochemical characterization of polyunsaturated fatty acid synthesis in Schizochytrium: release of the products as free fatty acids. *Plant Physiol. Biochem.,* **2009**, *47*(6), 472-478. [http://dx.doi.org/10.1016/j.plaphy.2009.02.002] [PMID: 19272783]

[115] Metz, J.G.; Roessler, P.; Facciotti, D.; Levering, C.; Dittrich, F.; Lassner, M.; Valentine, R.; Lardizabal, K.; Domergue, F.; Yamada, A.; Yazawa, K.; Knauf, V.; Browse, J. Production of polyunsaturated fatty acids by polyketide synthases in both prokaryotes and eukaryotes. *Science,* **2001**, *293*(5528), 290-293. [http://dx.doi.org/10.1126/science.1059593] [PMID: 11452122]

[116] Qiu, X. Biosynthesis of docosahexaenoic acid (DHA, 22:6-4, 7,10,13,16,19): two distinct pathways. *Prostaglandins Leukot. Essent. Fatty Acids,* **2003**, *68*(2), 181-186.

[http://dx.doi.org/10.1016/S0952-3278(02)00268-5] [PMID: 12538082]

[117] Ghasemnezhad, A.; Honermeier, B. Yield, oil constituents, and protein content of evening primrose (*Oenothera biennis* L.) seeds depending on harvest time, harvest method and nitrogen application. *Ind. Crops Prod.,* **2008**, *28*, 17-23. [http://dx.doi.org/10.1016/j.indcrop.2007.12.006]

[118] Jiang, Y.; Chen, F.; Liang, S.Z. Production potential of docosahexaenoic acid by the heterotrophic marine dinoflagellate *Crypthecodinium cohnii. Process Biochem.,* **1999**, *34*, 633-637. [http://dx.doi.org/10.1016/S0032-9592(98)00134-4]

[119] Kyle, D.J. Microbial oil mixtures and uses thereof US Patent 5,374,657, 1992.

[120] De Swaaf, M.E.; De Rijk, T.C.; Eggink, G.; Sijtsma, L. Optimisation of docosahexaenoic acid production in batch cultivations by *Crypthecodinium cohnii. J. Biotechnol.,* **1999**, *70*, 185-192. [http://dx.doi.org/10.1016/S0168-1656(99)00071-1]

[121] de Swaaf, M.E.; Pronk, J.T.; Sijtsma, L. Fed-batch cultivation of the docosahexaenoic-acid-producing marine alga Crypthecodinium cohnii on ethanol. *Appl. Microbiol. Biotechnol.,* **2003**, *61*(1), 40-43. [http://dx.doi.org/10.1007/s00253-002-1118-1] [PMID: 12658513]

[122] Dick, M.W. *Straminipolous fungi: Systematics of the Peronosporomyletes*; Kluwer Academic Publishers: Dordrecht, NL, **2001**. [http://dx.doi.org/10.1007/978-94-015-9733-3]

[123] Gladyshev, M.I.; Sushchik, N.N.; Makhutova, O.N. Production of EPA and DHA in aquatic ecosystems and their transfer to the land. Prostag. Oth. *Lipid M. Prostaglandins Other Lipid Mediat,* **2013**, *107*, 117-126. [http://dx.doi.org/10.1016/j.prostaglandins.2013.03.002]

[124] Gupta, A.; Barrow, C.J.; Puri, M. Omega-3 biotechnology: Thraustochytrids as a novel source of omega-3 oils. *Biotechnol. Adv.,* **2012**, *30*(6), 1733-1745. [http://dx.doi.org/10.1016/j.biotechadv.2012.02.014] [PMID: 22406165]

[125] Takeno, S.; Sakuradani, E.; Tomi, A.; Inohara-Ochiai, M.; Kawashima, H.; Shimizu, S. Transformation of oil-producing fungus, Mortierella alpina 1S-4, using Zeocin, and application to arachidonic acid production. *J. Biosci. Bioeng.,* **2005**, *100*(6), 617-622. [http://dx.doi.org/10.1263/jbb.100.617] [PMID: 16473770]

[126] Sakuradani, E.; Ando, A.; Shimizu, S.; Ogawa, J. Metabolic engineering for the production of polyunsaturated fatty acids by oleaginous fungus *Mortierella alpina* 1S-4. *J. Biosci. Bioeng.,* **2013**, *116*(4), 417-422. [http://dx.doi.org/10.1016/j.jbiosc.2013.04.008] [PMID: 23648102]

[127] Dyal, S.D.; Bouzidi, L.; Narine, S.S. Maximizing the production of γ-linolenic acid in *Mortierella ramanniana* var. *ramanniana* as a function of pH, temperature and carbon source, nitrogen source, metal ions and oil supplementation. *Food Res. Int.,* **2005**, *38*, 815-829. [http://dx.doi.org/10.1016/j.foodres.2005.04.002]

[128] Chen, H.C.; Liu, T.M. Inoculum effects on the production of γ-linolenic acid by the shake culture of *Cunninghamella echinulata* CCRC 31840. *Enzyme Microb. Technol.,* **1997**, *21*, 137-142. [http://dx.doi.org/10.1016/S0141-0229(96)00262-1]

[129] Čertík, M.; Adamechová, Z.; Guothová, L. Simultaneous enrichment of cereals with polyunsaturated fatty acids and pigments by fungal solid state fermentations. *J. Biotechnol.,* **2013**, *168*(2), 130-134. [http://dx.doi.org/10.1016/j.jbiotec.2013.03.016] [PMID: 23583333]

[130] Demir, M.; Turhan, I.; Kucukcetin, A.; Alpkent, Z. Oil production by *Mortierella isabellina* from whey treated with lactase. *Bioresour. Technol.,* **2013**, *128*, 365-369. [http://dx.doi.org/10.1016/j.biortech.2012.10.078] [PMID: 23201516]

[131] Bowles, R.D.; Hunt, A.E.; Bremer, G.B.; Duchars, M.G.; Eaton, R.A. Long-chain n-3 polyunsaturated fatty acid production by members of the marine protistan group the thraustochytrids: screening of isolates and optimization of docosahexaenoic acid production. *J. Biotechnol.,* **1999**, *70*, 192-202. [http://dx.doi.org/10.1016/S0168-1656(99)00072-3]

[132] Iida, I.; Nakahara, N.; Yokoshi, T.; Kamisaka, Y.; Yagi, H.; Yamaoka, M.; Suzuki, O. Improvement of docosahexaenoic acid production in a culture of *Thraustochytrium aureum* by medium optimization. *J. Ferment. Bioeng.,* **1996**, *81*, 76-78. [http://dx.doi.org/10.1016/0922-338X(96)83125-4]

[133] Wu, S.T.; Yu, S.T.; Lin, L.P. Effect of culture conditions on docosahexaenoic acid production by *Schizochytrium* sp. S31. *Process Biochem.,* **2005**, *40*, 3103-3108. [http://dx.doi.org/10.1016/j.procbio.2005.03.007]

[134] Huang, T.Y.; Lu, W.C.; Chu, I.M. A fermentation strategy for producing docosahexaenoic acid in *Aurantiochytrium limacinum* SR21 and increasing C22:6 proportions in total fatty acid. *Bioresour. Technol.,* **2012**, *123*, 8-14. [http://dx.doi.org/10.1016/j.biortech.2012.07.068] [PMID: 22929740]

[135] Unagul, P.; Assantachai, C.; Phadungruengluij, S.; Suphantharika, M.; Tanticharoen, M.; Verduyn, C. Coconut water as a medium additive for the production of docosahexaenoic acid (C22:6 n3) by *Schizochytrium mangrovei* Sk-02. *Bioresour. Technol.,* **2007**, *98*(2), 281-287. [http://dx.doi.org/10.1016/j.biortech.2006.01.013] [PMID: 16563749]

[136] Cheng, M.H.; Walker, T.H.; Hulbert, G.J.; Raman, D.R. Fungal production of eicosapentenoic and arachidonic acids from industrial waste streams and crude soybean oil. *Bioresour. Technol.,* **1999**, *67*, 101-110. [http://dx.doi.org/10.1016/S0960-8524(98)00113-8]

[137] Liang, Y.; Zhao, X.; Strait, M.; Wen, Z. Use of dry-milling derived thin stillage for producing eicosapentaenoic acid (EPA) by the fungus *Pythium irregulare. Bioresour. Technol.,* **2012**, *111*, 404-409. [http://dx.doi.org/10.1016/j.biortech.2012.02.035] [PMID: 22386467]

[138] Jang, H.D.; Yang, S.S. Polyunsaturated fatty acids production with a solid-state column reactor. *Bioresour. Technol.,* **2008**, *99*(14), 6181-6189. [http://dx.doi.org/10.1016/j.biortech.2007.12.024] [PMID: 18230414]

[139] Yu, L.J.; Qin, W.M.; Lan, W.Z.; Zhou, P.P.; Zhu, M. Improved arachidonic acids production from the fungus Mortierella alpina by glutamate supplementation. *Bioresour. Technol.,* **2003**, *88*(3), 265-268. [http://dx.doi.org/10.1016/S0960-8524(02)00312-7] [PMID: 12618051]

[140] Cui, Y.Q.; van der Lans, R.G.; Luyben, K.C. Effect of agitation intensities on fungal morphology of

submerged fermentation. *Biotechnol. Bioeng.,* **1997**, *55*(5), 715-726.
[http://dx.doi.org/10.1002/(SICI)1097-0290(19970905)55:5<715::AID-BIT2>3.0.CO;2-E] [PMID: 18636582]

[141] Higashiyama, K.; Yaguchi, T.; Akimoto, K.; Fujikawaa, S.; Shimizu, S. Effects of mineral addition on the growth morphology of and arachidonic production by *Mortierella alpina* 1S-4. *J. Am. Oil Chem. Soc.,* **1998**, *75*, 1815-1819.
[http://dx.doi.org/10.1007/s11746-998-0336-9]

[142] Chen, C.Y.; Yeh, K.L.; Aisyah, R.; Lee, D.J.; Chang, J.S. Cultivation, photobioreactor design and harvesting of microalgae for biodiesel production: a critical review. *Bioresour. Technol.,* **2011**, *102*(1), 71-81.
[http://dx.doi.org/10.1016/j.biortech.2010.06.159] [PMID: 20674344]

[143] Singhania, R.R.; Patel, A.K.; Soccol, C.R.; Pandey, A. Recent advances in solid-state fermentation. *Biochem. Eng. J.,* **2009**, *44*, 13-18.
[http://dx.doi.org/10.1016/j.bej.2008.10.019]

[144] López-Martínez, J.C.; Campra-Madrid, P.; Guil-Guerrero, J.L. γ-linolenic acid enrichment from *Borago officinalis* and *Echium fastuosum* seed oils and fatty acids by low temperature crystallization. *J. Biosci. Bioeng.,* **2004**, *97*(5), 294-298.
[http://dx.doi.org/10.1016/S1389-1723(04)70208-X] [PMID: 16233632]

[145] Yokoshi, T.; Usifa, M.T.; Kamisaka, Y.; Nakahara, T.; Suzuki, O. Increase in the γ-linolenic acid content by solvent winterization of fungal oil extracted from *Mortierella* genus. *J. Am. Oil Chem. Soc.,* **1990**, *67*, 846-851.
[http://dx.doi.org/10.1007/BF02540504]

[146] Kahveci, D.; Xu, X. Repeated hydrolysis process is effective for enrichment of omega-3 polyunsaturated fatty acids in salmon oil by *Candida rugosa* lipase. *Food Chem.,* **2011**, *129*, 1552-1558.
[http://dx.doi.org/10.1016/j.foodchem.2011.05.142]

CHAPTER 9

Biotechnological Production of Hydrocolloids

Cristine Rodrigues, Júlio C. de Carvalho*, Luciana P. S. Vandenberghe, Juliana de Oliveira, Rodrigo Malanski and **Carlos R. Soccol**

Bioprocess Engineering and Biotechnology Department, Federal University of Paraná, Curitiba, Brazil

Abstract: Hydrocolloids are an heterogeneous group of long chain polymers which have a variety of structures. Hydrocolloids are used in several industrial sectors. In foods, they are used to control and regulate a colloidal state and help in modifying the food sensory properties. Hydrocolloids are obtained from various natural sources such as seaweeds, plant seeds, tubers, plants, microorganisms, and animals. Due to their importance primarily as food additives, the annual use of hydrocolloids is constantly increasing. In this regard, studies on the use of cheaper sources and production processes are always relevant – this includes the production of hydrocolloids by microorganism cultures. This chapter presents the main hydrocolloids, both natural and biotechnologically produced, their main characteristics, and legal aspects of their application in food.

Keywords: Food additives, Functions of hydrocolloids, Hydrocolloids, Hydrocolloids production.

INTRODUCTION

As a definition, 'hydrocolloid' means particles of 10 to 1000 nm in diameter dispersed in water as a continuous phase [1]. The term can be used to group of polysaccharides and proteins characterized by forming viscous dispersions and/or gels when dispersed in water [2]. Hydrocolloids have different structures with influenced in their behavior in solutions; that are related to differences in relation

* **Address of correspondence to Julio Carvalho:** Department of Bioprocess Engineering and Biotechnology, Universidade Federal do Paraná, Caixa Postal 19011 - ACF Centro Politécnico. 81531-990 Curitiba - Brazil; E-mail: jccarvalho@ufpr.br.

Juliano Lemos Bicas, Mário Roberto Maróstica, Glaucia Maria Pastore (Eds.)

to flexibility, branchings, molecular weight and ionic charge [3]. Besides, the nature of the functions and interactions between protein and polysaccharide may have great variations which affects the interactions in food systems. [4, 5].

Hydrocolloids are used in a variety of industrial sectors. In foods are used to control and regulate a colloidal state, as thickening and gelling, to improve texture and/or viscosity, and stability of food system [6, 7].

Most food additives as hydrocolloids are sold in powdered form and must be dissolved in a liquid before use. Their solubility and dissolution kinetics are thus important factors in the development, formulation and quality control of food hydrocolloids [8]. The characterization of these additives may involve rheological, structural, microscopic and molecular aspects. The rheological aspect is the most important characterization being used as a determinant analysis of the product texture that is directly connected to sensory characteristics and acceptability of the consumer market [2].

One of the main functions of hydrocolloids in food is in the composition and stability of emulsions. Several factors such as temperature, pH, ionic strength and salts, can alter the viscosity of hydrocolloids solutions [9]. Most hydrocolloids can act as stabilizers but only a few can act as emulsifying as this requires an ability to facilitate the formation and stabilization of fine droplets during and after emulsification [10, 11].

Hydrocolloids are obtained from various natural sources such as agar, carrageenan and alginate are from seaweeds, guar gum, locust bean gum, tara gum and tamarind gum from plant seeds, konjac mannan from tubers, starch, pectin and cellulose from plants, xanthan gum, gellan gum, curdlan, dextran and cellulose from microorganisms, gelatin, caseinate, whey protein, egg white protein and chitin and chitosan from animals [12]. Hydrocolloids such as guar gum, locust bean gum, and tara gum are not pure polysaccharides [3]. Commercial carrageenan preparations are often mixed to provide characteristics for specific applications. This occurs because the structures of all naturally occurring polysaccharides in plants vary according to the growth conditions of these plants [13].

The most significant hydrocolloids used in food are pectins, xanthan and alginates [14]. So, this chapter deals with the main hydrocolloids used in food and its principals characteristics.

FUNCTIONS OF HYDROCOLLOIDS IN FOOD PRODUCTS

Food additives are utilized in food formulations for various purposes to enhance properties like flavor, color, shelf life and rheological properties. Hydrocolloids modify the food sensory properties, and are used to improve specific purposes [2]. These functional ingredients are present in dairy and bakery products, canned foods, salad dressings, beverages, sauces, soups and other processed foodstuffs [15].

At relatively low concentrations, hydrocolloids provides viscosity enhancement and/or prevention of sedimentation of dispersed particles. Proteins and hydrocolloids contribute to the microstructural properties of foods, because they can confer structure to the continuous phase of the medium [16, 17].

The functional properties of hydrocolloids include thickening, gelling, emulsifying, stabilization and controlling of the crystal growth of ice and sugar. For thickening, the hydrocolloids commonly are starch, xanthan, guar, locust bean, karaya, tragacanth, arabic gum and cellulose derivatives. The gelling are alginate, pectin, carrageenan, gelatine, gellan and agar [2].

The most important hydrocolloids for foods are starch and its derivatives, pectin, galactomannans, carrageenan, alginate, agar, arabic gum, cellulose and its derivatives [18], gelatin and xanthan gum. For each system, the selection of suitable hydrocolloid depends on functions and desirable properties in foods [19].

In many dairy products, hydrocolloids are adding to control texture interacting with the casein [20, 21]. For applications fermented dairy products, the choice of the proper type and level of hydrocolloid are important factors. In yogurt, suitable hydrocolloids often include carboxymethyl cellulose, pectin, alginate and xanthan gum, the hydrocolloids do not mask the natural flavor of the product [22].

Arozarena *et al.* [23] studied, in yellow cakes, the substitution of egg proteins with use of optimum leavening agent, emulsifiers and xanthan gum. Miller and

Hoseney [24] showed that after the inclusion of xanthan gum, the cakes obtained similar or better characteristics, in terms of volume, height and shrinkage, than those of the control cakes. Muthukumaran [25] studied the effects stabilizers on the stability of egg white foam and a satisfactory egg white foam was found using 0.125% xanthan gum. For enhancing the cloudy stability of juices for prolonged periods, pectin, galactomannans and arabic gum have been added [26 - 28].

In fish products, the capability of gel formation is important and it can be promote by use of gums and starches. However, carbohydrates affect the mechanical and gelling properties because they can modify the capacity of salt to solubilize myofibrillar proteins and the non-compability with the muscle protein [29, 30].

Hydrocolloids have been utilized in fried food in order to reduce the amount of oil absorbed [31, 32], however the purpose of the use of hydrocolloid includes to improve others properties like film-forming, viscosity control, adhesion, pickup control, mechanical resistance and freeze-thaw stability [33].

Cho, Shim and Lee [34] found that hydrocolloids together with appropriate amylase/amylopectin ratio would improve freeze-thaw stability of wheat dough, and Silva *et al.* [35] found that the hydrocolloids with high water binding capability could regulate dough rheology. Similarly to research conducted by Oishi *et al.* [36] who reported that hypoallergenic wheat flour noodles with sodium alginate and curdlan possessed better rupture strength and hardness.

Hydrocolloids have low calorie and high nutritional value, as dietary fibers could improve consumers' health, because it can decrease blood pressure and cholesterol level [37]. Specially, hydrocolloids containing β-glucan, a soluble dietary fiber found in high concentration in oats and barley, can reduce blood cholesterol level and the risk of heart-related diseases [38], and replace fat and/or flour in foods to reduce calories [39, 40].

Hydrocolloid gels are frequently used as semi- and soft-solid food models [41, 42]. Indeed, varying mixtures of hydrocolloid gels provide a wide range of textures that can be used to mimic the texture of real foods, despite their compositional simplicity and homogeneity [43].

NATURAL HYDROCOLLOIDS AND COMMERCIAL ASPECTS

Natural hydrocolloids referred in this chapter may be obtained from plants, animals or seaweeds. The most used in food applications are alginate, carrageenan, celluloses, galactomannans, gelatin, gum arabic, konjac gum, pectin and starches.

Alginates

Alginate is a polyssacharide extracted from brown seaweeds that has important properties for industry food, textile, paper and pharmaceutical. It is used as thickeners, stabilizers, gelling agents and emulsifiers [44, 45]. Besides its biocompability allows the use in medical applications including wound management [46] and drug delivery systems [47]. Chemical structure of alginate is composed by different copolymers, (1 → 4)-linked β-D-mannuronic acid (M) and α-L-guluronic (G) units (Fig. **1**), which are arranged in blocks the chain with homopolymeric of M regions, G regions and regions of alternating structure MG [48, 49]. MG regions are more flexible more soluble at lower pH, while M blocks can be immunostimulating and G blocks form stiff chains. Divalent cations (*e.g.* Ca^{2+}, Ba^{2+}, Sr^{2+}) can be crosslink in two G blocks of more than 6 residues each [45].

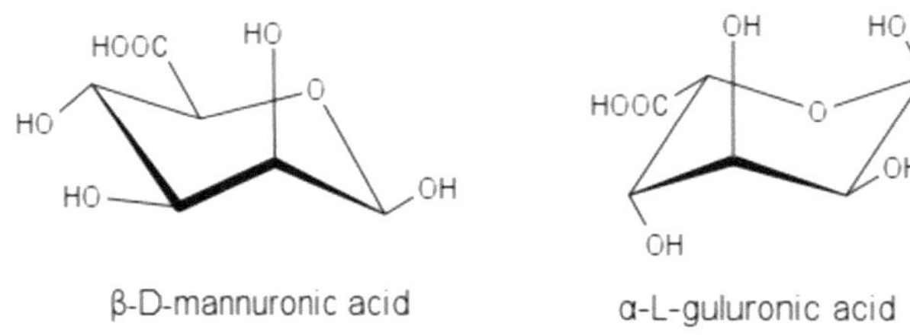

Fig. (1). Chemical structure of β-d-mannuronic acid (M) and α-l-guluronic acid (G).

Commercial alginates are obtained by extraction from brown algae (*Laminaria digitata*, *Laminaria hyperborea* and *Macrocystis pyrifera*). About 26,500 ton of alginate were commercialized in 2009 [50]. Biosynthesized alginate can be obtained by bacteria such as *Azotobacter vinelandii* and *Pseudomonas aeruginosa* [44, 45, 51]. These bacteria produce alginate utilizing a similar molecular mechanism, however they secrete alginate with different material properties [52].

Carrageenan

Carrageenans are used in food and pharmaceutical industries. As gelling and stabilizing agents, it can bind water, promote gel formation, thicken, stabilize and improve palatability. They are extracted from red seaweed [50, 53, 54]. In 2009, the sales volume of carrageenan was 50,000 ton [50].

Carrageenans are strongly anionic polymers [55]. The primary structure of carrageenan is based on alternating of 1,3-linked D-galactose and 1,4-linked D-galactose, with difference in degrees of sulfation. These carrageenans can be obtained by selective extraction techniques in order to have a purified form. The precursor structures are mu and nu carrageenans and a chemical modification by alkali produce kappa (κ), iota (ι), and lambda (λ)-carrageenans (Fig. **2**). [56].

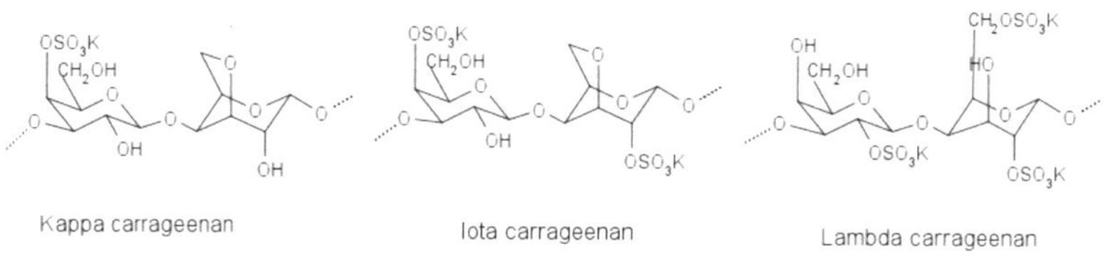

Fig. (2). Carrageenan structures.

Commercial carrageenans, kappa (κ), iota (ι), and lambda (λ)-carrageenans, differ only in the number of sulfate groups per disaccharide: one, two and three sulfate ester groups, resulting in calculated sulfate content of approximately 20%, 33% and 41% (w/w), for κ, ι and λ respectively [55, 57].

Cellulose

Cellulose, a polysaccharide present in plants, is one of the most abundant biopolymers. Algae, tunicates and some bacteria can synthesize cellulose too [58]. The difference is that cellulose produced by microorganisms does not contain lignin, pectin, or araban, it is a pure form of cellulose with different chemical and physical properties [59, 60].

Cellulose is a polymer composed by glucose units joined through β-1,4 glucosidic linkages. Cellulose used in food additives is modified cellulose. Cellulose pulp is dispersed in alkali solution with appropriate reagents in order to substitute the anhydroglucosc monomers of the cellulose chain. Modified celluloses are methyl cellulose, hydroxypropyl cellulose, hydroxypropylmethyl cellulose, methylethyl cellulose and carboxymethyl cellulose (Fig. **3**) [61].

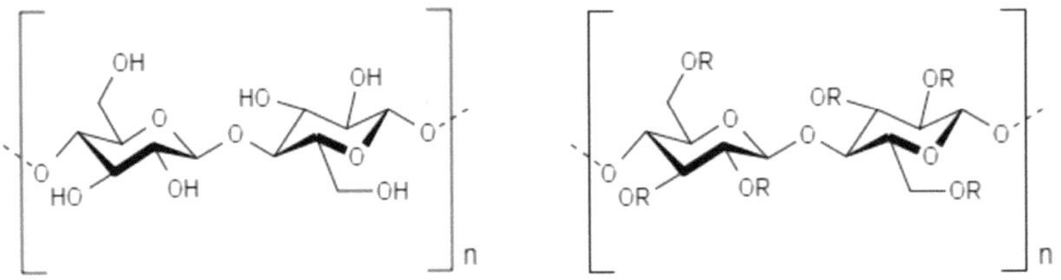

Cellulose

R = H or CH_3 - methyl cellulose
R = H or $CH_2CH(OH)CH_3$ - hydroxypropyl cellulose
R = H or CH_3 or $CH_2CH(OH)CH_3$ - hydroxypropylmethyl cellulose
R = H or CH_3 or C_2H_5 - methylethyl cellulose
R = H or CH_2COOH - carboxymethyl cellulose

Fig. (3). Cellulose structure and modified cellulose.

Cellulose derivatives, such as carboxymethyl cellulose, produced on an industrial scale are used in food industry as hydrocolloid [62]. Modified cellulose has diverse applications in foods as the regulation of rheological properties, emulsification, stabilization of foams, modification of ice crystal and water holding capacity [63].

The properties of modified celluloses are influenced by three main factors according to Murray [61]: type of substitution, degree of polymerization and the degree of the chain substitution. Particle size and powder bulk density affect the dissolving characteristics, moreover all modified celluloses, powder or granular form, are capable of absorbing water from the atmosphere.

Galactomannans

Galactomannans are storage polysaccharides isolated from the seed endosperm of some Leguminosae. Their properties as thickening, binding, and stabilizing become galactomannas important for industry. The consumption of galactomannans is around 90-100 thousand tons per year (70–80 thousand tons of guar gum and 12–14 thousand tons of locust bean gum) [64].

Consist of mannose and galactose at different ratios, β-(1,4)-D-mannose and a α-D-galactose (position C6) (Fig. **4**). This different ratio offer different galatomannans such as locust bean gum, guar gum, fenugreek gum and tara gum [65]. Among the different galatomannans, locust bean and guar gums are importants for industrial, tara and fenugreek gums is used limited due to availability and price [66].

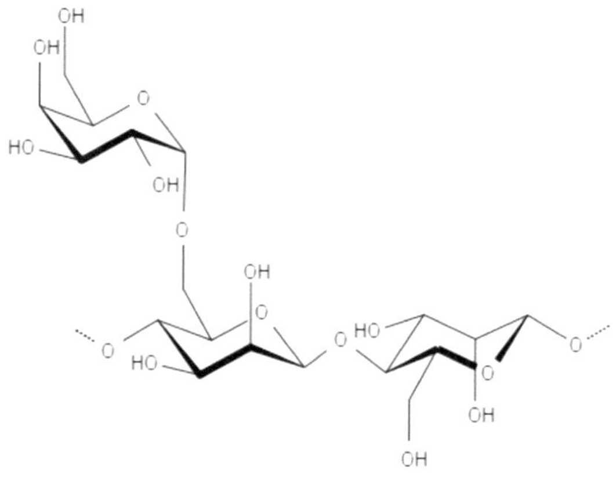

Fig. (4). Example of structure of galactomannans.

Locust bean gum (LBG) is a neutral galactomannan formed by mannose and galactose units ratio averaging 4 to 1. LBG is used as a thickener, stabilizer and emulsifier, and a characteristic of LBG is that its combination with other components to form synergistic interactions [65, 67]. The combination with κ-carrageenan increases significantly gel strength and changes the gel character from brittle to elastic [68].

Gelatin

Gelatin is derived of collagen (from cartilages, bones, tendons and skins of animals), obtained by hydrolysis process under controlled conditions of temperature and pH [69 - 71]. It features a complex molecular organization based on its special aminoacid composition, 33% glycine and 22% proline, and structure triple extended helix [72].

Gelatin used in the food and pharmaceutical industries derived from cows is a significant amount. The interest in gelatin derived from fish and poultry has been increased. There is no association between marine gelatin sources and risk of outbreaks of Bovine Spongiform Encephalopathy [73].

It can be commercialized in sheets, flakes, shreds, grains, fine powder, pale yellow or amber. Gelatin dry is stable in air, however in presence of moist or in solution is subject to microbial decomposition [74]. The global demand for gelatin is nearly 326,000 tons [73]. Because of its well-known gelling properties, gelatin is one of the most used hydrocolloids in food and pharmaceutical industries. It has been used to improve emulsion and foam stability in a series of food products, to provide defines mouth feel, fat reduction and texture [70, 73].

Gelatin offers many special properties including melt-in-the-mouth property or melting slightly below the physiological temperature of humans. Besides, gelatin is thermally reversible gel, versatile multi-functional hydrocolloid, has availability in different gel strengths and particle sizes for specific applications and it is easy to use because does not require the addition of salts or sugars to set [72].

Gum Arabic

Gum arabic is derived from *Acacia senegal* and *Acacia seyal* trees (stems and branches), which are abundant in the central Sudan, central Africa and in West Africa. The total production of gum arabic ranges about 60,000 - 80,000 tons [75, 76] and approximately 90% of the production comes from Sudan [77].

The chemical composition of gum arabic can vary according to the trees (source, age, climatic conditions and soil environment). Gum arabic is slightly acidic (pH 4.5 -5.0) and consisting mainly of a highly branched polysaccharide (galactose linked branches of arabinose and rhamnose, and glucuronic acid), can be found in a salt form (calcium, magnesium and potassium) of a polysaccharidic acid (arabic acid) [78, 79]. Consists of mainly three fractions, the major fraction consist of galactose backbone with linked branches of arabinose and rhamnose, which terminate in glucuronic acid found in nature as magnesium, potassium and calcium salt. A fraction is an arabinogalactan-protein complex, and the smallest fraction having the highest protein content is a glycoprotein which differs in its amino acids composition [75]. The low protein content adsorbs on the surface of oil droplets [78].

It is commonly used as emulsifier, thickener, encapsulation and stabilizier in food and pharmaceutical industry [75, 76, 80]. Gum Arabic is used in confectionery, bakery, dairy, beverage and encapsulation [76, 78]. In pharmaceutical industry, it is used as a suspending agent, emulsifier, adhesive and binder in tabletting and in demulcent syrups. In cosmetics as a stabilizer in lotions and protective creams. Other industrial applications include adhesives, textiles, lithography, printing and paper [78].

Konjac Gum

Konjac gum (konjac mannan) is derived from the tubers of *Araceae amorphophallus*, a perennial plant from Southeast Asia, which contains 60–80% konjac glucomannan [81, 82].

It is a heteropolysaccharide consisting of D-glucose and D-mannose, a mixture of 2,3,4-tri-O-methyl-D-glucose, 2,3,4-tri-O-methyl-D-mannose and 2,3,6-tri-O-methyl-D-mannose [83].

The uses of konjac are as a gelling agent, thickener, film former and emulsifier. Besides food application, konjac have been extended to pharmaceutical, chemical industry, petroleum, *etc.* [84]. Konjac gum is generally used with other hydrocolloids to increase the viscosity of blended systems and improve food quality due to its high viscosity [85, 86].

Pectin

Pectin is used to contribute to textural quality of fruit and vegetable products and, because of it can form viscoelastic solutions, it is usually used as a gelling agent in jams, jellies and confections [87 - 89].

Pectin is extracted from the plant cell wall, citrus peels, apple pomace and sugar beet pulps. The main feedstocks for commercial pectin production are apple pomace and citrus peel. Industrially, pectin is produced at acidic conditions with elevated temperature [90].

The structure of pectin is very complex and depends on source and the extraction process. The polysaccharide is formed by α-D-galacturonic acid containing1-4 linkages and the side chain contains neutral sugars (L-rhamnose, L-arabinose, and D-galactose) [87, 90]. Linear homopolymer of α-1,4-linked galacturonic acid, homogalacturonan (HG), comprises 65% of pectin [91].

Commercial pectins vary according to molecular weight, total charge, neutral sugar side chains, degree of acetylation and presence of ferulic acid and protein. They are rich in α-D-galacturonic acid region, which can be used to make high or low sugar jellies and stabilizer, depending on degree of esterification [89].

Starch

Starch is composed by two polymers of glucose, amylose and amylopectin (Fig. **5**), designed by nature as a plant energy reserve. Amylose is a linear molecule formed by glucose units α-1,4 linkages, comprising 20–30% of normal starch. Amylopectin comprises 70–80% and is branched molecule of the glucose units joined by α-1,6 linkages [92].

Native starch is insoluble and in general do not have the properties desired for use in processed foods and chemical, physical or biochemical treatment has been used to improve the functionally of starches [13, 93]. It has been found that certain hydrocolloids in combination with certain starches provide some improvement of properties without chemical modification [94].

On heating the starch in water, the granule starts to swell, the starch solubilizes and viscosity decreases. On cooling, a precipitate form gel. By this process is achieved a chemical modification of the starch. Modified Starches become tolerant of extremes of heat and cold and of high or low levels of pH [92, 95].

Starch has a great versatility application in the food industry. Some properties improved are to increase their tolerance to processing conditions, to improve paste and gel textures, to provide cold storage stability and to control water mobility [94].

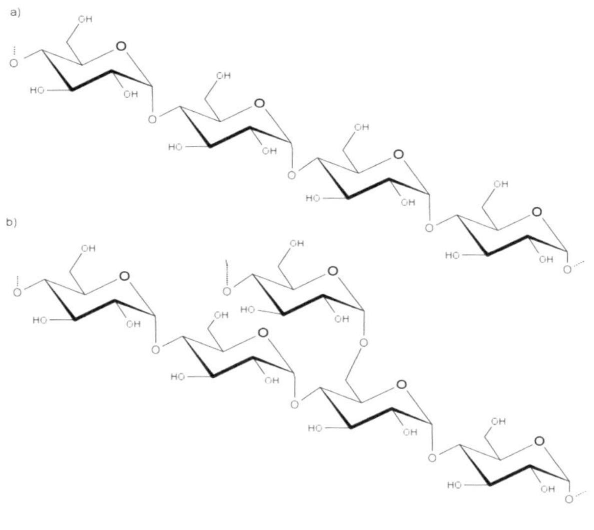

Fig. (5). Starch structure of amylose (**a**) and amylopectin (**b**).

BIOTECHNOLOGICALLY-PRODUCED HYDROCOLLOIDS

The principal hydrocolloids produced by microorganisms and used in food applications are gellan gum, xanthan gum and curdlan. Other hydrocolloids also used in the food industry can be cited as levan and pullulan.

Gellan Gum

It is a polysaccharide, produces outside the cells by the microorganism *Sphingomonas*. The original polysaccharide is formed by an L-glyceryl on O(2) of the 3-linked glucose from the tetrasaccharide series, and in some parts, an acetyl group can also be found [96]. In the industrial production, the fermentation broths go through a treatment with hot akali, in which these two substituents are taken out. The name "Gellan Gum" is given to the resulted polimer [97]. The majority of the acyl groups, with a milder extraction technique, can be maintained, which creates the "high acyl" gellan, that is already accessible commercially [97]. Different productions of the gellan gum are showed in Table **1**.

Table 1. Gellan gum production by submerged fermentation.

Substrate	Microorganism	Production	Reference
Soluble Starch	*Sphingomonas paucimobilis* ATCC 31461	43.6 g/L	[98]
Soy Bagasse	*Sphingomonas paucimobilis* NK 2000	7.5 g/L	[99]
Soluble Starch	*Sphingomonas paucimobilis* ATCC 31461	35.7 g/L	[100]

Also known as heteropolysaccharide-60 (PS-60), the gellan gum consists in an anionic polysaccharide which is formed by monosaccharides β-D-glucose, β-D-glucuronic acid and α-(1,4)-L-rhamnose. It has a wide variety of uses, given its high number of possible structures and properties, such as pharmaceutical, food, thickening, stabilizing, emulsifying, gelling agents and texturizing industries [99 - 101]. In the US and the EU, the gellan gum is accepted to be utilized in foods and in the medical area [102].

Xanthan Gum

Xanthan gum is a polysaccharide with many uses, but the main one is in food industry due to its capability to work as an emulsion stabilizer, and thickener, and as gelling agent when combined with different polysaccharides [103]. Xanthan gum exhibits pseudoplastic behaviour and is extensively used as stabilizer in sauces and dressings [104]. It consists in a main chain of trisaccharides and, along side, some small chains of glucuronic acid, mannose, pyruvil and acetyl residues [101]. The weight of this hydrocolloid varies from 500 to 2000 kDa, according to the microorganism that generated it [101]. The basic structural of xanthan gum showed in Fig. (**6**).

Table 2. Xanthan production by submerged fermentation.

Substrate	Micro-organism	Production	Reference
Beet molasses	*Xanthomonas campestris*	28 g/L	[110]
Palm date juice	*Xanthomonas campestris* NRRL B-1459	43,35 g/L	[105, 111]
Cheese whey and sucrose	*Xanthomonas campestris.*	25 g/L	[35]

Commercially, the hydrocolloid is produced by fermentation using bacterial *Xanthomonas campestris* and the global xanthan market has progressively increased, at an annual rate of 5–10% [105]. Different productions of the xanthan are showed in Table **2**.

Some studies aiming to increase the production of xanthan demonstrated that the elements the greater impact on its concentration are the carbon source, phosphate concentration and pH [106 - 108]. The best temperature range to produce xanthan is between 30 and 33 °C [1, 109]. The optimized pH range is between 6 and 8. The most common carbon sources used commercially are sucrose and glucose. The nitrogen sources are added in the form of glutamate [109]. Kalogiannis *et al.* [18] demonstrated that *Xanthomonas campestris*, when fermented in molasses with KH_2PO_4, can produce 53 g L^{-1} of xanthan. And when glucose is utilized the production decreased to 40 g L^{-1}. *Xanthomonas campestris* is a strictly aerobic bacterium, so the concentration of the oxygen is very important and can be improved with agitation [1].

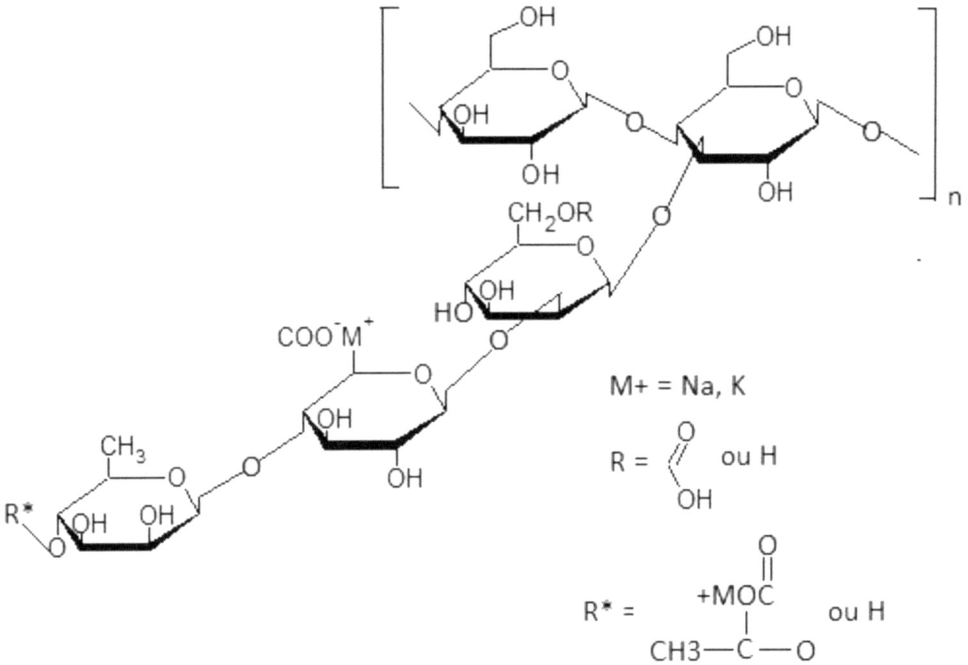

Fig. (6). Chemical structure of xanthan gum.

Curdlan

Curdlan consists in a linear chain of homopolysaccharides (Fig. 7). Its size varies from 400 to 500 D-glucoses joined by β-(1,3)-glucosidic linkages [111]. This biopolymer has many distinct thermal gelation properties therefore it has many application such as in food and pharmaceutical industries [112].

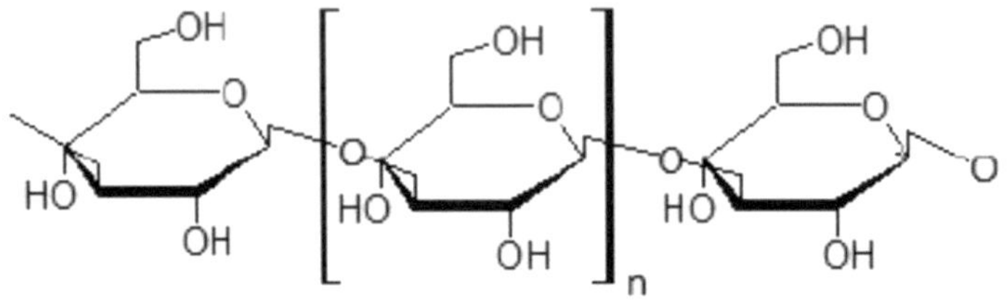

Fig. (7). Chemical structure of curdlan.

Curdlan's industrial production by fermentation only loses in quantity to xanthan and gellan gums. It can be produced by bacteria such as *Rhizobium radiobacter*, *Agrobacterium* species and *Alcaligenes faecalis* under low concentrations of nitrogen [15, 112]. The elements that have the greater impact on the curdlan's production are the concentration of nutrients such as nitrogen, phosphate and mainly the carbon, the pH and the aeration [1, 113, 114].

Salah *et al.* [15] studied the production of curdlan using hydrolyzed palm and bacteria *Rhizobium radiobacter* ATCC 6466 and obtained 22.83 g/L. A study on the effect of nitrogen source on production of curdlan was made by Jiang [112], using *Alcaligenes faecalis* ATCC 31749. In this study, the maximum production of curdlan obtained was 28.16 g/L.

Jin *et al.* [115] shown that for *Agrobacterium* sp. the best pH is 7 for culturin, but for curdlan production it is 4.5, and the optimal range of temperatures from 30 to 32°C. A low concentration of nitrogen has been proven to induce the production of curdlan, since the sulfate and the phosphate concentrations are fixed [116]. When increased the amount of ammonium concentration in the medium it has been shown that the microbial growth rate is reduced [1]. Another factor that appears to have a good influence in the curdlan production is addin uracil, to the medium.

Pullulan

Pullulan is a linear chain of glucoses, formed usually by a sequence of maltotriose trimers bound by the α-(1,4)-Glu-α-(1,4)-Glu-α-(1,6) linkages. But that is not the only possible configuration for the pullulan, another example can be the chain of maltotetraose units bound by (α-(1,4)-Glu-α-(1,4)-Glu-α-(1,6)-Glu-α-(1,6)) linkages [117]. Differently from other hydrocolloids pullulan is synthesized inside the cell and then transported to the environment. An example of a microorganism that is a good producer of pullulan is the *Aureobasidium pullulans* which is capable of producing up to 52.5 g L^{-1} of pullulan [65, 118], but its production varies widely among *A. pullulans* strains ranging from 1.3 g L^{-1} to 52.5 g L^{-1} [5, 14, 55, 65, 118 - 120]. One of the most important elements to take in consideration in the pullulan production is the carbon source, and its concentration to be utilized. It can be for example glucose, sucrose, dextran among others [55, 117, 119]. Revella *et al.* [14] demonstrated that the best carbon source to induce the production of pullulan is sucrose compared with xylose, glucose, fructose and cellobiose.

Levan

Levan is a chain of D-fructofuranosyl bound by by β-(2,6) linkages (Fig. **8**), and is biosynthesized outside the cell. Levan is produced by many microorganisms such as the bacterias *Zymomonas mobilis* or *Bacillus subtilis* [114, 121, 122] and also can be obtained by enzymatic syntheses both using sucrose as carbon source. An extracellular enzyme is responsible for the production of levan, and its usual name is levansucrase, but can also be named as 6-fructosyltransferase, β-(2,6--fructosyltransferase, and β-(2,6)-fructan:D-glucose 1-fructosyltransferase (EC 2.4.1.10) [123]. The best temperature to produce this enzyme varies according to the microorganism strain used, like for *Zymomonas mobilis*, the optimal temperature is 0 ºC, but for *Bacillus subtilis*, it is around 10 °C. The optimal production of levansucrase is obtained at acidic pHs. And its production is reduced by glucose ant at temperatures greater than 45 °C.

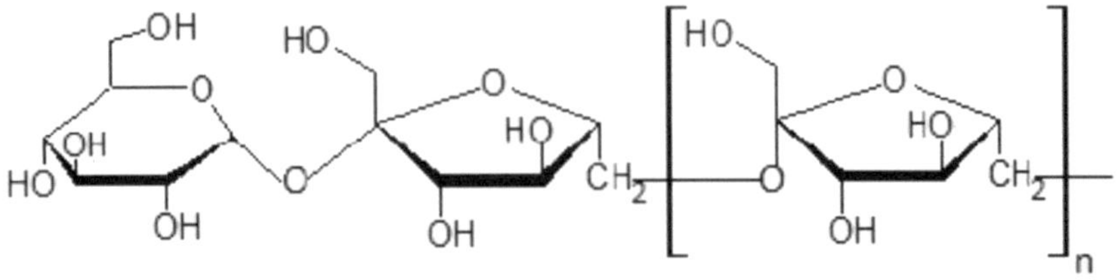

Fig. (8). Chemical structure of levan.

Dextran

Dextrans are a homopolysaccharides and consists in a group of chains connected to a main chain of glucoses bound by α-(1,6)-linkages. The final structure of the dextran can vary, because each microrganism will create its specific dextran [125].

Dextrans are not clearly allowed to be used in the food industries in the Europe and in the USA [124, 125]. But they still have many others uses for the dextrans, partially degraded dextrans and their derivatives, therefore have a significant commercial importance [125 - 127]. The specific properties of the dextrans such as stability, hydrophilicity, purity, and the ability to form clear and stable solution, are continuously increasing its importance [128].

LEGAL ASPECTS OF APPLICATION IN FOOD

The hydrocolloids are classified as food additives. There is not a category for hydrocolloids and they are classified according to their functional use such as gelling agent, thickener, emulsifier and others (Table **3**).

Table 3. Classification of hydrocolloids according to Codex General Standard for Food.

Component	Classification	INS number	Application
Alginate	Stabilizer, thickener, gelling agent, emulsifier	Sodium alginate – 401 Potassium Alginate – 402 Ammonium alginate – 403 Calcium alginate – 404 Propylene Glycol Alginate – 405	Ice cream, meat products, onion rings, pimento olive fillings, crabsticks and cocktail berries, jams, jellies e fruit fillings [130].

(Table 3) contd.....

Component	Classification	INS number	Application
Carrageenan	Thickener, gelling agent, stabilizer, emulsifier	407	dairy products, water-based foods, meat products, beverages, condiments, infant formula and pet food [132].
Cellulose	Emulsifier, stabilizer, anticaking agent, dispersing agent, thickener, binder, suspension agent, film coating, tableting agent, Carrier, glazing agent, Tableting aid, binder, filler, diluent of color and other food additives	Microcrystalline Cellulose - 460(i) Powdered Cellulose – 460 (ii) Methyl Cellulose – 461 Ethyl Cellulose - 462 Hydroxypropyl Cellulose – 463 Hydroxypropylmethyl cellulose – 464 Methyl Ethyl Cellulose - 465 Sodium Carboxymethyl Cellulose – 466 Ethyl Hydroxyethyl Cellulose - 467 Cross-Linked Sodium Carboxymethyl Cellulose – 468 Sodium Carboxymethyl Cellulose, enzymatically hydrolyzed – 469	vegetable products, alcoholic beverages, toppings, mousses, batters, instant products, frozen products, table sauces, soft drinks, bakery products and low pH milk products [61].
Galactomannan	Thickener, stabilizer, emulsifier, gelling agent	Locust bean gum – 410 Guar Gum – 412	Dairy products, frozen products (ice cream), soft drinks fruit juices, bread and pastry, fruit preserves, baby food, puddings, flans, pudding powder, dietary fibers and pet foods [131].
Gellan gum	Thickening agent, gelling agent, stabilizer	418	Dessert jellies, dairy products, sugar confectionery, fruit preparations, breadings and batters [132].
Gelatin	Stabilizer, gelling agent, emulsifying agent, crystallization inhibitor	-	Desserts, fruit gummies, marshmallows, pastilles, caramels, yogurt, meat and sausages, broths and canned meats [133].
Gum Arabic	Emulsifier, stabilizer, thickener	414	Confectionery industry (gums, pastilles, marshmallows and toffees), beverages, flavor encapsulation and dietary fibers [76].

(Table 3) contd.....

Component	Classification	INS number	Application
Pectin	Gelling agent, thickener, stabilizer, emulsifier	440	Jams, jellies, marmalades, industrial fruit preparations, fruit preparations for bakery products, dairy products, sugar confectionery, dough for bread and pastry, frozen pre-proofed dough and frozen bread [134].
Starch	Thickener, stabilizer, binder, emulsifier	Dextrin roasted starch - 1400 Acid treated starch - 1401 Alkaline treated starch - 1402 Bleached starch - 1403 Oxidized starch - 1404 Enzyme-treated starch - 1405 Monostarch phosphate - 1410 Distarch phosphate - 1412 Phosphated distarch phosphate - 1413 Acetylated distarch phosphate - 1414 Starch acetate - 1420 Acetylated distarch adipate - 1422 Hydroxypropyl starch - 1440 Hydroxypropyl distarch phosphate - 1442 Starch sodium octenylsuccinate - 1450 Acetylated oxidized starch - 1451	baked goods, batters and breadings (meat, poultry, seafood and vegetables, including even mushrooms or French fries), beverage emulsions and flavor encapsulation, confectionery, dairy products, fruit preparations, gravies, soups and sauces mayonnaise and salad dressings, meat products and savory snacks [135].
Xanthan gum	Thickener, stabilizer, emulsifier, foaming agent	415	Baked goods, bakery and pie fillings, dairy products, culinary products, frozen foods and dry mixes [136].

Codex Alimentarius is the international system that regulates the use of additives in foods according to "Codex General Standard for Food Additives" [129].

FUTURE DEVELOPMENTS

The annual use of hydrocolloids is constantly increasing, therefore efforts to find new sources of hydrocolloids, which are more efficient and economic, are still performing. Also, studies of the conditions of production, extraction and formulation including mixtures of different hydrocolloids are being conducted worldwide. This will certainly contribute to the reduction of costs and improvement of the final product to the consumer.

CONFLICT OF INTEREST

The authors confirm that they have no conflict of interest to declare for this publication.

ACKNOWLEDGEMENTS

Declared none.

REFERENCES

[1] Olphen, H.v. *Physical chemistry: Enriching topics from colloid and surface science*; Theorex: La Jolla, Calif., **1975**.

[2] Saha, D.; Bhattacharya, S. Hydrocolloids as thickening and gelling agents in food: a critical review. *J Food Sci Technol,* **2010**, *47*(6), 587-597. [http://dx.doi.org/10.1007/s13197-010-0162-6] [PMID: 23572691]

[3] *Carbohydrate chemistry for food scientists*; Inc (AACC): American Association of Cereal Chemists, **2007**.

[4] Dickinson, E. Stability and rheological implications of electrostatic milk protein–polysaccharide interactions. *Trends Food Sci. Technol.,* **1998**, *9*(10), 347-354. [http://dx.doi.org/10.1016/S0924-2244(98)00057-0]

[5] Tolstoguzov, V. Some thermodynamic considerations in food formulation. *Food Hydrocoll.,* **2003**, *17*(1), 1-23. [http://dx.doi.org/10.1016/S0268-005X(01)00111-4]

[6] BeMiller, J.N. *Gums and hydrocolloids analytical aspects*. In: *Carbohydrates in food*; CRC Press London: UK, **2006**; pp. 209-231.

[7] Funami, T. Next target for food hydrocolloid studies: Texture design of foods using hydrocolloid technology. *Food Hydrocoll.,* **2011**, *25*(8), 1904-1914. [http://dx.doi.org/10.1016/j.foodhyd.2011.03.010]

[8] Kravtchenko, T.P. A novel method for determining the dissolution kinetics of hydrocolloid powders. *Food Hydrocoll.,* **1999**, *13*(3), 219-225. [http://dx.doi.org/10.1016/S0268-005X(99)00002-8]

[9] Karazhiyan, H. Rheological properties of *Lepidium sativum* seed extract as a function of concentration, temperature and time. *Food Hydrocoll.,* **2009**, *23*(8), 2062-2068.

[http://dx.doi.org/10.1016/j.foodhyd.2009.03.019]

[10] Dickinson, E. Hydrocolloids at interfaces and the influence on the properties of dispersed systems. *Food Hydrocoll.,* **2003**, *17*(1), 25-39. [http://dx.doi.org/10.1016/S0268-005X(01)00120-5]

[11] Phillips, G.; Williams, P. *Handbook of hydrocolloids*; , **2000**.

[12] Dickinson, E. Effect of hydrocolloids on emulsion stability. In: *Gums and stabilizers for the food industry d12*; Williams, P.A.; Phillips, G.O., Eds.; Royal Society of Chemistry: Cambridge, UK, **2004**; pp. 394-404. [http://dx.doi.org/10.1039/9781847551214-00394]

[13] J.N., Pasting, paste, and gel properties of starch–hydrocolloid combinations. *Carbohydr. Polym.,* **2011**, *86*(2), 386-423. [http://dx.doi.org/10.1016/j.carbpol.2011.05.064]

[14] Quemener, B. Quantitative analysis of hydrocolloids in food systems by methanolysis coupled to reverse HPLC. Part 1. Gelling carrageenans. *Food Hydrocoll.,* **2000**, *14*(1), 9-17. [http://dx.doi.org/10.1016/S0268-005X(99)00042-9]

[15] Sahin, H.; Ozdemir, F. Effect of some hydrocolloids on the rheological properties of different formulated ketchups. *Food Hydrocoll.,* **2004**, *18*(6), 1015-1022. [http://dx.doi.org/10.1016/j.foodhyd.2004.04.006]

[16] Baeza, R.; Gugliotta, L.M.; Pilosof, A.M. Gelation of β-lactoglobulin in the presence of propylene glycol alginate: kinetics and gel properties. *Colloids Surf. B Biointerfaces,* **2003**, *31*(1–4), 81-93. [http://dx.doi.org/10.1016/S0927-7765(03)00045-6]

[17] Giboreau, A.; Cuvelier, G.; Launay, B. Rheological behaviour of three biopolymer/water systems, with emphasis on yield stress and viscoelastic properties. *J. Texture Stud.,* **1994**, *25*(2), 119-138. [http://dx.doi.org/10.1111/j.1745-4603.1994.tb01321.x]

[18] Kalogiannis, S. Optimization of xanthan gum production by Xanthomonas campestris grown in molasses. *Process Biochem.,* **2003**, *39*(2), 249-256. [http://dx.doi.org/10.1016/S0032-9592(03)00067-0]

[19] Sadar, L.N. *Rheological and textural characteristics of copolymerized hydrocolloidal solutions containing curdlan gum.,* **2004**.

[20] Syrbe, A.; Bauer, W.J.; Klostermeyer, H. Polymer Science Concepts in Dairy Systems—an Overview of Milk Protein and Food Hydrocolloid Interaction. *Int. Dairy J.,* **1998**, *8*(3), 179-193. [http://dx.doi.org/10.1016/S0958-6946(98)00041-7]

[21] Thaiudom, S.; Goff, H.D. Effect of κ-carrageenan on milk protein polysaccharide mixtures. *Int. Dairy J.,* **2003**, *13*(9), 763-771. [http://dx.doi.org/10.1016/S0958-6946(03)00097-9]

[22] Williams, P.A.; Phillips, G.O. GUMS | Food Uses. In: *Encyclopedia of Food Sciences and Nutrition,* 2nd ed; Caballero, B., Ed.; Academic Press: Oxford, **2003**; pp. 3001-3007. [http://dx.doi.org/10.1016/B0-12-227055-X/00574-5]

[23] Arozarena, I. Study of the total replacement of egg by white lupine protein, emulsifiers and xanthan gum in yellow cakes. *Eur. Food Res. Technol.,* **2001**, *213*(4-5), 312-316.

[http://dx.doi.org/10.1007/s002170100391]

[24] Miller, R.A.; Hoseney, R. The role of xanthan gum in white layer cakes. *Cereal Chem.,* **1993**. [USA].

[25] Muthukumaran, A. *Foam-mat freeze drying of egg white and mathematical modeling*; McGill University, **2007**.

[26] Silva, J.L.; Gonçalves, M.; Rao, M. Rheological Properties of High Methoxyl Pectin and Locust Bean Gum Solutions in Steady Shear. *J. Food Sci.,* **1992**, *57*(2), 443-448. [http://dx.doi.org/10.1111/j.1365-2621.1992.tb05513.x]

[27] Genovese, D.B.; Lozano, J.E. The effect of hydrocolloids on the stability and viscosity of cloudy apple juices. *Food Hydrocoll.,* **2001**, *15*(1), 1-7. [http://dx.doi.org/10.1016/S0268-005X(00)00053-9]

[28] Paraskevopoulou, A.; Boskou, D.; Kiosseoglou, V. Stabilization of olive oil – lemon juice emulsion with polysaccharides. *Food Chem.,* **2005**, *90*(4), 627-634. [http://dx.doi.org/10.1016/j.foodchem.2004.04.023]

[29] Ramírez, J.A. Food hydrocolloids as additives to improve the mechanical and functional properties of fish products: A review. *Food Hydrocoll.,* **2011**, *25*(8), 1842-1852. [http://dx.doi.org/10.1016/j.foodhyd.2011.05.009]

[30] Gómez-Guillén, C.; Borderías, A.J.; Montera, P. Thermal gelation properties of two different composition sardine (Sardina pilchardus) muscles with addition of non-muscle proteins and hydrocolloids. *Food Chem.,* **1997**, *58*(1–2), 81-87. [http://dx.doi.org/10.1016/S0308-8146(96)00217-8]

[31] Akdeniz, N.; Sahin, S.; Sumnu, G. Functionality of batters containing different gums for deep-fat frying of carrot slices. *J. Food Eng.,* **2006**, *75*(4), 522-526. [http://dx.doi.org/10.1016/j.jfoodeng.2005.04.035]

[32] Garmakhany, A.D. Study of oil uptake and some quality attributes of potato chips affected by hydrocolloids. *Eur. J. Lipid Sci. Technol.,* **2008**, *110*(11), 1045-1049. [http://dx.doi.org/10.1002/ejlt.200700255]

[33] Varela, P.; Fiszman, S.M. Hydrocolloids in fried foods. A review. *Food Hydrocoll.,* **2011**, *25*(8), 1801-1812. [http://dx.doi.org/10.1016/j.foodhyd.2011.01.016]

[34] Cho, Y.; Shim, J.; Lee, H. Characteristics of wheat flour dough and noodles with amylopectin content and hydrocolloids. *Korean Journal of Food Science and Techno.,* **2007**.

[35] Silva, E. Controlling rheology and structure of sweet potato starch noodles with high broccoli powder content by hydrocolloids. *Food Hydrocoll.,* **2013**, *30*(1), 42-52. [http://dx.doi.org/10.1016/j.foodhyd.2012.05.002]

[36] Oishi, Y. Preparation of hypoallergenic wheat flour noodles and evaluation of their physical properties. *Food Sci. Technol. Res.,* **2009**, *15*(1), 39-44. [http://dx.doi.org/10.3136/fstr.15.39]

[37] Alam, N.H.; Meier, R.; Rausch, T.; Meyer-Wyss, B.; Hildebrand, P.; Schneider, H.; Bachmann, C.; Minder, E.; Fowler, B.; Gyr, K. Effects of a partially hydrolyzed guar gum on intestinal absorption of carbohydrate, protein and fat: a double-blind controlled study in volunteers. *Clin. Nutr.,* **1998**, *17*(3),

125-129.
[http://dx.doi.org/10.1016/S0261-5614(98)80006-X] [PMID: 10205329]

[38] Mälkki, Y.; Virtanen, E. Gastrointestinal Effects of Oat Bran and Oat Gum: A Review. *LWT -. Food Sci. Technol. (Campinas.),* **2001**, *34*(6), 337-347.

[39] Lee, S.; Inglett, G.E. Rheological and physical evaluation of jet-cooked oat bran in low calorie cookies. *Int. J. Food Sci. Technol.,* **2006**, *41*(5), 553-559.
[http://dx.doi.org/10.1111/j.1365-2621.2005.01105.x]

[40] Lee, S.; Kim, S.; Inglett, G.E. Effect of shortening replacement with oatrim on the physical and rheological properties of cakes. *Cereal Chem.,* **2005**, *82*(2), 120-124.
[http://dx.doi.org/10.1094/CC-82-0120]

[41] Ishihara, S. Compression test of food gels on artificial tongue and its comparison with human test. *J. Texture Stud.,* **2013**, *44*(2), 104-114.
[http://dx.doi.org/10.1111/jtxs.12002]

[42] Moritaka, H.; Nakazawa, F. Flow velocity of a bolus in the pharynx and rheological properties of agar and gelatin. *J. Texture Stud.,* **2010**, *41*(2), 139-152.
[http://dx.doi.org/10.1111/j.1745-4603.2010.00218.x]

[43] Hayakawa, F. Characterization of eating difficulty by sensory evaluation of hydrocolloid gels. *Food Hydrocoll.,* **2014**, *38*(0), 95-103.
[http://dx.doi.org/10.1016/j.foodhyd.2013.11.007]

[44] Chèze-Lange, H. Production of microbial alginate in a membrane bioreactor. *Enzyme Microb. Technol.,* **2002**, *30*(5), 656-661.
[http://dx.doi.org/10.1016/S0141-0229(02)00030-3]

[45] Ertesvåg, H.; Valla, S. Biosynthesis and applications of alginates. *Polym. Degrad. Stabil.,* **1998**, *59*(1–3), 85-91.
[http://dx.doi.org/10.1016/S0141-3910(97)00179-1]

[46] Qin, Y. Alginate fibres: an overview of the production processes and applications in wound management. *Polym. Int.,* **2008**, *57*(2), 171-180.
[http://dx.doi.org/10.1002/pi.2296]

[47] Tønnesen, H.H.; Karlsen, J. Alginate in drug delivery systems. *Drug Dev. Ind. Pharm.,* **2002**, *28*(6), 621-630.
[http://dx.doi.org/10.1081/DDC-120003853] [PMID: 12149954]

[48] Pawar, S.N.; Edgar, K.J. Alginate derivatization: a review of chemistry, properties and applications. *Biomaterials,* **2012**, *33*(11), 3279-3305.
[http://dx.doi.org/10.1016/j.biomaterials.2012.01.007] [PMID: 22281421]

[49] Donati, I.; Holtan, S.; Mørch, Y.A.; Borgogna, M.; Dentini, M.; Skjåk-Braek, G. New hypothesis on the role of alternating sequences in calcium-alginate gels. *Biomacromolecules,* **2005**, *6*(2), 1031-1040.
[http://dx.doi.org/10.1021/bm049306e] [PMID: 15762675]

[50] Bixler, H.; Porse, H. A decade of change in the seaweed hydrocolloids industry. *J. Appl. Phycol.,* **2011**, *23*(3), 321-335.
[http://dx.doi.org/10.1007/s10811-010-9529-3]

[51] Moresi, M.; Bruno, M.; Parente, E. Viscoelastic properties of microbial alginate gels by oscillatory dynamic tests. *J. Food Eng.,* **2004**, *64*(2), 179-186. [http://dx.doi.org/10.1016/j.jfoodeng.2003.09.030]

[52] Hay, I.D.; Ur Rehman, Z.; Moradali, M.F.; Wang, Y.; Rehm, B.H. Microbial alginate production, modification and its applications. *Microb. Biotechnol.,* **2013**, *6*(6), 637-650. [PMID: 24034361]

[53] Cha, Su Antimicrobial Films Based on Na-alginate and κ-carrageenan. *LWT -. Food Sci. Technol. (Campinas.),* **2002**, *35*(8), 715-719.

[54] Prajapati, V.D.; Maheriya, P.M.; Jani, G.K.; Solanki, H.K. Carrageenan: a natural seaweed polysaccharide and its applications. *Carbohydr. Polym.,* **2014**, *105*(0), 97-112. [http://dx.doi.org/10.1016/j.carbpol.2014.01.067] [PMID: 24708958]

[55] Santo, V.E.; Frias, A.M.; Carida, M.; Cancedda, R.; Gomes, M.E.; Mano, J.F.; Reis, R.L. Carrageenan-based hydrogels for the controlled delivery of PDGF-BB in bone tissue engineering applications. *Biomacromolecules,* **2009**, *10*(6), 1392-1401. [http://dx.doi.org/10.1021/bm8014973] [PMID: 19385660]

[56] Imeson, A.P. 7 - Carrageenan and furcellaran. In: *Handbook of Hydrocolloids,* 2nd ed; Phillips, G.O.; Williams, P.A., Eds.; Woodhead Publishing, **2009**; pp. 164-185. [http://dx.doi.org/10.1533/9781845695873.164]

[57] Li, L.; Ni, R.; Shao, Y.; Mao, S. Carrageenan and its applications in drug delivery. *Carbohydr. Polym.,* **2014**, *103*(0), 1-11. [http://dx.doi.org/10.1016/j.carbpol.2013.12.008] [PMID: 24528694]

[58] Siró, I.; Plackett, D. Microfibrillated cellulose and new nanocomposite materials: a review. *Cellulose,* **2010**, *17*(3), 459-494. [http://dx.doi.org/10.1007/s10570-010-9405-y]

[59] Shi, Z. Utilization of bacterial cellulose in food. *Food Hydrocoll.,* **2014**, *35*(0), 539-545. [http://dx.doi.org/10.1016/j.foodhyd.2013.07.012]

[60] Lin, S-P. Biosynthesis, production and applications of bacterial cellulose. *Cellulose,* **2013**, *20*(5), 2191-2219. [http://dx.doi.org/10.1007/s10570-013-9994-3]

[61] Murray, J.C. 25 - Cellulosics. In: *Handbook of Hydrocolloids,* 2nd ed; Phillips, G.O.; Williams, P.A., Eds.; Woodhead Publishing, **2009**; pp. 710-723. [http://dx.doi.org/10.1533/9781845695873.710]

[62] Klemm, D.; Heublein, B.; Fink, H.P.; Bohn, A. Cellulose: fascinating biopolymer and sustainable raw material. *Angew. Chem. Int. Ed. Engl.,* **2005**, *44*(22), 3358-3393. [http://dx.doi.org/10.1002/anie.200460587] [PMID: 15861454]

[63] Stephen, A.M.; Phillips, G.O. *Food Polysaccharides and Their Applications*; CRC Press, **2006**. [http://dx.doi.org/10.1201/9781420015164]

[64] Sharma, B.R. *A Glimpse of Galactomannans*; India, **2008**.

[65] Wu, Y. An investigation of four commercial galactomannans on their emulsion and rheological

properties. *Food Res. Int.,* **2009**, *42*(8), 1141-1146.
[http://dx.doi.org/10.1016/j.foodres.2009.05.015]

[66] Michael, N.A.; Shinya, I.; Steve, W.C. Seed Polysaccharide Gums. In: *Functional Food Carbohydrates*; CRC Press, **2006**; pp. 127-165.

[67] Haddarah, A. The structural characteristics and rheological properties of Lebanese locust bean gum. *J. Food Eng.,* **2014**, *120*(0), 204-214.
[http://dx.doi.org/10.1016/j.jfoodeng.2013.07.026]

[68] Bayarri, S.; Durán, L.; Costell, E. Influence of sweeteners on the viscoelasticity of hydrocolloids gelled systems. *Food Hydrocoll.,* **2004**, *18*(4), 611-619.
[http://dx.doi.org/10.1016/j.foodhyd.2003.10.004]

[69] Badii, F.; Howell, N.K. Fish gelatin: Structure, gelling properties and interaction with egg albumen proteins. *Food Hydrocoll.,* **2006**, *20*(5), 630-640.
[http://dx.doi.org/10.1016/j.foodhyd.2005.06.006]

[70] Díaz-Calderón, P. Molecular configuration of gelatin–water suspensions at low concentration. *Food Hydrocoll.,* **2014**, *39*(0), 171-179.
[http://dx.doi.org/10.1016/j.foodhyd.2013.12.019]

[71] Jaswir, I. Extraction and characterization of gelatin from different marine fish species in Malaysia. *International Food Research Journal,* **2009**, *16*, 381-389.

[72] Karim, A.A.; Bhat, R. Gelatin alternatives for the food industry: recent developments, challenges and prospects. *Trends Food Sci. Technol.,* **2008**, *19*(12), 644-656.
[http://dx.doi.org/10.1016/j.tifs.2008.08.001]

[73] Karim, A.A.; Bhat, R. Fish gelatin: properties, challenges, and prospects as an alternative to mammalian gelatins. *Food Hydrocoll.,* **2009**, *23*(3), 563-576.
[http://dx.doi.org/10.1016/j.foodhyd.2008.07.002]

[74] Nations, F.a.A.O.o.t.U., Edible gelatin

[75] Chandraju, S.; Mythily, R.; Chidan Kumar, C. Isolation of simple sugar from a hydro colloid: gum arabic. *Recent Research in Science and Techno.,* 4(11). **2013**, *4*(11)

[76] Williams, P.A.; Phillips, G.O. 11 - Gum arabic. In: *Handbook of Hydrocolloids,* 2nd ed; Phillips, G.O.; Williams, P.A., Eds.; Woodhead Publishing, **2009**; pp. 252-273.
[http://dx.doi.org/10.1533/9781845695873.252]

[77] Nussinovitch, A.; Hirashima, M. *Cooking Innovations: Using Hydrocolloids for Thickening, Gelling, and Emulsification*; Taylor & Francis, **2013**.
[http://dx.doi.org/10.1201/b15547]

[78] Dauqan, E.; Abdullah, A. Utilization of gum Arabic for industries and human health. *Am. J. Appl. Sci.,* **2013**, *10*(10), 1270.
[http://dx.doi.org/10.3844/ajassp.2013.1270.1279]

[79] Ali, B.H.; Ziada, A.; Blunden, G. Biological effects of gum arabic: a review of some recent research. *Food Chem. Toxicol.,* **2009**, *47*(1), 1-8.
[http://dx.doi.org/10.1016/j.fct.2008.07.001] [PMID: 18672018]

[80] Lelon, J. Assessment of physical properties of gum arabic from Acacia senegal varieties in Baringo District, Kenya. *African Journal of Plant Science,* **2010**, *4*, 95-98.

[81] Takigami, S. 32 - Konjac mannan. In: *Handbook of Hydrocolloids,* 2nd ed; Phillips, G.O.; Williams, P.A., Eds.; Woodhead Publishing, **2009**; pp. 889-901. [http://dx.doi.org/10.1533/9781845695873.889]

[82] Nishinari, K.; Williams, P.A.; Phillips, G.O. Review of the physico-chemical characteristics and properties of konjac mannan. *Food Hydrocoll.,* **1992**, *6*(2), 199-222. [http://dx.doi.org/10.1016/S0268-005X(09)80360-3]

[83] Katsuraya, K. Constitution of konjac glucomannan: chemical analysis and 13C NMR spectroscopy. *Carbohydr. Polym.,* **2003**, *53*(2), 183-189. [http://dx.doi.org/10.1016/S0144-8617(03)00039-0]

[84] Zhang, Y-q.; Xie, B-j.; Gan, X. Advance in the applications of konjac glucomannan and its derivatives. *Carbohydr. Polym.,* **2005**, *60*(1), 27-31. [http://dx.doi.org/10.1016/j.carbpol.2004.11.003]

[85] He, X.J. Textural and rheological properties of hydrolyzed Konjac Glucomannan and Kappa-Carrageenan: Effect of molecular weight, total content, pH and temperature on the mixed system gels. *Emirates Journal of Food and Agriculture,* **2012**, *24*(3)

[86] Phillips, G.O.; Williams, P.A., Eds. Front matter. In: *Handbook of Hydrocolloids,* 2nd ed; Woodhead Publishing: Sawston, **2009**; pp. i-iii. [http://dx.doi.org/10.1533/9781845695873.frontmatter]

[87] Akhtar, M. Emulsion stabilizing properties of depolymerized pectin. *Food Hydrocoll.,* **2002**, *16*(3), 249-256. [http://dx.doi.org/10.1016/S0268-005X(01)00095-9]

[88] Wang, X.; Chen, Q.; Lü, X. Pectin extracted from apple pomace and citrus peel by subcritical water. *Food Hydrocoll.,* **2014**, *38*(0), 129-137. [http://dx.doi.org/10.1016/j.foodhyd.2013.12.003]

[89] Wicker, L. Pectin as a bioactive polysaccharide - Extracting tailored function from less. *Food Hydrocolloids 42, Part ,* **2014**, *2*(0), 251-259.

[90] Leroux, J. Emulsion stabilizing properties of pectin. *Food Hydrocoll.,* **2003**, *17*(4), 455-462. [http://dx.doi.org/10.1016/S0268-005X(03)00027-4]

[91] Mohnen, D. Pectin structure and biosynthesis. *Curr. Opin. Plant Biol.,* **2008**, *11*(3), 266-277. [http://dx.doi.org/10.1016/j.pbi.2008.03.006] [PMID: 18486536]

[92] Jobling, S. Improving starch for food and industrial applications. *Curr. Opin. Plant Biol.,* **2004**, *7*(2), 210-218. [http://dx.doi.org/10.1016/j.pbi.2003.12.001] [PMID: 15003223]

[93] *J.N. and R.L. Whistler, Starch: chemistry and technology*; Academic Press, **2009**.

[94] Kim, H-S.; Patel, B.; BeMiller, J.N. Effects of the amylose-amylopectin ratio on starch-hydrocolloid interactions. *Carbohydr. Polym.,* **2013**, *98*(2), 1438-1448. [http://dx.doi.org/10.1016/j.carbpol.2013.07.035] [PMID: 24053825]

[95] Copeland, L. Form and functionality of starch. *Food Hydrocoll.,* **2009**, *23*(6), 1527-1534. [http://dx.doi.org/10.1016/j.foodhyd.2008.09.016]

[96] Kuo, M-S.; Mort, A.J.; Dell, A. Identification and location of l-glycerate, an unusual acyl substituent in gellan gum. *Carbohydr. Res.,* **1986**, *156*(0), 173-187. [http://dx.doi.org/10.1016/S0008-6215(00)90109-5]

[97] Morris, E.R.; Nishinari, K.; Rinaudo, M. Gelation of gellan – A review. *Food Hydrocoll.,* **2012**, *28*(2), 373-411. [http://dx.doi.org/10.1016/j.foodhyd.2012.01.004]

[98] Bajaj, I.B.; Saudagar, P.S.; Singhal, R.S.; Pandey, A. Statistical approach to optimization of fermentative production of gellan gum from Sphingomonas paucimobilis ATCC 31461. *J. Biosci. Bioeng.,* **2006**, *102*(3), 150-156. [http://dx.doi.org/10.1263/jbb.102.150] [PMID: 17046526]

[99] Jin, H. Production of gellan gum by Sphingomonas paucimobilis NK2000 with soybean pomace. *Biochem. Eng. J.,* **2003**, *16*(3), 357-360. [http://dx.doi.org/10.1016/S1369-703X(03)00076-7]

[100] Nampoothiri, K.M. Fermentative production of gellan using Sphingomonas paucimobilis. *Process Biochem.,* **2003**, *38*(11), 1513-1519. [http://dx.doi.org/10.1016/S0032-9592(02)00321-7]

[101] Freitas, F.; Alves, V.D.; Reis, M.A. Advances in bacterial exopolysaccharides: from production to biotechnological applications. *Trends Biotechnol.,* **2011**, *29*(8), 388-398. [http://dx.doi.org/10.1016/j.tibtech.2011.03.008] [PMID: 21561675]

[102] Fialho, A.M.; Moreira, L.M.; Granja, A.T.; Popescu, A.O.; Hoffmann, K.; Sá-Correia, I. Occurrence, production, and applications of gellan: current state and perspectives. *Appl. Microbiol. Biotechnol.,* **2008**, *79*(6), 889-900. [http://dx.doi.org/10.1007/s00253-008-1496-0] [PMID: 18506441]

[103] Kiosseoglou, A. Functionality of medium molecular weight xanthan gum produced by Xanthomonas Campestris ATCC 1395 in batch culture. *Food Res. Int.,* **2003**, *36*(5), 425-430. [http://dx.doi.org/10.1016/S0963-9969(02)00175-8]

[104] Palaniraj, A.; Jayaraman, V. Production, recovery and applications of xanthan gum by Xanthomonas campestris. *J. Food Eng.,* **2011**, *106*(1), 1-12. [http://dx.doi.org/10.1016/j.jfoodeng.2011.03.035]

[105] Ben Salah, R. Optimisation of xanthan gum production by palm date (Phoenix dactylifera L.) juice by-products using response surface methodology. *Food Chem.,* **2010**, *121*(2), 627-633. [http://dx.doi.org/10.1016/j.foodchem.2009.12.077]

[106] Casas, J.A.; Santos, V.E.; García-Ochoa, F. Xanthan gum production under several operational conditions: molecular structure and rheological properties. *Enzyme Microb. Technol.,* **2000**, *26*(2-4), 282-291. [http://dx.doi.org/10.1016/S0141-0229(99)00160-X] [PMID: 10689089]

[107] Chantaro, P.; Pongsawatmanit, R. Influence of sucrose on thermal and pasting properties of tapioca starch and xanthan gum mixtures. *J. Food Eng.,* **2010**, *98*(1), 44-50.

[http://dx.doi.org/10.1016/j.jfoodeng.2009.12.006]

[108] Remaud-Simeon, M. Glucansucrases: molecular engineering and oligosaccharide synthesis. *J. Mol. Catal., B Enzym.,* **2000**, *10*(1–3), 117-128. [http://dx.doi.org/10.1016/S1381-1177(00)00119-3]

[109] García-Ochoa, F.; Santos, V.E.; Casas, J.A.; Gómez, E. Xanthan gum: production, recovery, and properties. *Biotechnol. Adv.,* **2000**, *18*(7), 549-579. [http://dx.doi.org/10.1016/S0734-9750(00)00050-1] [PMID: 14538095]

[110] Mabrouk, M.E. Xanthan production by a novel mutant strain of Xanthomonas campestris: Application of statistical design for optimization of process parameters. *Life Sci. J.,* **2013**, *10*(1), 1660-1667.

[111] Shih, I-L.; Chen, L-D.; Wu, J-Y. Levan production using Bacillus subtilis natto cells immobilized on alginate. *Carbohydr. Polym.,* **2010**, *82*(1), 111-117. [http://dx.doi.org/10.1016/j.carbpol.2010.04.030]

[112] Jiang, L. Effect of nitrogen source on curdlan production by Alcaligenes faecalis ATCC 31749. *Int. J. Biol. Macromol.,* **2013**, *52*(0), 218-220. [http://dx.doi.org/10.1016/j.ijbiomac.2012.10.010] [PMID: 23085490]

[113] McIntosh, M.; Stone, B.A.; Stanisich, V.A. Curdlan and other bacterial (1-->3)-β-D-glucans. *Appl. Microbiol. Biotechnol.,* **2005**, *68*(2), 163-173. [http://dx.doi.org/10.1007/s00253-005-1959-5] [PMID: 15818477]

[114] Seo, H-P.; Son, C.W.; Chung, C.H.; Jung, D.I.; Kim, S.K.; Gross, R.A.; Kaplan, D.L.; Lee, J.W. Production of high molecular weight pullulan by Aureobasidium pullulans HP-2001 with soybean pomace as a nitrogen source. *Bioresour. Technol.,* **2004**, *95*(3), 293-299. [http://dx.doi.org/10.1016/j.biortech.2003.02.001] [PMID: 15288272]

[115] Jin, L-H.; Um, H.J.; Yin, C.J.; Kim, Y.H.; Lee, J.H. Proteomic analysis of curdlan-producing Agrobacterium sp. in response to pH downshift. *J. Biotechnol.,* **2008**, *138*(3-4), 80-87. [http://dx.doi.org/10.1016/j.jbiotec.2008.08.010] [PMID: 18824044]

[116] Kim, D.Y.; Jung, S.B.; Choi, G.G.; Kim, Y.B.; Rhee, Y.H. Biosynthesis of polyhydroxyalkanoate copolyester containing cyclohexyl groups by Pseudomonas oleovorans. *Int. J. Biol. Macromol.,* **2001**, *29*(3), 145-150. [http://dx.doi.org/10.1016/S0141-8130(01)00144-1] [PMID: 11589966]

[117] Silva, M.F. Production and characterization of xantham gum by Xanthomonas campestris using cheese whey as sole carbon source. *J. Food Eng.,* **2009**, *90*(1), 119-123. [http://dx.doi.org/10.1016/j.jfoodeng.2008.06.010]

[118] Jiang, L. Optimization of fermentation conditions for pullulan production by Aureobasidium pullulan using response surface methodology. *Carbohydr. Polym.,* **2010**, *79*(2), 414-417. [http://dx.doi.org/10.1016/j.carbpol.2009.08.027]

[119] Duan, X.; Chi, Z.; Wang, L.; Wang, X. Influence of different sugars on pullulan production and activities of α-phosphoglucose mutase, UDPG-pyrophosphorylase and glucosyltransferase involved in pullulan synthesis in Aureobasidium pullulans Y68. *Carbohydr. Polym.,* **2008**, *73*(4), 587-593. [http://dx.doi.org/10.1016/j.carbpol.2007.12.028] [PMID: 26048225]

[120] Yaseen, E.I. Rheological properties of selected gum solutions. *Food Res. Int.,* **2005**, *38*(2), 111-119.

[http://dx.doi.org/10.1016/j.foodres.2004.01.013]

[121] Monsan, P. Homopolysaccharides from lactic acid bacteria. *Int. Dairy J.,* **2001**, *11*(9), 675-685. [http://dx.doi.org/10.1016/S0958-6946(01)00113-3]

[122] Castillo, E.; López-Munguía, A. Synthesis of levan in water-miscible organic solvents. *J. Biotechnol.,* **2004**, *114*(1-2), 209-217. [http://dx.doi.org/10.1016/j.jbiotec.2004.06.003] [PMID: 15464614]

[123] Meng, G.; Fütterer, K. Structural framework of fructosyl transfer in Bacillus subtilis levansucrase. *Nat. Struct. Biol.,* **2003**, *10*(11), 935-941. [http://dx.doi.org/10.1038/nsb974] [PMID: 14517548]

[124] SCO. *Opinion of the Scientific Committee on Food on a Dextran Preparation, Produced Using Leuconostoc mesenteroides, Saccharomyces cerevisiae and Lactobacillus spp, as a Novel Food Ingredient in Bakery Products.,* **2000**. Available from: http://ec.europa.eu/food/fs/sc/scf/out75_en.pdf

[125] Vandamme, E.; De Baets, S.; Steinbuchel, A. *Polysaccharides I: Polysaccharides and Prokaryotes*; Biopolymers Series, **2002**.

[126] de Belder, D. *Medical application of dextran and its derivatives. Polysaccharides in Medicinal Applications*; Marcel Dekker: New York, NY, **1996**, pp. 505-524.

[127] Ravella, S.R. Extracellular polysaccharide (EPS) production by a novel strain of yeast-like fungus Aureobasidium pullulans. *Carbohydr. Polym.,* **2010**, *82*(3), 728-732. [http://dx.doi.org/10.1016/j.carbpol.2010.05.039]

[128] Naessens, M. Leuconostoc dextransucrase and dextran: production, properties and applications. *J. Chem. Technol. Biotechnol.,* **2005**, *80*(8), 845-860. [http://dx.doi.org/10.1002/jctb.1322]

[129] Alimentarius, C. *General Standard for Food Additives*; GSFA, Codex STAN, **2008**, pp. 192-1995.

[130] Draget, K.I. 29 - Alginates. In: *Handbook of Hydrocolloids,* 2nd ed; Phillips, G.O.; Williams, P.A., Eds.; Woodhead Publishing, **2009**; pp. 807-828. [http://dx.doi.org/10.1533/9781845695873.807]

[131] Prajapati, V.D.; Jani, G.K.; Moradiya, N.G.; Randeria, N.P.; Nagar, B.J.; Naikwadi, N.N.; Variya, B.C. Galactomannan: a versatile biodegradable seed polysaccharide. *Int. J. Biol. Macromol.,* **2013**, *60*(0), 83-92. [http://dx.doi.org/10.1016/j.ijbiomac.2013.05.017] [PMID: 23707734]

[132] Sworn, G. 9 - Gellan gum. In: *Handbook of Hydrocolloids,* 2nd ed; Phillips, G.O.; Williams, P.A., Eds.; Woodhead Publishing, **2009**; pp. 204-227. [http://dx.doi.org/10.1533/9781845695873.204]

[133] Haug, I.J.; Draget, K.I. 6 - Gelatin. In: *Handbook of Hydrocolloids,* 2nd ed; Phillips, G.O.; Williams, P.A., Eds.; Woodhead Publishing, **2009**; pp. 142-163. [http://dx.doi.org/10.1533/9781845695873.142]

[134] Endreß, H.U.; Christensen, S.H. 12 - Pectins. In: *Handbook of Hydrocolloids,* 2nd ed; Phillips, G.O.; Williams, P.A., Eds.; Woodhead Publishing, **2009**; pp. 274-297. [http://dx.doi.org/10.1533/9781845695873.274]

[135] Taggart, P.; Mitchell, J.R. 5 - Starch. In: *Handbook of Hydrocolloids,* 2nd ed; Phillips, G.O.; Williams, P.A., Eds.; Woodhead Publishing, **2009**; pp. 108-141. [http://dx.doi.org/10.1533/9781845695873.108]

[136] Sworn, G. 8 - Xanthan gum. In: *Handbook of Hydrocolloids,* 2nd ed; Phillips, G.O.; Williams, P.A., Eds.; Woodhead Publishing, **2009**; pp. 186-203. [http://dx.doi.org/10.1533/9781845695873.186]

CHAPTER 10

Natural Antimicrobial Compounds

Sidnei Emilio Bordignon, Júlio Cesar de Carvalho, Jane Mary L. N. Gelinski, André Luís Lopes da Silva, Vanete Thomaz Soccol and **Carlos Ricardo Soccol**[*]

Bioprocess Engineering and Biotechnology Department, Federal University of Paraná, Curitiba, Brazil

Abstract: Processed foods depend on conservation methods to ensure biological stability of the product until consumption. Foods in general are a rich source of nutrients, and therefore may support the proliferation of opportunistic microorganisms. Most of these microorganisms only reduce the nutritional value and sensorial quality of the product; but in some cases, pathogenic contaminants may also grow. Traditional antimicrobial additives are decreasing in use due to some disadvantages related to physicochemical and sensory aspects, such as undesirable interactions with the food matrix, accumulation into consumer organism and even possible allergic reactions. Besides food traditional methods for preservation such as acidification or reduction of water activity, a possible strategy for increasing food shelf life is the use of natural antimicrobial compounds. Nowadays, the great challenge of the food industry is to make better use of these additives ensuring product integrity and, at the same time, generate minimal residual effects, avoiding undesirable physicochemical and sensorial modifications. In this context, there is an increasing preference of antimicrobial compounds from natural sources (microbial, animal or plant) targeting a wide use in the production of technologically advanced foods which, besides high quality and healthy, must be biologically friendly, meeting the demands of 21st century consumers.

Keywords: Additives, Bacteria, Bacteriocin, Biotechnology, Chelators, Microbial Control, Food Industry, Fungi, Legislation, Lysozyme, Membrane, Microbial Control, Nisin, Organic acids, Pediocin, Preservation, Resistance, Spoilage, Stability, Technology, Yeast.

[*] **Address correspondence to Carlos Ricardo Soccol:** Department of Bioprocess Engineering and Biotechnology, Universidade Federal do Paraná, Caixa Postal 19011, ACF Centro Politécnico, 81531-990 Curitiba, Brazil; E-mail: soccol@ufpr.br.

Juliano Lemos Bicas, Mário Roberto Maróstica, Glaucia Maria Pastore (Eds.)

INTRODUCTION

Food is susceptible to spoiling, and the need to preserve food quality is indissociable of the commercialization of processed foods. Microorganisms are ubiquitous, and processes guaranteed to reduce the microbial count in foods involve thermal processing, or other sterilization methods to a lesser extent. This means that products such as those minimally processed (*e.g.* vegetables, fruits, meat, dairy products, juices *etc.*) are *expected* to have low levels of contaminants, which may be controlled but not completely avoided during processing. Several traditional foods are stabilized, resulting in controlled development of microorganisms (*e.g.* wine, beer, cheese, yogurt), or results of physicochemical conservation methods (*e.g.* salting, concentrating, drying). With the development of food technology, thermal processing (pasteurizing, sterilizing, or freezing) became popular. However, most of these methods may dramatically alter food flavor or structure. Thus, in the perspective of the hurdle technology [1], food safety can be improved by combination of different food preservation methods, including the use of natural antimicrobial compounds. The combined use of those compounds with adequate methods of food processing can also contribute to reduce the risks of contamination by foodborne pathogens.

Antimicrobial compounds are produced by animals, plants and microorganisms. They occur naturally in foods or may be used as food additives. The first antimicrobial compounds used were spices and extracts from plants such as cinnamon, clove, garlic, mustard and onion [2], followed by fermentation products such as vinegar. With the development of chemistry, even inorganic substances such as hydrofluoric acid, fluorides, chlorates and other chemical compounds were used as preservatives [3]. But the development of efficient food additives was possible only with the advent of antibiotics, – which are highly effective and specific antimicrobial agents.

The use of antimicrobial compounds in foods is tightly regulated, for good reason: besides the usual problems associated with any additive – toxicity in high doses, allergies, other bioactivities – antimicrobial compounds may dramatically alter microbiota, and also may bioaccumulate in the environment. In addition, in the food chain, the selective pressure exerted mainly by inadequate use of antibiotics in primary production is responsible by appearance and spread of bacterial resistance [4]. Therefore, direct and indirect factors, such as cross-contamination during food processing, intentional addition of probiotic or others technological process (*e.g.* fermented foods), are issues for concern.

Actually, it was already known in 1948 that milk from cows treated for mastitis contained enough antibiotics to inhibit cheese-starter cultures [5], and regulation was issued to prevent or alert for the possibility of antibiotics in milk reaching the public at large. There are more than 30 permitted antimicrobial additives for food uses in EU, most of them organic acids or their salts (17), sulfites (8), nitrites (4) or phenols (3). Part of these preservatives is not permitted in processed food itself, but rather as a post-harvest treatment, as is the case of biphenyl or thiabendazole. Table **1**, presents the additives listed by the Codex Alimentarius and their typical dosage.

Table 1. Food additives with antimicrobial (preservative) action. The list covers most additives permitted by regulating agencies worldwide[1,2].

INS	Additive name	Typical maximal dosage (mg/kg)[3]	Applications other than as preservative
200	Sorbic acid	15-3000	Antioxidant, Stabilizer
201	Sodium sorbate		
202	Potassium sorbate		
203	Calcium sorbate		
209	Heptyl para-hydroxybenzoate	12 (EUA, beverages)	
210	Benzoic acid	15-5000	
211	Sodium benzoate		
212	Potassium benzoate		
213	Calcium benzoate		

(Table 1) contd.....

INS	Additive name	Typical maximal dosage (mg/kg)[3]	Applications other than as preservative
214	Ethyl para-hydroxybenzoate	2-1500	
215	Sodium ethyl para-hydroxybenzoate		
216	Propyl para-hydroxybenzoate		
217	Sodium propyl para-hydroxybenzoate		
218	Methyl para-hydroxybenzoate		
219	Sodium methyl para-hydroxybenzoate		
220	Sulfur dioxide	2-20	Antioxidant, Bleaching agent, Sequestrant, Stabilizer, Acidity regulator, Firming agent (calcium salts)
221	Sodium sulfite		
222	Sodium hydrogen sulfite		
223	Sodium metabisulfite		
224	Potassium metabisulfite		
225	Potassium sulfite		
226	Calcium sulfite		
227	Calcium hydrogen sulfite		
228	Potassium bisulfate		
230	Diphenyl	Just post-harvest, (Japan)	
234	Nisin	3-12	
235	Natamycin (Pimaricin)	2 (surface, 5mm depth)	
242	Dimethyl dicarbonate	250	
243	Lauric arginate ethyl ester	200	
249	Potassium nitrite	Viewed as color fixative in most legislations. 150-300 in Brazil	Color fixative
250	Sodium nitrite		
251	Sodium nitrate		
252	Potassium nitrate		
260	Acetic acid	GMP[4]	Acidity regulator, Acidulant
261	Potassium acetates		Acidity regulator
262	Sodium acetates		Acidity regulator, Sequestrant
263	Calcium acetate		Acidity regulator, Stabilizer
266	Sodium dehydroacetate	500 (Japan)	
270	Lactic acid	2000 to QSP (Brazil)	
280	Propionic acid	GRAS[4], GMP[4]	Flavoring agent

(Table 1) contd.....

INS	Additive name	Typical maximal dosage (mg/kg)[3]	Applications other than as preservative
281	Sodium propionate	GRAS[4], GMP[4]	
282	Calcium propionate		
283	Potassium propionate		
290	Carbon dioxide	GMP[4]	Carbonating agent, Packaging gas, Propellant
296	Malic acid	QSP[4]	Acidity regulator, Sequestrant
297	Fumaric acid	QSP[4]	Acidity regulator, Stabilizer
300	Ascorbic acid	GRAS[4], GMP[4]	Antioxidant, Color stabilizer, Flour enhancer
315	Erithorbic acid	GRAS[4], QSP[4]	Antioxidant
339(iii)	Trisodium phosphate	GRAS[4], QSP[4]	Acidity regulator, Emulsifier, Humectant, Stabilizer, Sequestrant, Thickener
354	Calcium tartrate	QSP[4]	Acidity regulator
385	Calcium disodium EDTA	75-200	Antioxidant synergist, Color retention agent, Sequestrant
386	Disodium EDTA		Antioxidant synergist, Color retention agent, Sequestrant, Stabilizer
1105	Lisozyme	500, GMP[4]	

[1]Data from FAO [7], FDA (EUA) [8], EC (Europe), MHLW (Japan) [9], ANVISA (Brazil) [10].

[2]Substances listed in the Codex Alimentarius, but lacking provision, and not listed in the legislation of EUA, EU, Japan or Brazil were omitted *e.g.* INS391, phytic acid, and INS240, formaldehyde.

[3]The maximal dosage varies with the kind of food or beverage.

[4]GMP: good manufacturing practices. GRAS: Generally recognized as safe (and thus not strictly an additive - FDA regulation). QSP: *Quantum Satis Para*, or the amount necessary for the desired effect (used in Brazilian and European legislation).

Although the list of permitted additives is different in USA and Japan, the same trend is observed – most additives are organic acid derivatives, while the most toxic substances are permitted only for specific uses. Also, because this specific use may be a result of the traditional form of preparation of the food, as is the case of nitrites – used for color fixation in ham, sausages and other meat products [6] – these may be listed in other categories, with the preserving action being regarded as secondary. Finally, several additives may have a weak preserving action, as is the case for acidulants (which are not biocidal or bacteriostatic, but simply maintain a low pH), anti-caking agents (which also maintain low a_w), and chelating agents (which bind to cofactors such as iron), all of these impairing microorganism growth.

TOWARDS NATURAL ANTIMICROBIAL COMPOUNDS

Independent from the source, any good bioactive compound idealized for use in preserving foods against pathogenic or spoilage microorganisms should have some desirable properties, such as: a) a broad spectrum of antibacterial and antifungal activity; b) antioxidant action; c) low toxicity; d) effectivity in the native environment; e) more effective action when interacting with other bioactive molecules present in the food; f) do not mask the aroma and flavor of the product, and g) effectiveness at low concentrations.

Control of the conditions for microbial growth in processed foods may be done by careful formulation and judicious use of other additives, thus guaranteeing minimal counts of microorganisms. Food processing and packaging should be able to maintain microbial contamination at lowest levels, but for foods with relatively high water activity, the interaction of factors is essential to restrict the microbial growing and ensuring a longer shelf life. Moreover, even when antimicrobial additives use is permitted, they should be used in the lowest level possible.

Artificial, broad-spectrum antimicrobial compounds tend to be unspecific and, therefore, too toxic (if effective) or too mild (if less toxic to humans). Natural antimicrobial compounds, on the other hand, are highly effective and more selective. Since competition among species in nature is fierce, there is an enormous amount of these compounds produced naturally by countless species. However, most of these antimicrobial compounds have limitations regarding the potential activity required for use in small quantities so that it does not cause changes in aroma and flavor. There is a trend towards the development of antimicrobial additives which derive from edible vegetables and animals, as well as GRAS microorganisms: the reasoning is that such additives would be well tolerated and quickly approved, because they are traditionally consumed in foods. Organic acids such as acetic, propionic and butyric may be produced by microorganisms (being natural, in that sense), but are also easily obtainable with high purity *via* chemical synthesis. The following sections briefly address the main antimicrobial additives from microorganisms currently in use and present the prospects of future natural additives.

Traditional Antimicrobial Compounds

There are several mechanisms proposed for the antimicrobial action of the currently used natural additives, which are not always totally understood. The activity depends on the properties and also on the concentration of the antimicrobial compound, which may act as a bacteriostatic in lower concentrations, and a bactericidal agent in higher concentrations. In general, many additives have multiple targets and are relatively mild and nonspecific in low concentrations. They inhibit bacterial and fungi growing through various actions, such as lowering pH, changing the proton gradient *via* cell membrane, acidifying the intracellular environment, and others.

Organic Acids: the short chain organic acids (*e.g.* lactic, acetic, propionic, butyric, citric) and their salts have bacteriostatic action against a broad spectrum of microorganisms. These acids can pass through the cytoplasmic membrane, reduce internal pH, altering the membrane potential and the electron transport chain; they may also bind to nucleic acids [11]. The pH of substrate has influences on the preservative action of organic acids and their salts and should be taken in account during food formulation. One possible explanation to the potential disruption across membranes is that the undissociated molecules of the acid cross the membrane, then dissociate (in a higher pH), generating protons and anions [12]. Upon dissipation of the proton motive force, the enzymatic synthesis is suppressed together with the interruption of the nutrient transport system, amino acid and energy metabolism, and synthesis of nucleic acids. Besides this, the use of sodium and potassium lactates in high concentrations decrease water activity, extending the shelf life of manufactured foods. Acetic acid and their derivatives (sodium acetate, calcium diacetate; dihydroacetic acid) remains extensively applied due to its wide spectrum of inhibition, since only *Acetobacter,* lactic acid bacteria and butyric bacteria are tolerant to it. The list of sensitive bacteria and fungi comprises: *Bacillus* sp., *Clostridium* sp., *Listeria* sp., *Salmonella* sp., *Staphylococcus aureus, Pseudomonas* sp., *E. coli, Campylobacter* sp., *Aspergillus* sp., *Rhizopus* sp., *Penicillium* sp. and *Saccharomyces* sp. The propionates, derived from propionic acid, have more specific applications, mainly to avoid the fungi proliferation in bakery and cheese surfaces [13].

The natural acidification in foods, mainly from metabolic activity from lactic and acetic acid bacteria, is useful in fermented beverages and dairy products. Non-fermented products have been preserved mainly by sorbic and benzoic acid in food processing, *i.e.* fruit juices, soft drinks. Benzoic acid is one of the most widely used preservatives in food industry; it is not cumulative in the human organism because it is converted, in the presence of glycine, into hippuric acid (which is then excreted in the urine) [14]. The main form of use is as sodium or potassium benzoate, which are the more soluble forms than the acid. The effective bactericidal and fungicidal action of benzoate salts happens in acidic conditions (pH lower than 4.2), and the compound is generally applied in a final concentration of 0.05% to 0.1%. Similarly, sorbic acid is more powerful in low pH and inhibitory to yeast and molds, but it is not effective to bacteria spoilage. The usual dosage of potassium sorbate is 0.05% to 0.2%, which can be applied inside or on surface by pulverization, *i.e.* into cheese, meat products, confectioneries and bread. An extend revision about biotechnological production of organic acids is presented in the Chapter 4.

Chelators: substances such as citric, phytic or even phosphoric acids may bind to cations, such as Ca^{2+}, Mn^{2+}, Mg^{2+} or Fe^{2+}, which are cofactors for essential enzymes. These additives may potentiate the action of other substances, for example, the lack of Mn^{2+} can reduce the resistance of lactic acid bacteria to oxidation. Metallo-proteins such as lactoferrin and ovotransferrin, with its high affinity for iron, are very effective chelators which their use may be intensified in the near future. However, very effective chelators may have an anti-nutritional action, as is the case for phytate [15], and for this its use should be properly evaluated.

Other binding molecules: There are several proteins capable of reversibly binding coenzymes. These are not called chelators since, by definition, chelators bind to a single metallic ligand. However the mechanism of action as an antimicrobial agent is similar: the protein binds very efficiently to a molecule which is essential for cell metabolism, therefore reducing its bioavailability. There are several natural compounds of this kind, and, just as chelators, they have anti-nutritional, as well as an antimicrobial, effect. The most important member of this group of molecules is avidin, an egg tetrameric protein that binds very efficiently to biotin. Some carbohydrates also have antimicrobial action, such as chitosan. This polyaminossaccharide is a commercial product obtained from modification on chitin from crab and shrimp shells. Chitosan is active against several bacteria (due to its polycationic structure), and can also suppress sporulation and spore germination in fungi. Currently, chitosan is used in food products mainly in edible bio-packaging, which can be produced by dipping or spraying the product. These systems with chitosan have been promising due to its high biodegradability and low toxicity [16 - 18].

Natamycin: This polyene antibiotic (also known as pimaricin) is used as an antifungal agent with a broad spectrum of activity, and is widely applied in the food industry. Natamycin is a 26-membered ring macrocyclic polyketide that is synthesized by a common pathway in which units originated from acetate, propionate or butyrate are condensed onto the growing chain by a polyketide synthase (PKS) in a process that resembles the long-chain fatty acid biosynthesis [19, 20]. It is produced by *Streptomyces natalensis, S. chattanoogensis, S. gilvosporeus* and *S. lydicusin* and it binds to ergosterol in fungal and yeast membranes, presumably impeding the sterol of properly distributing in the membrane, and killing the cell. The formation of pores seems to not occur with this polyene [21], but this antimycotic interferes with endocytosis [22]. Since bacteria do not have ergosterol in their membranes, natamycin has no antibacterial activity. Natamycin has been approved as a GRAS (Generally Recognized as Safe) product by FDA (Food and Drug Administration, USA) for use in food manufacturing.

Lysozyme: This is a bacteriolytic enzyme which hydrolyzes the peptidoglycan of bacterial cell walls and can also bind polymers of N-acetyl glucosamine. Although there are several lysozymes, the only used in commercial applications is hen egg white lysozyme [23]. This enzyme is active against Gram-positive bacteria, especially spores from *Clostridium* sp. (*Cl. tyrobutyricum, Cl. thermosac-charolyticum*), *Bacillus stearothermophilus* and *Listeria monocytogenes,* but

Gram-negative bacteria may be resistant because of the outer membrane, which acts as a physical barrier [24]. The efficacy of lysozyme over Gram-negative bacteria may be enhanced by the concomitant use of chelators or other permeabilizing agents [25].

Nisin: This small, ribosomally synthesized peptide is effective in the formation of pores in cell membranes, causing a rapid leakage of cytoplasmic content and depletion of proton motive force, which leads to cell death [26]. Nisin is produced by *Lactococcus lactis* and occurs naturally in some dairy products, and was the first bacteriocin to be permitted for food use by the FDA. More details about bacteriocins, especially nisin, are described in the following sections.

Novel Natural Compounds

Although few bioactive substances are effectively permitted as food or feed additives, there are several molecules being researched as potential new additives, mainly due to global concern about the emergence of drug-resistant bacteria, fungi, and viruses. Natural antimicrobial compounds are evaluated regarding their toxicity against humans and their effectiveness against bacteria, yeasts, filamentous fungi and viruses. Besides microorganisms, which already have been sources of novel antimicrobial compounds [27 - 29], some vegetable, such as herbs and fruits, holds a promise as sources for antimicrobial substances [30, 31]. Although phytochemicals of plants have varying degrees of antimicrobial activity, they are on the top list of compounds with wide possibilities of food application in relation to other antimicrobial substances. The main groups of phytochemicals are phenolic compounds (with antimicrobial and antioxidant activity) in essential oils, polypeptides, lectins, and alkaloids. The antimicrobial activity has been attributed mainly to the phenolic component of the essential oil fraction. The well-known spice oregano has a broad spectrum of antimicrobial activity probably due to its two main volatile constituents, thymol and carvacrol [32]. The use of oregano by the food industry tends to increase, since new research shows that combinations of oregano and other essential oils may have different degrees of effectiveness on different pathogens and spoilage microorganisms. Combinations of carvacrol, thymol and eugenol were evaluated by García-García, López-Malo and Palou [33] and they found synergistic activity against *Listeria inocua* compared to each isolated component. Other typical plant antimicrobial compounds are the aromatic medicinal plant, as flavonoids, thiosulfinates (allicin), polyphenol cathechins (green tea) and allyl isothiocyanate, component of mustard, with effective antibacterial and antifungal activities. These compounds may become important food additives in the future (Table **2**).

Similarly, it is known that several edible mushrooms produce a variety of bioactive substances [34], and that may provide new additives for food use. For example, aqueous, ethanolic and methanolic extracts of *Pleurotus florida* were effective against bacteria, and its analysis demonstrated the presence of phenolic compounds, glycosides and flavonoids [35]. It is possible that cultured macromycetes may provide a batch-to-batch consistent source of antimicrobials for food.

Table 2. Potential bioactive antimicrobial compounds in development.

Compound	Description / Mode of action	Main applications	Ref.
ANIMAL ORIGIN			
Lactoferrin	Iron-binding glycoprotein from milk which binds with specific target on cell surface of Gram-negative bacteria	Sports drinks, dietary supplements with immune properties claims; control of spoilage microorganisms and shelf-life extension of cheeses	[36] [37]
Ovotransferrin	Peptides classified as metalloproteinases from egg white albumin which inhibits bacterial growth by iron deprivation	Antimicrobial activity has been reported against Gram-positive and Gram-negative bacteria, specially foodborne pathogens and spoilage bacteria, but the food application is limited by allergenic reactions from egg white	[38]
Chitosan	Biopolymer of deacetylated chitin present in the exoskeletons of crustaceans and arthropods with low toxicity against mammalian cells	Direct application to reduce contamination in egg or as a packaging film on food as cheese, carrots and meat products surface	[39]
Pleurocidin	Heat-stable peptide, salt-tolerant, noncytotoxic to human cells but inhibited by divalent ions. Recombinant expression has been reported for large-scale production	Several important foodborne pathogens are sensitive to pleurocidin, as *Listeria monocytogenes, Vibrio parahemolyticus, E. coli* O157:H7, and *Penicillium expansum*	[40]
ANIMAL ORIGIN			

(Table 2) contd.....

Compound	Description / Mode of action	Main applications	Ref.
Protamine	Cationic peptide present in salmon with inhibitory effect of cellular growth via electrostatic interaction	The activity of protamine is dependent to linkage with negatively charged particles but non-specific binding is a limitation to food application. *Clostridium botulinum* and *E. coli* are sensitive to protamine	[41] [42]
Defensins	Cationic peptides with broad-spectrum antimicrobial activity obtained specially from protogynous hermaphroditic oranges potted grouper species and present in many animals	Antimicrobial activity against fungi, Gram-positive, Gram-negative bacteria. Antiviral activity by recombinant β-defensins	[43] [44]
Lipids	Free fatty acids isolated from mucosal surface of animal and milk lipids, such as caprylic acid, eicosapentaenoic acid – EPA, and docosahexaenoic acid – DHA	Wide-spectrum of antimicrobial activity against *Salmonella* Enteritidis, *Salmonella* Typhimurium, *E. coli* O157:H7, *L. monocytogenes,* and *Staphylococcus aureus*	[45] [46]
PLANT ORIGIN			
Saponins	Compounds responsible for plant protection from biotic stress by formation of complexes with cholesterol. Saponins acts through interaction with sterols and fatty acids on microbial membranes, promoting their solubilization	The antibacterial, antifungal and antiviral properties of saponins have potential for use in food preservation, its foam-like characteristics allows for the application as a natural bio-surfactant to stabilizing beverages and emulsions	[47] [32]
Flavonoids	Secondary metabolites from plants that can also improve sensory and nutritional quality of foods due to their antioxidant action provided by the phenolic structure which binds with different target components in the microbial cell. Flavonoids can promote inhibition by blocking nucleic acid synthesis, or altering the fluidity of the cytoplasmatic membrane, or by interference in the energy metabolism of the respiratory electron transport chain	Flavonoids compounds present a broad-spectrum antimicrobial activity including antifungal, antibacterial and antiviral properties, but the most useful application remains on beverage and fruit juices industries to control the fungi spoilage	[48] [45]
Thiosulfinates	Organosulfur compounds, mainly obtained from garlic fresh extract, promote inhibition of certain thiol-containing enzymes in the microorganisms by the rapid reaction of thiosulfinates with thiol groups, *e.g.* alcohol dehydrogenase, RNA polymerase	Allicin has multifunctional use as food additive by flavor and antimicrobial roles against bacteria (Gram positive and negative), some fungi and virus, but reports indicate possible allergic reactions associated to garlic consumption	[49] [45]

(Table 2) contd.....

Compound	Description / Mode of action	Main applications	Ref.
Catechins	Polyphenolic substances with antioxidant activity - isolated from green tea which exhibits antimicrobial properties suggested by binding with peptidoglycan on cell wall or by the generation of hydrogen peroxide after reaction with oxygen species in presence of superoxide dismutase	Catechins have not been used in the food industry as an antimicrobial agent, its use in oils and fats reduce the lipid oxidation. The *in vitro* assays show a broad-spectrum antimicrobial activity against important food pathogens, *e.g.* *Salmonella* sp., *Staphylococcus* spp., *Escherichia coli*, and pathogens in humans, *e.g. Helicobacter pylori, Candida albicans*	[50] [51]
Glucosinolates	Bioactive compounds from cruciferous vegetables (broccoli, cabbage) with antimicrobial action mediated by volatile substances, such as isothiocyanates and thiocyanates	Antibacterial and antifungal activities can be useful on packaged foods through gaseous glucosinolates, but certain studies with purified glucosinolates have indicated toxicity in some animal models	[52] [45]

BACTERIOCINS IN FOOD

Nisin was the first bacteriocin extensively studied for its antimicrobial role as a food preservative. It consists of a natural peptide with molecular size 3,500 Da produced by *Lactococcus lactis* subsp. *lactis*. It was discovery in 1928, in England, by researchers Rogers and Whittier, and since then numerous studies suggest its use in the food industry, strengthened by the GRAS certification granted by the FDA, in 1969. Due to the widely accepted claims of safety and the susceptibility by intestinal proteases and the fact the nisin does not alter sensory properties of the foods, this bacteriocin is a well-accepted additive for food technology, mainly in dairy products.

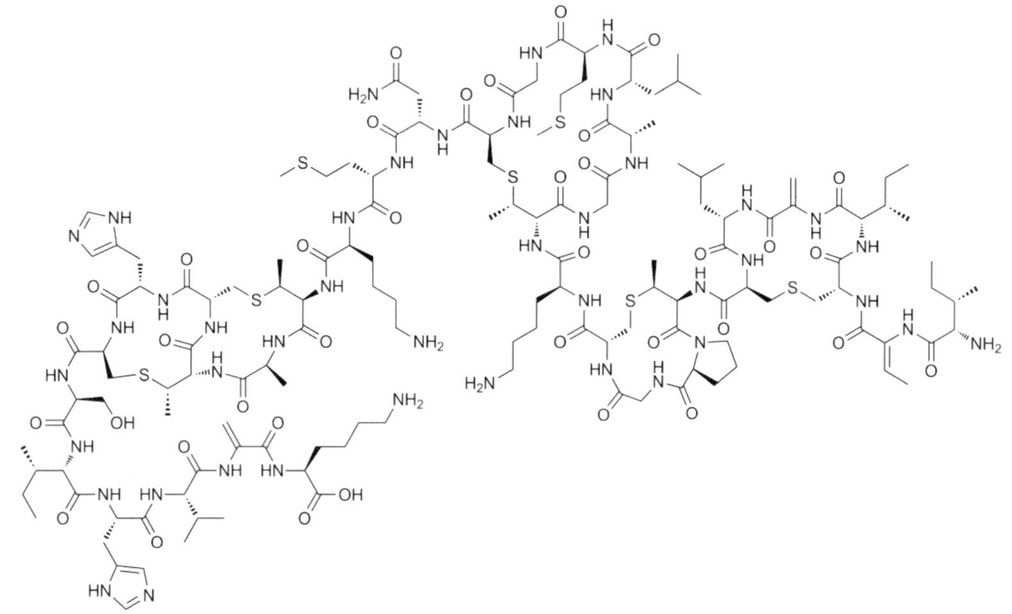

Fig. (1). Nisin structure.

Biochemically, nisin consists of a linear chain of 34 amino acid residues (Fig. **1**), with cationic and hydrophobic characteristics. Two forms, named variants A and Z, are widely known and studied and differ only in the amino acid on the 27th position, where histidine in nisin A is replaced by asparagine is in nisin Z. The antimicrobial spectrum is not affected by this substitution, however the Z variant has higher solubility and intermolecular diffusion and these characteristics are important for food technology. Other variants (termed nisin Q and F) have been reported more recently and obtained *via* substitution of others amino acids [53, 54].

Synthesis and Biotechnological Production of Nisin

The biosynthesis of nisin occurs extracellularly and mostly during the exponential growth cellular phase, although release occurs during stationary phase. In the ribosomes occurs the precursors or pre-peptides synthesis, which are enzymatically modified after translation by changing multiple amino acid residues. Then, the modified peptide is exported across the cytoplasmic membrane to be activated after a proteolytic cleavage, generating the final molecule (34 amino acids) with antimicrobial activity recognized [55, 56].

The main mode of action of nisin on vegetative cells occurs *via* depletion of the proton motive force, resulting from the release of intracellular ions through pores formed on the cell membrane. However, others mechanisms have been reported: (1) interference on the cell wall biosynthesis due to ability of nisin to bind peptidoglycan precursor (lipid II), using the same mechanism which allow the membrane permeabilization [57]; and (2) induction of cell autolysis by the release of hydrolytic enzymes (N-acetylmuramoyl-l-alanine amidase and N-acetylglucosaminidase) in the medium [58 - 62]. Nisin further demonstrates the ability to reduce thermal resistance and rate of germination of bacterial spores [63]. In food processing, this feature favor the combined use of nisin after high thermal treatment, reducing the risk of possible thermo-resistant spores.

The spectrum of bactericidal activity of nisin essentially comprises Gram-positive bacteria taxonomically related to the producer strain, thus, other lactic acid bacteria are the most sensitive, including the genera: *Lactococcus* spp., *Lactobacillus* spp., *Streptococcus* spp., *Pediococcus* spp., *Micrococcus* spp., but inhibition of important food pathogens, such as *Listeria monocytogenes* and *Staphylococcus aureus,* also occurs. At the research level, important reports have demonstrated the inhibitory effect of nisin on other microorganisms, including Gram-negative bacteria that are naturally resistant due to the denser layer of lipopolysaccharide (LPS) that acts as a protective barrier for the cytoplasmatic membrane. In these cases, chelating agents such as EDTA, or hydrolytic enzymes such as lysozyme, are usually applied together with nisin to increase permeability of target-cell membranes [64, 65].

The biotechnological production of nisin by *L. lactis* is well-known and extensively optimized for M17 and MRS synthetic media, where the cell growth and peptides expression are maximal. A strict control of temperature and pH are required to regulate nisin biosynthesis, besides rich protein composition in the fermentation medium, a typical requirement of lactic acid bacteria. Sucrose is usually the main source of carbon, and reports show significant increases in production by asparagine supplementation in MRS medium [66]. For industrial scale production, natural and cheapest substrates are required to the fermentation, such as fluid milk, skimmed milk or cheese whey, a common dairy industry by-product. The product available on the market is a mixture of milk proteins, salts and pure nisin at 2.5% concentration, standardized to an activity of 1 x 10^6 IU g^{-1}, sold by various companies worldwide. The high cost of production even with natural substrates, such as skimmed milk, has motivated studies to evaluate the different wastes from the dairy industry. Routinely, most of those industries discard large volumes of food-grade liquids resulting from pressing and washing steps, especially in the cheese manufacturing, as new substrates to submerse fermentation. The main steps after fermentation process are: separation of cells from broth, concentration, dehydration, and thus, the solid residue is standardized as a commercial formulation typically composed by sodium chloride and other denatured milk solids. Approaches seek to improve nisin downstream processing in order to obtain higher purity of the biomolecule, but generally, they are still too expensive for industrial implementation.

For research purposes, nisin can be purified by ion exchange chromatography, immunoaffinity or HPLC. Other alternative techniques can be used to increase concentration of the peptides, such as precipitation using organic solvents (ethanol or methanol), salts (ammonium sulfate) or acidic solutions (pH 2.0) [55].

The use of nisin as a biopreservative is allowed in more than 50 countries, as regulated by FAO/WHO since 1969, when it became recognized as a food additive for cheese up to the concentration of 12.5 mg of pure nisin per kilogram of product. Currently, each country establishes different rules for use by national legislation, including other foods such as meat derivatives, different cheeses, vegetables and tomatoes, and different maximum concentration. Some countries, such as France, Australia, Peru and the United Kingdom, do not regulate the maximum dosage, however, its use is restricted to acidic products because their activity is strongly reduced above pH 7.0. Furthermore, foods with high fat content tend to reduce the nisin efficacy because the peptides have high binding affinity to lipids, causing an unproductive adsorption that limits interaction of the peptide with membrane phospholipids. The presence of proteolytic enzymes in food is another limiting factor for use of the bacteriocin inactivation, since these enzymes, especially those present in meat products, cause inactivation of the bacteriocin.

Other Bacteriocins

Several other bacteriocins have been studied for use as food additives to control bacterial contamination. These bacteriocins can differ from nisin in terms of the producer strain, molecular size and strength, or by small differences in inhibitory spectrum. Table **3** shows the different classes of bacteriocins based on the initial grouping of Klaenhammer (1993) [67, 68] with new data from various authors.

In general, the bacteriocins have the same prospects of application in meat foods and dairy products. For example, sakacins, peptides produced by *Lactobacillus sakei*, are usually applied in the industry by introduction of bacteriocin-producing starters, *e.g.* for the processing of fermented milk or cured meats. In these cases the antimicrobial effect will dependent on cellular development and on the interaction between the bacteriocin released in the medium with the chemical composition of food.

Pediocins are also commercially available as natural antimicrobial compounds. These bacteriocins are metabolites synthesized by *Pediococcus* spp. strains, characterized by small molecular size (< 5,000 Da), thermal resistance and antibacterial function in a wide pH range. Pediocins PA-1/AcH also represent a large group of bacteriocins which include pediocin-like peptides having an N-terminal consensus sequence -Tyr-Gly-Asn-Gly-Val-Xaa-Cys. They are classified as Class IIa according to the traditional Klaenhammer classification [68] and their main characteristic is the anti-listerial activity. They are clustered based on the amino acid sequence similarity in the group, which ranges between 40 and 60%.

Table 3. Related bacteriocins based on Klaenhammer´s classification.

Class	Bacteriocins	Producer
I - Type A: Lantibiotics as short peptides (<5KDa) containing the unusual amino acids lanthionine (Lan), β-methyllanthionine (MeLan), dehydroalanine, and dehydrobutyrine.activity through the formation of pores in bacterial membranes.	nisin lactocin S epidermin gallidermin lacticin 481	*Lactococcus lactis* *Lactobacillus sake* *Staphylococcus epidermidis* *Staphylococcus gallinarum* *L. lactis*

(Table 3) contd.....

Class	Bacteriocins	Producer
I - Type B: Lantibiotics as smaller globular peptides and with a negative or no net charge; antimicrobial activity is related to the inhibition of specific enzymes.	mersacidin cinnamycin ancovenin duramycin actagardin	*Bacillus subtilis* *Streptomyces innamoneus* *Streptomyces* ssp. *S. cinnamoneus* *Actinoplanes* spp.
II - Class IIa: small (<10 kDa), heat-stable, non-lanthionin--containing peptides. Active against Listeria spp. includes pediocin-like peptides having an N-terminal consensus sequence -Tyr-Gly-Asn-Gly-Val-Xaa-Cys.	pediocin PA-1/AcH sakacin A sakacin P leucocin A-UAL 187 mesentericin Y105 enterocin A divercin V41 lactococcin MMFII	*Pediococcus acidilactici* *L. sake* *L. sake* *Leuconostoc gelidum* *Leuconostoc mesenteroides* *Enterococcus faecium* *Carnobacterium divergens* *L. lactis*
II - Class IIb: contains bacteriocins requiring 2 different peptides for activity.	lactococcin G lactococcin M lactacin F plantaricin A plantaricin S plantaricin EF plantaricin JK	*L. lactis* *L. lactis* *Lactobacillus johnsonii* *Lactobacillus plantarum* *L. plantarum* *L. plantarum* *L. plantarum*
II – Class IIc: contains the remaining peptides of this class, including sec-dependent secreted bacteriocins.	acidocin B carnobacteriocin A divergicin A enterocin P enterocin B	*Lactobacillus acidophilus* *Carnobacterium piscicola* *C. divergens* *E. faecium* *E. faecium*
III – Class III: (>30 kDa) heat-labile proteins	helveticin J	*Lactobacillus heleveticus*
IV – Class IV: complex molecules that require carbohydrate or lipid moieties for activity.	No bacteriocin has been purified.	---

(Adapted from Chen & Hoover, 2003) [67].

The pediocins are able to permeabilize the cytoplasmic membrane of target cells by forming pores inserted through the C-terminal chain in the membrane region. The cell lysis is induced after loss of the permeability barrier and loss of the membrane potential, similarly to the action mechanism of of nisin. Gram-positive bacteria are largely sensitive to pediocins, however Gram-negative pathogens, such as *Salmonella enterica* subsp. *enterica* Enteritidis, show no susceptibility to treatment it, similarly to most bacteriocins, because these bacteria do not adsorb the pediocin on these outer membrane [69].

The deeper knowledge about pediocins is concentrated on the by-products synthesized by *Pediococcus* strains, which comprises *P. acidilactici, P. pentosaceus* and *P. damnosus*. These strains are usually present spontaneously in fermentative processes of foods and beverages, such as in sauerkraut, cured meats and cheese manufacturing, and represent contaminant microorganisms in brewing industry and winery. *Pediococcus* spp. share similar growing conditions and metabolism for bacteriocin production, being able to grow and produce metabolites in a wide range of temperature, commonly between 28 to 40 °C. However, the most favorable temperature related to pediocin metabolism is 30 °C. In regard to oxygen concentration, the higher pediocin amounts generally are obtained in semi-aerobic fermentations (about 60% oxygen saturation), because the anaerobic or fully aerobic conditions tend to accumulate more biomass and lactate, and consequently a lower pediocin level [69]. Therefore, controlling the dissolved oxygen tension is an important factor for pediocin production.

The pH in the medium during the pediocin synthesis is similar to that of other bacteriocins: initially the pH is greater than 6.0 in the culture medium; followed by a sharp decrease (to pH 3.7 - 3.5) after approximately 24 hours of cell cultivation. This low pH is necessary to active the enzymes responsible for the post-translational modification promoted in the prepediocin, synthesized as a primary metabolite in the exponential growth phase [70].

Pediocin is preferentially obtained using glucose as carbon source, though other carbohydrates, *e.g.* sucrose, galactose and xylose, and natural substrates can be used in the process. The pediocin production have been successfully reported in some low cost substrates, such as whey [71], mussel-processing waste [72], acid grape broth [73], octopus protein waste [74] and protein waste from fish viscera [75].

Bacteriocin production in large scale is usually difficult, the fermentation tends to be unstable due to problems related to poor adaptability of the producer strain, low productivity and genetic instability. These drawbacks still limit a wider production and application of bacteriocins. Heterologous production could overcome these problems, since this strategy would allow for the use of industrial cultures more stable and productive, but it only remains as a promising strategy in research. Recombinant enterocin, pediocin and plantaricin have been more frequently related in the last years, especially by production in *Escherichia coli* [76 - 78], besides *Saccharomyces cerevisiae* [79] and other lactic acid bacteria [80]. The heterologous production opens the possibility to obtain multibacteriocinogenic strains, allowing a broader spectrum of antimicrobial activity [81]. However, generally, these proposals have limited applicability since the use of genetically modified organisms as food additives is not well-accepted by consumers and by "green industries" which propose new natural compounds to food applications.

Biotechnological Production of Natamycin

The industrial production of natamycin occurs *via* submerged fermentation, followed by extraction with organic solvent (methanol or butanol). Crystals are dried recovered after precipitation, generating a white power containing 50% pure natamycin. This commercial product presents weak stability to heat, light and oxidative compounds, and the activity tends to decrease in presence of high concentration of sodium chlorite, citric acid and sucrose. The photo-oxidation by UV irradiation is the most significant cause to loss of biological activity [82]. However, natamycin is well-accepted as food preservative since it is an effective inhibitor of all target organisms without causing negative effects on product quality. Nowadays, this product has been proposed as preservative to cheese, sausage, fruits and beverages (juices, beer, wine and yogurt).

Recent progress in the biotechnological natamycin production involves the use of precursors to increase the biosynthesis. The precursors already known are short-chain fatty acids and alcohols (*e.g.* sodium propionate, sodium acetate, n-butyl alcohol, ethanol and isobutyl alcohol) [83]. However, these precursors present different efficacies for the production of natamycin and different species of *Streptomyces* demonstrate different results for each precursor.

Natamycin production using the *S. gilvosporeus* strain XM-172 (*i.e.* high-yield mutant derived from *S. gilvosporeus* ATCC 13326) was significantly increased by feeding propionate (6 $g.L^{-1}$ by 24h at early log phase) in cultivation process under the optimal condition reached 6.72 $g.L^{-1}$ and increased to 85% as compared to the control [83]. Propanol (0.2 % by 28 h culture) was found to be the most effective additive for the natamycin production using the *S. natalensis* strain F4-245 under optimal conditions. In this case the maximal natamycin titer obtained was 10.38 $g.L^{-1}$, which was 17 % higher than that the cultures that where no added the alcohol [84].

The culture medium used for natamycin production is glucose basal medium with slight modifications performed due to different requirements to growth in different cultivation steps. These media are modified for different functions: (1) slant medium (for culture maintenance), (2) medium for the spore production (*e.g.* seed medium) and (3) production medium. The main components used to make these media are: glucose, malt extract, yeast extract, peptone, NaCl, soybean cake meal, beef extract, $CaCO_3$, $MgSO_4$, among others. But, each culture medium can be adjusted for optimal culture conditions according to each *Streptomyces* species or strain used. Nevertheless, a study of medium optimization for *S. natalensis* concluded that the major factors on natamycin fermentation were corn starch, glucose and NaCl [85].

Albeit recent progress in optimization of the culture conditions, other strategies are necessary to promote gains in production, such as the genetic improvement of the *Streptomyces* strains. Several approaches had been performed with good results, such as induction of mutations during space-flight and regular genetic transformation. The combination of the optimization of the culture conditions with strains genetically improved allows highest rate if natamycin production with industrial competitively.

Lysozyme Production

The biotechnological production of lysozyme has been obtained by cloning from different genes in many organisms genetically modified, especially in the yeast *Pichia pastoris*, among others yeasts, bacteria, rice, cattle, *e.g.* cloning of a cDNA for bovine lysozyme c2 originated from cow stomach. The use of recombinant lysozyme can be an efficient strategy to overcome bacterial resistance against wild lysozyme. High concentrations of this protein were achieved by secretion in semi-continuous and continuous fermentation processes; however these technologies are not evaluated in large scale yet [86]. Conventional ultrafiltration systems are not been suitable for lysozyme purification, because generally results in low purity and poor recovery. A four-stage cascade of continuous ultrafiltration is suggested as an option to recovery and concentrate lysozyme at 97% purity with buffer recycling system [87].

FURTHER DEVELOPMENTS

A great range of natural antimicrobial compounds are targets of innovative technologies to improve quality and food safety, but many of them have some restriction use due their relatively narrow activity spectra and moderate antibacterial effect. Thus, current investigations at bioprospecting lead us to some challenges: obtaining biocompounds of interest from non-conventional sources such as new microorganisms (endophytic fungi and bacteria), rhizobacteria, edible mushrooms, higher plants (medicinal or edible), fish, insects and etc. These sources drive us to search for bioactive substances (flavonoids like proanthocyanidin, tannins or tannins hydrolysable with catechin derivatives, phenolic compounds, carvacrol essential oil and of course, bacteriocins) with a variety of intrinsic properties (*e.g.* antioxidant and antimicrobial activity, and anti-tumor properties).

The antimicrobial potential of natural antimicrobial compounds may be improved by combined use of one or more agents of traditional food additives and food preservation methods, based on hurdle technology. Although it is known that many of natural antimicrobial and antioxidant compounds have great potential as food additive, some of their sensorial properties (aroma and flavor) may be too intense for adequate use in foods besides reduction of antimicrobial/antioxidant activities during shelf life of the food. Thus, innovative technologies for food products aiming to improve the action of these substances with antimicrobial potential have been evaluated by companies in the food sector: nanoencapsulation in which occurs gradual release of antimicrobials, essential oils or other compounds in food matrix. An example is the development of innovative bio packaging (biodegradable) in which natural active compounds are incorporated into the barrier and is gradually released. This can also be developed as an edible packaging.

The ultimate goal in food sector is the development of new products with food safety and quality, following the market trends with consumers that demand healthier foods. Thus, other challenges are also focused on the transformation of such compounds that have technological, industrial and commercial feasibility.

CONFLICT OF INTEREST

The authors confirm that they have no conflict of interest to declare for this publication.

ACKNOWLEDGEMENTS

Declared none.

REFERENCES

[1] Leistner, L. Basic aspects of food preservation by hurdle technology. *Int. J. Food Microbiol.,* **2000**, *55*(1-3), 181-186.
[http://dx.doi.org/10.1016/S0168-1605(00)00161-6] [PMID: 10791741]

[2] Tajkarimi, M.; Ibrahim, S. Antimicrobial herb and spice compounds in food. *Food Contr.,* **2010**, *21*(9), 1199-1218.
[http://dx.doi.org/10.1016/j.foodcont.2010.02.003]

[3] Lück, E.; Jager, M. *Antimicrobial food additives: Characteristics, Uses, Effects. 2nd*; Springer: Berlin, **1997**.
[http://dx.doi.org/10.1007/978-3-642-59202-7]

[4] Capita, R.; Alonso-Calleja, C. Antibiotic-resistant bacteria: a challenge for the food industry. *Crit. Rev. Food Sci. Nutr.,* **2013**, *53*(1), 11-48.
[http://dx.doi.org/10.1080/10408398.2010.519837] [PMID: 23035919]

[5] Welch, H. Problems of antibiotics in food as the Food and Drug Administration sees them. *Am. J. Public Health Nations Health,* **1957**, *47*(6), 701-705.
[http://dx.doi.org/10.2105/AJPH.47.6.701] [PMID: 13424815]

[6] Cammack, R.; Joannou, C. Nitrite and nitrosyl compounds in food preservation. *Biochim. Biophys. Acta. Bioenerg.,* **1999**, *1411*(2), 475-488.

[7] JECFA. *Compendium of Food Additive Specifications. FAO*; Rome, **2010**, 10, p. Food and Agriculture Organization of teh United Nations154.

[8] Food Additives Status List F.D.. *United States Food and Drug Administration*; , **2013**.

[9] FFCR. *Standards for use of food additives*; Ministry of Health, Labour and Welfare, **2014**.

[10] ANVISA. *Lista de Aditivos e Respectivas Concentrações Autorizadas*; Agência Nacional de Vigilância Sanitária: Brasília, **2011**, p. 163.

[11] Maillard, J.Y. Bacterial target sites for biocide action. *J. Appl. Microbiol.,* **2002**, *92*(s1) Suppl., 16S-27S.
[http://dx.doi.org/10.1046/j.1365-2672.92.5s1.3.x] [PMID: 12000609]

[12] Brul, S.; Coote, P. Preservative agents in foods. Mode of action and microbial resistance mechanisms. *Int. J. Food Microbiol.,* **1999**, *50*(1-2), 1-17.
[http://dx.doi.org/10.1016/S0168-1605(99)00072-0] [PMID: 10488839]

[13] Suhr, K.I.; Nielsen, P.V. Effect of weak acid preservatives on growth of bakery product spoilage fungi at different water activities and pH values. *Int. J. Food Microbiol.,* **2004**, *95*(1), 67-78.
[http://dx.doi.org/10.1016/j.ijfoodmicro.2004.02.004] [PMID: 15240076]

[14] Nair, B. Final report on the safety assessment of Benzyl Alcohol, Benzoic Acid, and Sodium Benzoate. *Int. J. Toxicol.,* **2001**, *20* Suppl. 3, 23-50.
[http://dx.doi.org/10.1080/10915810152630729] [PMID: 11766131]

[15] Connelly, P. Nutritional Advantages and Disadvantages of Dietary Phytates: Part 1. *J. Aust. Tradit.-Med. Soc.,* **2011**, *17*(1), 21.

[16] Dutta, P.K.; Tripathi, S.; Mehrotra, G.K.; Dutta, J. Perspectives for chitosan based antimicrobial films in food applications. *Food Chem.,* **2009**, *114*, 1173-1182.
[http://dx.doi.org/10.1016/j.foodchem.2008.11.047]

[17] Kong, M.; Chen, X.G.; Xing, K.; Park, H.J. Antimicrobial properties of chitosan and mode of action: a state of the art review. *Int. J. Food Microbiol.,* **2010**, *144*(1), 51-63.
[http://dx.doi.org/10.1016/j.ijfoodmicro.2010.09.012] [PMID: 20951455]

[18] Lucera, A.; Costa, A.; Conte, A.; Del Nobile, M.A. Food application of natural antimicrobial compounds. *Front. Microbiol.,* **2012**, *3*, 1-13.
[http://dx.doi.org/10.3389/fmicb.2012.00287] [PMID: 22275914]

[19] Hopwood, D.A. Genetic contributions to understanding polyketide synthases. *Chem. Rev.,* **1997**, *97*(7), 2465-2498.
[http://dx.doi.org/10.1021/cr960034i] [PMID: 11851466]

[20] Aparicio, J. F.; Colina, A. J.; Ceballos, E.; Martín, J. F. The biosynthetic gene cluster for the 26-membered ring polyene macrolide pimaricin. *J. Biol. Chem,* **1999**, *274*(15), 10133-39.
[PMID: 10187796]

[21] te Welscher, Y.M.; ten Napel, H.H.; Balagué, M.M.; Souza, C.M.; Riezman, H.; de Kruijff, B.; Breukink, E. Natamycin blocks fungal growth by binding specifically to ergosterol without permeabilizing the membrane. *J. Biol. Chem.,* **2008**, *283*(10), 6393-6401.
[http://dx.doi.org/10.1074/jbc.M707821200] [PMID: 18165687]

[22] Van Leeuwen, M.R.; Golovina, E.A.; Dijksterhuis, J. The polyene antimycotics nystatin and filipin disrupt the plasma membrane, whereas natamycin inhibits endocytosis in germinating conidia of

Penicillium discolor. *J. Appl. Microbiol.,* **2009**, *106*(6), 1908-1918. [http://dx.doi.org/10.1111/j.1365-2672.2009.04165.x] [PMID: 19228256]

[23] Salazar, O.; Asenjo, J.A. Enzymatic lysis of microbial cells. *Biotechnol. Lett.,* **2007**, *29*(7), 985-994. [http://dx.doi.org/10.1007/s10529-007-9345-2] [PMID: 17464453]

[24] Proctor, V.A.; Cunningham, F.E. The chemistry of lysozyme and its use as a food preservative and a pharmaceutical. *Crit. Rev. Food Sci. Nutr.,* **1988**, *26*(4), 359-395. [http://dx.doi.org/10.1080/10408398809527473] [PMID: 3280250]

[25] Masschalck, B.; Michiels, C.W. Antimicrobial properties of lysozyme in relation to foodborne vegetative bacteria. *Crit. Rev. Microbiol.,* **2003**, *29*(3), 191-214. [http://dx.doi.org/10.1080/713610448] [PMID: 14582617]

[26] Ruhr, E.; Sahl, H.G. Mode of action of the peptide antibiotic nisin and influence on the membrane potential of whole cells and on cytoplasmic and artificial membrane vesicles. *Antimicrob. Agents Chemother.,* **1985**, *27*(5), 841-845. [http://dx.doi.org/10.1128/AAC.27.5.841] [PMID: 4015074]

[27] Olasupo, N.A.; Fitzgerald, D.J.; Gasson, M.J.; Narbad, A. Activity of natural antimicrobial compounds against *Escherichia coli* and *Salmonella* enterica serovar Typhimurium. *Lett. Appl. Microbiol.,* **2003**, *37*(6), 448-451. [http://dx.doi.org/10.1046/j.1472-765X.2003.01427.x] [PMID: 14633097]

[28] Sandberg, M.; Määttänen, A.; Peltonen, J.; Vuorela, P.M.; Fallarero, A. Automating a 96-well microtitre plate model for *Staphylococcus aureus* biofilms: an approach to screening of natural antimicrobial compounds. *Int. J. Antimicrob. Agents,* **2008**, *32*(3), 233-240. [http://dx.doi.org/10.1016/j.ijantimicag.2008.04.022] [PMID: 18640013]

[29] Plaza, M.; Santoyo, S.; Jaime, L.; García-Blairsy Reina, G.; Herrero, M.; Señoráns, F.J.; Ibáñez, E. Screening for bioactive compounds from algae. *J. Pharm. Biomed. Anal.,* **2010**, *51*(2), 450-455. [http://dx.doi.org/10.1016/j.jpba.2009.03.016] [PMID: 19375880]

[30] Kabuki, T.; Nakajima, H. Characterization of novel antimicrobial compounds from mango *Mangifera indica* L. kernel seeds. *Food Chem.,* **2000**, *71*, 61-66. [http://dx.doi.org/10.1016/S0308-8146(00)00126-6]

[31] Puupponen-Pimiä, R.; Nohynek, L.; Alakomi, H.L.; Oksman-Caldentey, K.M. Bioactive berry compounds-novel tools against human pathogens. *Appl. Microbiol. Biotechnol.,* **2005**, *67*(1), 8-18. [http://dx.doi.org/10.1007/s00253-004-1817-x] [PMID: 15578177]

[32] Naidu, A.S. *Natural Food Antimicrobial Systems*; CRC Press: Boca Raton, FL, **2000**. [http://dx.doi.org/10.1201/9781420039368]

[33] García-García, R.; López-Malo, A.; Palou, E. Bactericidal action of binary and ternary mixtures of carvacrol, thymol, and eugenol against *Listeria innocua*. *J. Food Sci.,* **2011**, *76*(2), M95-M100. [http://dx.doi.org/10.1111/j.1750-3841.2010.02005.x] [PMID: 21535780]

[34] Barros, L.; Calhelha, R.C. Antimicrobial activity and bioactive compounds of Portuguese wild edible mushrooms methanolic extracts. *Eur. Food Res. Technol.,* **2007**, *225*, 151-156. [http://dx.doi.org/10.1007/s00217-006-0394-x]

[35] Menaga, D.; Mahalingam, P. Evaluation of phytochemical characteristics and antimicrobial activity of

pleurotus florida mushroom. *Asian J. Pharm. Clin. Res,* **2012**, *5*(4)

[36] Barbiroli, A.; Bonomi, F. Antimicrobial activity of lysozyme and lactoferrin incorporated in cellulose-based food packaging. *Food Contr.,* **2012**, *26*, 387-392. [http://dx.doi.org/10.1016/j.foodcont.2012.01.046]

[37] Balcao, V.M.; Costa, C.I. Nanoencapsulation of bovine lactoferrin for food and biopharmaceutical applications. *Food Hydrocoll.,* **2013**, *32*, 425-431. [http://dx.doi.org/10.1016/j.foodhyd.2013.02.004]

[38] Moon, S.H.; Lee, J.H. Antioxidant, antimicrobial, and cytotoxic activities of ovotransferrin from egg white. *Korean J. Food Sci. An,* **2012**, *32*, 612-617. [http://dx.doi.org/10.5851/kosfa.2012.32.5.612]

[39] Simões, A.D.; Tudela, J.A.; Allende, A.; Puschmann, R.; Gil, M.I. Edible coatings containing chitosan and moderate modified atmospheres maintain quality and enhance phytochemicals of carrot sticks. *Postharvest Biol. Technol.,* **2009**, *51*, 364-370. [http://dx.doi.org/10.1016/j.postharvbio.2008.08.012]

[40] Burrowes, O.J.; Hadjicharalambous, C.; Diamond, G.; Lee, T.C. Evaluation of antimicrobial spectrum and cytotoxic activity of pleurocidin for food applications. *J. Food Sci.,* **2004**, *69*, 66-71.

[41] Potter, R.; Truelstrup Hansen, L.; Gill, T.A. Inhibition of foodborne bacteria by native and modified protamine: importance of electrostatic interactions. *Int. J. Food Microbiol.,* **2005**, *103*(1), 23-34. [http://dx.doi.org/10.1016/j.ijfoodmicro.2004.12.019] [PMID: 16084263]

[42] Hansen, L.T.; Austin, J.W.; Gill, T.A. Antibacterial effect of protamine in combination with EDTA and refrigeration. *Int. J. Food Microbiol.,* **2001**, *66*(3), 149-161. [http://dx.doi.org/10.1016/S0168-1605(01)00428-7] [PMID: 11428574]

[43] Anderson, R.C.; Wilkinson, B. Ovine antimicrobial peptides: new products from an age-old industry. *Aust. J. Agric. Res.,* **2004**, *55*, 69-75. [http://dx.doi.org/10.1071/AR03064]

[44] Jin, J-Y.; Zhou, L.; Wang, Y.; Li, Z.; Zhao, J.G.; Zhang, Q.Y.; Gui, J.F. Antibacterial and antiviral roles of a fish β-defensin expressed both in pituitary and testis. *PLoS One,* **2010**, *5*(12), e12883. [http://dx.doi.org/10.1371/journal.pone.0012883] [PMID: 21188147]

[45] Juneja, V.K.; Dwivedi, H.P.; Yan, X. Novel natural food antimicrobials. *Annu. Rev. Food Sci. Technol.,* **2012**, *3*, 381-403. [http://dx.doi.org/10.1146/annurev-food-022811-101241] [PMID: 22385168]

[46] Shin, S.Y.; Bajpai, V.K.; Kim, H.R.; Kang, S.C. Antibacterial activity of bioconverted eicosapentaenoic (EPA) and docosahexaenoic acid (DHA) against foodborne pathogenic bacteria. *Int. J. Food Microbiol.,* **2007**, *113*(2), 233-236. [http://dx.doi.org/10.1016/j.ijfoodmicro.2006.05.020] [PMID: 16860896]

[47] Cheok, C.Y.; Salman, H.A.; Sulaiman, R. Extraction and quantification of saponins: A review. *Food Res. Int.,* **2014**, *59*, 16-40. [http://dx.doi.org/10.1016/j.foodres.2014.01.057]

[48] Cushnie, T.P.; Lamb, A.J. Antimicrobial activity of flavonoids. *Int. J. Antimicrob. Agents,* **2005**, *26*(5), 343-356.

[http://dx.doi.org/10.1016/j.ijantimicag.2005.09.002] [PMID: 16323269]

[49] Ankri, S.; Mirelman, D. Antimicrobial properties of allicin from garlic. *Microbes Infect.,* **1999**, *1*(2), 125-129.
[http://dx.doi.org/10.1016/S1286-4579(99)80003-3] [PMID: 10594976]

[50] Yilmaz, Y. Novel uses of catechins in foods. *Trends Food Sci. Technol.,* **2006**, *17*, 64-71.
[http://dx.doi.org/10.1016/j.tifs.2005.10.005]

[51] Bansal, S.; Choudhary, S.; Sharma, M.; Kumar, S.S.; Lohan, S.; Bhardwaj, V.; Syan, N.; Jyoti, S. Review- Tea: A native source of antimicrobial agents. *Food Res. Int.,* **2013**, *53*, 568-584.
[http://dx.doi.org/10.1016/j.foodres.2013.01.032]

[52] Vig, A.P.; Rampal, G.; Thind, T.S.; Arora, S. Bio-protective effects of glucosinolates – A review. *LWT-. Food Sci. Technol. (Campinas.),* **2009**, *42*, 1561-1572.

[53] Zendo, T.; Fukao, M.; Ueda, K.; Higuchi, T.; Nakayama, J.; Sonomoto, K. Identification of the lantibiotic nisin Q, a new natural nisin variant produced by *Lactococcus lactis* 61-14 isolated from a river in Japan. *Biosci. Biotechnol. Biochem.,* **2003**, *67*(7), 1616-1619.
[http://dx.doi.org/10.1271/bbb.67.1616] [PMID: 12913315]

[54] de Kwaadsteniet, M.; Ten Doeschate, K.; Dicks, L.M. Characterization of the structural gene encoding nisin F, a new lantibiotic produced by a Lactococcus lactis subsp. lactis isolate from freshwater catfish (Clarias gariepinus). *Appl. Environ. Microbiol.,* **2008**, *74*(2), 547-549.
[http://dx.doi.org/10.1128/AEM.01862-07] [PMID: 18039827]

[55] Arauz, L.J.; Jozala, A.F.; Mazzola, P.G.; Vessoni Penna, T.C. Nisin biotechnological production and application: a review. *Trends Food Sci. Technol.,* **2009**, *20*, 146-154.
[http://dx.doi.org/10.1016/j.tifs.2009.01.056]

[56] Nes, I.F.; Yoon, S-S.; Diep, D.B. Ribosomally Synthesiszed Antimicrobial Peptides (Bacteriocins) in Lactic Acid Bacteria: A Review. *Food Sci. Biotechnol.,* **2007**, *16*, 675-690.

[57] Bauer, R.; Dicks, L.M. Mode of action of lipid II-targeting lantibiotics. *Int. J. Food Microbiol.,* **2005**, *101*(2), 201-216.
[http://dx.doi.org/10.1016/j.ijfoodmicro.2004.11.007] [PMID: 15862882]

[58] Héchard, Y.; Sahl, H-G. Mode of action of modified and unmodified bacteriocins from Gram-positive bacteria. *Biochimie,* **2002**, *84*(5-6), 545-557.
[http://dx.doi.org/10.1016/S0300-9084(02)01417-7] [PMID: 12423799]

[59] Wiedemann, I.; Benz, R.; Sahl, H-G. Lipid II-mediated pore formation by the peptide antibiotic nisin: a black lipid membrane study. *J. Bacteriol.,* **2004**, *186*(10), 3259-3261.
[http://dx.doi.org/10.1128/JB.186.10.3259-3261.2004] [PMID: 15126490]

[60] Reisinger, P.; Seidel, H.; Tschesche, H.; Hammes, W.P. The effect of nisin on murein synthesis. *Arch. Microbiol.,* **1980**, *127*(3), 187-193.
[http://dx.doi.org/10.1007/BF00427192] [PMID: 6255884]

[61] Bierbaum, G.; Sahl, H.G. Induction of autolysis of staphylococci by the basic peptide antibiotics Pep 5 and nisin and their influence on the activity of autolytic enzymes. *Arch. Microbiol.,* **1985**, *141*(3), 249-254.
[http://dx.doi.org/10.1007/BF00408067] [PMID: 4004448]

[62] Deegan, L.H.; Cotter, P.D.; Hill, C.; Ross, P. Bacteriocins: biological tools for bio-preservation and shelf-life extension. *Int. Dairy J.,* **2006**, *16*, 1058-1071. [http://dx.doi.org/10.1016/j.idairyj.2005.10.026]

[63] Thomas, L.V.; Ingram, R.E.; Bevis, H.E.; Davies, E.A.; Milne, C.F.; Delves-Broughton, J. Effective use of nisin to control *Bacillus* and *Clostridium* spoilage of a pasteurized mashed potato product. *J. Food Prot.,* **2002**, *65*(10), 1580-1585. [PMID: 12380742]

[64] Helander, I.M.; Mattila-Sandholm, T. Permeability barrier of the Gram-negative bacterial outer membrane with special reference to nisin. *Int. J. Food Microbiol.,* **2000**, *60*(2-3), 153-161. [http://dx.doi.org/10.1016/S0168-1605(00)00307-X] [PMID: 11016605]

[65] Gelinski, J.M.; Baratto, C.M.; Scaratti, D.; Franco, B.D. *Lactobacillus sakei* 2a and its concentrated acid extract on inhibition of food-borne *Salmonella strains. J. Pure Appl. Microbiol.,* **2012**, *6*, 21-30.

[66] Penna, T.C.; Moraes, D.A. Optimization of nisin production by *Lactococcus lactis. Appl. Biochem. Biotechnol.,* **2002**, *98-100*, 775-789. [http://dx.doi.org/10.1385/ABAB:98-100:1-9:775] [PMID: 12018301]

[67] Chen, H.; Hoover, D.G. Bacteriocins and their food applications. *Compr. Rev. Food Sci. Food Saf.,* **2003**, *2*, 83-97.

[68] Klaenhammer, T.R. Genetics of bacteriocins produced by lactic acid bacteria. *FEMS Microbiol. Rev.,* **1993**, *12*(1-3), 39-85. [http://dx.doi.org/10.1111/j.1574-6976.1993.tb00012.x] [PMID: 8398217]

[69] Papagianni, M.; Anastasiadou, S. Pediocins: The bacteriocins of *Pediococci*. Sources, production, properties and applications. *Microb. Cell Fact.,* **2009**, *8*, 3. [http://dx.doi.org/10.1186/1475-2859-8-3] [PMID: 19133115]

[70] Anastasiadou, S.; Papagianni, M.; Filiousis, G.; Ambrosiadis, I.; Koidis, P. Pediocin SA-1, an antimicrobial peptide from *Pediococcus acidilactici* NRRL B5627: production conditions, purification and characterization. *Bioresour. Technol.,* **2008**, *99*(13), 5384-5390. [http://dx.doi.org/10.1016/j.biortech.2007.11.015] [PMID: 18093831]

[71] Guerra, N.P.; Pastrana, L. Dynamics of pediocin biosynthesis in batch fermentation on whey. *Electron J. Environ. Agric. Food Chem,* **2002**, *1*, 96-106.

[72] Guerra, N.P.; Pastrana, L. Nisin and pediocin production on mussel-processing waste supplemented with glucose and five nitrogen sources. *Lett. Appl. Microbiol.,* **2002**, *34*(2), 114-118. [http://dx.doi.org/10.1046/j.1472-765x.2002.01054.x] [PMID: 11849506]

[73] Nel, H.A.; Bauer, R.; Vandamme, E.J.; Dicks, L.M. Growth optimization of *Pediococcus damnosus* NCFB 1832 and the influence of pH and nutrients on the production of pediocin PD-1. *J. Appl. Microbiol.,* **2001**, *91*(6), 1131-1138. [http://dx.doi.org/10.1046/j.1365-2672.2001.01486.x] [PMID: 11851822]

[74] Vazquez, J. A.; Gonzalez, M. P.; Murado, M. A. Nisin and pediocin production by Lactococcus lactis and Pediococcus acidilactici using waste protein sources from octopus. *Electron J. Environ. Agric. Food Chem,* **2004**, *3*(2)

[75] Vázquez, J.A.; González, M.P.; Murado, M.A. Peptones from autohydrolysed fish viscera for nisin and pediocin production. *J. Biotechnol.,* **2004**, *112*(3), 299-311. [http://dx.doi.org/10.1016/j.jbiotec.2004.04.011] [PMID: 15313007]

[76] Nigutová, K.; Serencová, L.; Piknová, M.; Javorský, P.; Pristaš, P. Heterologous expression of functionally active enterolysin A, class III bacteriocin from *Enterococcus faecalis*, in *Escherichia coli. Protein Expr. Purif.,* **2008**, *60*(1), 20-24. [http://dx.doi.org/10.1016/j.pep.2008.03.006] [PMID: 18448357]

[77] Moon, G-S.; Pyun, Y-R.; Kim, W.J. Expression and purification of a fusion-typed pediocin PA-1 in *Escherichia coli* and recovery of biologically active pediocin PA-1. *Int. J. Food Microbiol.,* **2006**, *108*(1), 136-140. [http://dx.doi.org/10.1016/j.ijfoodmicro.2005.10.019] [PMID: 16403586]

[78] Gutiérrez, J.; Criado, R.; Citti, R.; Martín, M.; Herranz, C.; Nes, I.F.; Cintas, L.M.; Hernández, P.E. Cloning, production and functional expression of enterocin P, a sec-dependent bacteriocin produced by *Enterococcus faecium* P13, in *Escherichia coli. Int. J. Food Microbiol.,* **2005**, *103*(3), 239-250. [http://dx.doi.org/10.1016/j.ijfoodmicro.2004.11.035] [PMID: 16099309]

[79] Van Reenen, C.A.; Chikindas, M.L.; Van Zyl, W.H.; Dicks, L.M. Characterization and heterologous expression of a class IIa bacteriocin, plantaricin 423 from *Lactobacillus plantarum* 423, in *Saccharomyces cerevisiae. Int. J. Food Microbiol.,* **2003**, *81*(1), 29-40. [http://dx.doi.org/10.1016/S0168-1605(02)00164-2] [PMID: 12423916]

[80] Borrero, J.; Jiménez, J.J.; Gútiez, L.; Herranz, C.; Cintas, L.M.; Hernández, P.E. Protein expression vector and secretion signal peptide optimization to drive the production, secretion, and functional expression of the bacteriocin enterocin A in lactic acid bacteria. *J. Biotechnol.,* **2011**, *156*(1), 76-86. [http://dx.doi.org/10.1016/j.jbiotec.2011.07.038] [PMID: 21839785]

[81] Rodríguez, J.M.; Martínez, M.I.; Horn, N.; Dodd, H.M. Heterologous production of bacteriocins by lactic acid bacteria. *Int. J. Food Microbiol.,* **2003**, *80*(2), 101-116. [http://dx.doi.org/10.1016/S0168-1605(02)00153-8] [PMID: 12381397]

[82] Stark, J. Permitted preservatives - Natamycin. In: *Encyclopedia of Food Microbiology*; Academic Press: San Diego, **2000**; Vol. 3, pp. 1776-1781.

[83] Jing, K.; Hao, X.; Lu, Y. Effect of propionate on the production of natamycin with *Streptomyces gilvosporeus* XM-172. *Afr. J. Biotechnol.,* **2011**, *10*, 10030-10033.

[84] Li, M.; Chen, S.; Li, J.; Ji, Z. Propanol addition improves natamycin biosynthesis of *Streptomyces natalensis. Appl. Biochem. Biotechnol.,* **2014**, *172*(7), 3424-3432. [http://dx.doi.org/10.1007/s12010-014-0766-9] [PMID: 24532463]

[85] Yan, Y-Z.; Ke, N.; Wei, X-D.; Jin, Y-Y.; Bo, Z.; Zhao, W-J. Optimization of fermentation conditions for high production of natamycin by *Streptomyces natalensis. Chin. J. Antibiot.,* **2013**, *38*, 424-429.

[86] Digan, M.E.; Lair, S.V.; Brierley, R.A.; Siegel, R.S.; Williams, M.E.; Ellis, S.B.; Kellaris, A.; Provow, S.S.; Craig, W.S.; Veliçelebi, G.; Harpold, M.M.; Thill, G.P. Continuous production of a novel lysozyme *via* secretion from the yeast, *Pichia pastoris. Nat. Biotechnol.,* **1989**, *7*, 160-164. [http://dx.doi.org/10.1038/nbt0289-160]

[87] Mayani, M.; Filipe, C.D.; Ghosh, R. Cascade ultrafiltration systems - Integrated processes for purification and concentration of lysozyme. *J. Membr. Sci.,* **2010**, *347*, 150-158. [http://dx.doi.org/10.1016/j.memsci.2009.10.016]

CHAPTER 11

Use of (Bio) Surfactants in Foods

Jenyffer M. Campos[a], **Leonie A. Sarubbo**[b], **Juliana M Luna**[b], **Raquel Diniz Rufino**[b] and **Ibrahim M. Banat**[c,*]

[a] *Departamento de Nutrição, Universidade Federal de Pernambuco, Programa de Pós-graduação em Nutrição, Av. Prof. Moraes Rego, 1235 - Cidade Universitária, Recife, PE,Brazil, CEP: 50670-901*

[b] *Centro de Ciências e Tecnologia — Universidade Católica de Pernambuco, Rua do Príncipe,526 - Boa Vista - Recife, PE, Brazil, CEP: 50050-900*

[c] *School of Biomedical Sciences, Faculty of Life and Health Sciences, University of Ulster, BT52 1SA, Northern Ireland, UK*

Abstract: Surfactants are surface-active microbial products with an amphiphilic structure that allows these to act as emulsifiers, surfactants, foaming and dispersing agents. Biosurfactants have advantages over their chemically synthesized counterparts, such as biodegradability, lower toxicity, activity at extreme temperatures and pH values, besides the ability to be produced using sustainable by-products. Thus, biosurfactants have potential applications in the chemical, petroleum, environmental and pharmaceutical industries. Biosurfactants may also be used as food additives (which are compounds added to enhance the characteristics of food), acting as thickeners, emulsifiers and stabilizers. This chapter discusses the importance of the microbial biosurfactants, their characteristics and potential industrial applications, particularly as food additives.

Keywords: Amphiphilic, Bioemulsifiers, Biosurfactants, Dispersing, Emulsion, Foaming, Food additives, Interface, Solubilizes, Surface-active, Surfactants, Thickeners.

* **Address correspondence to Ibrahim M. Banat:** School of Biomedical Sciences, Faculty of Life and Health Sciences, University of Ulster, BT52 1SA, N. Ireland, UK; E-mail: im.banat@ulster.ac.uk.

Juliano Lemos Bicas, Mário Roberto Maróstica, Glaucia Maria Pastore (Eds.)

INTRODUCTION

Surfactants are surface active agents with amphiphilic molecules that have a distinct hydrophilic head moiety, often a sugar or amino acid, and a hydrophobic hydrocarbon/lipid tail moiety. The general ability of surfactants to be located at the interfaces between liquid phases with different degrees of polarity and hydrogen bridges, such as air-water, oil-water or water-solid interfaces, leads to a reduction in surface and interfacial tensions. The presence of both hydrophilic and lipophilic moiety provide the peculiar properties of the surfactants. The hydrocarbon chains are often the non-polar portions, while the polar portions can be ionic, non-ionic or amphoteric [1]. Biological origin surfactants have an advantage over synthetic chemical ones in terms of lower toxicity and biodegradability.

Therefore, surfactants change the surface and interfacial properties of a liquid [2]. These characteristics allow the formation of a microemulsion (oil-in-water or water-in-oil microbubbles) [3, 4]. This triggers industrial applications such as the formation of emulsions, wetting, foaming, solubilisation moisture retention, lubrication, detergency, phase dispersion and viscosity reduction [5 - 8]. Surface tension in an aqueous environment is reduced when the concentration of the surfactant is increased, forming micelles, which are aggregated to the amphipathic molecules. Within aqueous environments, the hydrophilic portion of a micelle is turned toward the outer portion of the molecule and the hydrophobic portion is positioned towards the inner part. The critical micelle concentration (CMC) is defined as the minimum surfactant concentration needed in order to initiate the formation of micelles [9 - 12]. At concentrations higher than the CMC, no further reduction in the surface or interfacial tension is detected. At the CMC, surfactant monomers are spontaneously incorporated into the micelle vesicles or bilayer-structured aggregates, which are produced due to the several weak physicochemical (hydrophobic, van der Waals and hydrogen bonding) interactions [13, 14]. The CMC is the most common measurement of the activity of a surfactant, especially regarding solubility in the aqueous phase [12]. The efficiency of a surfactant is determined by the CMC and this effectiveness reflects the ability of a given surfactant to reduce surface and interfacial tensions [2].

Surfactants have applications in the petrochemical, food, beverage, cosmetic, pharmaceutical, mining, paper, metallurgical, textile and agrochemical industries as well as in the bioremediation of contaminated soil and water [5, 15 - 19]. Moreover, they are used in soaps and detergents [20]. In 2012, the worldwide surfactant production exceeded three million tons a year, with the majority of these products being employed in domestic detergents.

This chapter describes the importance of biosurfactants (mainly those ones with microbial origin), their properties, characteristics and industrial applications particularly as potential food additives.

SURFACTANTS IN FOOD PRODUCTS

Food additives (vitamins, flavours, enzymes, antioxidants, *etc.*) play an important role in the functional properties, stability and quality of foods. For instance, amphiphilic molecules affect the physical and textural stability of the microstructure of a product. Moreover, many types of multifunctional amphiphilic macromolecules are very important as surface-active agents. A large number of researchers have recognised proteins, glycolipids, polysaccharides, lipoproteins and polar lipids as suitable stabilisers, emulsifiers or foaming agents in manufactured and natural foods [21].

Surfactants have been used in foods for centuries [22]. It can be noticed a growing awareness regarding the importance of colloid and surface science in solving technological problems and properties related to food formulas. Foods are mainly comprised of lipids, proteins and carbohydrates in a variety of complex mixtures. Colloidal systems in foods have an infinite number of combinations and are organized in complex microstructures in the form of dispersal agents, foams, emulsions, gels, *etc.* [23]. Such aggregated structures are composed by particles or by the combination of polymers and surfactants and are determined by attraction (van der Waals forces) and repulsion forces. The repulsion force can be electrostatic or steric depending on the food formulation composition. An electrostatic force is caused by charged interfaces. An example of this is seen in ionic surfactants. Steric repulsion is due to adsorbed polymers or vesicles in the continuous phase. An example of such event is a surfactant with a polymeric polar group. If this group is charged as well, it may also have an electrostatic effect. Understanding the interfacial properties and rheology of these systems requires knowledge about the interaction between proteins and hydrocolloids [24].

According to the Codex Alimentarius harmonised international food standards, guidelines and codes [25], food additives generally have no nutritional purpose and are used to change the biological, sensory and physicochemical aspects of food products as well as to facilitate production, storage and handling. The following is a list of the categories of additives:

- **Thickener**: substance that enhances viscosity;
- **Gelling agent**: substance that alters the texture of a food product through the formation of a gel;
- **Stabilizer**: substance that allows the maintenance of the uniform dispersal of immiscible substances;
- **Emulsifier**: substance that allows the creation or maintenance of a uniform blend of immiscible phases.

The emulsification of food additives (*i.e.* the dispersions/emulsions of immiscible ingredients) is important to the consistency and texture of the product as well as to the solubilisation of aromas through phase dispersion. Such functions are maintained by stabilizing aerated systems and controlling food globules clustering [26, 27]. An emulsion is a heterogeneous system consisting of at least one immiscible liquid dispersed in another one under the form of droplets [28]. Three main types of emulsion are known to be important in food:

1. Oil-in-water - oil droplets suspended in a continuous aqueous phase. This is the most versatile type of emulsion and is found in mayonnaise, creamer, ice cream, whipped toppings and cream liqueur. The properties are determined by the components in the water phase as well as the surfactants used.
2. Water-in-oil (margarine, butter and fat-based spreads). The stability of this type of emulsion depends on the fat or oil properties as well as on the surfactant employed in the water phase.
3. Water-in-oil-in-water - An oil-in-water emulsion is the droplets that contain water [24, 29 - 31].

Surfactants decrease the interfacial tension in such systems, leading to a reduction in the surface energy between the phases through steric and electrostatic barriers formations that prevent the coalescence of particles and enhance the stability of the emulsion [32].

SURFACTANTS USED IN FOOD INDUSTRY

Emulsifiers are essential ingredients in the majority of food products containing oils and fats. However, their use extends to other fields as well. To a certain extent, the reduction in interfacial tension provided by these substances reduces the effort required to overcome the surface energy leading to the dispersion of the phases into each other. It also allows the stabilization of the dispersion by preventing the separation into two immiscible phases, coalescence and flocculation, [33 - 36].

Thus, emulsifiers are indispensable to the food industry, in which foams and emulsions are often employed. Their most important functions in food systems are the reduction of surface tension at the oil-water interface, stabilization of the emulsion, control of the clustering of fat globules, improvement of texture and increase of the products' shelf life [37].

Emulsions and food colloids are complex systems with several components and/or mixtures of oils as well as other ingredients. The role of low molecular mass amphiphilic molecules is crucial for the stability of beverages, alcoholic emulsions, dressings and other manufactured products. Naturally-occurring surfactants are currently available as fine chemicals, such as liposaccharides, saponin, rosin, shellac, soaps and cholic acid conjugates. Plant extracts, pectin, tamarind seeds and gum tragacanth are other traditional emulsifiers and emulsion-stabilizing agents [38]. Monoglycerides (polar lipids) have also recently been described as emulsifiers for food products [22].

Surfactants employed in the food industry are divided into those with a low molecular mass, such as fatty acids, glycolipids and fatty alcohols, and those with a high molecular mass (macromolecular) extracted from proteins, polysaccharides and naturally-occurring materials (Fig. **1**). A better understanding of the adsorption, denaturation and coagulation stages in the stabilization of interfaces has led to considerable progress in the use of proteins as food emulsifiers [39]. Overcoming the required activation energy at each stage results in the progressive reduction of interfacial energy and greater stabilised emulsion. Low molecular mass surfactants decrease the equilibrium surface concentration in protein-stabilized dispersions. The greater energy absorption of low molecular mass surfactants leads to the displacement of macromolecules from the interface [31].

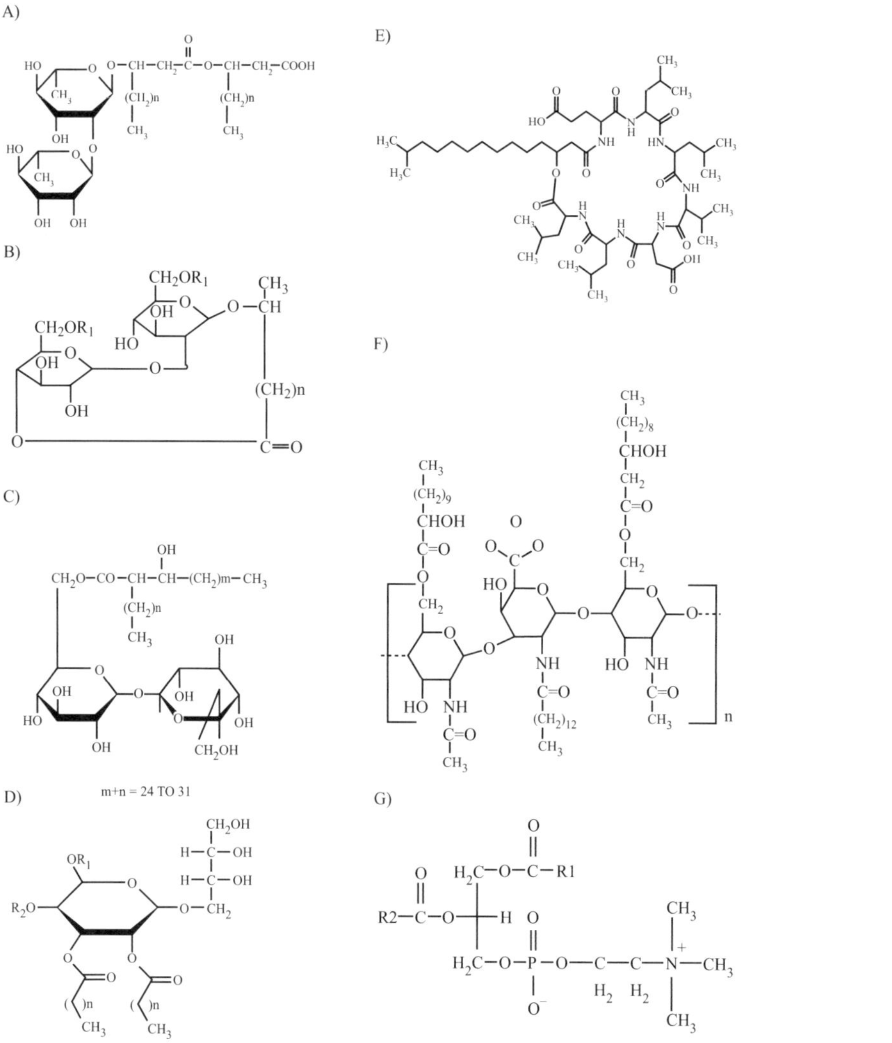

Fig. (1). The basic structure of the four main types of low molecular weight glycolipids (**A** - dirhamnolipid; **B** - lactonic sophorolipid; **C** - trehalose monomycolates; **D** - mannosylerythritol lopids) and two high molecular weight structure of the cyclic lipopeptide (**E** - surfactin; **F** - emulsan; **G** - lecithin).

The mobility of low molecular mass surfactants within aqueous solutions allows these substances to be efficient at reducing interfacial tension. Thus, these surfactants, quickly cover the newly established oil-water interface during the emulsification process. High molecular mass surfactants cover protein and polysaccharide groups. Proteins can penetrate the lipid phase to some degree. For anionic surfactants, which can be either of low or high molecular weight, saturation binding does not depend on the pH and appears to be controlled by hydrophobic interactions. Low molecular weight surfactants and polysaccharides are the predominant amphiphilic groups investigated for the purposes of emulsion stabilization. Polymer surfactant systems are able to control the stability and rheology in a broad range of formulations [31].

Carboxymethyl cellulose and glyceryl monostearate are extremely efficient synthetic emulsifiers. However, these additives are subjected to restrictions in the food industry, especially from the consumers, due to demands aiming to reduce the use of chemically synthesized products. Thus, growing consumer awareness has led to an increase in the demand for natural additives and ingredients [40].

Monoglycerides from fatty acids are commonly used as emulsifiers in water-in-oil emulsions. When transesterified with glycerol, hydrogenated oils or natural fats, they yield monoester and diesters blends of fatty acids which are considered "generally recognized as safe" (GRAS) for application in food, pharmaceuticals and cosmetics [41].

The hydrophile-lipophile balance (HLB) determines the emulsifying capability and ranges from 1 to 20 in commercial emulsifying agents. A low HLB (3 to 6) favours water-in-oil emulsions (propylene glycol fatty acid esters, glycerol esters, sorbitol fatty acid esters, polyglycerol esters, *etc.*), whereas an HLB ranging from 8 to 16 favours oil-in-water emulsions (phospholipids, proteins, hydrocolloids, sodium and potassium salts, alginates, polyoxyethylene fatty acid, guar gum, esters, *etc.*) [42]. These colloids are composed of complex blends of proteins and hydrocolloids, as well as of monomeric and polymeric materials. Over time, a surfactant replaces the less active material [43], leading to the development of different amphiphilic structures [31].

Lecithins, a very important food ingredient surfactant does not have microbial origin and is composed by hydrophobic molecules with a complex structure that can be useful in aqueous solutions. All oils contain small amounts of phospholipids, particularly the derivatives of phosphonyl-3-glycerol, whose hydroxyl groups in positions 1 and 2 are esterified with fatty acids. In contrast, the phosphoric residue is esterified by either a polyol or an amino alcohol. About 60% of phospholipids are found in the cell membranes of living organisms [44] and have amphiphilic characteristics, which explains the sensitivity of the bilayer created to hydration, pH, salinity, cholesterol, temperature, *etc.*

The intermolecular forces of amphiphilic molecules have been illustrated and quantified, such as the electrostatic forces, hydrophobic interactions, lipophilic interactions, the Van der Waals interactions, the interactions between charged and induced electric dipoles and the interactions between induced dipoles [31].

Cocoa powder is a common ingredient in dry mixtures for desserts and beverages. Compared to ready-to-eat products, dry mixes have a longer shelf life and are easier to pack, transport and store. However, the solubility of cocoa powder in water consists in a slow process. Thus, several synthetic surfactants, emulsifiers and instantizing agents have been investigated in order to find substances that can enhance the solubility of instant mixes. Due to concerns from consumers regarding the chemical substances found in food, natural ingredients have become a key issue. Lecithin is a natural ingredient derived from soybeans that provides excellent wetting, emulsifying and dispersing properties. When combined with naturally hydrophobic substances, such as cocoa powder, it is necessary to find the right set of ingredients in dry mix desserts and drinks to formulate instant products of high quality [23]. Cocoa powder is often used with cold milk or cold water. In both cases, the cocoa butter repels water molecules and prevents the cocoa particles to get wet. Thus, the hydrophobic powder must be converted into a hydrophilic powder in order to improve its wettability and dispensability. Lecithin is added exactly for this purpose, since the molecule is a combination of phosphatides comprised with surface-active hydrophilic and hydrophobic parts. The hydrophobic portion unites with the cocoa butter and the hydrophilic portion is positioned outside the cocoa particle, thereby attracting water and leading to the wetting, dispersion and suspension of cocoa powder in cold milk or water [23].

PRODUCTION AND SUSTAINABILITY OF SURFACTANTS

Currently, there are no microbial surfactants that are widely used in food industries. Safety and cost are key issues for additives in food products. Thus, prior to receiving an authorization for application, an additive must be submitted to a toxicological evaluation in order to determine its synergic, accumulative and protective effects [45, 46]. The worldwide production of surfactants was from the order of 17 million tons in 2000, with an expected increase of 3 to 4% a year. Due to their diverse properties and structures, biosurfactants have applications in different industrial processes and new employments for these biomolecules are likely to be discovered. It is therefore widely believed that they will be remarked as "multifunctional materials" [47, 48].

Health and food authorities in many countries have progressively imposed restrictions on the use of synthetic emulsifiers in food ingredients. Thus, researchers are continuously seeking natural alternative sources for amphiphilic molecules that could be suitable and employed in new food formulations. In the food industry, raw materials are processed with food additives containing emulsifiers to acquire the desired texture and consistency. The application of surface-active compounds in baking and meat processing is due to the ability of these substances to alter the rheological characteristics of flour and emulsify fat tissue [49]. Naturally occurring and commercially available plant-derived food emulsifiers, such as lecithin and Arabic gum, are viable alternatives instead of chemical additives [50]. However, problems may arise from the fact that some of these products have to be imported from countries subjected to climatic and political upheavals. To overcome such setbacks, large-scale microbial cultivation has been suggested for the production of food emulsifiers [51]. The large-scale industrial production of natural food-grade additives has been successfully achieved with gellan gum, nisin and xanthan gum [52 - 54].

Lecithins with sphingolipids (sphingosine) and phospholipids (phytosphingosine) are extracted from soybeans, eggs, oat, wheat, *etc.* and are cheap materials. Many manufacturers plant crude lecithin in a variety of ways in order to extract products with different degrees of purity and specificity, such as de-oiled lecithin, plastic lecithin, phosphatidyl choline, phosphatidyl choline-enriched lecithin, phosphatidyl ethanolamine, phosphatidyl inositol, phosphatidyl serine and phosphatidic acid [21]. In lipid fractions (oleoresins), the monoglyceride content is higher (2 to 3% by weight). Many oleoresins extracted from flowers, fruit, leaves, spices, *etc.*, contain triglycerides, monoglyceride derivatives and nonsaponifiable fats (waxes). These fractions can be "self-emulsifiable" and form water-in-oil emulsions *in situ*. Saturated fatty acid monoglycerides are combined forming lamellar liquid crystalline phases at the oil-water interface at and above the critical temperature for emulsification. Recently, oleoresins have been extracted and the active component (antioxidants, such as lutein in marigolds and carnosic acid in rosemary, lycopene in tomatoes *etc.*) is removed in order to enrich the residual oil phase through further extraction with monoglycerides, followed by the use of these naturally occurring self-emulsifying oils in water-in-oil emulsions.

Amphiphilic non-phosphorous molecules, especially diglyceride ethers with a mono-saccharide or disaccharide, follow the phospholipids. Ceramides and phytoceramides are fatty-acid amides with either sphingosine or phytosphingosine that are often united with the galactose in animals or glucose in plants. Attempts have been made in order to extract molecules from wheat, soybeans and oats to investigate their surface properties. Improved separation methods are expected to reduce extraction costs, allowing digalactosyl-diglycerides and trigalactosyl-diglycerides to become viable, natural oil-in-water and water-in-oil emulsifiers. The unsaponifable portion of fat has low solubility in water. The components of this part are different hydrocarbons, fatty alcohols terpenes and fat-soluble vitamins. Studies have reported stable "surface complexes" at oil-in-water interfaces composed of saponins and other monomeric emulsifiers capable of enhancing the stability of emulsions [55]. This has led to a category of products denominated "nutraceuticals" [31].

BIOSURFACTANTS OF MICROBIAL ORIGIN AND POSSIBLE ROLES IN REPLACING TRADITIONAL SURFACTANTS

Increasing environmental concern regarding the use of chemical surfactants has led researchers to turn towards microbial-derived surface-active compounds denominated biosurfactants, which are biodegradable and have a low degree of toxicity. Biosurfactants are the metabolic by-product of bacteria, yeasts and filamentous fungi. As mentioned above, the occurrence of hydrophilic and hydrophobic groups allows a surfactant to be distributed at the interface between liquid phases with different polarities, thereby reducing surface and interfacial tensions. This property makes biosurfactants appropriate for industrial applications [56].

Biosurfactants with emulsifier properties are also found in animals, such as lipoproteins from chicken eggs, soluble proteins from poultry meat or casein and milk fat globule membrane from bovine milk [57]. Advanced techniques in recent years have led to a better understanding of the rheological properties, texture and microstructures of food colloids. Emulsifying and dispersing agents in food products do not necessarily need to reduce surface tension [58]. For instance, liposan does not reduce the surface tension of water, but has been successfully employed for the emulsification of vegetal oils [59]. In contrast, sophorolipids from *Turulopsis bombicola* are able to reduce surface and interfacial tensions, but are not considered efficient emulsifiers.

Several natural food emulsifiers derived from vegetal material , such as Arabic gum and lecithin, hold considerable acceptance within the market. However, lecithin has functional limitations when employed in products submitted to modern processing conditions, such as microwave cooking and irradiation. On the other hand, the production of emulsifiers from microbial cultures is an alternative to existing additives and offers products that are tolerant to modern food processing methods [28]. Butter, cream, mayonnaise, margarine, hotdogs and salad dressing for example are processed foods in emulsion form. Other applications for emulsifiers include the enhancement of shelf life and the texture increase of products containing starch (due to the formation of complexes with the components of these products), changes in the rheological properties of wheat flour through interactions with gluten and improvements in the consistency and texture of fat-based products through the control of polymorphisms and the crystalline structures of fats [20, 28, 60].

Surface-active compounds synthesized from the petrochemical industry account for 70 to 75% of all surfactants employed in industrialized countries [61] and biological concerns have led to the search for more environmentally positive or friendlier products [8, 15, 62] in order to replace the non-biodegradable branched alkyl benzenes and to enhance their specificity [63]. Microbiological products with surfactant properties have been obtained from microorganisms [64 - 66], such as *Pseudomonas aeruginosa, Arthrobacter, Bacillus subtilis, Acinetobacter calcoaceticus, Torulopsis bombicola* and *Candida lipolytica* [10, 67, 68]. Microorganisms produce biosurfactants with different molecular structures [3]. While some biosurfactants are produced on soluble substrates or a combination of substrates, most are produced on water-insoluble substrates, such as fats, oils and liquid or solid hydrocarbons [69].

The majority production of biosurfactants is achieved with bacteria, with reports dating from 1949, when a new biologically active compound produced by the *Bacillus subtilis*, denominated surfactin, was discovered [2]. Surfactin is considered one of the best biosurfactants, reducing surface tension at concentrations under 0.005%. The genera *Pseudomonas* and *Bacillus* are described as major biosurfactant-producing bacteria.

Some biosurfactants produced from *Bacillus subtilis* bacterial strains are considered safe, with some cell products receiving GRAS status, such as α-amylase and pullulanase derived from *Bacillus licheniformis* and pectate lyase from *B. subtilis*. However, there are reports that indicate both these bacteria as being involved in some cases of fatal food poisoning [2].

The primary criteria for identifying biosurfactant-producing bacteria is the ability to reduce surface or interfacial tension, but some exceptions can be found in the screening of microrganisms for use in the food industry [51]. Bacteria are not the only microorganisms investigated for biosurfactant production. The yeasts *Candida valida, Candida utilis, Rhodospiridium diobovatum, Hansenula anomala* and *Rhodotorula graminis*, the red alga *Porphiridium cruentum* and the bacteria *Acinetobacter calcoaceticus* and *Klebsiella spp* have also been investigated for the production of high molecular weight emulsifiers [51].

The genus *Candida* has been used in the emulsification and solubilisation of hydrocarbons, with microbial growth on such substrates facilitated by the production of biosurfactants. Moreover, biosurfactants can be produced from low-cost, renewable sources [70 - 78]. *Yarrowia* is another genus of the yeast that has been employed in the production of biosurfactants. Both genera have GRAS status, which demonstrates that such microorganisms do not represent risk of toxicity or pathogenicity, thereby allowing their use in foods. Such yeasts also include *Kluyveromyces lactis, Saccharomyces cerevisiae* and *Yarrowia lipolytica* [76].

A biosurfactant isolated from *S. cerevisiae* has demonstrated great emulsifying properties. The compound contains mainly proteins and some carbohydrates and appears to come from phosphomannoprotein of the yeast cell wall [79]. Another bioemulsifier extracted from *S. cerevisiae* cell walls (mannoprotein) has been tested for its emulsifying abilities. The product is extracted by autoclaving in citrate buffer (pH 7) and consists of 380 to 410 g/kg of protein and 210 g/kg of carbohydrates. Its function is similar to the one of sodium caseinate. Analyses have demonstrated that emulsions of 60% oil-in-water, 8 g/L of bioemulsifier and 5 to 50 g/L of sodium chloride are stable for three months at 4 °C and with a pH ranging from 3 to 11. Practical applications of this bioemulsifier indicate that it can be used with carboxymethyl cellulose rather than xanthan for mayonnaise production and employed to stabilize oil-in-water emulsions for the production of crackers, cookies, cakes, ice cream, sausages and other food products [80].

Using local substrates as the carbon source, *Candida lipolytica* can produce biopolymers with considerable emulsifying activity [70, 81]. Cirigliano and Carmam [59] isolated a bioemulsifier generated by *C. lipolytica* cultivated in an environment containing n-hexadecane and demonstrated the potential application of this product in the food industry. *C. utilis* produces a biosurfactant with excellent emulsification properties, low viscosity and carbohydrate content between 79 and 98%. These features point out that this agent could be employed in the formulation of low-fat food, such as mayonnaise kinds of salad dressings. Some authors compared dressings prepared using 0.2% and 0.8% m/V of a bioemulsifier from *C. utilis* with dressings prepared using Mayodan, which is a commercially available stabilizer composed of guar and xanthan gum. Although the results were only preliminary and not completely reproducible, the bioemulsifier from *C. utilis* indicated both emulsification activity and stability when stored at 4 °C for one week [51].

ADVANTAGES OF BIOSURFACTANTS OVER SYNTHETIC SURFACTANTS

Biosurfactants occupy a special chemical position in industries interested in their use. These substances can be applied in oil recovery processes, in the biodegradation of water-insoluble pollutants and as components in pharmaceuticals, detergents and food industries [82 - 84]. Extremes of temperature, pH and ionic strength are often found in industrial processes. Therefore, surfactants need to demonstrate adequate stability under such conditions [2, 85]. Despite the diversity in properties and chemical composition of biosurfactants, certain features are common to the majority of such substances and offer significant advantages over chemically derived conventional surfactants [26, 86]:

- **Surface and interfacial activity**: lower concentrations of biosurfactants are needed to reduce the surface tension of liquids;

- **Application under extreme conditions**: The lipopeptide from *Bacillus licheniformis* JF-2 is stable at temperatures around 75 °C for up to 140 hours and with a pH range from 5 to 12. Biosurfactants can take salt concentrations up to 10%, whereas 2% of NaCl is enough to inactivate conventional surfactants;

- **Biodegradability**: bacteria and other microorganisms in water and soil can easily degrade biosurfactants, making these substances suitable for the treatment of waste and bioremediation processes;

- **Low toxicity**: Concerns regarding the allergic effect of artificial products favour the application of biosurfactants in cosmetics, pharmaceuticals and food products;

- **Availability**: Widely available raw materials and even industrial waste can be used in the production of biosurfactants;

- **Specificity**: Biosurfactants are complex organic molecules with specific actions that are compatible with the decontamination of pollutants or application at cosmetics, foods and pharmaceuticals;

- **Compatibility and digestibility**: These properties are particularly important to the cosmetic, food and pharmaceutical industries as they can be easily digested.

Biosurfactants have potential employment as emulsifiers in the processing of food products with applications in baked goods through changes in the rheological characteristics of flour and in meat-based products through the emulsification of animal fat. *Candida utilis* produces a bioemulsifier that can be used in salad dressing [40]. Rhamnolipids have been employed to enhance the properties of butter and pastry products [47].

S. cerevisiae produces mannoprotein in a low-cost, large-scale process, which stabilizes water-in-oil emulsions used in desserts and mayonnaise. Besides being stable in a broad pH range, mannoprotein is also employed in animal rations [20, 87]. *Saccharomyces lipolytica* CCT-0913 generates a biosurfactant with stability in a wide range of temperatures and pH values [88]. The yeasts *Candida valida, Candida utilis, Rhodospiridium diobovatum,Hansenula anomala* and *Rhodotorula graminis*, the red alga *Porphiridium cruentum* and the bacteria *Acinetobacter calcoaceticus* and *Klebsiella spp.* produce extracellular bioemulsifiers that offer better stabilizing activity in comparison to carboxymethyl cellulose and Arabic gum [2].

Candida bombicola creates a biosurfactant that has been employed for the partial solubilisation of coal [89]. Sophorolipids, surfactin and rhamnolipids have also been used in the removal of heavy metals from sediments [10]. Furthermore, biosurfactants have been applied in uranium processing as well as the textile, ceramic and paper industries [3].Finally, insect repellents, antacids, contact lens solutions, deodorants, toothpaste, fingernail products and other products require surfactants [5, 90, 91].

Despite the several potential applications, biosurfactants are not yet employed as food additives on a large scale. Due to government regulations, approval is required for such use. However, the identification of compounds with a low degree of toxicity and satisfactory surface-active properties is of considerable interest. Quantitative structure-activity relationship (QSAR) models constitute a tool that can be used for this purpose by relating physicochemical aspects (solubility and molecular weight) to biological responses (pharmacological effects, toxicity and carcinogenicity). The immune response is a bioindicator of toxic substances, as low concentrations of lead, cadmium, benzopyrene, nitrosamines, saturated fatty acids and nicotine can reduce the immune response over time. No long-term studies have been carried out on low concentrations of toxic agents found in foods [91].

A legislation for the allowed toxicological standard is required for the use of surfactants in foods. Such a norm is the threshold limit regarding the content of a certain substance in water, food, air or soil. A toxicological standard is the result of a detailed analysis of the risks and benefits. For such, political and socioeconomic interests are balanced against toxicological thresholds, health-based recommendations and technical feasibility. For food additives, the Joint Food and Agriculture Organization of the United Nations and the World Health Organization Expert Committee on Food Additives have long set the establishment of acceptable daily intake values to provide "an adequate margin of safety in order to reduce to a minimum any health hazard for all consumers groups". For such, it is crucial to distinguish nongenotoxic substances, for which a threshold may be established, from genotoxic substances, for which no such limit exists [92].

The threshold dose is the highest level of exposure that does not offer any important toxicological effect, which is denominated the no-observed-adver-e-effect level (NOAEL). In a long-term toxicity study, a high dose may cause poisoning without mortality and a low dose should not affect normal growth, development or longevity. Therefore the NOAEL is the highest dose that has no toxic effect in the most sensitive animals. The determined dose is then divided by a safety factor to establish the acceptable daily intake, which determines the amount of a food additive that can be ingested on daily bases over a lifetime without constituting any noticeable health risk. Thus, substances that can be accumulated in the body are not suitable to be employed as additives [31].

Biosurfactants also show several physiological functions (cell adherence/release, hydrocarbon transport, antimicrobial and antibiofilm activity) that have many useful applications in the food and other industries [93 - 95]. However, the mechanisms of action are not yet fully understood.

CONCLUDING REMARKS

There has been increasing interest in the isolation and identification of new microbial polysaccharides and surfactants. Such issue has been fuelled by the possibility of discovering a new gum or emulsifier with properties that allow the application as a gelling, emulsification or dispersing agent. The high production costs related to biosurfactants stem primarily from inefficient processing methods and the use of expensive substrates. The prerequisites that will allow biosurfactants to gain a significant niche of the market include the following: The use of cheap substrates that can provide high yields; improvements in processing technologies to facilitate the recovery of the product; greater knowledge regarding the metabolism of biosurfactant-producing microorganisms; and the selection of biosurfactants for specific applications. Moreover, the development of fermentation technologies will enhance the possibility of modifying the structure and function of biopolymers in a controllable tendency, allowing the development of structural binding models and "designer biopolymers".

New ingredients capable of tolerating modern food processing methods (ultra-high temperatures, extrusion microwave heating, *etc.*) that can function adequately in new low-salt, low-fat and low-calorie formulas will be developed. However, the success of food products and ingredients produced through biotechnology depends on their acceptance from consumers. Thus, governments, representatives of the food industry, educators, media, scientists and technicians directly involved in these processes should educate consumers regarding the potential benefits of biotechnology. Discoveries stemming from biotechnological processes are surely effective on offering solutions to the challenges faced by the food industry.

CONFLICT OF INTEREST

The authors confirm that they have no conflict of interest to declare for this publication.

ACKNOWLEDGEMENTS

The editors acknowledge Espaço da Escrita (Coordenadoria Geral - State

University of Campinas) for the English corrections in in this text.

REFERENCES

[1] Singh, A.; Van Hamme, J.D.; Ward, O.P. Surfactants in microbiology and biotechnology: Part 2. Application aspects. *Biotechnol. Adv.,* **2007**, *25*(1), 99-121. [http://dx.doi.org/10.1016/j.biotechadv.2006.10.004] [PMID: 17156965]

[2] Barros, F.F.; Quadros, C.P.; Maróstica, M.R.; Pastore, G.M. Surfactina: propriedades químicas, tecnológicas e funcionais para aplicações em alimentos. *Quim. Nova,* **2007**, *30*, 409-414. [http://dx.doi.org/10.1590/S0100-40422007000200031]

[3] Deleu, M.; Paquot, M. From renewable vegetables resources to microorganisms: new trends in surfactants. *C. R. Chim.,* **2004**, *7*, 641-646. [http://dx.doi.org/10.1016/j.crci.2004.04.002]

[4] Nguyen, T.T.; Edelen, A.; Neighbors, B.; Sabatini, D.A. Biocompatible lecithin-based microemulsions with rhamnolipid and sophorolipid biosurfactants: formulation and potential applications. *J. Colloid Interface Sci.,* **2010**, *348*(2), 498-504. [http://dx.doi.org/10.1016/j.jcis.2010.04.053] [PMID: 20471022]

[5] Penfold, J.; Chen, M.; Thomas, R.K.; Dong, C.; Smyth, T.J.; Perfumo, A.; Marchant, R.; Banat, I.M.; Stevenson, P.; Parry, A.; Tucker, I.; Grillo, I. Solution self-assembly of the sophorolipid biosurfactant and its mixture with anionic surfactant sodium dodecyl benzene sulfonate. *Langmuir,* **2011**, *27*(14), 8867-8877. [http://dx.doi.org/10.1021/la201661y] [PMID: 21644533]

[6] Marchant, R.; Banat, I.M. Microbial biosurfactants: challenges and opportunities for future exploitation. *Trends Biotechnol.,* **2012**, *30*(11), 558-565. [http://dx.doi.org/10.1016/j.tibtech.2012.07.003] [PMID: 22901730]

[7] Desai, J.D.; Banat, I.M. Microbial production of surfactants and their commercial potential. *Microbiol. Mol. Biol. Rev.,* **1997**, *61*(1), 47-64. [PMID: 9106364]

[8] Chen, M.L.; Penfold, J.; Thomas, R.K.; Smyth, T.J.; Perfumo, A.; Marchant, R.; Banat, I.M.; Stevenson, P.; Parry, A.; Tucker, I.; Grillo, I. Solution self-assembly and adsorption at the air-water interface of the monorhamnose and dirhamnose rhamnolipids and their mixtures. *Langmuir,* **2010**, *26*(23), 18281-18292. [http://dx.doi.org/10.1021/la1031812] [PMID: 21028852]

[9] Płaza, G.A.; Chojniak, J.; Banat, I.M. Biosurfactant mediated biosynthesis of selected metallic nanoparticles. *Int. J. Mol. Sci.,* **2014**, *15*(8), 13720-13737. [http://dx.doi.org/10.3390/ijms150813720] [PMID: 25110864]

[10] Mulligan, C.N. Environmental applications for biosurfactants. *Environ. Pollut.,* **2005**, *133*(2), 183-198. [http://dx.doi.org/10.1016/j.envpol.2004.06.009] [PMID: 15519450]

[11] Chen, M.L.; Penfold, J.; Thomas, R.K.; Smyth, T.J.; Perfumo, A.; Marchant, R.; Banat, I.M.; Stevenson, P.; Parry, A.; Tucker, I.; Grillo, I. Mixing behavior of the biosurfactant, rhamnolipid, with

a conventional anionic surfactant, sodium dodecyl benzene sulfonate. *Langmuir,* **2010**, *26*(23), 17958-17968.
[http://dx.doi.org/10.1021/la1031834] [PMID: 21043468]

[12] Chen, M.; Dong, C.; Penfold, J.; Thomas, R.K.; Smyth, T.J.; Perfumo, A.; Marchant, R.; Banat, I.M.; Stevenson, P.; Parry, A.; Tucker, I.; Campbell, R.A. Adsorption of sophorolipid biosurfactants on their own and mixed with sodium dodecyl benzene sulfonate, at the air/water interface. *Langmuir,* **2011**, *27*(14), 8854-8866.
[http://dx.doi.org/10.1021/la201660n] [PMID: 21657229]

[13] Maier, R.M. Biosurfactants: evolution and diversity in bacteria. *Adv. Appl. Microbiol.,* **2003**, *52*, 101-121.
[http://dx.doi.org/10.1016/S0065-2164(03)01004-9] [PMID: 12964241]

[14] Raza, Z.A.; Khalid, Z.M.; Khan, M.S.; Banat, I.M.; Rehman, A.; Naeem, A.; Saddique, M.T. Surface properties and sub-surface aggregate assimilation of rhamnolipid surfactants in different aqueous systems. *Biotechnol. Lett.,* **2010**, *32*(6), 811-816.
[http://dx.doi.org/10.1007/s10529-010-0216-x] [PMID: 20155387]

[15] Makkar, R.S.; Cameotra, S.S.; Banat, I.M. Advances in utilization of renewable substrates for biosurfactant production. *AMB Express,* **2011**, *1*(1), 5.
[http://dx.doi.org/10.1186/2191-0855-1-5] [PMID: 21906330]

[16] Das, P.; Mukherjee, S.; Sivapathasekaran, C.; Sen, R. *Biosurfactants. Advances in Experimental Medicine and Biology*; Sen, R., Ed.; Springer: Berlin, **2010**, Vol. 672, pp. 88-101.

[17] Fracchia, L.; Cavallo, M.; Allegrone, G.; Martinotti, M.G. *Current Research*; Méndez Vilas, A., Ed.; Technology and Education Topics in Applied Microbiology and Microbial Biotechnology: Spain, **2010**, pp. 827-837.

[18] Campos, J.M.; Stamford, T.L.; Sarubbo, L.A.; de Luna, J.M.; Rufino, R.D.; Banat, I.M. Microbial biosurfactants as additives for food industries. *Biotechnol. Prog.,* **2013**, *29*(5), 1097-1108.
[http://dx.doi.org/10.1002/btpr.1796] [PMID: 23956227]

[19] Sachdev, D.P.; Cameotra, S.S. Biosurfactants in agriculture. *Appl. Microbiol. Biotechnol.,* **2013**, *97*(3), 1005-1016.
[http://dx.doi.org/10.1007/s00253-012-4641-8] [PMID: 23280539]

[20] Banat, I.M.; Makkar, R.S.; Cameotra, S.S. Potential commercial applications of microbial surfactants. *Appl. Microbiol. Biotechnol.,* **2000**, *53*(5), 495-508.
[http://dx.doi.org/10.1007/s002530051648] [PMID: 10855707]

[21] Garti, N. What can nature offer from an emulsifier point of view: trend and progress? *Colloids Surf. A Physicochem. Eng. Asp.,* **1999**, *152*, 125-146.
[http://dx.doi.org/10.1016/S0927-7757(98)00621-9]

[22] Xu, W.; Nikolov, A.; Wasan, D.T.; Gonsalves, A.; Borwankar, R.P. Fat particle structure and stability of food emulsions. *J. Food Sci.,* **1998**, *63*, 183-188.
[http://dx.doi.org/10.1111/j.1365-2621.1998.tb15705.x]

[23] Kourkoutas, Y.; Banat, I.M. Biosurfactant production and application. In: *The Concise Encyclopedia of Bioresource Technology*; Pandey, A.P., Ed.; Haworth Reference Press, **2008**; pp. 505-515.

[24] Tadros, T.F. *Applied Surfactants, Principles and Application*; Wiley-VCH: Weinheim, Germany, **2005**.
[http://dx.doi.org/10.1002/3527604812]

[25] *Codex General Standard for Food Additives"*, (GSFA, Codex STAN 192-1995)

[26] Nitschke, M.; Pastore, G.M. Biossurfactantes: propriedades e aplicações. *Quim. Nova,* **2002**, *25*, 772-776.
[http://dx.doi.org/10.1590/S0100-40422002000500013]

[27] Rodríguez Patino, J.M.; Carrera Sánchez, C.; Rodríguez Niño, M.R. Implications of interfacial characteristics of food foaming agents in foam formulations. *Adv. Colloid Interface Sci.,* **2008**, *140*(2), 95-113.
[http://dx.doi.org/10.1016/j.cis.2007.12.007] [PMID: 18281008]

[28] Sarubbo, L.A. Otimização da produção de biopolímeros por *Candida lipolytica* utilizando óleo de babaçu como fonte de carbono. *Master's Thesis , Universidade Federal de Pernambuco, Brasil,* **1997**.

[29] Benichou, A.; Aserin, A.; Garti, N. Double emulsions stabilized with hybrids of natural polymers for entrapment and slow release of active matters. *Adv. Colloid Interface Sci.,* **2004**, *108-109*, 29-41.
[http://dx.doi.org/10.1016/j.cis.2003.10.013] [PMID: 15072926]

[30] Morais, J.M.; Santos, O.D.; Nunes, J.R.; Zanatta, C.F.; Rocha-Filho, P.A. W/O/W multiple emulsion obtained by one-step emulsification method and evaluation of the involved variables. *J. Dispers. Sci. Technol.,* **2008**, *29*, 63-69.
[http://dx.doi.org/10.1080/01932690701688391]

[31] Kralova, I.; Sjoblon, J. Surfactant used in food industry: a review. J. *Dispersion Sci. Technol.,* **2009**, *30*, 1363-1383.
[http://dx.doi.org/10.1080/01932690902735561]

[32] Velikonja, J.; Kosaric, N. *Biosurfactants: production properties, applications*; Kosaric, N., Ed.; Marcel Dekker Inc: New York, **1993**, p. 16.

[33] Brennan, J.G. *In Food Processing Handbook*; Wiley-VCH: Weinheim, Germany, **2006**.

[34] Fiechter, A. Biosurfactants: moving towards industrial application. *Trends Biotechnol.,* **1992**, *10*(6), 208-217.
[http://dx.doi.org/10.1016/0167-7799(92)90215-H] [PMID: 1368395]

[35] Lin, S.C.; Sharma, M.M.; Georgiou, G. Production and deactivation of biosurfactant by *Bacillus licheniformis* JF-2. *Biotechnol. Prog.,* **1993**, *8*, 138-145.
[http://dx.doi.org/10.1021/bp00020a004]

[36] Campos, J.M.; Stamford, T.L.; Sarubbo, L.A. Production of a bioemulsifier with potential application in the food industry. *Appl. Biochem. Biotechnol.,* **2014**, *172*(6), 3234-3252.
[http://dx.doi.org/10.1007/s12010-014-0761-1] [PMID: 24504690]

[37] Sharma, S.C. Gums and hydrocolloids in oil-water emulsions. *Food Technol.,* **1981**, *35*, 59-67.

[38] Ishigami, Y.; Suzuki, S. Development of biochemicals—functionalization of biosurfactants and natural dyes. *Prog. Org. Coat.,* **1997**, *31*, 51-61.
[http://dx.doi.org/10.1016/S0300-9440(97)00018-0]

[39] Darling, D.F.; Birkett, R.J. *Foods and Foams*; Dickinson, E., Ed.; Royal Society of Chemistry: London, **1987**.

[40] Shepherd, R.; Rockey, J.; Sutherland, I.W.; Roller, S. Novel bioemulsifiers from microorganisms for use in foods. *J. Biotechnol.,* **1995**, *40*(3), 207-217.
[http://dx.doi.org/10.1016/0168-1656(95)00053-S] [PMID: 7632394]

[41] Krogh, N.J.; Sparto, F.V. *In Food Emulsions*; Marcel Dekker: New York, **2004**.

[42] Dickinson, E.; McClemens, D.J. *Advances in Food Colloids*; Blackie: Glasgow, **1996**.
[http://dx.doi.org/10.1007/978-1-4613-1223-9]

[43] Hasenhuettl, G.L.; Hartel, R.W. *Food Emulsifiers and Their Applications*; Chapman and Hall: New York, **1997**.
[http://dx.doi.org/10.1007/978-1-4757-2662-6]

[44] Karleskind, A. *Oils and Fats Manual*; Intercept: Hampshire, UK, **1996**.

[45] BRASIL. *Portaria nº 540 - Secretaria de Vigilância em Saúde - Ministério da Saúde, de 27 de outubro de,* **1997**.

[46] BRASIL. *Resolução RDC nº 234 - Agência Nacional de Vigilância Sanitária - Ministério da Saúde, de 19 de agosto de,* **2002**.

[47] Muthusamy, K.; Gopalakrishnan, S.; Ravi, T.K.; Sivachidambaram, P. Biosurfactants: properties, commercial production and application. *Curr. Sci.,* **2008**, *94*, 736-747.

[48] Banat, I.M.; Franzetti, A.; Gandolfi, I.; Bestetti, G.; Martinotti, M.G.; Fracchia, L.; Smyth, T.J.; Marchant, R. Microbial biosurfactants production, applications and future potential. *Appl. Microbiol. Biotechnol.,* **2010**, *87*(2), 427-444.
[http://dx.doi.org/10.1007/s00253-010-2589-0] [PMID: 20424836]

[49] Fiechter, A. Biosurfactants: moving towards industrial application. *Trends Biotechnol.,* **1992**, *10*(6), 208-217.
[http://dx.doi.org/10.1016/0167-7799(92)90215-H] [PMID: 1368395]

[50] Whistler, R.L. Exudate gums. , **1993**.
[http://dx.doi.org/10.1016/B978-0-08-092654-4.50016-4]

[51] Larsson, K.; Friberg, E. *Food Emulsions,* 2nd ed; Friberg, S.E.; Larsson, K.; Sjoblom, J., Eds.; Marcel Dekker: New York, **1990**.

[52] Sutherland, I.W. *Biotechnology of Microbial Polysaccharides*; Cambridge University Press: Cambridge, **1990**.
[http://dx.doi.org/10.1017/CBO9780511525384]

[53] Roller, S. The biotechnological development of new food preservatives. *Biotechnol. Genet. Eng. Rev.,* **1991**, *9*, 183-206.
[PMID: 1797014]

[54] Roller, S.; Dea, I.C. Biotechnology in the production and modification of biopolymers for foods. *Crit. Rev. Biotechnol.,* **1992**, *12*, 261-277.
[http://dx.doi.org/10.3109/07388559209069195]

[55] Mitra, S.; Dungan, S.R. Micellar properties of quillaja saponin. 2. Effect of solubilized cholesterol on solution properties. *Colloids Surf. B Biointerfaces,* **2000**, *17*, 117-133. [http://dx.doi.org/10.1016/S0927-7765(99)00088-0]

[56] Makkar, R.S.; Cameotra, S.S. Utilization of molasses for biosurfactant production by two *Bacillus* strains at thermophilic conditions. *J. Am. Oil Chem. Soc.,* **1997**, *74*, 887-889. [http://dx.doi.org/10.1007/s11746-997-0233-7]

[57] Zajjc, J.E.; Panchal, C.J. Bio-emulsifiers. *CRC Crit. Rev. Microbiol.,* **1976**, *5*(1), 39-66. [http://dx.doi.org/10.3109/10408417609102309] [PMID: 793775]

[58] Cooper, D.G.; Paddock, D.A. Production of a biosurfactant from *Torulopsis bombicola. Appl. Environ. Microbiol.,* **1984**, *47*(1), 173-176. [PMID: 16346455]

[59] Cirigliano, M.C.; Carman, G.M. Purification and characterization of liposan, a bioemulsifier from *Candida lipolytica. Appl. Environ. Microbiol.,* **1985**, *50*(4), 846-850. [PMID: 16346917]

[60] Nitschke, M.; Costa, S.G. Biosurfactants in food industry. *Trends Food Sci. Technol.,* **2007**, *18*, 252-259. [http://dx.doi.org/10.1016/j.tifs.2007.01.002]

[61] Breckenridge, C.R.; Polman, J.K. Solubilization of coal by biosurfactants derived from *Candida bombicola. Geomicrobiol. J.,* **1994**, *12*, 285-288. [http://dx.doi.org/10.1080/01490459409377996]

[62] Georgiou, G.; Lin, S-C.; Sharma, M.M. Surface-active compounds from microorganisms. *Biotechnology (N. Y.),* **1992**, *10*(1), 60-65. [http://dx.doi.org/10.1038/nbt0192-60] [PMID: 1368190]

[63] Urum, K.; Pekdemir, T.; Copur, M. Surfactants treatment of crude oil contaminated soils. *J. Colloid Interface Sci.,* **2004**, *276*(2), 456-464. [http://dx.doi.org/10.1016/j.jcis.2004.03.057] [PMID: 15271574]

[64] Sen, R.; Swaminathan, T. Characterization of concentration and purification parameters and operating conditions for the small-scale recovery of surfactin. *Process Biochem.,* **2005**, *40*, 2953-2958. [http://dx.doi.org/10.1016/j.procbio.2005.01.014]

[65] Trindade, J.R.; Freire, M.G.; Amaral, P.F.; Coelho, M.A.; Coutinho, J.A.; Marrucho, I.M. Aging mechanisms of oil-in-water emulsions based on a bioemulsifier produced by *Yarrowia lipolytica. Colloids Surf. A Physochem. Eng. Asp,* **2008**, *324*, 149-154. [http://dx.doi.org/10.1016/j.colsurfa.2008.04.009]

[66] Satpute, S.K.; Banpurkar, A.G.; Dhakephalkar, P.K.; Banat, I.M.; Chopade, B.A. Methods for investigating biosurfactants and bioemulsifiers: a review. *Crit. Rev. Biotechnol.,* **2010**, *30*(2), 127-144. [http://dx.doi.org/10.3109/07388550903427280] [PMID: 20210700]

[67] Beopoulos, A.; Cescut, J.; Haddouche, R.; Uribelarrea, J-L.; Molina-Jouve, C.; Nicaud, J-M. *Yarrowia lipolytica* as a model for bio-oil production. *Prog. Lipid Res.,* **2009**, *48*(6), 375-387. [http://dx.doi.org/10.1016/j.plipres.2009.08.005] [PMID: 19720081]

[68] Freitas, F.; Alves, V.D.; Pais, J.; Costa, N.; Oliveira, C.; Mafra, L.; Hilliou, L.; Oliveira, R.; Reis, M.A. Characterization of an extracellular polysaccharide produced by a Pseudomonas strain grown on glycerol. *Bioresour. Technol.,* **2009**, *100*(2), 859-865. [http://dx.doi.org/10.1016/j.biortech.2008.07.002] [PMID: 18713662]

[69] Van Hamme, J.D.; Singh, A.; Ward, O.P. Physiological aspects. Part 1 in a series of papers devoted to surfactants in microbiology and biotechnology. *Biotechnol. Adv.,* **2006**, *24*(6), 604-620. [http://dx.doi.org/10.1016/j.biotechadv.2006.08.001] [PMID: 16979315]

[70] Sarubbo, L.A.; Porto, A.L.; Campos-Takaki, G.M. The use of babassu oil as substrate to produce bioemulsifiers by *Candida lipolytica. Can. J. Microbiol.,* **1999**, *45*(5), 423-426. [http://dx.doi.org/10.1139/w99-025] [PMID: 10446719]

[71] Sarubbo, L.A.; Marçal, M.C.; Neves, M.L.; Silva, M.P.; Porto, A.L.; Campos-Takaki, G.M. Bioemulsifier production in batch culture using glucose as carbon source by *Candida lipolytica. Appl. Biochem. Biotechnol.,* **2001**, *95*(1), 59-67. [http://dx.doi.org/10.1385/ABAB:95:1:59] [PMID: 11665807]

[72] Sarubbo, L.A.; Luna, J.M.; Campos-Takaki, G.M. Production and stability studies of the bioemulsifier obtained from a new strain of *Candida glabrata* UCP 1002. *Electron. J. Biotechnol.,* **2006**, *9*, 400-406.

[73] Sarubbo, L.A.; Farias, C.B.; Campos-Takaki, G.M. Co-utilization of canola oil and glucose on the production of a surfactant by *Candida lipolytica. Curr. Microbiol.,* **2007**, *54*(1), 68-73. [http://dx.doi.org/10.1007/s00284-006-0412-z] [PMID: 17171462]

[74] Rufino, R.D.; Sarubbo, L.A.; Neto, B.B.; Campos-Takaki, G.M. Experimental design for the production of tensio-active agent by Candida lipolytica. *J. Ind. Microbiol. Biotechnol.,* **2008**, *35*(8), 907-914. [http://dx.doi.org/10.1007/s10295-008-0364-3] [PMID: 18563463]

[75] Rufino, R.D.; Sarubbo, L.A.; Campos-Takaki, G.M. Enhancement of stability of biosurfactant produced by *Candida lipolytica* using industrial residue as substrate. *World J. Microbiol. Biotechnol.,* **2007**, *23*, 734-741. [http://dx.doi.org/10.1007/s11274-006-9278-2]

[76] Fontes, G.C.; Amaral, P.F.; Coelho, M.A. Produção de Biossurfactante por levedura. *Quim. Nova,* **2008**, *31*, 2091-2099. [http://dx.doi.org/10.1590/S0100-40422008000800033]

[77] Sobrinho, H.B.; Rufino, R.D.; Luna, J.M.; Salgueiro, A.A.; Campos-Takaki, G.M.; Leite, L.F.; Sarubbo, L.A. Utilization of two agroindustrial by-products for the production of a surfactant by *Candida sphaerica* UCP0995. *Process Biochem.,* **2008**, *43*, 912-917. [http://dx.doi.org/10.1016/j.procbio.2008.04.013]

[78] Luna, J.M.; Sarubbo, L.A.; Campos-Takaki, G.M. A new biosurfactant produced by *Candida glabrata* UCP 1002; characteristics of stability and application in oil recovery. *Braz. Arch. Biol. Technol.,* **2009**, *52*, 785-793. [http://dx.doi.org/10.1590/S1516-89132009000400001]

[79] Barriga, J.A.T.; Cooper, D.G.; Idziak, E.S.; Cameron, D.R. Components of the bioemulsifier from S.

cerevisiae. *Enzyme Microb. Tech.,* 25, 1-2, 96-102 **1999**, *25*(1-2), 96-102.

[80] Torabizadeh, H.; Shojaosadati, S.A.; Tehrani, H.A. Preparation and characterization of bioemulsifier from *Saccharomyces cerevisiae* and its application in food products. *Lebenson. Wiss. Technol.,* **1996**, *29*, 734-737.
[http://dx.doi.org/10.1006/fstl.1996.0114]

[81] Marçal, M. do C. R. Produção de biopolímeros por *Candida lipolytica* em meios suplementados por óleos vegetais (Babaçu, Coco, Dendê). *Master's Thesis. Universidade Federal de Pernambuco, Brasil,* **1991**.

[82] Franzetti, A.; Gandolfi, I.; Raimondi, C.; Bestetti, G.; Banat, I.M.; Smyth, T.J.; Papacchini, M.; Cavallo, M.; Fracchia, L. Environmental fate, toxicity, characteristics and potential applications of novel bioemulsifiers produced by *Variovorax paradoxus* 7bCT5. *Bioresour. Technol.,* **2012**, *108*, 245-251.
[http://dx.doi.org/10.1016/j.biortech.2012.01.005] [PMID: 22277206]

[83] Fracchia, L.; Cavallo, M.; Martinotti, M.G.; Banat, I.M. Biosurfactants and Bioemulsifiers Biomedical and Related Applications - In: *Present Status and Future Potentials, Biomedical Science, Engineering and Technology*; Ghista, Dhanjoo N, Ed.; , **2012**; Chapter 14, pp. 325-370.

[84] Pacheco, G.J.; Ciapina, E.M.; Gomes, Ede.B.; Junior, N.P. Biosurfactant production by *rhodococcus erythropolis* and its application to oil removal. *Braz. J. Microbiol.,* **2010**, *41*(3), 685-693.
[http://dx.doi.org/10.1590/S1517-83822010000300019] [PMID: 24031544]

[85] Ilori, M.O.; Amobi, C.J.; Odocha, A.C. Factors affecting biosurfactant production by oil degrading *Aeromonas spp* . isolated from a tropical environment. *Chemosphere,* **2005**, *61*(7), 985-992.
[http://dx.doi.org/10.1016/j.chemosphere.2005.03.066] [PMID: 15878609]

[86] Bueno, S.M. *Bactérias produtoras de biossurfactantes: isolamento, produção, caracterização e comportamento num sistema modelo. Doctoral Thesis , Universidade Estadual Paulista, Brasil,* **2008**, 99f.

[87] Torabizadeh, H.; Shojaosadati, S.A.; Tehrani, H.A. Preparation and characterization of bioemulsfier from *Saccharomyces cerevisiae* and its application in food products. *Lebenson. Wiss. Technol.,* **1996**, *29*, 734-737.
[http://dx.doi.org/10.1006/fstl.1996.0114]

[88] Lima, A.S.; Alegre, R.M. Evaluation of emulsifier stability of biosurfactant produced by *Saccharomyces lipolytica* CCT-0913. *Braz. Arch. Biol. Technol.,* **2009**, *52*, 285-290.
[http://dx.doi.org/10.1590/S1516-89132009000200004]

[89] Lin, S.C.; Sharma, M.M.; Georgiou, G. Production and deactivation of biosurfactant by *Bacillus licheniformis* JF-2. *Biotechnol. Prog.,* **1993**, *9*, 138-145.
[http://dx.doi.org/10.1021/bp00020a004]

[90] Kleckner, V.; Kosaric, N. Biosurfactants for cosmetics. In: *Biosurfactants. Surfactant science series*; Kosaric, N., Ed.; Dekker: New York, **1993**; Vol. 48, pp. 373-389.

[91] Stickl, H.A. *International Journal of Hygiene and Environmental Medicine,* **1991**, *191*(2-3), 232-240.

[92] Van Leeuwen, F.X. *Food Safety and Toxicity*; de Vries, J., Ed.; CRC Press: New York, **1997**.

[93] Quinn, G.A.; Maloy, A.P.; Banat, M.M.; Banat, I.M. A comparison of effects of broad-spectrum antibiotics and biosurfactants on established bacterial biofilms. *Curr. Microbiol.,* **2013**, *67*(5), 614-623.
[http://dx.doi.org/10.1007/s00284-013-0412-8] [PMID: 23783562]

[94] Banat, I.M.; De Rienzo, M.A.; Quinn, G.A. Microbial biofilms: biosurfactants as antibiofilm agents. *Appl. Microbiol. Biotechnol.,* **2014**, *98*(24), 9915-9929.
[http://dx.doi.org/10.1007/s00253-014-6169-6] [PMID: 25359476]

[95] De Rienzo, M.A.; Banat, I.M.; Dolman, B.; Winterburn, J.; Martin, P.J. Sophorolipid biosurfactants: Possible uses as antibacterial and antibiofilm agent. *N. Biotechnol.,* **2015**, *32*(6), 720-726.
[http://dx.doi.org/10.1016/j.nbt.2015.02.009]

CHAPTER 12

Production and Applications of Food Enzymes

Gabriela Alves Macedo*, **Camilo Barroso Teixeira**, **José Valdo Madeira Júnior** and **Paula Speranza**

Bioprocess Laboratory, School of Food Engineering, University of Campinas, Campinas, Brazil

Abstract: Enzymes important biocataçysts widely used in food processing and are also useful for generating food ingredients. They may be recovered from animals, plants and mainly microorganisms, which represent the main source of industrial enzymes. This chapter seeks to link the development of new enzymes for food processes involving the requirements of the food industry, including: main companies producing enzymes; main enzymes used for food processing; food enzyme applications, emerging enzymes and production for food processing.

Keywords: L-asparaginase, Baking, Biotechnology, Dairy products, Enzymes, Food processing, Market, Phytase, Starch, Tannase.

INTRODUCTION

Enzyme technology has passed through a maturation phase in which research and development discovered new applications for biocatalysts in chemical, pharmaceutical and food industries. The advances in biotechnology, mainly in genetics, protein engineering and direct evolution have inaugurated a new era for enzymes and its use increased in research and industrial scale due to variety of reactions in environmental conditions. Moreover, the capacity to replace harmful chemical reactions and the trend for developing cleaner technologies enhances the importance of enzymes for the green chemistry and the biobased economy society towards the creation of new processes and product innovation in the market. There are many advantages in using enzymes, including higher product quality,

* **Address Correspondence to Gabriela Alves Macedo:** School of Food Engineering, University of Campinas (UNICAMP). Rua Monteiro Lobato, 80. 13083-862. Campinas, Brazil; E-mail: gmacedo@fea.unicamp.br.

Juliano Lemos Bicas, Mário Roberto Maróstica, Glaucia Maria Pastore (Eds.)

lower manufacturing cost, less waste and lower energy consumption [1].

Enzymes have been applied in food fermentation since earliest times for the manufacturing of cheese, sourdough, beer, wine and vinegar, besides non-food products, such as leather, indigo and linen. For their empirical usages, enzymes were generated by spontaneously growing microorganisms or by animal and vegetable preparations, such as calves' rumen or papaya fruit [2].

The commercial production of enzymes for food use have been reported since 1874, when the scientist Christian Hansen manufactured cheese using chymosin obtained from calves' stomachs [3]. This enzyme is currently produced using the recombinant DNA technique with the bovine chymosin gene expressed in *Escherichia coli* K-12. This was the first recombinant enzyme approved by the US Food and Drug Administration (FDA) for use in food [4]. In the 1930s, pectinases were used for clarification of juices, and some years later, invertase was also used for sacarose hydrolysis to produce inverted sugar syrup applying a pioneer process with immobilized enzymes. The application of enzymes in large scale in the food industry began in 1960s, when amyloglucosidase was used for starch hydrolysis to produce glucose syrups [5, 6]. This process was used to substitute the acid hydrolysis due to many advantages, *e.g.* higher product yields, increased degree of purity and easier crystallization. In 1973, the production of fructose syrup became feasible using immobilized glucose isomerase. After that, the industrial application of immobilized enzymes became true.

In the meanwhile, developments of recombinant DNA techniques permitted an improvement in enzyme production by microbial fermentation, increasing microbial stability and altered specificity and selectivity of enzymes. Those techniques showed excellent results to modelling fungi and bacteria for specific enzyme production and are continuing to broaden the applications of enzymes in food technology and different sectors of industry [7 - 10]. The objective of this chapter is to provide an overview on the current enzyme global market and the food applications of enzymes, emphasizing the traditional commercial applications, enzyme production processes and emerging enzymes.

MAIN COMPANIES, COMMERCIAL AND MARKET ASPECTS OF ENZYMES

The necessity of product innovation as a trend for a sustainable market, inspired the technological development stimulating the creation of new applications for enzymes in different industrial sectors in recent years. There are four major sectors of industrial enzymes, *i.e.* detergent (household care), technical (textile, leather, pulp and paper, ethanol and others), food and feed enzymes. The most important enzymes in industrial market are protease, amylase, lipase, ligase, phytase, cellulase, xylanase *etc.* [11]. To illustrate the enzyme market, 75% of industrial enzymes are consumed by food and brewing, feed and detergent industries [12]. Food enzymes constitute the major market share of industrial enzymes. Nevertheless, the segment of technical enzymes has increased fast in last 4-5 years due to the fuel ethanol enzymes (c.a. 11% of the global market in 2009), whose market has increased 15%-20% each year in the last years [11]. In fact, enzyme market has been expanding in recent years. In 2011, this market reached US$ 3.5 billion, with an annual increase of 6.1% [12].

The global industrial enzyme market operates as an oligopoly market, with the presence of three major suppliers: Novozymes, Genencor International (DuPont) and DSM [11]. In 2011, Novozymes and Genencor, presented market share of 47% and 21%, respectively. After several years of development, industrial enzyme market from China has become an important production base with about 10% of the global market. The expectations from the industrial market from China, considering its economic situation and domestic demand, indicates a faster growth rate of 10% or the global 6-8% for the next three years. The Chinese R&D delay and the foreign companies patent monopoly in China market have let the Chinese enterprises with a low competitiveness. Therefore, Novozymes, DuPont (Genencor) and other foreign brands still dominates the Chinese market. However, the Chinese VTR occupy a significant position, with 16.5% (in 2011) in market segments, *e.g.* feedstuff share in Chinese feed enzyme market [12].

Novozymes reported in 2012 that this company still retained its global market share of 47% and that total sales increased by 7% compared to 2011, with an estimated enhancement of 7% for global market for industrial enzymes. The major contributors for this increase were household care enzymes and microorganisms, with a growth of 14% and 23%, respectively. The market from Latin America enhanced 13% in sales, pushed by the increase of household care, feed and other technical enzymes. In terms of food enzymes, those used for preparing healthy foods were the main responsible for the growth of this sector. Enzymes for food and beverages are the second largest product area for Novozymes, corresponding to 28% of sales in 2012 (Fig. **1**), and globally these sector accounts to an anual value higher than US$1 billion. Even with the use of enzymes in many different products, producers and countries, the development of processed foods segment is still considered critical to the long-term growth of the company. The demand for innovative products by manufacturers of foods and drinks have been increasingly the necessity to meet the efficiency, health and the high regulatory standards [13].

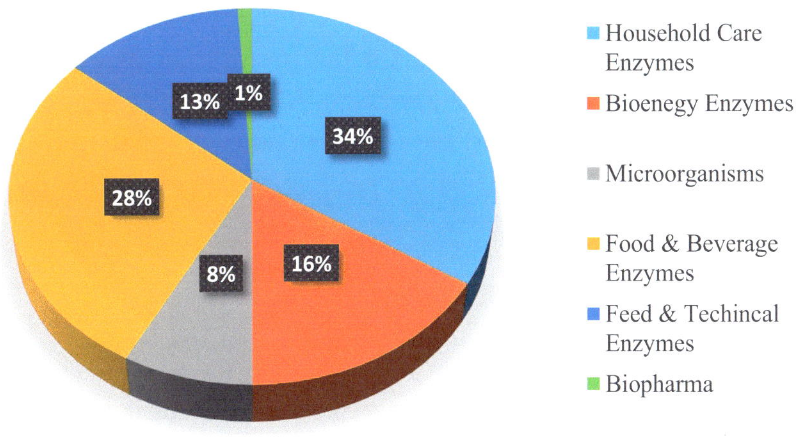

Fig. (1). 2012 enzyme business and bio business sales [13].

Enzymes for Food Industry: General Characteristics and Food Applications

Enzymes are ideal catalysts for the food industry. When compared to chemical synthesis, few by-products influencing the taste and color are formed in enzymatic reactions. Moreover, the reaction may be efficiently controlled and can be easily interrupted when the reaction reaches the desired degree of conversion [14].

Food and feed industries are possibly the sectors where processing is more dependent on biological agents. Most of the advances in these fields are related to the scope for more efficient biocatalysts, which may be achieved by a good screening for biocatalysts, structural modification and immobilization of enzymes produced by fermentation with fungi or bacteria. More precisely, enzyme optimization techniques aim at the development of biocatalysts with enhanced stability (*e.g.* temperature), specific activity and product specificity. Protein engineering and enzyme immobilization are the main tools used for this biocatalyst optimization, although developments in protein expression and microbial cell culture together with high-throughput screening techniques have been widely considered for improving enzymes. In terms of microbial screening, the use of extremophiles (*e.g.* extremes of temperature, salt concentration) and has also being useful to identify and develop more efficient biocatalysts [6].

Enzyme preparations used in food processing typically contains the enzyme of interest as well as diluents, preservatives and stabilizers besides certain impurities, *i.e.* other types of enzymes and metabolites from the producing microorganism, besides residues of raw materials used for the preparation of the fermentation media or for isolating and purifying the enzyme. All materials must be appropriate with high purity degree and current Good Manufacturing Practice (cGMP). The pathogenic and toxigenic potential of the production strain is the main target in evaluating enzyme safety [15, 16].

APPLICATIONS OF FOOD ENZYME

Many enzyme applications are consolidated in food industry, such as brewing, high fructose syrup from corn or from sucrose, juice clarification, confectionary and others [17]. We will discuss some of the most important and significant enzyme processing in food industries in the next topics.

Starch Processing

High Fructose Corn Syrup (HFCS) is a sweetener derived from corn starch widely used in soft drinks, candies, baking, jams and jellies and many other foods. It was the first food ingredient made industrially with an enzymatic process. Many advantages are associated to this process with regard to chemical process to hydrolyze corn sugar into fructose syrup, including higher quality products, energy efficiency, safer working environment and milder conditions. This process has several environmental benefits such as reducing the use of strong acids and bases, lower power consumption (less greenhouse gases), less toxic and easier to treat waste and a safer environment for workers in production [17].

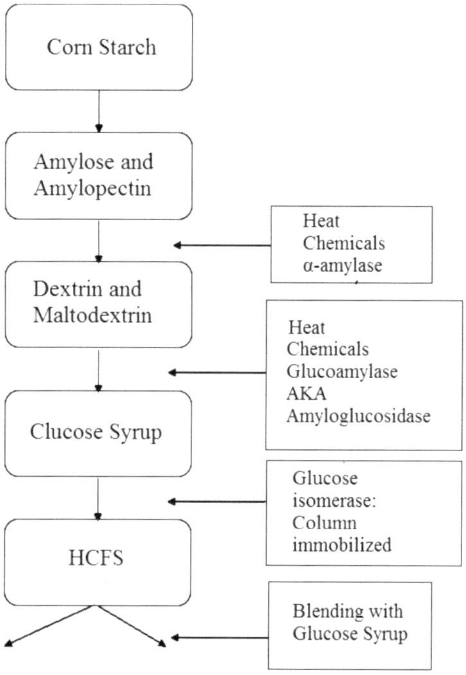

Fig. (2). Flowchart of HFCS productions from Cornstarch. Adapted from Kay Parker [18].

Corn starch, a polymer of glucose, consists of branched amylose and linear amylopectin. For breaking it down, heat, caustic soda and/or hydrochloric acid besides three different enzymes are applied resulting in simple sugars glucose and fructose present in HFCS. The flowchart of the process is illustrated on Fig. (**2**). The first enzyme, *Bacillus* spp. α-amylase, randomly hydrolyzes α-1,4 glycosidic bonds releasing short chain dextrins and oligosaccharides. The second enzyme, glucoamylase (or amyloglucosidase), obtained from fungi sources such as *Apergillus*, hydrolyses α-1,4 (and α-1,6 to a lesser extent) glycosidic bonds from non-reducing ends of dextrins and oligosaccharides to produce glucose monomers. After treatment with these two enzymes, the product consists of glucose syrup. In the third step, glucose isomerase catalyses the isomerization of glucose to fructose. In such process, α-amylase and glucoamylase are used as free enzymes, while glucose isomerase is immobilized in package columns for reacting with the glucose syrup. The final product ispurified by liquid chromatography to form a mixture of 90% fructose and 10% glucose [18].

Dairy Applications

Cheeses

The purified chymosin is an important enzyme originally found in rennet which is already produced by genetically modified microorganisms, avoiding the need to sacrifice the calves. Therefore, the cheese industry does dependent of animal enzymes for the production of rennet anymore. Moreover, the current demand for this enzyme in cheese production could hardly be supplied just by calf rennet. This ensures an abundant supply of high quality enzyme (chymosin) at a reasonable price [17]

Chymosin presents as principal role in cheesemaking, by coagulating milk after hydrolysing the Phe105-Met106 peptide bond of the micelle-stabilizing protein, κ-casein, [19, 20]. This coagulation is divided in three principal steps: the first comprises the enzymatic hydrolysis of the casein micelles; the second is the aggregation of the casein particles and the third one is the modification in properties and structure of the coagulum [21].

Cheese Flavors

There are a wide variety of cheese options with characteristic flavor, which is developed during ripening, and most of it is due to the action of lipase enzymes. Lipases act on fat cheese to release short chain fatty acids, which are responsible for the characteristic aroma or are precursors of flavoring molecules of different types of cured cheese (Romano, Provolone, blue cheese *etc.*). These enzymes are produced by different microorganisms, and in the case of cheese, one of the most remarkable example are the fungi from the genus *Penicillium* [17].

Lipases from indigenous, endogenous and /or exogenous sources are responsible for triglycerides hydrolysis in all cheese varieties resulting in the fatty acids releasing during cheese ripening. The short-chain fatty acids from triglycerides of ruminant milk have low flavor thresholds when released, contributing to impart characteristic flavors of different cheese types. For instance, a more extensive lipolysis is typical in some hard Italian varieties and in blue cheese [22]. These free fatty acids may contribute to cheese flavor either directly (short-chain fatty acids) or indirectly (precursors of volatile compounds, such as methyl ketones in blue cheese) [22].

Lactose-free Dairy Products

A significant portion of the adult population presents lactose intolerance, a disorder resulting from the lack of lactase (enzyme responsible for lactose hydrolysis), naturally present in the intestinal tract of children, and responsible for causing gastrointestinal disturbances, bloating, diarrhea or a combination of gastrointestinal symptoms. Lactose is present in dairy products such as milk, ice cream, yogurt and cheese. It may be hydrolyzed by lactase forming two easily digestible sugars: glucose and galactose. Without this enzyme, lactose is fermented in the food in the gut producing undesirable side effects just cited above.

Lactose-intolerant people (approximately 20-30 percent of American adults) can now enjoy dairy products thanks to the commercial availability of lactase. Currently, There are many "lactose-free" products on the market and they are prepared by a pretreat milk or the final product with the enzyme lactase. Moreover, the lactase is available for household consumers with lactose intolerance [17].

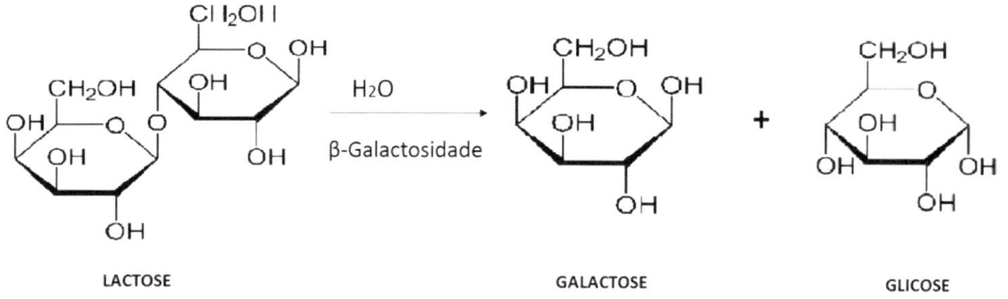

Fig. (3). Enzymatic hydrolysis of lactose by β-galactosidase.

The enzymatic treatment of dairy products for the production of lactose-free food is based on the lactose hydrolysis by β-galactosidase generating glucose and galactose as products (Fig. **3**). The taste of milk become more sweet (glucose and galactose are four times sweeter than the lactose) and crystallization in refrigerated products, such as ice creams, are avoided. Another advantage may be the side reaction synthesis of galacto-oligosaccharides, which are prebiotic carbohydrates [23, 24].

Baking Applications

Bromate Replacers

In the production of bread, oxidizing compounds that assist the formation of consistent dough are used. These chemical oxidants (bromate, azodicarbonamide) have been widely used for strengthening the gluten network. Although potassium bromate has been used for many years in breadmaking, its use is now banned in many countries. Indeed, some studies evaluate the toxic effects of such compound [25 - 27].

The dough and breadmaking quality of wheat flour is affected by the quality and quantity of gluten proteins. Arabinoxylan contributes to the primary gluten network by cross-linking ferulic acid and gluten proteins. Glucose-oxidase reaction (with glucose) generates H_2O_2 which is considered responsible for the molecular mechanism of dough and bread improvement through catalyzing the oxidative cross-linking of gluten proteins and ferulic acid present in arabinoxylan [28].

Bakery Products

Some enzymes are used to modify the starch, which keeps the softer bread for a long time. This change in the structure of the starch prevents the hardening of bread, which otherwise would lose moisture over time when the starch molecules revert from a soluble form to the insoluble form (retrogradation). When starch loses its interaction with water, also loses flexibility and the texture of the bread become hard and brittle. This results in loss of quality and reduces shelf life. The starch modifying enzymes could help in retarding retrodegradation. The bread becomes softer and more flavorful for 3-6 days [17].

The enzyme application in breadmaking is well established for upgrade the processing and quality of flour, dough and bread. Besides the traditional enzymes available, new enzymes (oxidizing and crosslinking enzymes) are also being developed for this application [29].

Amilolitic enzymes, such as α-amylases, act in starch substrate hydrolyzing glycoside bonds which results in flour stabilization, increased loaf volume and improved crumb texture and crust color. Endo-protease act in gluten substrate hydrolyzing peptide bonds resulting in improved dough handling and machinability. Endo-xylanase act in arabinoxylan substrate degrading and solubilizing arabinoxylans resulting in increase of loaf volume and improved crumb texture. Lipase act in flour lipid and glycolipids releasing fatty acids, generating mono and di-glycerides, resulting in increased dough strength, loaf volume and crumb texture. Lypoxygenase oxidizes polyunsaturated fatty acids generating hydroxyperoxides and cooxidation of proteins and carotenoids resulting in crumb bleaching. Transglutaminase act in proteins catalyzing the cross-linking of proteins by formation of covalent isopeptide bonds [29].

Fruit Juice

Pectin is a polysaccharide present in large concentrations in juices. Pectin gives a turbid appearance to the juice and results in a mouth feel not very palatable to consumers. Pectinases are enzymes that hydrolyze pectin and generate clear juices with more desirable characteristics (appearance, stability, mouth feel and taste).

The majority of industrially processed fruit juices and wines are clarified to avoid turbidity. The visual perception of this turbidity is the result of suspended pectin particles coming from the plant cell walls [30 - 32]. The pectin hydrolysis by pectinase results in pectin–protein complexes which flocculates, precipitates and decant. The final juice presents a minor concentration of pectins with lower viscosity, which facilitates the downstream filtration process [32].

Pectinases occur naturally in most fruits but the juice processing industries often add external enzymes to accelerate the settling velocity process traditionally used to clarify juices. The use of these enzymes in juice production also increases extraction yield, thus reducing waste, increasing the efficiency of utilization of raw materials and therefore reducing the production costs. These clear juices are usually more esthetically pleasing and are widely available at retail [17].

Meat Tenderizing

Meat is mainly composed of a very complex set of proteins with defined structure. The primary responsible for meat softness are the myofibrillar proteins and connective tissue proteins. Protease enzymes are used to reduce this complex network of proteins, hydrolyzing them into smaller fragments. Proteases, such as papain and bromelain, have already been used to soften tough meat cuts. The use of these enzymes generates less waste and better use of raw materials enables the consumer to enjoy the ability to tenderize tougher cuts of meat [17].

Papain, bromelain and ficin are proteases from vegetable origin and are the most widely used enzymes for meat tenderization [33, 34]. Microbial proteases from *Aspergillus* species are also approved by U.S.D.A for meat tenderization. The application of these enzymes is made by various forms, such as marinades, addition in brine and diverse spices as meat tenderizers [33, 34].

Papain is a cysteine protease obtained from the latex of *Carica papaya* and has a physiological function in plant protection against insects [35, 36]. Ficin is extracted from the latex of *Ficus glabrata, Ficus anthelmintica* and *Ficus laurifolia* [36, 37,]. Bromelain is a mixture of cysteine proteases, extracted from the stem and fruit of the pineapple (*Ananas comosus*) plant [35, 38, 39]. They have some differences between each other due to specificity of protease. Sullivan and Calkins [40] evaluated the difference effect of papain, bromelain and ficin on meat tenderization of beef muscle of high and low-connective tissue and concluded that papain showed the best results to improve tenderness in meat although presented juiciness and textural changes as negative aspects. Bromelain increased tenderness and degraded collagen more than the contractile proteins. Ficin had the most equilibrated degradation of both myofibrillar and collagen proteins.

Confectioneries

Sweets often have short shelf life due to the presence of sucrose, which begins to crystallize after cooking. Similar behavior is observed for soft cookies and other sweet bakery. The enzyme invertase hydrolyses sucrose into two simple sugars, glucose and fructose (Fig. **4**), which no longer undergo crystallization, thereby increasing product's shelf life. Enzymes replace the use of chemicals in the manufacturing process, reducing the risk to the environment and maintaining the desirable sensory characteristics to the product for extended periods [17].

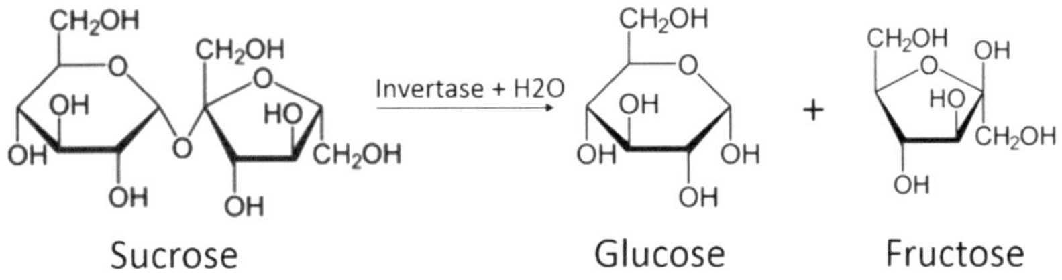

Fig. (4). Sucrose hydrolysis by invertase.

EMERGING ENZYMES IN FOOD INDUSTRY

The food industry is interested in finding and developing new applications for enzymes that have desirable properties, which are not easily obtained in nature. These enzymes have to present some important features such as the ability to act in specific reactions and the ability to remain stable under extreme conditions of temperature, pressure, pH and other conditions employed in these processes. Microbial enzymes are those that exhibit these characteristics, and genetic improvement of these microorganisms has allowed achieving these goals more easily.

In this section are presented some new enzymes that have been used and / or tested by the food industry. These enzymes, due to the specificity they present, are of great interest nowadays.

Phytase

Food products made from plants, like cereals and legumes, are some of the main staple for humans being in many parts of the world. They constitute an important source of carbohydrates, protein, dietary fibers and vitamins [41] . However, many of these cereals and legumes have a high concentration of antinutrients, compounds which are capable of blocking the use of other substances, or have toxic properties. Among these antinutrients, phytic acid (myo-inositol-1, 2, 3, 4, 5, 6-hexakisphosphates) or phytate (Fig. **5**) is extremely important for human nutrition [42].

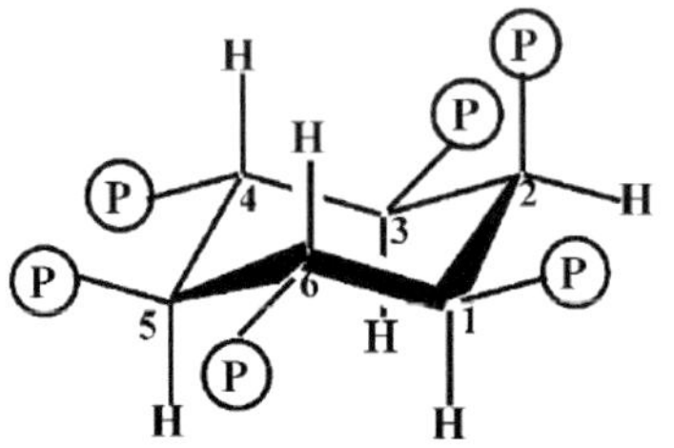

Fig. (5). Chemical structure of phytic acid [43].

Phytate is the main storage form of both phosphate and inositol in seeds and grains, and it accumulates in these parts during ripening [44] . Its antinutrient activity is due to its chelation of various indispensable metals such as calcium, zinc, magnesium and iron, forming insoluble chelate complexes [45, 46]. Phytates can also bind with proteins and vitamins, forming insoluble complexes, which reduce the efficiency of utilization, digestibility and activity of these compounds [47, 48]. The dephosphorylation of phytate is a strategy for improving nutritional value of cereals and legumes, removal of phosphate groups from the inositol ring decreases the mineral binding strength of phytate [42, 49].

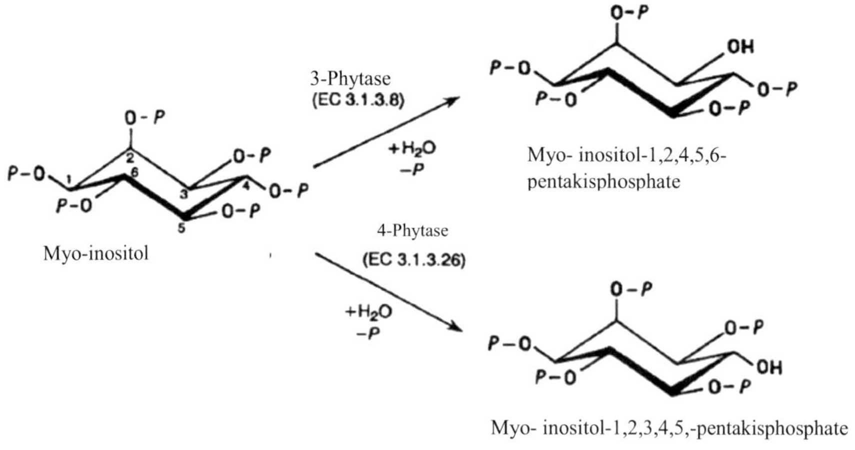

Fig. (6). Hydrolysis of phytic acid [51].

As a solution, phytase (EC 3.1.3.8) is an emerging enzyme for degrading phytate during food processing or digestion. This enzyme catalyzes the sequential release of phosphate from phytate [42]. Phytase removes orthophosphate groups from inositol ring of phytic acid, producing free inorganic phosphorus and another myo-inositol phosphate intermediate (Fig. **6**) [50].

Most phytases were isolated from filamentous fungi, such as *Aspergillus ficuum*, *Mucor piriformis* and *Cladosporium species* [52] . This enzyme has several applications in food industries. The addition of phytase in processing bread improves its quality, since it improves their nutritional value, reducing the phytate content and promoting the activation of endogenous α-amylase enzyme, providing more calcium [53].

The addition of phytase in soymilk that is rich in phytate, may potentially improve mineral bioavailability and absorption [54]. The study of the biological activity of rats treated with sorghum hydrolyzed by phytase showed improvement in phosphorus digestibility, besides the inhibition of aspartate aminotransferase and alanine aminotransferase [55].

L-Asparaginase

The thermal processing of food, primarily those derived from plant origin such as potato and cereal products, leads to the formation of acrylamide [56, 57]. The acrylamide formation occurs along with Maillard reaction, through the reaction of reducing sugars and asparagine at temperatures higher than 120 °C in a variety of low-moisture heated foods (*e.g.,* potato derivatives, bakery products and roasted coffee) [58, 59]. Acrylamide induces tumors in mice and rats, and exerts reproductive and neurotoxic damage [60]. As this compound has negative impact on human health, many efforts have been made to reduce their presence in food.

L-Asparaginase (E.C. 3.5.1.1) has been claimed to reduce effectively acrylamide levels in food, since L-asparagine is considered to be one of the main precursors for the acrylamide formation in foods [56, 61]. This enzyme catalyzes the hydrolysis of aspartic acid and ammonia, thereby specifically consuming a key precursor for acrylamide formation. The application of L-asparaginase was confirmed to be one of the successful ways of acrylamide mitigation in potato products.

The enzyme was able to eliminate one of the precursors of acrylamide - amino acid L-asparagine (without compromising the sensory properties of processed products [58].

Asparaginase is also used to reduce the formation of acrylamide in biscuits without altering the final color of the product [62] . Studies of the use of L-asparaginase in roast and ground coffee has been able to reduce the concentration of acrylamide in these products, however, often occurs loss of organoleptic properties in these products. Optimization studies on the reaction conditions attempt to prevent these losses [63].

Tannase

Tannins are natural polyphenolic compounds with varying molecular weights, coming from the secondary metabolism of plants [64, 65]. Tannins are classified into two groups: hydrolysable tannins and condensed tannins [66, 67]. The hydrolysable tannins are joined by esteriac links (ester linkage between the aromatic ring group and the glucose residue) and depsidic (ester linkage between the aromatic rings), being readily hydrolyzed by acids and enzymes in their bases forming units [68] . The basic structural unit of this type of tannin is a polyol, usually glucose, with their hydroxyl groups esterified by gallic acid (gallotannins) or by hexahidroxidifênico (ellagitannins) (Fig. **7**). The condensed tannins may contain flavanoids two to fifty units. Flavonoids are the major class of polyphenols, composed of two aromatic rings linked by a heterocyclic ring. Can be divided into subclasses such as flavonols, flavones, flavanones [69]. The condensed tannins have a complex structure; are resistant to hydrolysis, but can be soluble in aqueous organic solvents, depending on their structure (Fig. **8**). Because of their chemical structures, without the presence of ester bonds and depsidic chemical linkages, proanthocyanidins are not susceptible to hydrolysis by tannase [70].

These compounds form strong complexes with minerals and some macromolecules (*e.g.* proteins, cellulose, starch and others) [72, 73]. By forming these complexes, digestion of these compounds may be compromised.

Furthermore, tannins inhibit digestive enzymes including pectinase, amylase, lipase, protease, cellulase and β-galactosidase in the saliva and rumen, thereby reducing the absorption of nutrients [65, 74].

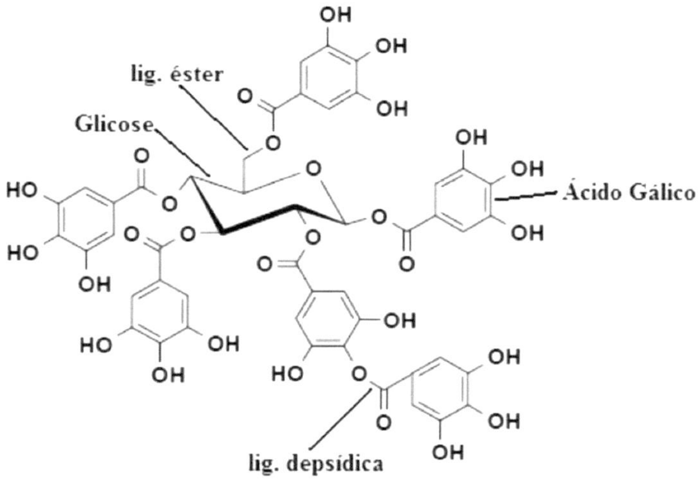

Fig. (7). Chemical structure of hydrolyzable tannin [71].

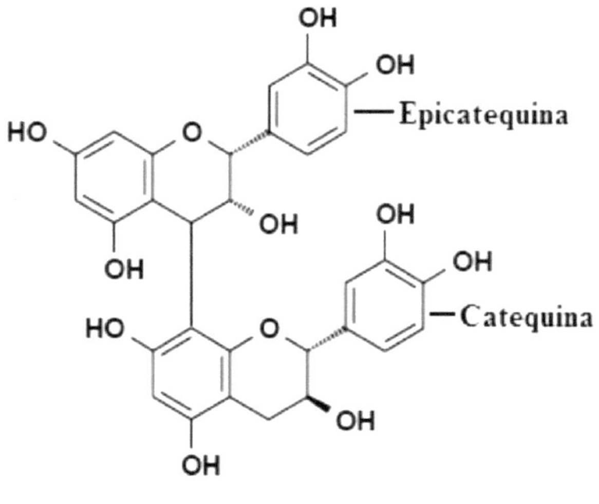

Fig. (8). Chemical structure of condensed tannin [71].

Tannin acyl hydrolase (E.C. 3.1.1.20), or simply tannase, catalyzes the hydrolysis reaction of the ester bonds present in (hydrolysable tannins) the gallotannins, complex tannins, and gallic acid ester to release gallic acid [64, 73] . One of the main industrial applications of tannase is employment in instant tea processing, in order to remove unwanted insoluble complexes known as "tea cream", resulting from the polymerization of polyphenols (tannins) and proteins interact with caffeine. The use of tannase avoids the use of chemicals to eliminate the "tea cream", ensuring a final product of excellent quality, water soluble and is characterized by a high content of aromatic components and appropriate color [75].

The application of tannase in wine also favors the removal of phenolic compounds, stabilizing and increasing their quality [73]. About 50% of the coloring of wine is due to the presence of tannins and, the oxidation of these components in contact with air (generating quinones) can cause an undesirable turbidity and consequently loss of quality of the final product. This turbidity can be avoided with the use of tannase, which prevents the oxidation reaction [73].

The tannase is also used as a clarifying agent in some fruit juices by removing phenolic compounds present in plant material, thereby preventing complexation and precipitation thereof [76]. In addition, tannase plays an important role in production of antioxidants such as epigallocatechin gallate, epicatechin and gallic acid [71, 76 - 78].

Thus, although these enzymes have great potential to be applied in the food industry, new technologies must be employed to facilitate their implementation on a large scale. The use of immobilization techniques and genetic improvement are successful examples to increase productivity and reduce the costs of these enzymes [79 - 81].

FOOD ENZYME PRODUCTION

The use of enzymes has a long history, especially in food processing. Currently, microbial enzymes represent the main source for industrial enzymes, since microorganisms are easily cultivated and manipulated, they may produce enzymes uninterruptedly in an optimized and automated process. However, microbial production of enzymes also presents some challenges, such as the high costs for culture medium (representing approximately 30% of the manufacturing costs). To overcome this obstacle, cheaper culture media have been tested in fermentation processes, such as agroindustrial residues (sugarcane bagass, citrus pomace, wheat bran). Furthermore, it has been studied alternatives to improve the performance of microorganism by genetic engineering [82 - 84].

Fermentation is a primary process for enzyme production. Both fungi and bacteria may be used, which may be grown under submerged (SmF) or solid-state fermentation (SSF). Submerged fermentation is generally considered for the production of enzymes from bacterial origin, since these organisms develop better in a liquid medium.

In contrast, microorganisms grown under solid fermentation can produce large amounts of other compounds which are of no interest to the final product, and scaling-up solid state fermentations remains one of its main challenges.

On the other hand, there are several advantages for using SSF instead of SmF, such as high productivities, low energy costs and low production costs. However, a few years ago this process was rarely implemented in biotechnology industry, because of the low reproducibility. Nowadays, with advances in biotechnology research, SSF has reached higher levels of standardization and reproducibility.

In food industry, extracellular enzymes are preferred, due to their simple downstreaming process. In submerged fermentation, the extracellular enzyme is diluted in the culture medium, which can be processed to yield the enzyme with different purity grades. In the case of SSF, the enzymes produced through this technique have, in most cases, greater molecular stability. Therefore, one should evaluate the type of fermentation for a better enzyme yield produced [85, 86].

For enzyme production, the culture media used in microbial fermentations should contain all necessary elements. In addition, they should satisfy the technical objectives of the process by offering a suitable environment for the growth and enzyme synthesis, and at the same time they must be economically profitable. The medium composition must present specific requirements are dependent on genetic information for each microbial species. Some species with a wide nutritional flexibility are able to synthesize many of their metabolites from simple precursors, while other species are more demanding and require complex nutrients for their growth and reproduction [87, 88].

Nutrients can be divided into two classes: macronutrients (carbon, oxygen, hydrogen, nitrogen, sulfur and phosphorus) and micronutrients (iron, magnesium, manganese, calcium, zinc, potassium, sodium, copper, chloride, selenium, *etc.*). Both types are essential, but the former are required in large quantities because they are the main constituents of cellular organic compounds [85, 86, 89].

Basically, the fermentation medium should have an ideal ratio of carbon and nitrogen to improve microbial development. This relationship is one of the most important to prevent the catabolic repression, a key mechanism in the regulation of extracellular enzymes that degrade complex substrates to which the microorganism is exposed [90]. Another important component is the inducing factor (for inducible enzymes), *i.e.* compounds that promote the production of a particular enzyme [91 - 93]. Metabolic repression is another important factor to consider when formulating a culture medium. In such case, some enzymes are not produced if a simpler source of energy is available. For instance, proteases are produced in the presence of proteins (inducing agent), but their synthesis is suppressed in the presence of glucose or free aminoacids (metabolic repressor) [94].

Table 1. Composition of medium culture for enzyme production.

enzyme	micro-organism	macronutrients	micronutrients	conditions	enzyme activity	reference
protease	*Bacillus, Aspergillus*	sucrose, citric acid, yeast extract, fosfate, sulphate	K, Mg, Ca, Cl	100-200 rpm, 25-35°C, 48-96 h	1,200-12,000 U/mL	[89, 93, 95]
lipase	*Rhizopus, Aspergillus, Geotrichum*	glucose, peptone, yeast extract, fosfate, oil-seed (olive, sunflower)	Mn, K, Cl	120-180 rpm, 30°C, 120 h	10-20 U/mL	[92, 96, 97]
amylase	*Bacillus, Aspergillus*	starch, peptone, sulphate	K, Mg	150 rpm, 30 °C, 48 h	3,000 U/mL	[91, 98, 99]
glucosidase	*Candida, Penincillium*	beef extract, peptone, ovalbumin, soluble cassava	Cl	30 °C, 120 h	5-20 U/mL	[100, 101, 102]

Additionally, other parameters should be controlled to reach optimized enzymatic production, such as incubation conditions and microorganism. Table **1** shows examples of the conditions employed for the production of major enzymes used in the food industry.

Considering the microorganism employed, one of the most important strategies consists of screening or genetic manipulation of microorganism [101, 103]. Genetically engineered microorganisms that show over-expression has been traditionally used to produce enzymes in the food industry [104, 105].

These genetically modified microorganisms have been considered in fermentations due to their simple handling, fast growth rate, high enzyme yield, better enzyme stability and others. However, many drawbacks are associated to this strategy, *e.g.* difficulties in scaling-up [105], thereby new genes and host are studied to overcome these problems.

Recently, the genomes of yeasts and some prokaryote organisms have been elucidated, opening the way to novel and potent modified in the industrial area such as an *Aspergillus niger* strain used as a host for over-expressing food enzymes, such as xylanase [106]. Xylanase belongs to a hydrolytic enzyme of the xylan, that is one of the component of hemicellulose, it found in various plant resources. This enzyme may be used in food, feed and papermaking industries [106]. Other examples of enzymes produced by genetically modified microorganisms is shown in Table **2**.

Table 2. Genetic manipulation on microorganisms producers of food enzymes.

enzyme	origin of gene(s)	Gene Code	Host microorganism	properties	reference
protease	*Bacillus subtillis*	degU32 allele	*Bacillus licheniformis*	hypersecretion of enzyme	[82]
lipase	*Escherichia coli*	LIP2	*Yarrowia lipolytica*	high productivity	[107]
β-glucosidase	*Saccharomycopsis fibuligera*	BGL1	*Saccharomyces cerevisiae*	non-limiting activity	[101]
amylase	*Bacillus sp.*	amy	*Bacillus subtilis*	increase rate production	[108]

CONCLUDING REMARKS

The large increase in recent years in the commercialization of enzymes, is due, among other factors, the development of more robust enzymes produced by fermentation processes that provide greater efficiency and the search for new enzymes with specificity that meet the current demand of the food industry. These new enzymes allow obtaining products that are not possible by traditional chemical processes. In the coming years, a further increase in the use of enzymes in the food industry will likely be observed.

CONFLICT OF INTEREST

The authors confirm that they have no conflict of interest to declare for this publication.

ACKNOWLEDGEMENTS

Declared none.

REFERENCES

[1] Binod, P. Review article - industrial enzymes present status and future perspectives for India. *J. Sci. Ind. Res.,* **2013**, *72*, 271-286.

[2] Kirk, O.; Borchert, T.V.; Fuglsang, C.C. Industrial enzyme applications. *Curr. Opin. Biotechnol.,* **2002**, *13*(4), 345-351.
[http://dx.doi.org/10.1016/S0958-1669(02)00328-2] [PMID: 12323357]

[3] Nielsen, P.H.; Malmos, H.; Damhus, T.; Diderichsen, B.; Nielsen, H.K. Simonsen, M., SchiV, H.E., Oestergaard, A., Olsen, H.S., Eigtved, P.; Nielsen, T. K. Enzyme applications (industrial). In: *Encyclopedia of Chemical Technology*; John Wiley & Sons, Inc: New York, **1994**; Vol. 9, pp. 567-620.

[4] Flamm, E.L. How FDA approved chymosin: a case history. *Biotechnology (N. Y.),* **1991**, *9*(4), 349-351.
[http://dx.doi.org/10.1038/nbt0491-349] [PMID: 1367006]

[5] Vasic-Racki, D. History of industrial biotransformations—Dreams and realities. In: *Industrial Biotransformations*; Wiley-VCH: Weinheim, Germany, **2006**; pp. 1-35.

[6] Fernandes, P. Enzymes in food processing: a condensed overview on strategies for better biocatalysts. *Enzyme Res.,* **2010**, *2010*, 862537.
[http://dx.doi.org/10.4061/2010/862537] [PMID: 21048872]

[7] Falch, E.A. Industrial enzymes--developments in production and application. *Biotechnol. Adv.,* **1991**, *9*(4), 643-658.

[http://dx.doi.org/10.1016/0734-9750(91)90736-F] [PMID: 14542053]

[8] Miguel, Â. S. M.; Martins-Meyer, T. S.; Figueiredo, É. V. d. C.; Lobo, B. W. P.; Dellamora-Ortiz, G. M. Enzymes in Bakery: *Current and Future Trends,* **2013**.

[9] Petersen, S. Protein Engineering: Design and Engineering on the Nano Scale. In: *Applied Biocatalysis*; Straathof AJJ, A. P., Ed.; Harwood Academic Publications: Amsterdam, **2005**; pp. 263-310.

[10] Poulsen, P.B. K. History of Enzymology with Emphasis on Food Production. In: *Handbook of Food Enzymology, Whitaker JR, V. A*; Wong, D.W., Ed.; New York, **2003**.

[11] ResearchandMarkets. Global Industrial Enzymes Market. *Analysis,* **2010**.

[12] Reportsnreports. *Global and China Industrial Enzyme Industry Report, 2012-2014,* **2012**.

[13] Novozymes. *The Novozymes Report,* **2012**.

[14] Olsen, H.S. Enzymes in Food Processing. In: *Biotechnology Set*; Wiley-VCH Verlag GmbH, **2008**; pp. 663-736.

[15] Olempska-Beer, Z.S.; Merker, R.I.; Ditto, M.D.; DiNovi, M.J. Food-processing enzymes from recombinant microorganisms--a review. *Regul. Toxicol. Pharmacol.,* **2006**, *45*(2), 144-158. [http://dx.doi.org/10.1016/j.yrtph.2006.05.001] [PMID: 16769167]

[16] Pariza, M.W.; Johnson, E.A. Evaluating the safety of microbial enzyme preparations used in food processing: update for a new century. *Regul. Toxicol. Pharmacol.,* **2001**, *33*(2), 173-186. [http://dx.doi.org/10.1006/rtph.2001.1466] [PMID: 11350200]

[17] ETA. *Enzymes: A Primer On Use and Benefits Today and Tomorrow.,* [accessed Nov 02, 2014]; Available at: http://www.enzymeassociation.org/wp-content/uploads/2013/09/benefits_paper.pdf

[18] Kay Parker, M. S.; Nwosu., Veronica High fructose corn syrup: Production, uses and public health concerns *Biotechnology and Molecular Biology Review,* **2010**, *5*(5), 71-78.**2010**, *5*(5) 71-.

[19] Fox, P.F.; Guinee, T.P.; Cogan, T.M.; McSweeney, P.L. *Fundamentals of Cheese Science*; Aspen Publishers: Gaithersburg Maryland, **2000**.

[20] Sousa, M.J.; Ardö, Y.; McSweeney, P.L. Advances in the study of proteolysis during cheese ripening. *Int. Dairy J.,* **2001**, *11*(4–7), 327-345. [http://dx.doi.org/10.1016/S0958-6946(01)00062-0]

[21] Hyslop, D.B. Enzymatic Coagulation of Milk. In: *Advanced Dairy Chemistry—1 Proteins*; Fox, P.F.; McSweeney, P.L., Eds.; Springer: US, **2003**; pp. 839-878. [http://dx.doi.org/10.1007/978-1-4419-8602-3_24]

[22] McSweeney, P.L. Biochemistry of cheese ripening. *Int. J. Dairy Technol.,* **2004**, *57*(2-3), 127-144. [http://dx.doi.org/10.1111/j.1471-0307.2004.00147.x]

[23] Ladero, M.; Perez, M.T.; Santos, A.; Garcia-Ochoa, F. Hydrolysis of lactose by free and immobilized β-galactosidase from Thermus sp. strain T2. *Biotechnol. Bioeng.,* **2003**, *81*(2), 241-252. [http://dx.doi.org/10.1002/bit.10466] [PMID: 12451560]

[24] Mahoney, R.R. Galactosyl-oligosaccharide formation during lactose hydrolysis: A review. *Food Chem.,* **1998**, *63*(2), 147-154. [http://dx.doi.org/10.1016/S0308-8146(98)00020-X]

[25] Ahmad, M.K.; Khan, A.A.; Mahmood, R. Alterations in brush border membrane enzymes, carbohydrate metabolism and oxidative damage to rat intestine by potassium bromate. *Biochimie,* **2012**, *94*(12), 2776-2782.
[http://dx.doi.org/10.1016/j.biochi.2012.09.001] [PMID: 22974983]

[26] Limonciel, A.; Wilmes, A.; Aschauer, L.; Radford, R.; Bloch, K.M.; McMorrow, T.; Pfaller, W.; van Delft, J.H.; Slattery, C.; Ryan, M.P.; Lock, E.A.; Jennings, P. Oxidative stress induced by potassium bromate exposure results in altered tight junction protein expression in renal proximal tubule cells. *Arch. Toxicol.,* **2012**, *86*(11), 1741-1751.
[http://dx.doi.org/10.1007/s00204-012-0897-0] [PMID: 22760423]

[27] Zhang, X.; De Silva, D.; Sun, B.; Fisher, J.; Bull, R.J.; Cotruvo, J.A.; Cummings, B.S. Cellular and molecular mechanisms of bromate-induced cytotoxicity in human and rat kidney cells. *Toxicology,* **2010**, *269*(1), 13-23.
[http://dx.doi.org/10.1016/j.tox.2010.01.002] [PMID: 20067818]

[28] Decamps, K.; Joye, I.J.; Rakotozafy, L.; Nicolas, J.; Courtin, C.M.; Delcour, J.A. The bread dough stability improving effect of pyranose oxidase from trametes multicolor and glucose oxidase from Aspergillus niger: unraveling the molecular mechanism. *J. Agric. Food Chem.,* **2013**, *61*(32), 7848-7854.
[http://dx.doi.org/10.1021/jf4021416] [PMID: 23875828]

[29] Goesaert, H.; Gebruers, K.; Courtin, C.M.; Brijs, K.; Delcour, J.A. Enzymes in Breadmaking. In: *Bakery Products*; Blackwell Publishing, **2007**; pp. 337-364.

[30] Binning, R.P. P. Apple juice. In: *Fruit Juice Processing Technology, S*; Nagy, C.S.; Shaw, P.E., Eds.; AgScience: Auburndale, FL, USA, **1993**; pp. 271-317.

[31] Schobinger, U., Ed. J., W. Fruit juice embellishment and clarification. In: *Fruit and Vegetable Juices, Handbook of Food Technology,* 2nd ed; Eugen Ulmer: Stuttgart, Germany, **1987**; pp. 168-189.

[32] Pinelo, M.; Zeuner, B.; Meyer, A.S. Juice clarification by protease and pectinase treatments indicates new roles of pectin and protein in cherry juice turbidity. *Food Bioprod. Process.,* **2010**, *88*(2-3), 259-265.
[http://dx.doi.org/10.1016/j.fbp.2009.03.005]

[33] Ashie, I.N.; Sorensen, T.L.; Nielsen, P.M. Effects of Papain and a Microbial Enzyme on Meat Proteins and Beef Tenderness. *J. Food Sci.,* **2002**, *67*(6), 2138-2142.
[http://dx.doi.org/10.1111/j.1365-2621.2002.tb09516.x]

[34] Dransfield, E.E. Enzymes in the tenderization of meat. In: *Enzymes and Food Processing., Birch GG, B. N*; Parker, K.J., Ed.; Applied Science Publishers: London, England, **1981**; pp. 177-194.
[http://dx.doi.org/10.1007/978-94-011-6740-6_10]

[35] Bekhit, A.A.; Hopkins, D.L.; Geesink, G.; Bekhit, A.A.; Franks, P. Exogenous proteases for meat tenderization. *Crit. Rev. Food Sci. Nutr.,* **2014**, *54*(8), 1012-1031.
[http://dx.doi.org/10.1080/10408398.2011.623247] [PMID: 24499119]

[36] Konno, K.; Hirayama, C.; Nakamura, M.; Tateishi, K.; Tamura, Y.; Hattori, M.; Kohno, K. Papain protects papaya trees from herbivorous insects: role of cysteine proteases in latex. *Plant J.,* **2004**, *37*(3), 370-378.

[http://dx.doi.org/10.1046/j.1365-313X.2003.01968.x] [PMID: 14731257]

[37] Gaughran, E.R. Ficin: History and Present Status. *Pharm. Biol.,* **1976**, *14*(1), 1-21. [http://dx.doi.org/10.3109/13880207609081900]

[38] Barrett, A.J.; Rawlings, N.D.; Woessner, J.F. *Handbook of Proteolytic Enzymes*; Elsevier Academic Press: Amsterdam, **2004**.

[39] Grzonka, Z.; Kasprzykowski, F.; Wiczk, W. Cysteine proteases. In: *Industrial Enzymes*; Polaina, J.; MacCabe, A.P., Eds.; Springer: NY, USA, **2007**; pp. 181-195. [http://dx.doi.org/10.1007/1-4020-5377-0_11]

[40] Sullivan, G.A.; Calkins, C.R. Application of exogenous enzymes to beef muscle of high and low-connective tissue. *Meat Sci.,* **2010**, *85*(4), 730-734. [http://dx.doi.org/10.1016/j.meatsci.2010.03.033] [PMID: 20416788]

[41] Katina, K.; Arendt, E.; Liukkonen, K-H.; Autio, K.; Flander, L.; Poutanen, K. Potential of sourdough for healthier cereal products. *Trends Food Sci. Technol.,* **2005**, *16*(1), 104-112. [http://dx.doi.org/10.1016/j.tifs.2004.03.008]

[42] Kumar, V.; Sinha, A.K.; Makkar, H.P.; Becker, K. Dietary roles of phytate and phytase in human nutrition: A review. *Food Chem.,* **2010**, *120*(4), 945-959. [http://dx.doi.org/10.1016/j.foodchem.2009.11.052]

[43] Raboy, V. myo-Inositol-1,2,3,4,5,6-hexakisphosphate. *Phytochemistry,* **2003**, *64*(6), 1033-1043. [http://dx.doi.org/10.1016/S0031-9422(03)00446-1] [PMID: 14568069]

[44] Loewus, F.A.; Reddy, N.; Sathe, S. *Biosynthesis of Phytate in Food Grains and Seeds*; CRC Press: Boca Raton, FL, USA, **2001**. [http://dx.doi.org/10.1201/9781420014419.ch4]

[45] Liu, B-L.; Rafiq, A.; Tzeng, Y-M.; Rob, A. The induction and characterization of phytase and beyond. *Enzyme Microb. Technol.,* **1998**, *22*(5), 415-424. [http://dx.doi.org/10.1016/S0141-0229(97)00210-X]

[46] Papatryphon, E.; Howell, R.A.; Soares, J.H. Growth and mineral absorption by striped bass Morone saxatilis fed a plant feedstuff based diet supplemented with phytase. *J. World Aquacult. Soc.,* **1999**, *30*(2), 161-173. [http://dx.doi.org/10.1111/j.1749-7345.1999.tb00863.x]

[47] Li, D.; Che, X.; Wang, Y.; Hong, C.; Thacker, P. Effect of microbial phytase, vitamin D and citric acid on growth performance and phosphorus, nitrogen and calcium digestibility in growing swine. *Anim. Feed Sci. Technol.,* **1998**, *73*(1), 173-186. [http://dx.doi.org/10.1016/S0377-8401(98)00124-2]

[48] Sugiura, S.; Gabaudan, J.; Dong, F.; Hardy, R. Dietary microbial phytase supplementation and the utilization of phosphorus, trace minerals and protein by rainbow trout [Oncorhynchus mykiss (Walbaum)] fed soybean meal based diets. *Aquacult. Res.,* **2001**, *32*(7), 583-592. [http://dx.doi.org/10.1046/j.1365-2109.2001.00581.x]

[49] Sandberg, A-S.; Brune, M.; Carlsson, N-G.; Hallberg, L.; Skoglund, E.; Rossander-Hulthén, L. Inositol phosphates with different numbers of phosphate groups influence iron absorption in humans. *Am. J. Clin. Nutr.,* **1999**, *70*(2), 240-246.

[PMID: 10426701]

[50] Debnath, D.; Sahu, N.; Pal, A.; Baruah, K.; Yengkokpam, S.; Mukherjee, S. Present scenario and future prospects of phytase in aquafeed. *Asian-australas. J. Anim. Sci.,* **2005**, *18*, 1800-1812. [http://dx.doi.org/10.5713/ajas.2005.1800]

[51] Dvoráková, J. Phytase: sources, preparation and exploitation. *Folia Microbiol. (Praha),* **1998**, *43*(4), 323-338. [Praha]. [http://dx.doi.org/10.1007/BF02818571] [PMID: 9821286]

[52] Haefner, S.; Knietsch, A.; Scholten, E.; Braun, J.; Lohscheidt, M.; Zelder, O. Biotechnological production and applications of phytases. *Appl. Microbiol. Biotechnol.,* **2005**, *68*(5), 588-597. [http://dx.doi.org/10.1007/s00253-005-0005-y] [PMID: 16041577]

[53] Haros, M.; Rosell, C.M.; Benedito, C. Fungal phytase as a potential breadmaking additive. *Eur. Food Res. Technol.,* **2001**, *213*(4-5), 317-322. [http://dx.doi.org/10.1007/s002170100396]

[54] Tang, A.L.; Wilcox, G.; Walker, K.Z.; Shah, N.P.; Ashton, J.F.; Stojanovska, L. Phytase activity from Lactobacillus spp. in calcium-fortified soymilk. *J. Food Sci.,* **2010**, *75*(6), M373-M376. [http://dx.doi.org/10.1111/j.1750-3841.2010.01663.x] [PMID: 20722939]

[55] Schons, P.F.; Ries, E.F.; Battestin, V.; Macedo, G.A. Effect of enzymatic treatment on tannins and phytate in sorghum (Sorghum bicolor) and its nutritional study in rats. *Int. J. Food Sci. Technol.,* **2011**, *46*(6), 1253-1258. [http://dx.doi.org/10.1111/j.1365-2621.2011.02620.x]

[56] Stadler, R.H.; Blank, I.; Varga, N.; Robert, F.; Hau, J.; Guy, P.A.; Robert, M-C.; Riediker, S. Acrylamide from Maillard reaction products. *Nature,* **2002**, *419*(6906), 449-450. [http://dx.doi.org/10.1038/419449a] [PMID: 12368845]

[57] Zyzak, D.V.; Sanders, R.A.; Stojanovic, M.; Tallmadge, D.H.; Eberhart, B.L.; Ewald, D.K.; Gruber, D.C.; Morsch, T.R.; Strothers, M.A.; Rizzi, G.P.; Villagran, M.D. Acrylamide formation mechanism in heated foods. *J. Agric. Food Chem.,* **2003**, *51*(16), 4782-4787. [http://dx.doi.org/10.1021/jf034180i] [PMID: 14705913]

[58] Ciesarová, Z.K. E; Boegl, P. Impact of L-asparaginase on acrylamide content in potato products. *J. Food Nutr. Res.,* **2006**, *45*(4), 141-146.

[59] Claeys, W.L.; De Vleeschouwer, K.; Hendrickx, M.E. Kinetics of acrylamide formation and elimination during heating of an asparagine-sugar model system. *J. Agric. Food Chem.,* **2005**, *53*(26), 9999-10005. [http://dx.doi.org/10.1021/jf051197n] [PMID: 16366686]

[60] Pedreschi, F.; Mariotti, M.S.; Granby, K. Current issues in dietary acrylamide: formation, mitigation and risk assessment. *J. Sci. Food Agric.,* **2014**, *94*(1), 9-20. [http://dx.doi.org/10.1002/jsfa.6349] [PMID: 23939985]

[61] Friedman, M. Chemistry, biochemistry, and safety of acrylamide. A review. *J. Agric. Food Chem.,* **2003**, *51*(16), 4504-4526. [http://dx.doi.org/10.1021/jf030204+] [PMID: 14705871]

[62] Anese, M.; Quarta, B.; Frias, J. Modelling the effect of asparaginase in reducing acrylamide formation

in biscuits. *Food Chem.,* **2011**, *126*(2), 435-440.
[http://dx.doi.org/10.1016/j.foodchem.2010.11.007]

[63] Guenther, H.; Anklam, E.; Wenzl, T.; Stadler, R. H. Acrylamide in coffee: review of progress in analysis, formation and level reduction. In: *Food Addit. Contam*; , **2007**; 24, pp. (sup1)60-70.
[http://dx.doi.org/10.1080/02652030701243119]

[64] Aguilar, C.N.; Rodríguez, R.; Gutiérrez-Sánchez, G.; Augur, C.; Favela-Torres, E.; Prado-Barragan, L.A.; Ramírez-Coronel, A.; Contreras-Esquivel, J.C. Microbial tannases: advances and perspectives. *Appl. Microbiol. Biotechnol.,* **2007**, *76*(1), 47-59.
[http://dx.doi.org/10.1007/s00253-007-1000-2] [PMID: 17530245]

[65] Chávez-González, M.; Rodríguez-Durán, L.V.; Balagurusamy, N.; Prado-Barragán, A.; Rodríguez, R.; Contreras, J.C.; Aguilar, C.N. Biotechnological advances and challenges of tannase: an overview. *Food Bioprocess Tech.,* **2012**, *5*(2), 445-459.
[http://dx.doi.org/10.1007/s11947-011-0608-5]

[66] Carneiro, A. d. C.; Vital, B. R.; Pimenta, A. S.; Mori, F. A. Reatividade dos taninos da casca de Eucalyptus grandis para produção de adesivos. In: *Cerne*; , **2001**; 7, pp. (1)001-009.

[67] Salunkhe, D.K.; Chavan, J. *Dietary tannins: consequences and remedies*; CRC Press, **1989**.

[68] Hagerman, A.E.; Butler, L.G. The specificity of proanthocyanidin-protein interactions. *J. Biol. Chem.,* **1981**, *256*(9), 4494-4497.
[PMID: 7217094]

[69] Faria, A.R. Natural compounds with anti-Trypanossoma cruzi activity. *Revista Brasileira de Farmacognosia,* **2007**, *17*(3), 455-465.

[70] Pinto, G.; Bruno, L.; Hamacher, M.; Tarzi, S.; Couri, S. In Increase of tannase production in solid state fermentation by Aspergillus niger 3T5B8 *25th Symposium on biotechnology for fuels and chemicals, poster presentation,* **2003**Breckenridge, CO, USA, pp. 3-68.

[71] Battestin, V.; Macedo, G.; De Freitas, V. Hydrolysis of epigallocatechin gallate using a tannase from Paecilomyces variotii. *Food Chem.,* **2008**, *108*(1), 228-233.
[http://dx.doi.org/10.1016/j.foodchem.2007.10.068]

[72] Field, J.A. L. G. Biodegradation of tannins. In: *Metal ions in biological systems, Degradation of environmental pollutants by microorganisms and their metaloenzymes, H., S*; Dekker, M., Ed.; New York, **1992**; Vol. 28, pp. 61-97.

[73] Lekha, P.K.; Lonsane, B.K. Production and application of tannin acyl hydrolase: state of the art. *Adv. Appl. Microbiol.,* **1997**, *44*, 215-260.
[http://dx.doi.org/10.1016/S0065-2164(08)70463-5] [PMID: 9311108]

[74] Frutos, P.; Hervas, G.; García, F.G.; Mantecón, A. Review. Tannins and ruminant nutrition. *Span. J. Agric. Res.,* **2004**, *2*(2), 191-202.
[http://dx.doi.org/10.5424/sjar/2004022-73]

[75] Sanderson, G.W.; Coggon, P. Green tea conversion using tannase and natural tea enzymes U.S. Patent 3,812,266, May 21, 1974.

[76] Lekha, P.; Lonsane, B. Comparative titres, location and properties of tannin acyl hydrolase produced by Aspergillus niger PKL 104 in solid-state, liquid surface an submerged fermentations. *Process*

Biochem., **1994**, *29*(6), 497-503.
[http://dx.doi.org/10.1016/0032-9592(94)85019-4]

[77] Battestin, V.; Macedo, G.A. Tannase production by Paecilomyces variotii. *Bioresour. Technol.,* **2007**, *98*(9), 1832-1837.
[http://dx.doi.org/10.1016/j.biortech.2006.06.031] [PMID: 17045475]

[78] Macedo, J.; Battestin, V.; Ribeiro, M.; Macedo, G. Increasing the antioxidant power of tea extracts by biotransformation of polyphenols. *Food Chem.,* **2011**, *126*(2), 491-497.
[http://dx.doi.org/10.1016/j.foodchem.2010.11.026]

[79] Cho, E-A.; Kim, E-J.; Pan, J-G. Adsorption immobilization of Escherichia coli phytase on probiotic Bacillus polyfermenticus spores. *Enzyme Microb. Technol.,* **2011**, *49*(1), 66-71.
[http://dx.doi.org/10.1016/j.enzmictec.2011.03.006] [PMID: 22112273]

[80] Iwamoto, K.; Tsuruta, H.; Nishitaini, Y.; Osawa, R. Identification and cloning of a gene encoding tannase (tannin acylhydrolase) from Lactobacillus plantarum ATCC 14917(T). *Syst. Appl. Microbiol.,* **2008**, *31*(4), 269-277.
[http://dx.doi.org/10.1016/j.syapm.2008.05.004] [PMID: 18653299]

[81] Vidya, J.; Vasudevan, U.M.; Soccol, C.R.; Pandey, A. Cloning, Functional Expression and Characterization of L-Asparaginase II from E. coli MTCC 739. *Food Technol. Biotechnol.,* **2011**, *49*(3), 286-290.

[82] Borgmeier, C.; Bongaerts, J.; Meinhardt, F. Genetic analysis of the Bacillus licheniformis degSU operon and the impact of regulatory mutations on protease production. *J. Biotechnol.,* **2012**, *159*(1-2), 12-20.
[http://dx.doi.org/10.1016/j.jbiotec.2012.02.011] [PMID: 22387916]

[83] Pleissner, D.; Kwan, T.H.; Lin, C.S. Fungal hydrolysis in submerged fermentation for food waste treatment and fermentation feedstock preparation. *Bioresour. Technol.,* **2014**, *158*(0), 48-54.
[http://dx.doi.org/10.1016/j.biortech.2014.01.139] [PMID: 24583214]

[84] Zhang, B-B.; Lu, L-P.; Xia, Y-j.; Wang, Y-L.; Xu, G-R. Use of agar as carrier in solid-state fermentation for Monacolin K production by Monascus: A novel method for direct determination of biomass and accurate comparison with submerged fermentation. *Biochem. Eng. J.,* **2013**, *80*(0), 10-13.
[http://dx.doi.org/10.1016/j.bej.2013.09.007]

[85] Bhavsar, K.; Buddhiwant, P.; Soni, S.K.; Depan, D.; Sarkar, S.; Khire, J.M. Phytase isozymes from Aspergillus niger NCIM 563 under solid state fermentation: Biochemical characterization and their correlation with submerged phytases. *Process Biochem.,* **2013**, *48*(11), 1618-1625.
[http://dx.doi.org/10.1016/j.procbio.2013.08.001]

[86] Prévot, V.; Lopez, M.; Copinet, E.; Duchiron, F. Comparative performance of commercial and laboratory enzymatic complexes from submerged or solid-state fermentation in lignocellulosic biomass hydrolysis. *Bioresour. Technol.,* **2013**, *129*, 690-693.
[http://dx.doi.org/10.1016/j.biortech.2012.11.135] [PMID: 23352484]

[87] Bauchop, T.; Elsden, S.R. The growth of micro-organisms in relation to their energy supply. *J. Gen. Microbiol.,* **1960**, *23*(3), 457-469.
[PMID: 13687855]

[88] Madigan, M.T.; Martinko, J.M.; Parker, J.; Brock, T.D. *Biology of microorganisms, prentice hall*; Upper Saddle River, NJ, **1997**.

[89] Belmessikh, A.; Boukhalfa, H.; Mechakra-Maza, A.; Gheribi-Aoulmi, Z.; Amrane, A. Statistical optimization of culture medium for neutral protease production by Aspergillus oryzae. Comparative study between solid and submerged fermentations on tomato pomace. *J. Taiwan Inst. Chem. E.,* **2013**, *44*(3), 377-385.
[http://dx.doi.org/10.1016/j.jtice.2012.12.011]

[90] Deutscher, J. The mechanisms of carbon catabolite repression in bacteria. *Curr. Opin. Microbiol.,* **2008**, *11*(2), 87-93.
[http://dx.doi.org/10.1016/j.mib.2008.02.007] [PMID: 18359269]

[91] Kuddus, M.; Ramteke, P.W. Recent developments in production and biotechnological applications of cold-active microbial proteases. *Crit. Rev. Microbiol.,* **2012**, *38*(4), 330-338.
[http://dx.doi.org/10.3109/1040841X.2012.678477] [PMID: 22849713]

[92] Salihu, A.; Alam, M.Z. AbdulKarim, M. I.; Salleh, H. M. Lipase production: an insight in the utilization of renewable agricultural residues. *Resour. Conserv. Recycling,* **2012**, *58*, 36-44.
[http://dx.doi.org/10.1016/j.resconrec.2011.10.007]

[93] Sarkar, N.; Ghosh, S.K.; Bannerjee, S.; Aikat, K. Bioethanol production from agricultural wastes: An overview. *Renew. Energy,* **2012**, *37*(1), 19-27.
[http://dx.doi.org/10.1016/j.renene.2011.06.045]

[94] Sumantha, A.; Larroche, C.; Pandey, A. Microbiology and industrial biotechnology of food-grade proteases: a perspective. *Food Technol. Biotechnol.,* **2006**, *44*(2), 211-220.

[95] Potumarthi, R.; Subhakar, C.; Jetty, A. Alkaline protease production by submerged fermentation in stirred tank reactor using Bacillus licheniformis NCIM-2042: Effect of aeration and agitation regimes. *Biochem. Eng. J.,* **2007**, *34*(2), 185-192.
[http://dx.doi.org/10.1016/j.bej.2006.12.003]

[96] Sánchez-Otero, M.G.; Ruiz-López, I.I.; Ávila-Nieto, D.E.; Oliart-Ros, R.M. Significant improvement of Geobacillus thermoleovorans CCR11 thermoalkalophilic lipase production using response surface methodology. *N. Biotechnol.,* **2011**, *28*(6), 761-766.
[http://dx.doi.org/10.1016/j.nbt.2011.01.010] [PMID: 21315194]

[97] Maldonado, R.R.; Burkert, J.F.; Mazutti, M.A.; Maugeri, F.; Rodrigues, M.I. Evaluation of lipase production by Geotrichum candidum in shaken flasks and bench-scale stirred bioreactor using different impellers. *Biocatal. Agric. Biotechnol.,* **2012**, *1*(2), 147-151.

[98] Hashemi, M.; Mousavi, S.; Razavi, S.; Shojaosadati, S. Mathematical modeling of biomass and α-amylase production kinetics by Bacillus sp. in solid-state fermentation based on solid dry weight variation. *Biochem. Eng. J.,* **2011**, *53*(2), 159-164.
[http://dx.doi.org/10.1016/j.bej.2010.09.017]

[99] Rajagopalan, G.; Krishnan, C. α-amylase production from catabolite derepressed Bacillus subtilis KCC103 utilizing sugarcane bagasse hydrolysate. *Bioresour. Technol.,* **2008**, *99*(8), 3044-3050.
[http://dx.doi.org/10.1016/j.biortech.2007.06.001] [PMID: 17644331]

[100] Gonçalves, F.A.; Leite, R.S.; Rodrigues, A.; Argandona, E.J.; Fonseca, G.G. Isolation, identification

and characterization of a novel high level β-glucosidase-producing Lichtheimia ramosa strain. *Biocatal. Agric. Biotechnol.,* **2013**, *2*(4), 377-384.

[101] Gurgu, L.; Lafraya, Á.; Polaina, J.; Marín-Navarro, J. Fermentation of cellobiose to ethanol by industrial Saccharomyces strains carrying the β-glucosidase gene (BGL1) from Saccharomycopsis fibuligera. *Bioresour. Technol.,* **2011**, *102*(8), 5229-5236. [http://dx.doi.org/10.1016/j.biortech.2011.01.062] [PMID: 21324680]

[102] Nimpiboon, P.; Nakapong, S.; Pichyangkura, R.; Ito, K.; Pongsawasdi, P. Synthesis of a novel prebiotic trisaccharide by a type I α-glucosidase from B. licheniformis strain TH4-2. *Process Biochem.,* **2011**, *46*(2), 448-457. [http://dx.doi.org/10.1016/j.procbio.2010.09.018]

[103] Jørgensen, S.; Skov, K.W.; Diderichsen, B. Cloning, sequence, and expression of a lipase gene from Pseudomonas cepacia: lipase production in heterologous hosts requires two Pseudomonas genes. *J. Bacteriol.,* **1991**, *173*(2), 559-567. [PMID: 1987151]

[104] Coughlan, M.P. The properties of fungal and bacterial cellulases with comment on their production and application. *Biotechnol. Genet. Eng. Rev.,* **1985**, *3*(1), 39-110. [http://dx.doi.org/10.1080/02648725.1985.10647809]

[105] Treichel, H.; de Oliveira, D.; Mazutti, M.A.; Di Luccio, M.; Oliveira, J.V. A review on microbial lipases production. *Food Bioprocess Tech.,* **2010**, *3*(2), 182-196. [http://dx.doi.org/10.1007/s11947-009-0202-2]

[106] Thomas, L.; Joseph, A.; Gottumukkala, L.D. Xylanase and cellulase systems of Clostridium sp.: an insight on molecular approaches for strain improvement. *Bioresour. Technol.,* **2014**, *158*(0), 343-350. [http://dx.doi.org/10.1016/j.biortech.2014.01.140] [PMID: 24581864]

[107] Fickers, P.; Fudalej, F.; Nicaud, J-M.; Destain, J.; Thonart, P. Selection of new over-producing derivatives for the improvement of extracellular lipase production by the non-conventional yeast Yarrowia lipolytica. *J. Biotechnol.,* **2005**, *115*(4), 379-386. [http://dx.doi.org/10.1016/j.jbiotec.2004.09.014] [PMID: 15639099]

[108] Yuki, S. On the gene controlling the rate of amylase production in Bacillus subtilis. *Biochem. Biophys. Res. Commun.,* **1968**, *31*(2), 182-187. [http://dx.doi.org/10.1016/0006-291X(68)90727-4] [PMID: 4968472]

SUBJECT INDEX

L

M

N

O

P

R

S

T

V

X

Y

Made in the USA
Las Vegas, NV
03 November 2026

57424109R00307